HANDBOOK OF PRINTED CIRCUIT MANUFACTURING

HANDBOOK OF PRINTED CIRCUIT MANUFACTURING

Raymond H. Clark

 VAN NOSTRAND REINHOLD COMPANY
——————————————————————— *New York*

Library of Congress Catalog Card Number: 84–13117
ISBN: 0-442-21610-6

Manufactured in the United States of America

Published by Van Nostrand Reinhold Company Inc.
135 West 50th Street
New York, New York 10020

Van Nostrand Reinhold Company Limited
Molly Millars Lane
Wokingham, Berkshire RG11 2PY, England

Van Nostrand Reinhold
480 Latrobe Street
Melbourne, Victoria 3000, Australia

Macmillan of Canada
Division of Gage Publishing Limited
164 Commander Boulevard
Agincourt, Ontario M1S 3C7, Canada

15   14   13   12   11   10   9   8   7   6   5   4   3   2

Library of Congress Cataloging in Publication Data
   Clark, Raymond H.
   Handbook of printed circuit manufacturing.

   Includes index.
   1. Printed circuits–Design and construction.
I. Title.
TK7868.P7C55   1985   621.381'74   84-13117
ISBN 0-442-21610-6

*To all people who make their living working
in the printed circuit industry.*

Preface

Of all the components that go into electronic equipment, the printed circuit probably requires more manufacturing operations—each of which must be performed by a skilled person—than any other.

As a shift supervisor early in my printed circuit career, I had to hire and train personnel for all job functions. The amount of responsibility delegated to my subordinates depended strictly on how well I had been able to train them. Training people can be a trying experience and is always a time-consuming one. It behooved me to help my workers obtain the highest degree of job understanding and skill that they and I were capable of. One hindrance to effective teaching is poor continuity of thought, for example, having to say to a trainee, "Wait a minute; forget what I just told you. We have to go back and do something else first." It was in trying to avoid pitfalls such as this that I undertook a detailed examination of the processes involved, what I thought each trainee had to know, and what questions they would most frequently ask. From this analysis I developed the various process procedures. Only after I had done so was I able to train effectively and with the confidence that I was doing the best possible job. Answers had to be at hand for all of their questions and in whatever detail they needed to know. The desire to train effectively, to help the trainee master the necessary skills, and to provide a ready reference for use as needed helped prompt the writing of the *Handbook of Printed Circuit Manufacturing*.

All of the many processes in printed circuit manufacturing are fairly to extremely complicated. Some of them, such as plating baths, constituted an open system that depends on the chemical and thermodynamic equilibrium of many factors. Plating baths and certain other printed-circuit processes have a very distinct personality. They may behave differently one day than they did the day before, and differently in the morning than in the afternoon. All platers and plating chemists know what I am talking about. Temperature, brightener, chloride level, contamination, anode level, electrical connections, cleanliness of bus bars, the type of pattern being plated—even the type of resist—are only a few of the factors affecting bath performance.

When a problem develops, the cause may not be immediately obvious. Any one effect can be caused by any number of things. Some problems are brought about by a number of factors coming to bear at one time. When scrap is being produced, down time mounts, shipments don't go out the door, people are standing around waiting, and nothing seems to have a positive effect on the difficulty, where are the plater, supervisor, chemist, or manager to turn? Since problems sometimes result from situations that may not be directly related to the area in which they occur, vendor technical service experience may not be of sufficient scope to resolve them directly. No one is going to see a greater variety of problems, nor be more dedicated to resolving them, than the people who work in a printed circuit shop. The desire to help others solve the same difficulties that I have had to face and to prevent their recurrence is another motivation behind writing *Handbook of Printed Circuit Manufacturing*.

One of the frequent drawbacks in our industry is that affecting the transfer of information and experience. Too many printed circuit shops do not formalize their processes by writing them down. Even if the process has been written down, it is often so out-of-date that it is never referred to. There being no standard procedure for any process, a hundred printed circuit manufacturers will produce the same board a hundred different ways. The consistency in quality, amount of scrap, amount of customer returns, and ability to deliver a job on time thus varies considerably. Also, since so little written information is available on the performance of each process—from the hands-on point of view needed by the process operator and supervisor (not to mention the chemists, engineers, and technical service people)—it is difficult for them to develop a complete understanding of their job function and of the technology with which they must deal. Since so little has been published on printed circuit manufacturing and since there are no industry-wide accepted procedures written down for all to learn from, and thus no ready sources of process explanation, it is difficult for process operators and supervisors to learn from those who have gone before them. Often, what exists in print has been written by a consultant or vendor technical service people. What they have to say may be important, but it may lack any real understanding of the actual situations encountered by the shop floor personnel. The desire to provide a detailed set of procedures and a related explanation of each process is another motivation behind writing this book.

This author believes that training will be most effective only when the supervisor fully understands all the processes he or she is in charge of and can answer all questions posed by trainees and even those by more experienced people. Such an understanding provides confidence to the person doing the training and leads to increased job satisfaction on the part of both the supervisor and the process operator. Supervisors and managers without full knowledge of their processes will tend to end up with "operator dependent processes" in which the person on the line must supply the expertise necessary to make things work. This all too common situation leads to poor product quality, a bad working environment, and a breakdown in supervision and management.

Too little time is spent developing the employees' full understanding of the printed circuit manufacturing process or even of their own specific jobs. The situation is often worse for printed circuit sales people and other sales people who service the printed circuit industry but are not out on the manufacturing floor all day every day, where they could at least learn by osmosis. It can scarcely be doubted that their ability to meet their company goals would also be enhanced by a complete understanding of printed circuit manufacturing.

The biggest challenge in printed circuit manufacturing—today and always—is to produce printed circuits that precisely conform to design requirements. If they are routinely produced in this way, they will require a shorter lead time, will be shipped on time, and will not be returned by the customer (except in the occasional instances when customers do not understand their own requirements). If this challenge can be met, the industry will remain strong and printed circuit companies profitable.

In sum, the *Handbook of Printed Circuit Manufacturing* has been written to meet the needs of those engaged in manufacturing, or who want to know more about manufacturing, printed circuits. It is meant to be a primary, comprehensive reference—a source book of hands-on manufacturing know-how, step-by-step procedures, trouble-shooting information, and guidelines explaining each process and technology in detail. It has been written by one who is still involved in manufacturing printed circuits and in training others to do so, and it is written for, and dedicated to, the people who are involved in the same tasks.

One last note: The information presented here is true and complete to the best of the author's knowledge. It is based on his own experience, observation, and education. He advises all who work in printed circuit manufacturing to (1) question, (2) experiment, (3) observe, and (4) evaluate. For a variety of reasons, certain processes work better in some houses than in others. Read all you can, and learn from your own experiences. It is the author's wish that what is presented here will help you benefit from the experience of one who has gone before.

RAYMOND H. CLARK
San Jose, California

Acknowledgments

I would like to acknowledge the help or guidance of the following people as well as of the many whom I may have forgotten to mention and who may not be aware of the contributions they have made, directly or indirectly, to my career and this book: Jerry Hertel, Jerry Banks, Donna Esposito, Elmer Hayes, Dr. Leo Roos, Gary Glines, Thelma Andre, Bill Collier, Debbie White, Harsh Kacchi, Jim Langan, Bernie McDermott, Kent Carter, Charles Simon, Zane Warner, Glen Selvedge, Dr. Paul Craven, Greg Eger, Britt Watts, Bob Grille, Tom Rourke, Don Wolf, Terry Rose, Tom Richards, Diane Harrington, Chuck Candelaria, Paul Zeb Jr., Pete Hepner, and Michael Gardner and Alberta Gordon of Van Nostrand Reinhold.

I would also like to acknowledge the patience and support of my wife, Cathi Ann Ege Clark, during the writing of this book.

Contents

HANDBOOK OF PRINTED CIRCUIT MANUFACTURING

SECTION ONE
DESIGN AND MANUFACTURE OF PRINTED CIRCUITS

1
How Printed Circuit Boards are Manufactured: Processes and Materials

Printed circuits are found in virtually all electronic equipment manufactured in the last twenty years. They are perhaps the most custom designed and manufactured component in that equipment, and very often the most expensive single component. Considering the fact that a multilayer printed circuit may cost more than a thousand dollars, and have that much more value added in other components, it behooves the buyer, designer and user of printed circuitry to have at least rudimentary understanding of how they are made. What follows here is a brief description of the steps followed by most printed circuit manufacturers.

WHAT IS A PRINTED CIRCUIT BOARD?

Printed circuits boards are dielectric substrates with metallic circuitry photochemically formed upon that substrate. There are three major classifications:

- *Single sided boards.* Dielectric substrate with circuitry on one side only. There may or may not be holes for components and tooling drilled into them.
- *Double sided boards.* Dielectric substrate with circuitry on both sides. An electrical connection is established by drilling holes through the dielectric and plating copper through the holes.
- *Multilayer boards.* Two or more pieces of dielectric material with circuitry formed upon them are stacked up and bonded together. Electrical connections are established from one side to the other, and to the inner layer circuitry by drilled holes which are subsequently plated through with copper.

WHAT IS THE DIELECTRIC SUBSTRATE MADE OF?

The dielectric substrate most commonly used is made from fiberglass sheets which have copper foil bonded onto both sides with epoxy resin; this is called epoxy/fiberglass laminate. Other materials are glass with polyimide, teflon, or triazine resins; and paper covered with phenolic resin. All of these substrates have copper foil bonded to both sides with the resin.

THE MANUFACTURING PROCESS

1. Planning receives all documentation from the customer: artwork, drawings, manufacturing specifications, sales order. Together with the Photo department, the planning engineer reviews all the documentation for completness and accuracy. If it is all acceptable the following happens:

- Planning issues a job traveler which details the manufacturing steps to be followed, and the requirements at each step.
- Laminate is blanked for the job.
- The drilling tape is programmed for the part, and a Drilling first article is drilled and bought off.
- The Photo department generates working film which registers to the first article.

2. Once the drilling tape has been bought off the complete job can be drilled. The panels are pinned together in stacks from one to four high, depending on panel thickness. These are loaded onto the drilling machine. The drilling program tape is fed into the numerical control (N/C) drilling machine memory. This tape commands the N/C drill to drill all the holes in the proper location with the correct drill size. Most modern drilling machines will automatically change drill bits when the hole size changes.

3. After drilling, the panels are deburred, then go to Plating to have copper chemically deposited in the holes; this process is called *electroless copper plating,* or *through hole plating.* The electroless copper deposit is only about 20 millionths of an inch thick. The deposit serves two purposes: (1) It provides the electrical connection between the sides of a panel and to the inner layers of multilayer boards. (2) It serves as a metal substrate upon which subsequent electrolytically plated copper may be deposited.

4. After the holes have been plated, the circuit pattern is imaged onto the panels. The image defines the circuit for plating and etching purposes. This image is commonly applied using one of the following methods:

- *Silk screening.* The image is screened on with an ink plating resist.
- *Dry film photoresist.* Both sides of the panel are coated with a thin layer (0.001 to 0.002 inch) of dry film photoresist. The circuit artwork is layed on top of the photoresist and exposed to ultraviolet light. The u.v. polymerizes the resist from a soft gelatinous state to a hard, chemically resistant plating or etch resist. The unexposed photoresist is washed away during development, to leave the circuit pattern.

5. After imaging, the circuitry is electroplated with copper to a thickness of 0.001 to 0.002 inch. The copper plating is then plated over by another metal,

usually tin-lead. This second metal performs two functions: (1) It preserves the solderability of the circuitry by protecting the copper from oxidation. (2) It serves as an etch resist for the subsequent copper foil etching operation. Tin-lead is not the only metal used for these purposes, but it is the most common and easiest to use. Nickel, tin, and tin-nickel are used for special purposes.

6. After the circuit has been plated to the correct thickness, the resist (dry film photoresist or screened on ink) must be stripped by a solvent. This will bare the unwanted copper foil which must be etched away to leave the plated circuit.

Copper foil is etched by spraying the etchant to both sides of the panel as it moves on a conveyor. The speed of the conveyor and the type of etchant used play major roles in yielding straight side walls with minimum undercut. Sulfuric acid/hydrogen peroxide, ammoniacal/chloride, cupric chloride, and ferric chloride are common etchants. After etching, the panels show tin-lead covered copper circuitry on the epoxy/fiberglass substrate.

7. Contact finger plating is the next operation in manufacturing the printed circuit. Contact fingers are the rows of tabs along one or more sides of the boards. These tabs fit into connectors in the electronic equipment. They must be durable and resistant to tarnishing and oxidation. A strip of plater's tape is applied to mask off the contact fingers from the rest of the board. Tin-lead is chemically stripped from the contact fingers, and nickel, then gold are plated. Nickel serves as a wear resistant barrier between the copper and the gold. It prevents atoms of gold and copper from migrating into each other; this preserves the electrical conductivity of the gold. Gold is used because of its excellent conductivity and resistance to oxidation.

8. Reflow (tin-lead fusing) is the next operation. Tin-lead is a dull gray metal, very porous and easily oxidized. The reflow operation melts the tin-lead for a few seconds; this fuses the two metals into the bright, shiny, corrosion resistant alloy known as *solder*. Tin-lead plating is sometimes called solder plating, because solder is an alloy of tin and lead. However, the actual alloy is not formed until reflow.

Reflow is accomplished by two common methods, infrared and hot oil. A conveyorized infrared oven is the fastest method of all. However, there are times when hot oil is easiest and more effective to use: small quantities, and thick panels with large ground shield areas. After being reflowed the panels will exhibit shiny traces and holes, with gold contact fingers, and begin to look like what most board users think of as a printed circuit.

9. Inspection is usually performed next. Broken traces or short circuits can be fixed easily at this stage. There is no substitute for quality workmanship and no one is more qualified to inspect the work being done than the operator and supervisor at each operation. However, accidents in processing and handling do occur. The in-process inspection affords an opportunity to take stock of quality factors and highlight areas needing attention.

10. The next operation is the application of soldermask and nomenclature/legend. Soldermask is an epoxy barrier applied to one or more sides of the panels. It prevents solder bridges from forming during the assembly wave soldering operation performed by the board user. It is silk screened on and baked to cure. After soldermask, an epoxy nomenclature, or legend, is silk screened on to one or both sides also. This silk screening operation is performed using artwork supplied by the board user to identify locations on the boards. It, too, is baked to cure.

11. Fabrication is the final operation performed in manufacturing printed circuits. Fabrication is the operation of cutting the board from the panel. A panel may contain one board (one up) or contain several boards (two up, three up, etc.) Each board has its own special shape and must be routed from the panels to meet tight dimensional requirements. Beveling and second drilling are also performed in Fabrication.

Printed circuit boards contain holes which are plated through, and some which are not to have any plating in them at all. Since the electroless copper process, will deposit copper inside all drilled holes, a second drilling step is needed. Second drilling is simply the operation of drilling holes which were not drilled when the panels were drilled the first time. Sometimes, because of extremely tight dimensional tolerances (± 0.002 inch), non-plated holes must be first drilled, then plugged during plating to prevent them from being plated. This is typically done for the case of tooling holes to align boards for automatic component insertion machines. It is extremely time consuming. The decision to second drill or to plug is made by the planning engineer, after reviewing the drawings for that printed circuit board.

Beveling, also called chamfer, puts a tapered edge on the contact fingers; this facilitates loading printed circuit boards into connectors.

12. If hardware is required by the board user—terminal lugs, standoffs, eyelets, connectors, etc.—it is added after fabrication.

13. The boards are given a final inspection, usually on the basis of an A.Q.L. (acceptable quality level) sampling plan. Each board manufacturer must determine the A.Q.L. which proves reliable for them. All that remains is to package and ship.

Multilayer printed circuits are manufactured in substantially the same way as double sided boards—once the inner layers have been laminated into a panel. The basic process is similar in the Planning and Photo departments. There are added considerations, since all the layers must register virtually perfectly:

1. Thin epoxy/fiberglass and copper laminate is blanked to panel size.
2. Special tooling holes are punched, or drilled, into the artwork and laminate. These tooling holes are used to align the artwork to the laminate, and to align the layers during lamination into panel form. They assure proper registration and orientation throughout the process, including drilling.

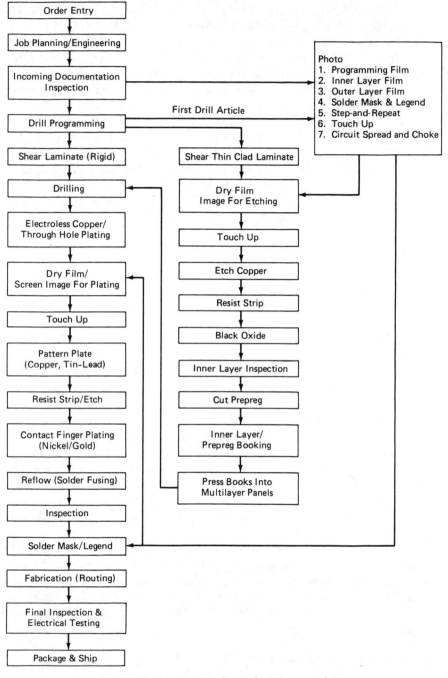

Fig. 1-1. Basic sequence of manufacturing operations.

3. The inner layer laminate is imaged, usually with dry film photoresist. The image is called a *print and etch pattern*. There will be no plating.
4. The layers are etched to remove unwanted copper foil, and the resist is stripped. This leaves copper circuitry on the epoxy/fiberglass substrate.
5. The etched layers are immersed in a hot caustic oxidizing bath. This bath forms dendritic crystals of black oxide. The black oxide improves bonding during multilayer lamination.
6. The layers are carefully sequenced and oriented during the stackup operation (also called *booking*). During booking, sheets of prepreg (fiberglass coated with epoxy resin) are placed between the inner layers. This will be used to bond them together. The number of sheets is determined by the thickness requirements for the board.
7. The inner layers and prepreg are pinned together between caul plates, and loaded into the multilayer press.
8. Pressure and temperature are applied to cure the epoxy and bond the layers into one panel.

After the panels are removed from the press, they are drilled and processed like a double sided board except for smear removal prior to electroless copper. Drill bits reach hundreds of degrees in temperature. This smears epoxy around the inside of the holes. If this happens on a multilayer board, the smear could prevent an electrical connection from being made during electroless copper plating. The smear must be removed by chemical means, such as concentrated sulfuric acid or chromic acid. Plasma is also used today for removing epoxy smear. For a summary of the entire procedure, see Fig. 1-1. A final word about tin-lead:

Tin-lead is preferred by board manufacturers and users for three reasons: (1) It is plated over the copper circuitry to serve as an etch resist during removal of unwanted copper foil. (2) When reflowed, tin-lead protects the circuitry, especially the pads and plated through holes, from oxidation; this preserves the future solderability of the printed circuit. (3) Solder is the easiest metal to solder to. Electrical components are soldered into printed circuit boards; if the boards are already covered by a layer of solder, a reliable joint can most readily be formed.

As mentioned before, other metals are used, including gold. None have the outstanding solderability of solder, though. A trend is underway in the electronics industry toward printed circuits which have solder only on the pads and holes. The rest of each board is soldermask over bare copper. The big advantage of this type board is greater reliability for fine line, tightly spaced circuitry.

BARE BOARD ELECTRICAL TESTING

One service increasingly offered by the board manufacturer is that of bare board electrical continuity testing. This is the only reliable way to determine electrical integrity. Once the bed of nails test fixture has been built, little expense is involved in testing. Unwanted short circuits or open circuits are detected and their location printed out on a tape. This makes it easy to find problems without depending on visual inspection methods. Considering that some printed circuit boards can cost more than a thousand dollars, and may have thousands more loaded as components, it makes little sense not to use bare board electrical testing for large board quantities.

DIELECTRIC SUBSTRATES

Printed circuits are built on a variety of substrates; the particular substrate used for any given design is chosen on the basis of function, operating environment, and cost. Functional considerations include electrical, mechanical, and flammability factors. Environmental operating considerations include temperature and humidity conditions, as well as vibration and thermal stresses. Cost must be considered, since it is desirable to use the least costly material which will fulfill the preceeding requirements.

Epoxy/Fiberglass

Probably the majority of printed circuits used in radio, television, computers, and telecommunication equipment are manufactured on this material; as are printed circuits for many other uses. Epoxy/fiberglass is the single most important laminate in the U.S. printed circuit industry. There are variations which enhance fire extinguishing properties, drilling, punching, and machining properties, properties of moisture absorption, chemical and heat resistance, and mechanical strength. These are listed below.

NEMA	MILITARY	DESCRIPTION
G10	GE	General purpose epoxy/fiberglass woven fabric
FR-4	GF	Same as G10, but fire retardant, FR-4 is much more widely used than is G10, and is used where the material requirement calls for G10. FR-4 is sometimes referred to as G10FR.
FR-5	GH	This laminate is a similar to FR-4 except that it is very strong and hard. FR-5 is used at higher operating temperatures than FR-4, where it is important to maintain electrical and mechanical properties. It is difficult to drill and machine because of its hardness.
	(HOWE 1102)	This is a modified-epoxy/fiberglass laminate for high temperature applications. It has poorer electrical and flame retardant properties than does FR-5. However, it is far easier to machine and to drill. Like FR-5 and polyimide, it is used in high temperature environments and is especially used for burn-in boards. It is a black laminate and will not show discoloration due to prolonged heating. It is manufactured only by Howe Laminates and has no military or NEMA designation.

Epoxy/Fiberglass/Paper Composites

These are manufactured to reduce cost, and to alter the mechanical properties. Typically fiberglass and epoxy resin are applied over a paper core. Laminates which are made on paper will tend to absorb moisture to a greater extent.

NEMA	MILITARY	DESCRIPTION
CEM-1	None	Both CEM-1 and CEM-3 are epoxy/fiberglass
CEM-3	None	over a paper core, differing only in the type of paper used. They are less expensive, more punchable substitutes for FR-4.
FR-3	PX	This laminate contains no fiberglass, only epoxy over paper, It, too is a less costly punchable substitute for FR-4.

Phenolic/Paper

This material is one of the cheapest laminates made. It has been widely used for circuits which are punched from the panel form. With greater need for enhanced chemical resistance, wave soldering, and greater overall reliability in electrical components, the use of phenolic resin over paper has diminished greatly in the U.S. electronics industry.

NEMA	MILITARY	DESCRIPTION
XXXPC	None	Phenolic resin over paper base.
FR-2	None	This is virtually the same as XXXPC, except that it is fire retardant. It is generally used when the specification calls for XXXPC.

Polyester/Fiberglass

This is a fairly cheap laminate, with good moisture absorption and electrical properties; although it is inferior to epoxy substrates, especially as regards mechanical strength and heat resistance. It is finding wider acceptance for circuits used in extremely high volume, such as electronic game cartridges and disk drives.

NEMA	MILITARY	DESCRIPTION
FR-6	None	Polyester/fiberglass has not gained wide acceptance as yet. It has been manufactured primarily with nonwoven fiberglass, which has an esthetically unappealing surface.

Polyimide/Fiberglass

This material has excellent electrical and mechanical properties, which are maintained at elevated operating temperatures. Polyimide is very hard and requires frequent drill bit changes during drilling. A major application is for burn-in boards, where electrical components are tested for reliability. It is expensive, and several epoxy/fiberglass substitutes have been developed. Polyimide is also used for multilayer construction to eliminate z-axis expansion in high reliability applications—usually for the military.

NEMA	MILITARY	DESCRIPTION
None	GI	Polyimide resin on woven fiberglass.

Teflon/Fiberglass

This material is used for printed circuits requiring very low dielectric constants. Teflon is soft and difficult to drill and machine cleanly. Prior to through-hole plating, the drilled holes must be etched with a metallic sodium compound to improve copper/teflon adhesion. Teflon/fiberglass is the most expensive of the widely used printed circuit substrates.

NEMA	MILITARY	DESCRIPTION
GX/GT	GX/GT	Both GX and GT are teflon resin over woven fiberglass. The only difference between them is the testing method and certification supplied. The dissipation factor and dielectric constants for GX must be known, fall within a narrow tolerance, and be so tested in the X-band frequency. Material labeled GX meets closer tolerances than does material labeled GT.
GR/GP	GR/GP	GR and GP are teflon over nonwoven glass fibers. The differences between GR and GP are analogous to the differences between GX and GT. The nonwoven glass offers slightly lower dissipation factors at a slightly more uniform dielectric constant.

Other Materials

1. Teflon over ceramic, for electronic applications requiring high dielectric constants of 10.2 or more.

TABLE 1-1. Most common copper clad laminates.

NEMA EQUIVALENT (LI 1-1971)[5,6]	LAMINATE COMPOSITION	LAMINATE CHARACTERISTICS	COLOR[3]	THICKNESS[3] RANGE	COPPER FOIL[3] CLADDING	MAXIMUM OPERATING TEMPERATURE	Underwriters Laboratories File No. E37002			SPECIFICATIONS AND MATERIAL DESIGNATIONS[5]				
							PLASTICS RECOGNITION	C.C.I.L.[4] RECOGNITION	FIRE RESISTANCE	MIL-P-13949E[6]	ISO R 1642	IEC 249-2	D.I.N. 40 802[6]	B.S. 4584[6]
XXXPC	phenolic/paper	Commercial electric and electronic applications. Stable electric properties in high humidity and temperature conditions. Good punching at room temperature. Good dimensional characteristics.	Tan, Natural (brown)	0.031" to 0.093" (0.80mm) (2.40mm)	¾, 1, 2, ounce	125°C	yes	yes	94 HB	—	PF CP 4	1-IEC-PF-CP-Cu	PF-CP-1	PF-CP-Cu-7
FR-2	phenolic/paper	Fire resistant. Very good punching qualities when warm.	Tan	0.031" to 0.093" (0.80mm) (2.40mm)	¾, 1, 2, ounce	75°C (<0.057") 105°C (≥0.057")	yes	yes	94 V-0	—	PF CP 5	—	PF-CP-2	PF-CP-Cu-8
G-10	epoxy/glass fabric	Superior mechanical strength, high electric properties over wide humidity and temperature ranges. Excellent chemical resistance, low warpage, and predictable machinability. Dimensionally stable.	Natural (very pale green)	0.020" to 0.125" (0.50mm) (3.20mm)	All thicknesses	130°C	yes	yes	94 HB	FL-GE	EP GC 1	4-IEC-EP-GC-Cu	EP-GC-1	EP-GC-Cu-2
FR-4	epoxy/glass fabric	Fire resistant version of G-10.	Natural (very pale green) Tan, Blue, Green	0.020" to 0.125" (0.50mm) (3.20mm)	All thicknesses	130°C	yes	yes	94 V-0	FL-GF	EP GC 2	5-IEC-EP-GC-Cu	EP-GC-2	EP-GC-Cu-3

CEM-1	epoxy/ glass fabric, cellulose paper core	Composite with electrical properties, moisture resistance and thermal stress equal to G-10. Good punching at room temperature, excellent drilling qualities.	Tan, Blue	0.031" to 0.093" (0.80mm) (2.40mm)	¾, 1, 2 ounce	130°C	yes	yes	94 V-0	—	—	—
CEM-3	epoxy/ glass fabric, glass paper core	Composite with electrical properties, moisture resistance, and thermal stress equal to G-10. Very good punching at room temperature, excellent drilling qualities	Natural (light brown), Tan	0.031" to 0.093" (0.80mm) (2.40mm)	All thicknesses	130°C	pending	pending	94 V-0	—	—	—
G-11	epoxy/ glass fabric	Similar to G-10. Retention of electrical and mechanical properties in heated environments.	Natural (pale green)	0.031" to 0.125" (0.80mm) (3.20mm)	¾, 1, 2 ounce	140°C (<0.057") 170°C (≥0.057")	yes	no	94 HB	FL-GB	EP GC 3	—
FR-5	epoxy/ glass fabric	Fire resistant version of G-11.	Natural (pale green)	0.031" to 0.125" (0.80mm) (3.20mm)	¾, 1, 2 ounce	140°C (<0.057") 170°C (≥0.057")	yes	no	94 V-1	FL-GH	EP GC 4	—
—	triazine/ glass fabric	Superior performance and excellent electrical stability at very high temperatures and humidity. Excellent resistance to hostile processing conditions, including heat and chemicals. Very high mechanical strength and very low dimensional change.	Natural (yellow-green),	0.020" to 0.125" (0.50mm) (3.20mm)	All thicknesses	220°	no	no	94 V-0	—	—	—
—	polyimide/ glass fabric	Most stable performance at highest temperatures. Unaffected by processing chemicals.	Natural (dark amber)	0.020" to 0.125" (0.050mm) (3.20mm)	All thicknesses	240°C	no	no	94 V-1	FL-GI	—	—

2. Polyethylene, for low dielectric applications similar to teflon/fiberglass.

3. Polystyrene, for low dielectric applications as a substitute for teflon/fiberglass.

4. Aluminum and aluminum/copper laminated substrates are used in microwave and other types of circuits where there is a need for rigidity or heat sinking. See Table 1-1.

5. OhmegaPly is the brand name of laminates, both rigid and thin clad, which contain a layer of resistor material. The resist is etched onto the board surface. Using a material such as this would eliminate surface mounted resistors to a large degree and reduce the overall size of the board.

6. Polyimide/Kevlar, for high temperature, low thermal expansion applications, such as chip carriers, where the silicon chip is bonded directly to the surface of the printed circuit. The polyimide resin is coated over DuPont Kevlar fibers, not over fiberglass.

2
Computer Aided Design and Design Automation

Vectron Graphic Systems, Inc.
Santa Clara, California

This is a concise introduction to the basics of computer aided design (CAD) and design automation for printed circuits. It is not "everything you ever wanted to know about computer aided design," but it will provide enough information for you to find out the rest. It will not tell you whether or not you should install a computer system in your company's printed circuit design facility or tell you which one to install if you decide you need one, but it will give you some general considerations to think about, which will help you make these decisions. After reading this, you can find more specific information about how these computer systems perform, and how much they cost, by discussing them with the sales people for computer aided design systems.

First, let us discuss what is meant by the two terms *computer aided design,* and *design automation.* A CAD system is one which makes printed circuit board artwork from a layout, which has already been designed. Specific information about the layout is entered into the computer system by one of two methods discussed below; and the computer draws the artwork which is used to manufacture the printed circuit. The computer aided design system is a substitute for the process of hand-taping the printed circuit board artwork. Such systems are referred to as *digitizing* systems, after the digital process of entering the layout into the system.

A *design automation* system, by contrast, is one in which the computer does all or part of the designing task for the printed circuit boards. The information from a schematic diagram (a list of what is connected to what) is entered into the computer, and the computer goes about the task of designing the printed circuit board. The computer may ask for help in placing certain packages, or it may leave some traces unrouted to be completed by hand; but, in general, the computer is doing the designer's job. Of course, when the designing job is finished, the design automation system draws the artwork, just as the computer aided design system does.

USING CAD

The Advantages of CAD Artwork

Those who work with printed circuit boards know that taped-up artwork is no joy. Computer aided design systems produce artwork that eliminates some of the chronic problems of tape-ups.

First, the artwork produced on a computer aided design system is more accurate than a hand-made tape-up artwork. The accuracy of a tape-up is limited by the steadiness of the hand of the taper. If the taper is unskilled, or negligent, or ill, the artwork quality suffers. By contrast, computer aided design artwork is accurate to within about half a mil, regardless of the skill level or attentiveness of the operator. With the computer drawing the artwork, lines that are supposed to be on 25 mil centers are on 25 mil centers, not 24 and not 26. If the computer is told to place 8 mil lines with 8 mil air gaps, that will be just the result.

More accurate artwork brings a number of advantages. The boards will be more manufacturable. The printed circuit manufacturer will have no problems due to artwork induced short or open circuits, there will be no layer-to-layer misregistration, or metal bridges. Also, the printed circuit assembler will be able to make use of auto-insertion equipment for loading IC's and other components into the board. In fact, the computer aided design system should be able to punch the tape which drives the auto-insertion machinery. The greater accuracy of the artwork will allow the designer to meet military specifications, which are impossible to meet with hand taping.

A second advantage of CAD artwork is improved stability over the hand tape-up. With time, the tape of a hand tape-up can creep along the Mylar base, leaving artwork which is even less accurate than it was initially. An even worse experience is that of removing a tape-up from its folder, only to find two or three pads and pieces of tape still in the bottom of the folder. Where were they on the artwork before falling off? Since CAD artwork is made from a single sheet of photographic film, these problems do not occur.

A third advantage of CAD artwork over taping is speed. Most people can enter a design into a computer aided design system faster than they can tape it. A very fast taper can keep up with a CAD operator, but only at the expense of accuracy. Even the fastest taper has no chance, however, if any section of the board pattern is repeated. While the taper must tape the repeated section, the computer aided design operator simply tells the computer to do it. In the extreme, a memory array which would require several days of taping, can be completed by a computer aided design operator in ten to fifteen minutes.

A related benefit of a computer aided design system is the ability to edit quickly, easily, and accurately. How many times have you lifted a piece of tape from artwork, only to have a tangle of other pieces come off with it? With a

CAD system, the computer always makes it clear which line is to be moved or removed.

By entering layouts into a computer aided design system, two other benefits can be had. First, the CAD system can punch a paper tape that will drive a numerically controlled drilling machine for drilling the printed circuit boards. This tape will direct the drill to the centers of the pads on the board. and ensure the use of correct drill bit sizes. Second, the CAD system can draw the fabrication and assembly drawings with a minimum of difficulty. Many CAD systems can draw on vellum, or Mylar, with a variety of pen types; documentation grade drawings are made available as fast as the artwork is completed. In addition to using less drafting time, a CAD system always produces drawings with perfect, even lettering, correctly drawn symbols, and information that matches the printed circuit artwork—because it came from the same data base.

What is the catch to computer aided design? Cost. The advantages of CAD artwork do not come unless you have a computer aided design system. To decide if you need a CAD system, you must compare the costs of the system with your costs of personnel time, defective boards from inaccurate artwork, and rework time. In contrast to hand taping, where only a light table, a stool, and some supplies are needed; the CAD system requires a large investment in equipment, procedures, and personnel training. The following section discusses the equipment and people needed for computer aided design.

Equipment Necessary for CAD

Computer aided design for printed circuit artwork requires a number of special pieces of equipment. The first of these pieces is the device that actually draws the finished artwork, the photoplotter. A photoplotter uses a light beam to make marks—pads and lines—on a piece of photographic film. It moves the light beam around the film on a mechanical arm, to draw the printed circuit artwork directly onto the film. The film is then developed, to yield the finished artwork. Most photoplotters get their data directly from a computer tape. The tape is mounted on the photoplotter with the film, the "start" button is pushed, the photoplotter reads the tape, then draws the artwork.

A photoplotter generates actual size artwork, rather than a two-to-one enlargement typical of tape-ups. The photoplotter must be able to plot different pad sizes and line widths; hence, it must be able to change the size of the light beam which it uses to draw upon the film. This change is accomplished by shining the light beam through an aperture; which makes the beam the shape and size of the desired pad or line width. The photoplotter contains a wheel, in which is mounted an entire set of apertures. The computer tape which tells the photoplotter where to plot must also tell it which aperture on the wheel to use. Although the wheel has only 24 positions on it, it can be replaced with another source of apertures. Wheels containing letters or logos are also available.

Photoplotters are notoriously temperamental pieces of equipment, requiring constant care and attention. The light beam must remain focused, and the mechanism clean, in order to maintain the desired accuracy. Light intensity must be varied by using filters, in order that exactly the correct amount of exposure may be achieved. Even computer aided design shops which maintain a full time professional to run their photoplotter expect to botch a few pieces of film now and then. Photoplotters must reside in photographic darkrooms, with attendant requirements for lightproofing, clean air, and constant temperature and humidity. Because of these drawbacks, many shops equipped with CAD systems do not bother with a photoplotter. Instead, they use their computer aided design system to make tapes which are sent out to photoplotting services.

The next piece of equipment needed is a computer. The computer makes the tape which is later given to the photoplotter. A computer system for CAD work needs the standard list of computer accessories: disk drives to store data, magnetic tape unit to make photoplotter tapes, a console for directing the computer, a printer for lists which must be printed, and so on.

The next piece of equipment needed is a digitizer, or digitizing terminal, for entering printed circuit layouts. The digitizing terminal has a keyboard, a digitizing tablet (or table), and a video screen. The digitizing tablet is a special type of drafting table with imbedded electronic sensors. These sensors allow the computer to detect the location, on the table, of a special stylus or cursor which comes with the table. Data about the circuit layout is entered into the computer by pointing to the layout with the stylus. The video screen draws a picture of the layout that has been entered. This is necessary for verifying the correctness of the data which has been entered.

Another required piece of equipment for a computer aided design system is a pen plotter. Pen plotters are similar to photoplotters, except that they draw on paper (or vellum or Mylar) with ink, instead of on film with light. The pen plotter is plugged directly into the computer for running, rather than being driven by a tape like the photoplotter. Pen plots are indispensable for checking correctness of the final circuit which has been entered. It is cheaper and faster to make a pen plot, and then make any needed corrections, before using the photoplotter.

The final and most important item of equipment requirement for a computer aided design system is the software. It is the software which transforms the equipment from just a computer system into a computer aided design system. Without the software, the computer will not know how to draw on the plotters. Without software, the digitizing terminal will not know how to use the tablet. The system will not put out photoplotting tapes without software. The software makes the system run well (or poorly). NOTE: It is the software that determines how useful the CAD system will be. Companies manufacturing computer aided design systems have devoted more time to developing their software than to anything else.

People Necessary for CAD

There are four job classifications for people working with CAD systems: two of these are already working in the printed circuit design shop, the two others will be new. The printed circuit designer is one of the existing classifications. The computer aided design system does not actually design printed circuits, it just draws the artwork; the designer is still required. The next type of person needed is the digitizer. It is the job of the digitizer to enter the designer's layouts into the computer system. The digitizer will take the place of the taper, already found in the printed circuit design shop. The skills required for digitizing are a little different than those required for taping. Although both need a rudimentary knowledge of printed circuit artwork, the taper also needs the manual dexterity to work with the tape. The digitizer needs to be trained in the use of the computer system; training which can usually be accomplished in a few weeks.

A new type of person who will be required is the computer operator. The computer operator is responsible for general system maintenance, performing data backups, telling the computer to make photoplot tapes, or plots, and so on. In very small operations, the computer operator may also be the same person as the digitizer, but a system of any size requires a separate person to take care of the computer operator's tasks. The computer operator needs about the same amount of training as the digitizer, if not a little more.

The fourth type of person needed for a computer aided design system is the manager. The manager keeps the work flowing smoothly through the system. He/she needs to know not only about the general management of a printed circuit design operation; but, also about the operation of the computer system, and the possibilities and limitations inherent in that system. The manager must keep the other people working as a well oiled machine.

The CAD Process

The following is a discussion of the process by which a printed circuit designer's layout is turned into finished artwork, with the use of a computer aided design system.

The first process which must occur is that of digitizing the layout, entering it into the computer. A layout of the circuit is taped onto the digitizing terminal, which includes the special electronic drafting table with the stylus (Fig. 2-1). Next, the digitizer uses the stylus to point to items of the layout. To enter a trace the digitizer points to one end, and then the other; the computer draws

Fig. 2-1. Digitizer using an electronic tablet to enter circuit layout into the computer.

a straight line between the two points. To insert a pad, the digitizer points to it, and the computer adds it to the board. To insert a package, the digitizer points to the first pin and tells the computer what kind of package to insert; and the computer inserts the package. To repeat some feature on the board— an array of memory traces, for example—the digitizer indicates what part of the layout is to be repeated, and tells the computer where on the layout it is to be duplicated. The computer can step and repeat even very complex patterns instantly. To work with a different trace width, the digitizer simply tells the computer what trace width he/she wants to work with now, and the computer complies. If the board contains a ground plane, the digitizer indicates this fact, digitizes an outline of the plane, draws in the tie bars; and the computer does the rest.

While the digitizer is busy entering the layout, the computer is cleaning up the entry automatically. One of the most valuable functions a CAD system performs is to evenly space lines and pads on the board. The digitizer tells the computer that, for example, everything on the board is to be on 25 mil centers. Subsequently, if the digitizer tries to insert a line that is 23 mils from the previous line, the computer will simply move the line 2 mils over to make the required spacing. The computer can also be told that all lines are to be verticle, horizontal, or at 45 degree diagonals. If the digitizer then indicates a line that sags a little at one end, the computer will straighten it out automatically. The computer is told what line width to use, and it uses just that width.

After the digitizer has completed entering the layout, the CAD system produces a pen plot of what the entered circuit will look like on film. This plot can be made in multiple colors, to look like the original layout, or the system can plot each side of the board separately. This plot is used to check that the data entry has been done properly. The digitizer, or a person other than the digitizer, must verify that all the lines and pads have been entered, and that all the traces go to the correct locations. Typically, the plot is checked against the original schematic diagram. It is not necessary to check spacing of the plot; the computer system will already have corrected that. If the digitizer knows his/her job, then lines that appear to be correctly spaced are exactly correct (after all, the computer drew them), pads are exactly the right size, and everything is in exactly the right location. There is no chance that something will be a few mils off one way or the other. With digitizing systems, things are either right, or they are very obviously wrong.

After the plot has been checked and corrected, it is given back to the digitizer, who goes back to the terminal and makes any required changes. The changes are made in a manner similar to the original data entry. The digitizer points to the offending line or pad, and indicates where it should be, or that it should be removed. The computer will move or remove discrepant items. The digitizer can check on the video screen that the changes have been made correctly. When changes are completed, the job is ready for photoplotting; a second pen plot may be run first, to spot check the corrections.

An alternative method of entering a layout is to enter the information from the keyboard, using a net list. The digitizer looks at the connections which are indicated by the schematic diagram (Fig. 2-2). The advantage of working with a net list is that layouts can be quickly entered. However, this is done at the expense of accuracy. Generating a net list increases the possibility of error.

When the tape is done, it is taken to the photoplotter—either in-house, or an outside photoplotting service—and the artwork is made. Some last minute changes can be made at this point, if desirable. The board is usually reduced from 2:1, the scale at which it is digitized, to 1:1 for photoplotting. However, 2:1 artwork can also be photoplotted if required. The sizes of pads and traces can also be changed en masse if necessary.

After photoplotting, the artwork is checked. It must be verified that the photoplotter was in focus, that the right pad sizes and line widths were loaded into the photoplotter, and that the film has been properly developed. If no problems come to light, the artwork is ready to be sent to the printed circuit board manufacturing shop. The CAD system draws the pen plots of the fabrication and assembly drawings as the artwork is being photoplotted so that these drawings are ready at the same time (Fig. 2-3).

```
*BSID U11 12 XA 1          *PA7 U5 28 J2 16
*CLK U1 37 U12 8           *PB0 U5 29 J2 17
     U5 3                  *PB1 U5 30 J2 18
     U17 1                 *PB2 U5 31 J2 19
*CLK.BAR U12 12 RP5 8 J1 49 *PB3 U5 32 J2 20
*HOLD U1 39 U16 2          *PB4 U5 33 J2 21
*HLDA U1 38 U12 9          *PB5 U5 34 J2 22
*HLDA.BAR U12 11 RP5 9 J1 41 *PB6 U5 35 J2 23
*IO.BAR/M U12 16 RP5 4 J1 33 *PB7 U5 36 J2 24
*IO/M.BAR U1 34 U13 9      *PC0 U5 37 J2 25
          U13 13 U11 11 U12 4 *PC1 U5 38 J2 26
          U5 7            *PC2 U5 39 J2 27
*IO/MB.BAR J1 34 RP4 9 U11 9 *PC3 U5 1 J2 8
*IN0 U6 26 J2 33           *PC4 U5 2 J2 7
*IN1 U6 27 J2 34           *PC5 U5 5 J2 6
*IN2 U6 28 J2 35           *RD.BAR U1 32 U11 2
*IN3 U6 1 J2 32                    U4 22
*IN4 U6 2 J2 31                    U5 9
*IN5 U6 3 J2 30                    U13 2
*IN6 U6 4 J2 29                    R4 2
*IN7 U6 5 J2 28                    U15 5
*INT U1 10 U12 18         *RDB.BAR J1 32 RP4 2 U11 18
*INTA.BAR U1 11 U11 15    *RESET U1 3 U12 6
*INTAB.BAR J1 43 RP4 5 U11 5       U5 4
*IRD.BAR U13 3 U10 1 U10 19 *RESET.BAR U12 14 RP5 6 J1 47
*IWR.BAR U13 6 U9 1 U9 19 *RST5.5 U1 9
*PA0 U5 21 J2 9                   XA 13
*PA1 U5 22 J2 10          *RST6.5 U1 8
*PA2 U5 23 J2 11                  XA 14
*PA3 U5 24 J2 12          *RST7.5 U1 7
*PA4 U5 25 J2 13                  XA 15
*PA5 U5 26 J2 14          *S0 U1 29 U12 17
*PA6 U5 27 J2 15          *S0.BAR U12 3 RP5 3 J1 40
```

(a)

Fig. 2-2. (a) Example of a net list, which can be used to enter circuit layout into the computer. (b) The schematic diagram is the beginning of a printed circuit. The circuit artwork must contain all the connections shown in the schematic.

How to Judge CAD Systems

Computer aided design systems for your printed circuit artwork cost a lot of money. Obviously, a lot of thinking and evaluation are needed to determine which one is best. In addition to complete stand alone systems, your options include smaller systems; timesharing services; service bureaus; and of course, continuing to perform hand taping of the artwork. Depending upon the specific requirements of your company, any of these might be the correct decision.

The first decision which must be made is to determine if a CAD system is needed at all. Compare the cost of a CAD system with the costs of continuing to cope with current artwork generation problems. The costs of the CAD system are the costs of the original installation, plus the cost of ongoing maintenance, and the cost of additional people which may be needed to run the system. These are all costs available from CAD salesmen.

The cost of not using CAD is more difficult to calculate, because most of the costs are indirect and intangible. Here is a list of considerations which have led

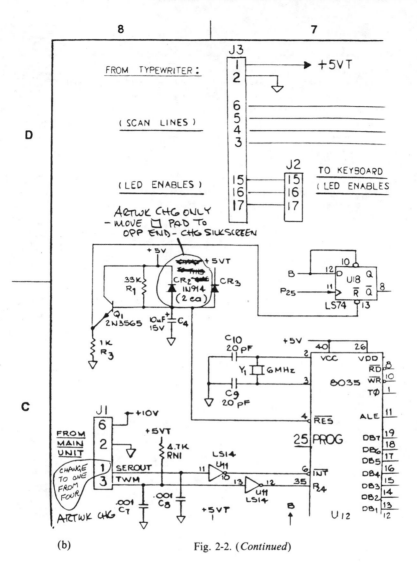

(b)

Fig. 2-2. (*Continued*)

many companies to conclude that a computer aided design system is cost effective for them:

1. How much is it worth to be able to get artwork faster—hence get your products to the market faster? The CAD system will speed up the process dramatically.
2. How much is it worth to avoid downstream manufacturing difficulties caused by inaccurate artwork? CAD systems help improve manufacturing yields and reduce field failures.

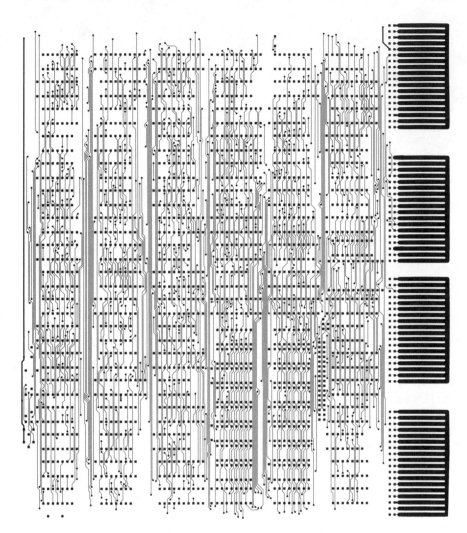

Fig. 2-3. Example of photo-plotted artwork.

3. How much is the ability to make design changes quickly worth, in order to be able to respond to changing market demands?
4. How much is it worth to be able to hire and train a relatively unskilled digitizer when you need one, rather than scour the countryside for a new taper when the work load increases? The CAD system requires less experienced people than does a manual taping operation.
5. It is possible for a CAD system to increase your department's efficiency enough that there will be no need to hire additional people to handle increasing work loads.

What are the critical factors which make one system better than another? Before answering this, it is important to dispell some common myths about the components of the system. The photoplotter does not determine how useful the system is. All photoplotters are, for the most part, very much alike in terms of accuracy. There are differences in speed of photoplotting, but the photoplotter will probably never be the bottleneck in any operation, as long as it is not out of commission. An exception to this would be when there are a dozen or so digitizers all working on different circuits at the same time. Also, the type of computer in the system should not be used to judge the system. A salesman may say that a given system is faster and more accurate because it contains a 32 bit minicomputer; but in truth, the speed and accuracy of a computer aided design system are more dependent upon the ingenuity of the people who designed and wrote the software than upon the particular hardware upon which the system runs. Lastly, the features, such as color graphics, should not be used to judge the system. Although color may be impressive, and useful in some applications, it is not that useful for a double sided printed circuit, and it is not that expensive. Do not assume that any particular feature is all that useful, until you have demonstrated it for yourself.

The prime consideration for a computer aided design system must be throughput. CAD systems are justified on the basis that they generate more artwork than competing methods. A system that cannot get the work through is worth nothing. Here is a list of considerations that can be used to help judge the value of a CAD system:

1. Decide which editing features are useful for your application. If your boards use silk screens, check on how easy it is to do silk screens. If you use buried power and ground planes, find out how much trouble they will be using a given system.

2. One feature of the CAD system that dramatically affects the speed at which work can be done is the speed at which the system can display the layout on the video screen of the digitizing terminal. Whenever a change has been made, the operator will want a display to check the correctness of the change.

3. It is also important to know how fast you can get pen plots. This is a function not only of the pen plotter, but also of how good a duty cycle you can expect. Will the plotter run all night unattended? Will the paper have to be changed after every plot? If you run a one or two shift operation, these two questions may be more important than the actual speed of the pen plotter.

4. How many digitizing terminals will be needed to get the work done? If the workload is large, the CAD system must be able to handle as many digitizing stations as is needed. Will it be necessary to have terminals at long distances from each other? Do not assume that a given system will handle it; not all of them will.

5. Make a determination, as best you can, of how much trouble you will have with a given system. All complicated machinery breaks down occasionally.

Find out from other users how reliable various CAD systems are. Find out how long before the CAD manufacturer gets the repairman out to fix it.

6. How much work will be lost if the system goes down? The last thing entered by the digitizer? Everything since lunch? A whole days work? Find out.

COMPARING DESIGN AUTOMATION TO CAD

A design automation system differs from a computer aided design system. The design automation system does the actual printed circuit board design, in addition to generating artwork from designs prepared by printed circuit designers. It does this with the addition of a large block of software written for that purpose. The software is intended to take the place, at least in part, of the printed circuit board designer. In other respects, design automation systems are similar to computer aided design systems: it takes about the same staff to run them, and they run with the same computer hardware, with the possible exception that a somewhat more powerful computer and a disk drive may be necessary. A design automation system usually has a CAD system embedded within it, so that the system can be used purely as a CAD system, which is necessary to make changes to digitized boards, or to input design automation into the system.

Advantages of Design Automation

Using a design automation system for designing printed circuits offers several advantages over the manual design process. Computers are well suited for this work, and function well in the printed circuit design environment. A design automation system offers fast design turnaround. Boards that might take a designer several weeks can be designed by the computer in a day or two. There are many reasons why computers are so much faster than a designer. First, the computer can check a large number of possibilities very quickly. Although most computers tend to bog down toward the end of a design (when the last few traces are being routed), in the beginning they are literally as fast as lightning, routing traces by following rote rules of good board design. Next, the computer spends no time thumbing back and forth among the pages of a large schematic. Design automation systems first compile a list of the interconnects to be made on a board; after this, the system needs only a few thousandths of a second to find out what is to be connected where. Finally, the computer spends no time with pencils. Once it decides where a trace is to go, it places it there almost instantly. There is no sharpening of pencils, no aligning of rulers, and no fuss with other mechanical aspects of drawing a layout.

A design automation system produces printed circuit designs that contain no rule violations. The computer is never guilty of squeezing just one too many traces through a tight spot on the board; nor does it ever carelessly run a wide trace a little too close to a large pad. The computer patiently checks every line it places on the board against every other line and pad for spacing violations. The design automation system also avoids careless mistakes in connecting things. The computer never looks at a 14 pin package and mistakes it for a 16 pin package. It never misreads the signal pin assignments out of the manufacturer's specification book.

For all these reasons, printed circuit board design departments can save money with design automation. Since the computer can design routine boards, the designers can turn to designing very complex boards, or boards in which signal shielding or signal length are of paramount importance—boards that the computer is likely to have trouble with. Manually designed boards must be exhaustively checked for correctness. With design automation, this is greatly simplified, since if the input to the computer is checked and is correct, then the design can be relied upon to be correct.

Design automation systems provide all the advantages of a CAD system. A design automation system photoplots the same accurate, stable artwork as a computer aided design system, yielding the same manufacturability and reliability in the artwork as boards from a CAD system. The design automation system also produces paper tapes for numerically controlled drills. Fabrication and assembly drawings are also provided from the same data base that creates the artwork.

The Design Automation Process

To use a design automation system for creating printed circuit designs, you must follow certain procedures. The net list of the board must be entered into the computer. This can be done by typing in a prepared net list, or by using the digitizing terminal to enter the information from a schematic diagram. If this latter method is used, then the design automation system must have a program that can produce a net list from a schematic diagram. If this latter method is used, then the design automation system must have a program that can produce a net list from a schematic diagram which has been entered.

Next, physical data about the board must be entered. The computer must be told the size of the board, the location of power and ground buses (if you want them located in a certain place on the board), the locations of connectors to the board, locations of areas where no traces should be routed, and so forth. Again, there are two ways to enter these data: the first is to type in the x- and y-coordinates of the various features of the board, and the second is to use the digitizing system to draw in the various features. If the digitizer is being used, it may be possible to design in a few critical traces before the computer gets started on the rest of the board.

Next, the package placement scheme must be decided. Not all design automation systems have package placement programming. When this is the case, packages must be placed by the designer, who types the information in, or uses the digitizer. Systems which do have package placement programming will consult the interconnect list then make decisions on how the packages are to be placed in relation to one another. Some systems request guidelines regarding how packages should be placed on the board; for example, whether the long or short axis of a chip should be placed along the horizontal direction, or how many rows and columns of ICs should be used. In situations where several electronic devices are packaged in one IC, the design automation system must also make decisions about logical groupings for the devices within the available ICs. Ideally, this package assignment should be done in conjunction with the package placement; but some systems do the assignment first, as a separate step.

After package assignment, the system must do pin assignment in situations where there is a choice of pins for use by certain signals. Although printed circuit designers typically do this as they are designing, computer systems usually do it first. This not only reflects the fact that it is easier to program this way, but also that computers do an adequate job of pin assignment in advance. The printed circuit designer must thumb through pages of schematics to make reasonable guesses about what pin assignments would be advantageous to minimize trace crossings. The computer can find out where each of the signals to a given IC are connected, and make pin assignments in a second or two.

Routing is the last phase of the automated design cycle. The computer now places traces on the board to make electrical connections. This program is usually the most time consuming part of the operation for the computer. The computer must look at interconnects one at a time, and insert each onto the board. It uses factors such as how crowded areas of the board are becoming, where previously placed traces are located, and which traces go in the same general direction, to optimize the route for the given trace it is trying to place. Most computers sort traces according to some criteria, such as routing the shortest one first, or perhaps the longest, or perhaps traces which run to a given connector on the board.

When the computer has finished with the board, a pen plot is run of the resulting design to see if any problems arose, and to print a list of traces that the computer could not insert, if any. If there is a list of "fails," the computer's design is turned over to a designer for completion. While computers are very good at placing ordinary traces in the early stages of designing a board, they are not very good at squeezing the last few traces onto a board. The more out of the ordinary a job is, the more difficulty the computer will have with it. Designers usually have little trouble placing the last few traces on the board, since their thinking is more flexible than the rules followed by computers. When the board has been completed, additions and changes which may be necessary are entered into the system by the computer aided design functions.

After the design has been checked for accuracy, it is photoplotted just as it would have been with a CAD system.

Judging Whether or Not a Design Automation System is Needed

Since design automation systems are a viable alternative to designing by hand, they must be given consideration. To do this, it must be determined whether the company can make economic use of a design automation system, by comparing the costs of the two methods.

The costs of a design automation system can be obtained from a salesman, and the criteria and numbers evaluated. The costs include the cost of the initial purchase price of the system, the costs of ongoing maintenance, and the cost of additional people that may be needed to run the system. The costs of designing printed circuit boards by hand is more difficult to determine. The cost of the designers' salaries and the overhead to sustain them are obvious. But, there are equally valid intangible costs to factor in. How much is it worth to a company to be able to get products to market several weeks ahead of the competition, by virtue of being able to save weeks on printed circuit design time? How much field maintenance will be saved by having printed circuit designs which are right, rather than ones which have had a few wires added to them? How much is it worth to be able to handle a surge in the work load by having the computer work overtime, instead of having to scour the countryside for extra designers, hire in job shoppers, or shed critical designs to an outside service?

Which Design Automation System Is Best?

There is so much information available on design automation systems from competing sources, that the problem is to figure out which information is important, and which is not. First, the statistics about computer speeds, disk drive speeds, programming techniques used on construction of the design automation programs, algorithm names, and so on, are unimportant. A design automation system for designing printed circuit boards requires a complex interaction of computer, disk drive, and software. There is little reason to believe that a design automation system with a "500 nanosecond cycle time" for its computer is any faster in designing printed circuits than a design automation system whose computer has a slower cycle time. There is no reason to believe that a router using "Lee's algorithm" will produce better routed boards than one using a nameless algorithm. The most important thing about a design automation system is whether or not it will design your boards effectively. Advertising puffery and sales rhetoric have never designed a printed circuit.

There is a great disparity of capabilities among the various design systems. The following is a list of some of the information which must be ascertained:

1. Is data entered by constructing a net list and then typing it in, or can a schematic be entered from which the computer will extract the net list?
2. Can physical data about the board size and location of connectors, and so forth, be entered through digitizing, or must the coordinates of each feature of the board be entered by typing?
3. Does the computer decide package location on its own, or must the designer do this?

The question which must always be in mind is whether or not the system can help design your boards. To help you answer, most design automation manufacturers tell you their "completion rates," which is the percentage of traces the computer designs before the job must be turned over to a designer. Most design automation systems cannot design every board to completion. The completion rate gives some idea of how close the system will come to the ideal of 100% completion.

There are two reasons to be skeptical of this number as a measure of how well the design automation system will design your boards. First, unless a benchmark is run on the design automation system, the completion rate told to you by the manufacturer will be the completion rate on someone else's boards. Since your boards may be totally different from someone else's, the completion rate on your boards may not be as great. It is not true that a router which does well on one kind of board will do equally well on another; very minor changes in board styles can markedly affect completion rates. Any of the following factors, for example, will affect completion rate:

1. Whether or not the board is long and skinny.
2. Whether or not it contains edge connectors.
3. Whether or not it contains buried power and ground.
4. Whether or not it contains chips rotated 90 degrees.

Second, what you are really interested in is the completability of the board, once the computer has finished with it. A board on which the computer has completed 95% of the traces, and used 100% of the available room, may not be as good as one where the computer has completed only 90% of the traces, but used only 85% of the space. How easily a design can be completed can only be judged by looking at sample boards which the computer has designed. Check to see how much channel space is left, and how indiscriminately the design automation system has placed feed through holes in the way of possible later traces.

There are a few other capabilitites of the router which should be determined. Find out if the router can handle varying sizes of pads. A router which does not understand that it must leave more clearance around larger pads can be aggravating to work with. Can the system use different widths of traces to route with? Ask about the methods used by the computer to design the boards, ignoring sales rhetoric:

1. Does it place feed throughs intelligently?
2. Does it conserve channel space wherever possible?
3. Does it minimize the number of feed throughs?
4. Does it save space around pins, to be used later for electrical connections to those pins?
5. Does it connect multiple pin nets well, or does it run parallel paths across the board carrying the same signal?
6. Are 45 degree angle traces used to snake connections through tight spots, or are 45 degree traces only added at the end for cosmetic purposes?

These sorts of considerations will help decide the utility of a design automation system for today's high density boards.

3
Digital Printed Circuit Design

Robert Gorge
Fermi National Laboratories

DEFINITIONS

Bill of Material. An itemized list of all components, hardware, wire, etc, used on a printed wiring board assembly

Crate (Module). A complete electronic unit which operates with one or more complete printed circuit card assemblies

Decoupling Capacitor. A capacitor mounted in close proximity to an integrated circuit, between its power and ground busses

DIP. Dual Inline Package; a common integrated circuit package; an equal number of pins arranged in two rows

Feedthrough (Via). A hole placed into a printed wiring card to accomplish a connection from one side of the card to the other

I.C. Integrated Circuit; any of various modular independent circuit devices encapsulated in plastic or ceramic packages

Mother Board. A printed wiring board which has one or more printed wiring board assemblies plugged into its connectors

Multilayer Board. A printed circuit board which has more than a top and bottom (2) layer; the additional layers are sandwiched between the two outer layers

PAL. Programmable Array Logic integrated circuit; a device which can be programmed to function in a large varity of circuit configurations

Parts List.—see Bill of Material

SIP. Single Inline Package; a device which has its pins arranged in one row

Terminal Area. The round copper area with a hole in its center into which a component is mounted and soldered

V_{cc}. A term used to denote a positive power buss going to an integrated circuit

Vias.—see Feedthrough

2X, 1X. A term used to denote scale, $2\times$ being two times size, $1\times$ full size

X-Y Coordinates. The vertical and horizontal grid coordinates used to layout printed circuit boards; all holes should fall on cross points of these coordinates whenever possible

An initial printed circuit board design must be developed upon sound principles, which take into account manufacturability, test, installation, and serviceability.

When all areas of interest concerning an original printed circuit board design are met, this will lead to an optimum level being achieved in the finished printed circuit board assembly. This will result in greater product reliability, and easier serviceability, which reduces down time.

In keeping with the orignal design intention for a printed circuit board, a procedure must be followed so that all areas of interest concerning the printed circuit board are considered and incorporated into the initial design layout for the board. To aid in the design and development of two layer and multilayer digital printed circuit boards, the following procedure is suggested. This procedure is intended as a guide, and as such must always be subject to modification to compensate for the constraints placed upon an initial design by time, cost, inventory, or other problems which are always encountered in the development of a printed circuit board layout.

The procedure follows a logical progression from the initial concept through to the installation of a complete circuit card. Therefore, the procedure must begin with the schematic diagram for the intended printed circuit card. The designer should receive, along with the schematic information, samples of any components which will be used on the circuit card and deviate from the "standard," or normally used components within a company, so that spacing of components is considered at the onset.

Since most schematic diagrams are not complete documents when the designer first sees them, a relatively finished drawing of the intended circuit must be generated. The designer can develop a finished printed circuit board much easier if he or she is working from a well organized schematic diagram which contains all pertinent information needed to transform the schematic diagram into usable hardware. With a well organized schematic document the geographic placement of components will be speeded up since the schematic will help dictate where they must be placed on the circuit board.

When time considerations are paramount, and there are long lead time items which will be incorporated in the finished printed circuit card, a preliminary bill of material can be generated from the schematic document. This will allow time for procuring the long lead time items, while the actual printed circuit board layout is under development. If time is not of the utmost importance, the bill of material will be the last document generated, as this will allow the designer freedom to change components as the board layout progresses and not cause useless stock to remain in inventory or force the designer to use a less than desirable component in the layout.

In a case where several people are involved with the development of a printed circuit board, the semi-formal schematic document will greatly aid in any discussions prior to the actual layout of the board. This can save a great deal of time which would be lost if changes are made after the actual layout has been started.

The schematic diagram is often viewed as one of those documents which can be "done later." This situation should be avoided for two reasons. First, the printed circuit board must be handled as a complete package to ensure that all pertinent information is generated, avoiding time delays later on in the manufacturing of the board. Secondly, the schematic diagram is a valuable aid in the geometric placemant of components, which again will save time during the development stages.

There are any number of things which could go wrong when there are several printed circuit boards being developed at the same time which will all be used in a common crate or module, often coming together in a "mother board" configuration. Add to this the fact that several designers may be working on boards, possibly in different departments or sections, and the risk of things going wrong increases. To eliminate this situation a board outline drawing can be generated, which eliminates many of the problems associated with multiple layouts for a common crate or module.

The board outline is simply that, a 2× outline of the intended printed circuit board, photographically reproduced on a 7 mil clear mylar sheet. Internal to the board outline is a gray precision grid which will drop out photographically when the finished artwork is reduced to 1×. Exterior to the board outline are photographic reduction targets and dimensions. Across the top, the sheet is punched for a standard, seven-round-hole pin bar.

The internal grid serves several purposes. It eliminates potential errors of misalignment when the designer must look through several layers. Should rework be necessary, which is often the case, the designer will not have to search for a matching grid, since it is always present. The grid and alignment holes are constant from one board outline to another, which eliminates checking board to board alignment prior to photography. Should front panel card edge components be used, which are common to several boards, alignment is much easier producing a much more uniform appearance in the finished product. Should additional designers be required for the project and outside personnel are involved, using outline formats ensures that no errors of this type will creep into the system design. This board outline system ensures a precision drawing from one printed circuit board layout to another, also ensuring proper fit into card guides, insertion depth into connectors, properly routed boards and standardized handling of artwork for photography.

With the use of photographic reduction targets on a basic layout sheet used in conjunction with the pin bar system, a large number of potential problems are eliminated. These positive results in time and effort greatly outweigh the cost of the board outlines. Often a relatively few board outline patterns will cover a large percentage of the designs generated within a year. The price of this system is further reduced in costing out a printed circuit artwork generation when redundant drafting time is eliminated.

The board outline format is the basic document upon which all the supporting artwork documents will be developed. Using the schematic diagram and the board outline diagram, the designer can now determine the minimum number of layers which will be required to achieve the desired results. To determine the number of layers, an IC (integrated circuit) count must be made so that the board density can be determined. The IC count must be reduced to its lowest possible number without effecting the basic function of the circuit.

With the density of printed circuit boards always increasing, the designer is asked to get more and more onto the board. The density of the board can be reduced in several ways which will aid the designer in developing the board. One method is to substitute IC packages; for example, using a PAL (programmable array logic) in place of existing packages on the board will reduce the overall count. Combining the logic functions of different IC packages in a common single package of the same function, using pin outs which are more compatable to layouts, will reduce the overall IC count. Discrete resistors can be replaced with resistor networks packaged in SIPs or DIPs which will aid both in layout techniques and in gaining needed board geography.

When you have reduced the IC count to its lowest possible count, along with the discrete components, these are then translated into units. A fourteen pin integrated circuit represents one unit.

The total usable board area, expressed in square inches, is divided by the unit count. Keep in mind this is the usable board area, so all areas which cannot be used, such as a connector area, heat sink, rail contact area, etc. must be deducted.

For general use in printed circuit boards which are developed manually, a realistic unit number is 0.6. Should the calculated number be lower, additional geography will be needed, and this can be accomplished by changing the decoupling method, using capacitors under the ICs in sockets, or using external power busses which are mounted to the board after assembly.

All these considerations have some negative factors within them such as manufacturing difficulty, longer assembly times, increased down times, and higher reject rates.

Another possibility is to break the schematic diagram into more than one printed circuit board, which will greatly increase the needed board geography.

If all efforts have been made to reduce the unit count, and still one cannot obtain a number the designer can work with, the next consideration is to move the power and ground into inner foils, producing a multilayer board. Should this not provide the necessary board geography, additional layers must be added to achieve the desired results.

Once the determination has been made as to how many layers the finished printed circuit board is to have, the designer can proceed with the geographic assignment of the various IC packages.

Since the two layer board is the "workhorse" of the industry, the remainder of the discussion will concern itself with this type of board.

To aid in the circuit layout and function, as well as manufacture and test, the IC packages are arranged in horizontal or vertical rows. All ICs are oriented in the same direction with a decoupling capacitor at the V_{cc} termination of each package.

Since there is always a strong similarity between a well organized schematic diagram and the finished printed circuit board, the placement of IC packages can be roughly determined from the schematic design. Those areas which will be of greatest difficulty in the development of the board can be worked out by the designer, using a pencil layout, to further determine the proper IC positions within the printed circuit board format. Keep in mind that the individual IC packages must yield their location to circuit function.

As the various packages are adjusted within the layout, using the combination of the schematic diagram and the pencil layout, the discrete components will also be added and checked for physical size to ensure that they will have adequate clearance on the finished board. The red and blue pencil layout can proceed to the point where the printed circuit designer has a high degree of confidence in the layout and switch over to the actual tape layout of the board. If it is intended that the actual tape layout will be done by someone else, the designer must take pains in the pencil layout to ensure that all pertinent information is on the layout. The person doing the taping should also be aware of manufacturing techniques of printed circuit boards as well as assembly procedures.

Final geographic placement can be made using the red and blue pencil layout and the schematic diagram. Should intermediate approvals be required, the schematic diagram and the red and blue pencil layout will supply all the information required for a decision to be made on further effort. With any and all approvals complete and any additions or deletions to the circuitry incorporated into the schematic diagram and preliminary layout, the actual taped layout of the board can proceed.

The assignment of signal paths follows a priority system which starts with the highest priority level of signal and descends to those signals which can tolerate greater movement on the layout. The ground signal is often run at the outer extremity of the printed circuit format. This allows for easy connection of internal ground busses to the main bus.

The ground and power signals are the first to be assigned to ensure a good clean path for the ground signals, decoupling and filtering, and of sufficient width to more than adequately carry the intended current for the board. No feedthroughs or vias are ever used in these signal paths. The ground and power busses are run between the rows of pins of the ICs on the component side of the board. This will allow for all cross traces to be on the wiring side of the board where they are clearly visible after assembly and the technician troubleshooting the board can easily follow a signal path. Only the ground and power busses will be hidden after assembly, but these terminations are understood and are usually not considered in troubleshooting a circuit card.

Depending upon the type of circuit involved in the layout, the remaining signal priorities would follow in a descending order such as: secondary grounds, secondary power, oscillators, high current, general signals, low level signals, test points, LEDs, lamps, and cosmetic signals.

Throughout the assignment of signal priorities, care must be exercised to eliminate the unnecessary use of feedthrough or vias as means of establishing higher and lower priority levels. Should the designer fall into this trap, the final signal traces to be placed on the board would be a study in feedthroughs. Rerouting a signal path is often more desirable than adding feedthroughs as long as this system is not over used to the extent of producing a "leggy" board. Either of these two conditions is a source of potential problems in the final card. Using a combination of geometric placement, proper routing and feedthroughs will produce a good, sound, well thought out design.

Conductor widths must be of sufficient size to be able to handle the current of the board. Ground and power are typically 0.100 mils at 2:1 with a minimum of 0.060 in most cases. General signal traces are 0.040 at 2:1 with the minimum spacing being 0.030. A minimum of 0.030 traces with a space of 0.020 can be attained with manual layouts, but beyond this taping becomes very difficult and drafting time increases dramatically.

Terminal areas are standard to minimum annular area to provide for sufficient pads for assembly as well as rework.

All terminal areas are placed on X- and Y- coordinates for drill tape and automatic insertion considerations. The X- and Y- coordinates are functions of the grid selected for your board outline drawing. Grids of 0.010 square, accented every 1 inch, are the most commonly used. Should minimum traces become the rule in the layouts being generated, consideration should be given to using a 0.005 square grid.

Pin one of each IC on the wiring side is marked with a round pad or a different pad configuration to aid in reading the completed assembly, making test and trouble shooting much easier.

When the printed circuit artwork is completed, the supporting artwork can be generated to complete the documentation package. Using a method of placing all pads on the format sheet, and all traces on another sheet, will produce a pad master drawing. The pad master drawing will double as a solder mask drawing, using standard photographic procedures to increase the pad areas sufficiently for a solder mask. A secondary generation of the pad master document will serve as the master or drilling drawing. A standard board configuration drawing, showing all manufacturing information, is normally generated at the time the board outline format is developed and would always accompany any printed circuit board layout on that document.

The final artwork document which will be generated is the silk screen. All pertinent information which will aid in assembly, test, field service, version modification, inventory, or general aesthetics of the finished printed circuit card should be incorporated on this artwork. The schematic number is always included on the silk screen artwork to tie all information together when the printed circuit card is complete.

The cost of the silk screen artwork and solder mask are outweighed by the positive benefits of these two documents.

The solder mask will enhance the circuit function, provide for a corporate identifying board color and serve as a good base for the silk screen.

The silk screen will aid anyone using the circuit card in all areas of its useful life. Where long periods of time may elapse between service of the boards, the silk screen information will serve to reduce service time considerably.

A secondary generation of the silk screen artwork drawing will serve as an assembly drawing which can be used in conjunction with the bill of material or parts list to provide full information necessary to assemble the circuit boards.

The total package is now complete and all persons concerned with the manufacture, assembly, test, installation and maintenance can have at their fingertips all pertinent information concerning that particular printed circuit assembly.

A good design will always take into account those problems encountered by the people who must fabricate the board, test it, as well as service the board.

SECTION TWO
PLANNING, DOCUMENT CONTROL, AND QUALITY

4
Definitions

Actinic. Regions of the light spectrum which are capable of producing chemical change.

Activation. Treatment to overcome a passive surface condition.

Acutance. Measure of sharpness between image and non-image areas in a phototool.

Addition Agent (Additives). A material added to a plating bath for the purpose of modifying the character of the deposit. The addition agent, which includes brighteners, is usually added in relatively small amounts and is often of organic or colloidal nature.

Additive Plating. The selective plating of electrical circuits and configurations directly upon a dielectric substrate. In the truest sense, no etching is required for additive circuits.

Adhesion. Chemical or mechanical forces that hold two surfaces together, as adhesion of photoresist to copper or electrodeposited copper's adhesion to electroless, etc.

Alloy Plate. An electrodeposit containing two or more metals so combined as to be indistinguishable with the unaided eye. Solder is an alloy plate of lead and tin.

Amp-Hour. An amount of electricity defined as one amp supplied for one hour. Of course, $\frac{1}{10}$ of an amp supplied for 10 hours is also one-amp hour. The general formula for amp-hour is

$$\text{Amp-hour} = \text{Amps} \times \text{Hours}.$$

This is a value platers need to know in order that additions of brightener can be added on a scheduled routine. A typical rate of brightener addition may call for 300 ml of brightener for each 1000 amp-hours of plating. Thus, 500 amps of plating over 2 hours is 500 amps $\times$ 2 hours = 1000 amp-hours.

Annular Ring. The pad of circuit metal surrounding a hole.

Anode. The electrode in a plating tank which has a positive charge placed on it. The anode may also be the source of metal ions for plating; this is called a *sacrificial anode.* A sacrificial anode corrodes during the normal plating operation; this supplies metal ions to the plating bath at the same rate at which they are removed (for 100% efficiency). An inert, or nonsacrifical anode, does not corrode during plating, and is typically made of titanium, platinized titanium, stainless steel, or carbon. The term *anode* also means the metal which is suspended in the electroplating bath, e.g., copper anode, or tin-lead anode.

Anodizing. Anodic treatment of metals, particularly aluminum, to form an oxide film of controlled properties.

Artwork. The design which is photographed to produce the master circuit pattern. Sometimes the negatives and positives used to image photoresist are called artwork; actually they are phototools.

Base. (1) The dielectric support for printed circuits. (2) The film support for photographic emulsions. (3) A substance which has a pH of greater than 7.

Bath Voltage. The total voltage between the anode and cathode in an electroplating bath.

Blistering. Swelling or separation between resist and the surface to which it is applied; may or may not be accompanied by underplating and more severe resist failure.

Board. (1) Printed circuit board. (2) The foil and base from which a printed circuit is fabricated.

Breakdown. Resist failure; degrees range from halo, to blisters, underplate, lifting, sloughing or "falling off."

Bridge. (*Synonyms:* Tent, Cap) The ability of photoresist to span across a metallized plated-through hole and protect it from etchants. This property is utilized to produce copper-through holes.

Bright Dip. Dip used to produce a bright surface on a metal.

Bright Plate. An electrodeposit which has a bright reflective surface.

Brightener. An addition agent which causes the formation of a bright plate, or which improves the brightness of the deposit.

Brush Plating. Plating solution (sometimes in gel form) held in, or fed to an absorbing pad, or brush, which also carries the anode (usually insoluble). Electrical connections are made to the anode and work, and the brush is moved back and forth over the work.

Build-Up. Undesirable overplating which occurs on corners and edges of the base metal during plating.

Burnt Deposit. A rough, nonadherent, or otherwise unsatisfactory electrodeposit produced by the application of excessive current density.

Bus Bar. The long copper bars which run the length of the plating tank, and from which the anodes and printed circuit panels (cathodes) are suspended. These bars carry the electricity used to electroplate.

Carryover. Drag-out of a solution which is introduced into yet another solution.

Chemical Milling or Machining. (Also called *photofabrication*) Fabrication of metal parts by chemical etching utilizing a photoresist material. Often used for precision or delicate parts.

Clad. A thin layer of metal foil bonded to a substrate.

Cleaning. The removal of grease, oxides and other foreign material from a surface.

Scrub Cleaning. Hand or automatic machines which use pumice slurry, brushes, or wetted fiber wheels to remove oxides abrasively and roughen the board surface.

Solvent Cleaning. Cleaning by means of organic solvents.

Alkaline Cleaning. Cleaning by means of alkaline (basic) solutions.

Electrocleaning. Cleaning by a current being passed through the solution.

Cathodic or Direct Cleaning. Electrolytic cleaning in which the work is the cathode (−).

Anodic or Reverse Cleaning. Electrolytic cleaning in which the work is the anode (+).

Soak Cleaning. Chemical cleaning without the use of current.

Collimation. The process whereby light rays are made parallel.

Component. Commonly used to describe resistors, capacitors, transistors, etc. Actually, these are component parts. A component is a functional unit of one or more circuits and devices which make up an assembly or system.

Composite Plate. An electrodeposit consisting of layers of two or more metals; sometimes called multiple-metal plate.

Conductor. (1) A current carrying line between terminal areas. Often used as a synonym for circuit line, trace, circuit path, etc. (2) A material capable of carrying an electrical charge or current.

Corrosion. (1) Gradual destruction of a material usually attributable to a chemical process. (2) (Of anodes in plating). Solution of anode metal by the electrochemical action in the plating bath.

Covering Power. The ability of a plating solution to produce a deposit at very low current densities.

Current Density. Total amps per square foot of plating area. This is an important number to know to determine plating amperage. The current density may apply also to the total amps per square foot of anode area: this is anode current density. Each plating bath has its own optimum current density, for both the anodes and the cathodes. Typical values for common electroplating baths are:

Copper:	30 amps per square foot (a.s.f.)
Tin-Lead:	20 a.s.f.
Nickel:	15 a.s.f.

Current Efficiency. In electroplating, the proportion of the current used to deposit or dissolve metal. The remainder of the current is usually consumed in the evolution of hydrogen and oxygen.

Deburring. Removal of burrs, sharp edges, etc. from work.

Deionized Water. Chemically pure water which is equivalent to distilled water. It is made by passing the water through resin beds which remove the minerals. Deionized water is usually plumbed throughout the plating shop in plastic pipes. It is used to make up fresh plating baths, and to add water to baths which have evaporated.

Definition. Fidelity of reproduction.

Delamination. Separation of resist from the metal substrate; also, separation of laminated layers in multilayer boards or of the metal foil from its dielectric support.

Delta D (ΔD). Difference in photographic density between areas. ΔD between steps of a $\sqrt{2}$ density tablet is 0.15.

Density. Concentration of matter; mass per unit volume. Photographic density is the degree of opaqueness. Circuit density is the number of circuits in the board area, etc.

Dielectric. The insulating support for circuits. The base that supports the copper foil for printed circuits.

Dimensional Stability. A relative measure of the change in dimensions as a function of temperature, humidity, etc.

Dip Soldering. Placing a circuit pattern in contact with the surface of a static pool of molten solder to cover the entire circuit in one operation.

Double-Sided Board. A printed board with conductive material on both sides.

Drag-In. The quantity of water or solution that adheres to the material introduced into a bath.

Drawing, or Print. The blueprint of the printed circuit board. It is an engineering drawing which contains the completely dimensioned drawing of the part. It also contains such items as:

1. The hole chart, a list of all holes, their size and tolerance, and whether or not they are to be plated or nonplated.
2. Metal plating requirements, i.e., the types of metals to be plated, and their thicknesses.
3. The part number.
4. The revision level.

Dummy. A cathode used for working the plating solution but which is not used after plating.

Dummying. Plating with dummy cathodes.

Electrode. A metal conductor through which current enters or leaves an electrolytic cell.

Electroforming. The production or reproduction of articles by electrodeposition.

Electroless Copper. This is copper which is chemically deposited inside the holes which have been drilled in printed circuit panels. It is also called *through hole plating, copper deposition* or *Cuposit* (Cuposit is the brand name which is given to electroless copper lines supplied by the Shipley Company). Electroless copper is used to make electrical connections between the sides of printed circuit boards, and to the inner layers of multilayer boards.

Electroless Plating. A type of immersion plating in which a chemical reducing agent is employed to reduce metal ions to metal on the surface of the work.

Electrolytic Plating or Electroplating This is the type of plating which results from using electricity to supply the energy for metal deposition.

Emulsion. The light sensitive silver halide coating of photographic films which produces an image after exposure and development.

Epoxy. A thermosetting material which is commonly used with fiberglass to produce the dielectric support for quality copper foil circuit boards.

Etch Factor. Ratio of depth of etch to undercut lateral etch. (Etch factor is calculated for only one side of a conductor.)

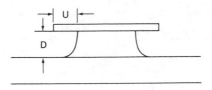

Circuit cross section

Etchant. A solution which removes unwanted metallic portions of a printed board by chemical action. Common etchants are $FeCl_3$, AP, etc.

Etchback or Smear Removal. This is the process of removing unwanted epoxy from plated through holes. The epoxy may have resulted from drilling (drill bits may reach 500 degrees and smear epoxy around the inside of holes), or it may be necessary to remove extra epoxy from around the annular ring of inner layer connections in the holes of multilayer boards. Epoxy smear will prevent adhesion of copper to the inside walls of holes, especially on 0.125 thick panels; and it will prevent good electrical connections from forming to inner layer connections in the holes of multilayer boards.

Etched Printed Circuit. A printed circuit formed by etching.

Etching. This is the removal of metal by chemical methods. The most common application is the etching of copper by alkaline ammonia, sulfuric acid–hydrogen peroxide, ferric chloride, or cupric chloride.

Exposure. Subjecting photoresist to UV light to cause polymerization.

External Layer. Outer surface of a multilayer circuit board.

Eyeballing. A registration technique where the phototool is aligned visually with drilled holes (opposed to pin registration). Term also includes drilling and other activities which depend upon visual alignment.

Fillet. Used interchangeably with *cove.* Also describes a build-up of solder and the rounding of circuit intersections and pads to avoid sharp corners.

Finger. That part of a circuit board used to provide electrical connection by pressure contacts.

Flash Plate. A very thin deposit of metal (but, often up to 0.5 mil in circuit manufacture).

Flexible Circuit. A circuit or circuits on or between flexible dielectric insulating material.

Flip Flop. This refers to running multiple circuits on a panel. When run flip flop, panels are processed with half the circuits oriented from the circuit side, and half from the component side. Running with a flip flop configuration requires that only one piece of stepped-and-repeated film be made; this piece will be used to image both sides of the panel.

Foil. A very thin sheet of metal.

Foot. A foot or fillet at the resist-substrate interface.

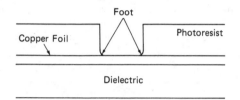

Gray Scale. A photographic density scale which consists of a series of known densities arranged in order of increasing opaqueness.

Ground Plane, or Land Area. Any large expanse of metal on a printed circuit boards may be called a ground plane, or land area. A ground plane may also be a metallic plane at ground potential.

Ground Plate. A continuous sheet of metal in a multilayer P.C. board, usually between layers.

Haloing. Change of resist color along image edges; may indicate loss of adhesion.

Hard Water. Water containing calcium or magnesium ions which will form insoluble curds with soaps.

Hole Size. The diameter of a hole. The pin used to measure the diameter may also be referred to as the hole size. Hole sizes, or diameters, are measured in thousandths of an inch, or in mils: 1 mil = 1 thousandth = $\frac{1}{1000}$ inch. Hole sizes are important to platers because they are used to determine the thickness of plated metal, especially copper.

Image. A likeness of an object produced on a light sensitive material. To image-expose.

Immersion Plating. Deposit by simple immersion without any outside source of current.

Inclusion. Foreign particles which may be included in metal. For example, lead inclusions may be present in copper foil or iron in anodes. The term also includes impurities in plated-through holes or plated surfaces resulting in defects or irregularities.

Internal Layer. A conductive pattern contained entirely within a multilayer board.

Jumper. An electrical connection between two points on a printed board, added after the board is fabricated.

Laminate. n. A product of two or more layers of material bonded together. A piece of copper foil bonded to a dielectric is called copper laminate. *v.* To unite layers of material, e.g., photoresist to a metal clad dielectric board; to bond layers of a multilayer board together.

Lamination. The process of applying photoresist to a surface, while removing the coversheet.

Latent Image. The stored up affect of light on a photosensitive material, when it is developed the true image appears where the latent image was. Some photoresist types darken in exposed areas, revealing the latent image.

Leveling Action. The ability of a plating solution to produce a surface smoother than that of the base metal.

Line. A single conductor on a printed circuit board.

Line Width. The width of a conductor on a printed circuit board. Etched conductors are measured at their narrowest cross section.

Mask. A covering to protect the circuitry pattern, or background during board manufacture.

Matte. A surface or finish that is dull and lackluster.

Mil. One-thousandth of an inch (0.001 inch).

Multilayer Printed Circuit. A printed circuit board with alternate layers of conductors and dielectric bonded together and interconnected with plated-through holes or posts.

Multiple Lamination. More than one layer of resist film applied to increase total film thickness.

Negative. The production of light and dark areas in the reverse order of the original. A configuration of clear lines on a black background. Photographic negatives are used with photoresist to "tent" and for print-and-etch boards.

Nodule. Wartlike protuberance generated on a board surface during electroplating. Sources of nodules are anode sludge and other particulate bath contamination.

Nonconductor. (Synonym: dielectric). A material such as glass or plastic not capable of conducting an electrical current.

Opaque. Not transparent or translucent. Also to retouch or paint over areas of a photographic film.

Ounces of Copper. This refers to the thickness of copper foil on the surface of the laminate; ½ ounce copper, 1 ounce copper, and 2 ounce copper are common thicknesses. One ounce copper foil contains 1 ounce of copper per square foot of foil. The foil on the surface of laminate may be designated for the copper thickness on both sides by: 1/1 = 1 ounce, two sides, 2/2 = 2 ounces, two sides, and 2/1 = 2 ounces on one side and 1 ounce on the other side.

$$\frac{1}{2} \text{ oz.} = 0.72 \text{ mil} = 0.00072 \text{ inch}$$
$$1 \text{ oz.} = 1.44 \text{ mils} = 0.00144 \text{ inch}$$
$$2 \text{ oz.} = 2.88 \text{ mils} = 0.00288 \text{ inch}$$

Overhang. The increase in conductor width caused by plating buildup.

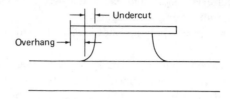

Overlay. A transparent film which includes the master pattern. It is placed over the printed circuit board to blank for inspection, exposure, etc.

Pad. Area of a printed circuit used for making connections to the pattern; also, the area around a drilled or plated-through hole.

Panel Plating. Plating which is performed to an unimaged panel, bare copper. It may be done to build up the copper on the surface of the panel, or to deposit additional copper in the holes of bare copper panels after they have been through electroless copper. This is done for panels which will be imaged with dry film photoresist. The photoresist leaves a contaminant on the surface which will cause a metal peeler if not etched off.

Pattern Plating. Plating which is done to panels which have a circuit image applied with plating resist. Pattern plating is usually copper/tin-lead.

Peeling. Detachment of a plated metal coating from the basis metal.

Periodic Reverse. A method of plating in which the current is reversed periodically from positive to negative.

pH. Measure of acidity or alkalinity. Inverse log of the hydronium concentration or

$$\frac{1}{\log [H^+]} = pH.$$

Neutral is 7 or $[H^+] = 1 \times 10^7$. As the superscript approaches 0, the solution becomes increasingly acidic; as it approaches 14 it becomes more basic or alkaline.

Photofabrication. The manufacture of precise shapes by print-and-etch technique.

Photographic Spread or Choke. This refers to the increasing (spread) or decreasing (choke) of circuitry line width. This is generally done to eliminate an airgap problem on the artwork.

Photoresist. A photosensitive, chemically resistant material used to mask areas of printed circuit board blanks. The areas not protected with photoresist are then etched away (print-and-etch) or plated (pattern plating).

Phototool. Photomask, photographic master, etc.

Pinhole. Small hole or imperfection; minute void.

Pit. Depression produced in a metal surface by nonuniform electrodeposition or corrosion, or contamination.

Plastisol. A mixture of resins and plasticizers converted to a continuous plastic film by the application of heat. The original mixture becomes a part of the film with no significant solvent evaporation. The plastisol coating is often used on plating racks to prevent the electrodeposit from building up on the rack.

Plated-Through Hole. Connections between insulated layers of circuit foil formed by holes. Hole walls are chemically metallized then electroplated to the desired thickness.

Plating. Chemical, galvanic, or electrochemical deposition of metal.

Plating Area. This is the portion of an imaged or unimaged panel which is copper; it is the area which will be plated:

$$\text{Plating Area} = \text{Total Panel Area} - \text{Area of resist.}$$

Plating Bar. Temporary circuit or margin on a P.C. board to connect fingers or circuits for electroplating.

Plating Bath. The electroplating solution or electrolyte.

Plating Resist. This is the resist used to image a printed circuit panel for plating, or for etching. The resist may be applied by silk screen printing, or by dry film or liquid photoresist.

Pores. Discontinuities in a metal coating which extend through it.

Porosity. The number of pores in a given area of electrodeposit.

Positive. Clear and opaque areas in the same order as the original. Positive phototools have black lines (circuits) on a transparent background and are used with photoresist to pattern plate.

Printing. Reproducing a pattern on a surface; exposing photoresist through a phototool, or silk screening.

Profile. Shape perpendicular to a surface; i.e., profile of resist line on copper panel in cross section.

Rack. Frame used to suspend and conduct current to boards (cathodes) during electrodeposition.

Rectifier. A device for converting alternating current into direct current for electroplating.

Registration. The position of circuit phototools with respect to their desired location on a board or panel.

Resist. A material (ink, paint, metallic plating, etc.) used to protect portions of a printed circuit from the etchant.

Resolution. The ability to reproduce artwork of various size lines and spaces. A near synonym is *resolving power,* or to define circuit pattern for plating.

Robber. Area designed around a printed circuit board to absorb unevenly distributed current.

Screen Printing. Putting an image onto base material by forcing ink through a stencil screen with a squeegee.

Scum. Areas in which the resist has not been completely removed or in which dissolved resist has been redeposited after development. Scum manifests itself as a stain, white powder, strings, or clumps of resist. These areas can act as a plating or etch resist.

Single-Sided Board. A circuit board which has conductors on only one side.

Solderability. Surface sustains an unbroken molten solder film, all holes fill during wave solder.

Specific Gravity. The density of a substance relative to water.

Strike. (1) Solution used to deposit a thin initial film of metal. (2) Thin initial film of metal followed by other coatings. (3) To plate for a short time at a high initial current density, after which the current density is reduced to its normal operating range.

Strip. Removal of the photoresist film from the etched or plated circuit board.

Stripper. Organic solvent to remove photoresist from etched or plated boards.

Substrate. The base metal; sometimes the dielectric base to which metal (usually copper) is laminated to produce circuit boards.

Tap Water. Tap water is the city water which comes from the faucet. It is not to be confused with deionized water.

Throwing Power. This is the ability of a plating bath to plate metal. As metal is deposited onto the surface of the panel, it corrodes off the anode. The concentration of metal (in a sacrificial anode plating system) should remain constant; this is called throwing metal from the anode to the cathode.

Tooling Holes. Holes placed in the PC base to accurately position the boards for drilling or exposure. *Synonym:* registration holes.

Traveler. This is the set of instructions which the planning engineer writes for manufacturing the printed circuit. The traveler lists all the steps which the panels will go through during manufacturing, and all the requirements at each step. The traveler should always be consulted and read before any work is performed on the panels. It should also be signed off when work has been completed, and the quantity of panels written in.

Trees. Irregular projections of plating which usually occur at high current density points (look like small trees), in tin-lead plating.

Undercut. To etch away material on underside of a deposit. (*See* Overhang.)

Ultraviolet. That portion of the light spectrum with wavelengths shorter than visible light and longer than X-rays.

Viscosity. The resistance to flow of liquids; a sort of internal friction.

Wettability. Surface will sustain an unbroken water film.

Wetting Agent. A substance which reduces the surface tension of a liquid causing it to spread more readily.

5
Planning

The minimum duty of the Planning Department is to issue a traveler for every job brought in by the Sales Department. It is the ideal department to serve as a central clearing house for all manufacturing documentation: customer specifications, artwork, drawings, engineering change notices, and any special notes or requirements. The manufacturing process begins in Planning with the issue of the job traveler. Since the traveler must list all materials and operations as manufacturing instructions, it is logical that changes in the drawings, artwork, materials or other requirements at any operation should be cleared through and documented by the Planner.

The Planner should develop a relationship with various personnel of the customer: buyer, quality, drafting and engineering. This will help avoid problems and delays. Discrepancies in customer and vendor are best served by rapid and friendly resolution and clarification of problems and requirements. It is critical that the Planner have an excellent understanding of printed circuit manufacturing and required documentation.

There are other ways in which the Planner can be of great value. The more planning that goes into each job, the better will be the result in terms of conformance to specification, and in manufacturing yield. As the customer requires changes in specifications it should be the responsibility of the Planner to issue notices to all operations affected: Photo, Drilling, Fabrication, etc. When a change is necessary in the manufacturing process to circumvent a problem: change in drill size, second drilling, or altering the artwork, the Planner should document the change and follow through.

As keeper of manufacturing documents, the Planner has another vantage point which should be utilized. Prior to issuing travelers for reordered part numbers, the Planner should review the documents in the Control File, including past travelers. This file should contain not only special notes on the part, but also past travelers, returned material notices, and vendor/customer documentation communications. By reviewing past travelers, areas of excessive loss can be spotted and the reason identified to prevent a reoccurrence; any notes written on the traveler which need attention can also be addressed before the panels reach the stage where the problem occured.

When it becomes necessary to use the services of other companies for artwork preparation, drill programming, or tool making, the Planner should be consulted, as he/she is most familiar with tooling and layout requirements of the customer and of the shop to utilize outside tooling.

The Planner may help utilize materials in a more cost effective manner if he/she is kept informed of inventory for laminate and other critical materials. If the warehouse is loaded with materials not commonly used (white elephants in some cases) the Planner can schedule less demanding jobs to run on those materials. In the case of multilayers, the number of thin clad laminates can be minimized, for most applications, by using existing, or standard sized, materials if they do not conflict with customer design specifications.

The following specification for running a printed circuit planning area is only one example of how a Planning area can be operated.

PLANNING RESPONSIBILITIES

A. It is the responsibility of the Planner to review the package coming in from Sales for completeness, and agreement for part numbers, revision numbers, board/panel quantity, panel size, number up per panel, material, and other requirement per the print and customer specification.

B. To review new revisions and determine what changes have been made in the part. The Planner must also determine if existing tooling may be used on the new revision.

C. To contact Sales, and/or the customer when problems need attention or more information is required to build the part, and to follow through until the problem has been resolved.

D. To maintain the integrity of the customer files, control files, and film storage files. The Planner shall maintain a log to track the movement of film, prints and other documentation of the files.

E. To plan the production of each part and write the traveler and other necessary documentation. The Planner is the only one authorized to write or make changes to a traveler. If another individual sees a need for a change in the way the job was planned, he must take it to the planner to discuss and make the actual change.

Definitions

Control File—(Also called Film File number.) This is the primary file for each part number. In it are old copies of travelers run on past jobs, prints, customer film inspection check list, and other documentation related to the part number.

Job Number—(Also called the Work Order Number.) This is the identifying lot number for each job run on a part number.

Number Up—This is the number of boards per panel.

Silk Screen—This is one method of imaging a board prior to plating. However,

Ⓐ WORK ORDER				
PROCESS	QTY	DATE	EMP	REJ.
Rel. Date				
Film Insp.				
Program				
Process Film				
Blank				
Bake				
IL Image				
IL Etch				
Black Oxide				
Press				
Bake				
Drill N/C				
De Burr				
Cuposit				
Image S/S D/F				
Touch-Up				
Plate (Cu)				
Plate (Ni)				
Plate (SnPb)				
Plate (Special)				
Etch				
Tip Plate				
Reflow				
Inspection				
Clean				
Solder Mask				
Legend				
Fabrication N/C				
Fabrication Manual				
Champher				
Final Clean				
Final Inspection				
Electrical Test				
Hardware				
Shipping	Qty	Date	Invoice #	

Multilayer (bracket spanning IL Image through Bake)

Dwg.# _____ Rev. _____
A/W # _____ Ⓑ Rev. _____
A/W # _____ Rev. _____
Mst Dwg.# _____ Rev. _____

MATERIAL

Type: FR-4 _____ Ⓒ Color: Nat. _____
Thk: .031 .062 .093 .125 _____
Oz Sides .5/.5 1/1 1/0 2/2 2/0 _____
Size _____ x _____ 1 2 3 4 Deep
Run _____ Panels Bds Per Panel= _____

Ⓓ **BOARD MARKINGS**

UL Date Code Logo Lot #
Etched Screened Stamped

Ⓔ **SCREENING**

Solder Resist Over: Cu SnPb Ni _____ Sides
Type: Green Clear _____
Legend : Blk White _____

Pattern Ⓕ **PLATING** Contacts

Cu Ni Sn Pb Au/Cu Au/Ni Au/Au
Cu Thickness _____ Ni Thickness _____
Ni Thickness _____ Au Thickness _____
SnPb Thickness _____ Au Thickness _____

HOLE SIZE Ⓖ

Min	Max	Drill Size	Size After Plate	**TESTING** Ⓙ
—	—	—		Micro Section
—	—	—		Porosity
—	—	—		Solder Sample
—	—	—		Source Insp.
—	—	—		Test Reports
—	—	—		Certs

NOTES: *I* _____
_____
_____
_____
_____
_____
_____

CUSTOMER:
QTY:

DUE DATE: P.O. NO.:

Ⓗ

PART NO:

ITEM:

REV: CONTROL #:

JOB NO:

Fig. 5-1. Typical job traveler.

when it is referred to on a print or in a customer specification, it is the nome, or legend, and is applied with an epoxy ink.

Solder Resist—This is soldermask, like PC-401 or SR1000.

Circuit Side—(Also called the solder side.) This is the side opposite the component side. Circuit side is the side which is hit by the wave solder.

Oz's Copper—This is the ounces per square foot of copper.

Fiscal Year—Begins in July.

Customer File—This is a small card file that contains a card on each part number. They are filed numerically for each customer. It is a quick reference for Control Number.

PLANNING PROCEDURES: REFER TO COPY OF TRAVELER

1. Select a job to be planned from the production control files. You must have:

   a. Sales Order.
   b. Quote Summary.
   c. Set of prints.
   d. Customer Artwork.

2. Assign a control number:

   a. Check customer file to see if the job has been run previously, or if an earlier revision has been run. *Note:* If the part number has been run under an earlier revision, see the section on New Revisions.
   b. If the job has not been run before, assign a control number. *Note:* Immediately fill out a customer file card. Make special notes as you plan the job here: data code, color of soldermask and legend, thickness of contact finger metals, unusual thickness/type of material.

3. Review Sales Order, Quote Summary, Customer Specification, and the print.
4. Fill out a traveler:

   - Blue—single sided.
   - White—double sided.
   - Green—multi-layer.

   A. Fill out the information on the right edge of traveler

      - Customer.
      - Purchase Order Number.
      - Item—write in New, Reorder, or New Revision.
      - Control Number.
      - Quantity (Qty).
      - Due date.
      - Part number.
      - Revision.
      - Job number: xx-yyyy   xx—Week due
                                     yyyy—Sales Order Number.

B. Drawing number and revision.
   Artwork number and revision.
C. Material:

- Circle FR-4.
- Circle natural.
- Circle thickness.
- Circle oz/side.

Note: If the material is anything other than FR-4, natural, and the thickness and oz/side is anything other than what can be circled, you must black out everything which does not apply, and write in the correct specification in the space.

- Size—Write in panel size. This must be from the list of standard panel sizes.
- Run—Write in panel quantity.
- Boards per panel—Write in the number up per panel.

D. Board Markings—See blue notebook.
E. Screening

- Solder Resist—This is soldermask; read print and specification carefully. Some customers want soldermask over bare copper.
- Legend—Circle or write in color and/or location.

F. Plating—Circle pattern, and contacts, if required.

- Pattern—Circle appropriate metals (usually copper and SnPb); Cu thickness, always use 0.0015 inch, unless thicker is requested.
- Contacts—Circle Au/Ni, gold/nickel, unless the customer asks for something else, like Au/Cu. Always put nickel under gold.

G. Hole Size—Fill in according to finished hole size and tolerance. Use the following as a guide:

- Plated Holes:

Dia. ± .003	Drill size is .002 over tolerance
+ .003 Dia. − .002	Drill size is .003 over tolerance
Dia. ± .005	Drill size is Dia. + .005
+ .005 Dia. − .000	Drill size is .003 over tolerance

Example

	Min.	Max	Drill Size
.031 ± .003	.028	.034	.036
.031 + .003 − .002	.029	.034	.037
.031 ± .005	.026	.036	.036

- Non-plated holes. Non-plated holes are generally drilled (opened) in Fabrication with no pilot drilling. There are some exceptions:

- Tooling Holes—Sometimes these are drilled to finished size, then plugged during plating. Try not to first drill more than two per board.

Example: If there are 20 non-plated holes 0.125 inch diameter. Drill two per board, plug during plating, and drill the rest in Fabrication.

Min	Max	Drill Size
Plug		0.125
Open in fab		0.125

Note: You must make a note under Plate (Cu) to plug these holes. You must use a red ink stamp to mark the plugged holes on the print.

- Auto Insertion Tooling Hole and other tightly (+0.003/−0.000) toleranced non-plated holes must be first drilled then plugged. They cannot be second drilled.
- Except for tooling and auto insert tooling holes, all non-plated holes should be drilled only in Fabrication.

H. Work Order:

- All steps not used must be blacked out with a felt tipped pen.
- New jobs will include steps 2 and 3, program and film inspection, but these steps will be blacked out for reorders.
- Obviously, steps 7 through 11 do not apply to single sided and double sided boards, as they apply only to multilayers.

- Notes—Special information will be written here. All special notes should be written in red ink.
- Special requirements may be circled here.

5. Make a label for the film envelope.
6. Make an envelope for the drill film.
7. Shoot the following film, or ask Photo to shoot it for you:

   a. Programming film: silver positive of side shown on print.
                        silver positive of drill/pad master.
   b. Drill film: silver negatives of both sides, or diazos of both sides, negative or positive.

8. Stack the following, until Photo has inspected and bought off the customer film:

   a. Film envelope.
   b. Drill film envelope.
   c. Control file.
   d. Sales Order and Quote Summary.
   e. Customer film.
   f. All prints.
   g. Programming and drill film.
   h. Customer film inspection check list.

9. When Photo has bought off the customer film:

   a. Make a copy of the traveler and place in the Photo In basket.
   b. Place extra copies of prints and other documentation in the Control File.
   c. Place a copy of the Customer Film inspection check list in the film envelope (the original to the Control File).
   d. Place the Sales Order, Quote Summary, Program and Drill film, and the travelers on the Production Control desk.

10. Log completed jobs taken to the Production Control desk, in the Planning Log.

Single Sided Boards

1. Use blue travelers.
   2. Fill out traveler in the same manner as for double sided, with some exceptions:

   a. *No Cuposit*—Cross CUPOSIT off the traveler, and make a note to the Platers to skip it.
   b. *All* holes are non-plated.

Multilayer Boards

1. Filling out the back of the traveler—see attached example.

    a. Nominal thickness includes the 0.0015 inch per side that will be plated on to the board, unless stated on blue print.

    b. All panel sizes must be from the standard panel size list for multi-layers:

        12 $\times$ 16
        14 $\times$ 18
        16 $\times$ 18
        18 $\times$ 24

    c. Under the Comments section, write in what is to be blanked.

2. At the top of each traveler, on the front side, write Multilayer.
3. Work Order, section I:

    a. Next to Cuposit, write in Smear Removal, and Micro-Section.
    b. Use correction fluid to white out Fabrication Manual, and write in Micro-Section.

4. Always use green travelers.

PLANNING ORDER OF PRIORITY

    1. Multilayers.
    2. Reorders.
    3. New jobs.

NEW REVISIONS

1. Determine the reasons for the new revision:

    a. Ask Sales.
    b. Compare prints and artwork.

2. Determine if existing tooling and artwork can be used with only minor changes. If not, the job will have to be programmed and the artwork reprocessed. It is better to determine if we can use the existing tooling.
3. If the existing tooling can be used, make a note of this on the traveler you will be filling out, and on a Customer Documentation form.
4. Write in the new revision information on the Customer File Card.
5. Generally, new revisions require purging of the Control File:

a. Check Control File and Production Control to see if any jobs on the old revision are currently running in house, or if there are jobs not yet released on the old revision.

b. If the old revision is not running, and there are no unreleased jobs:

- Examine contents of Control File for special notes or production problems. Make a note of anything which may affect the new revision also.
- Dispose of entire contents of Control File for that part number revision.

6. If there are unreleased job orders on the old revision:

a. Call the customer, they may wish to change to the new revision. If they do, make the change and notify Sales and Production Control.

b. Record the results of your phone call on a Customer Documentation Form.

c. If the customer wants to run with the old revision on the previously existing, but unreleased job order, see 7 below.

7. If the old revision is currently being run in the shop:

a. Write a note referring to the fact that we do have two revisions of a part number running in the house. The note should list the job numbers of each revision and say that when the last job has been run, the files will be purged, as it will not be run again.

b. Attach one copy of the note to the film envelope; and give a copy to Photo, Drilling, and Fabrication.

c. Bundle up all Control File information on the old revision, place the note on top, and wrap a rubber band around it. When the last job has been run, discard the file contents referring to the old revision.

REORDERS

Running a reorder is little more than reviewing the past travelers and other information in the Control File, and copying a traveler.

1. Select a job to be planned from the Production Control File.
2. Pull the Customer File Card. *Note:* this has the Control Number listed (Film Number).
3. Pull the Control File.

- Review the Documentation forms and past travelers for special notes and production problems.
- Review sales order.
- Review information on Customer File card and make sure all the above agree.

4. Fill out one traveler for each due date on the Sales Order. Place the traveler in a plastic folder with a copy of the print.
5. Pull the Drill Film.
6. Make a copy of the traveler. Place this in the In basket of Photo.
7. Take the traveler, Sales Order, and Drill Film and place them on the desk of Production Control.
8. Log out every job you take to Production Control in the Planning log.

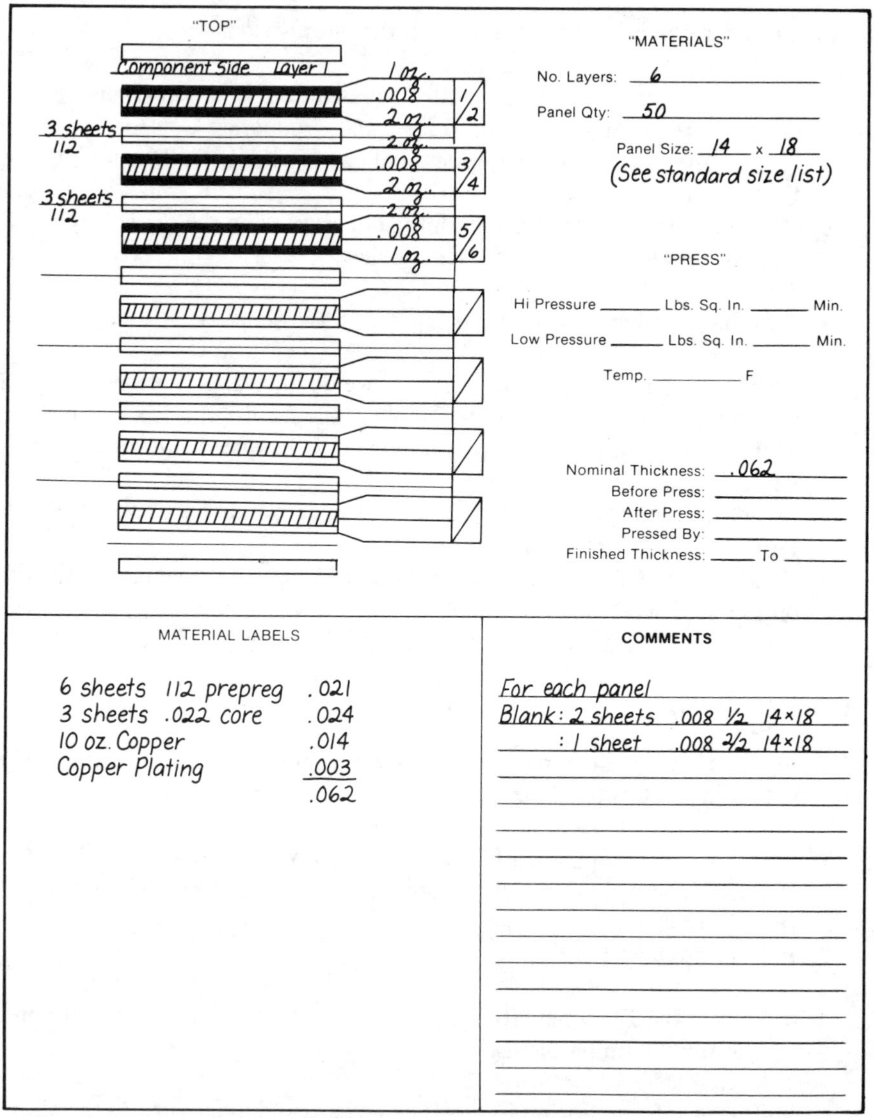
"TOP"

Component Side Layer 1

3 sheets 112

3 sheets 112

1 oz. .008 2 oz. — 1/2

2 oz. .008 2 oz. — 3/4

2 oz. .008 1 oz. — 5/6

"MATERIALS"

No. Layers: 6

Panel Qty: 50

Panel Size: 14 x 18

(See standard size list)

"PRESS"

Hi Pressure ______ Lbs. Sq. In. ______ Min.

Low Pressure ______ Lbs. Sq. In. ______ Min.

Temp. ______ F

Nominal Thickness: .062

Before Press: ______

After Press: ______

Pressed By: ______

Finished Thickness: ______ To ______

MATERIAL LABELS

6 sheets	112 prepreg	.021
3 sheets	.022 core	.024
10 oz. Copper		.014
Copper Plating		.003
		.062

COMMENTS

For each panel

Blank: 2 sheets .008 1/2 14×18

: 1 sheet .008 2/2 14×18

Fig. 5-2. Planning the construction of a multilayer.

6
Quality Assurance Program

Too often, in the printed circuit industry, there is a tendency to confuse Quality Assurance with Inspection. Inspection is but one area of concern for the Quality Assurance department. There are a number of functions which must be fullfilled by Quality Assurance for a modern printed circuit shop to deal effectively with the tasks of manufacturing and meeting the customers' requirements. A list of Quality Assurance functions must include the following:

1. Set up inspection points at one or more junctures for work in progress.
2. Provide written guidelines for purchasing critical materials, and for inspecting these materials (such as copper clad laminate) when they are delivered.
3. Establish acceptability criteria for work-in-progress, as well as for incoming documentation and incoming materials.
4. Work with manufacturing personnel to establish written process procedures and workmanship standards.
5. Work with manufacturing and quality assurance personnel to establish an ongoing training program.
6. Work with Manufacturing Planning to anticipate manufacturability and quality problems before a job is released from Planning.
7. Resolve the differing quality requirements of incoming customer documentation for running quick turnaround/prototype jobs, versus running major production runs.
8. Set up the procedural framework for identifying, resolving, and preventing reoccurrence of manufacturing errors and process problems.
9. Monitor critical processes through operation of a chemical/metallurgical laboratory.
10. Establish procedures for tracking such critical information as:

   - Product yield
   - Reasons for which product defects occurred
   - Rate of customer returns, and reasons for those returns
   - Dollar value of scrapped product
   - Job turn-around time, and percentage of jobs shipped complete
   - Quality problems, and trends in quality.

11. Dissemination of the data tracked above to all personnel in the company.
12. Providing immediate feedback to manufacturing when a quality problem is noticed.
13. Placing of the burden for quality on the shoulders of those actually doing the work: the individual and his/her supervisor.
14. Help to bring uniformity and consistency to all manufacturing operations, on all shifts, and ultimately to the quality of the product.
15. The Quality Assurance Manager must be effective in dealing with in-house quality problems, as well as in dealing with quality problems which a customer claims to be having as a result of the printed circuit.

There is a conceptual difference between Quality Control and Quality Assurance. Quality Assurance, as discussed in this book, is the attempt to achieve conformance to design requirements by optimizing the components of manufacturing, as listed above. Quality Control denotes the attempt to achieve a quality product by inspection of that product, and arm-twisting manufacturing into doing a better job. The concept of Quality Assurance opens up new dimensions to the creative individuals who are not afraid of a challenge.

The quality assurance manager must get involved with all the operations performed at the printed circuit manufacturing plant. He/she must understand the processes well enough to discuss problems that arise in the product. Also, since one of the goals of Quality Assurance must be to establish, along with manufacturing, written procedures and workmanship standards, the Quality Assurance manager must relate to the manufacturing people. Very often, simply talking with manufacturing people and supervisors will reveal a wealth of information about the process.

The smooth operation of the quality program, and the platform upon which quality begins is the written procedure. The sanctity of the written procedure must be taught to all employees from the day they are hired; and it must be demonstrated and respected by every employee from the President on down. There can be no uniformity of operation from one shift to another at any operation, and there can be no effective training program without the written procedure. Every process which occurs in printed circuit manufacturing must be identified; once this list has been compiled the processes can be scheduled for completion of the written procedure.

The first area which must be addressed is the procedural framework for setting up the quality assurance department. The types of procedures needed include the inspection guidelines, calibration program, processing reworked and returned product, inspection stamp control, material review boards, and a host of other situations normally dealt with in any Quality Assurance Manual. These procedures should thoroughly outline how the quality assurance department is organized. The Quality Assurance Manual must also contain copies of all the forms and tags used by the department. Examples of points to be covered are shown in the next section.

TYPICAL QUALITY ASSURANCE MANUAL

1. A general section entitled "Elements of Quality" provides the reader with a brief overall look at the goals and duties of the department. All other sections must reflect these goals and duties. This section also defines the authority of the Quality Assurance Manager.

2. The "Material Accountability" section lists only copper clad laminate as the material to be accounted for. This is the most critical of all the materials. The copper foil must be inspected for defects prior to use. Should there be no defects, the thickness of the material must be measured to ensure compliance with the military specification to which it should have been purchased: MIL-P-13949. If there are defects in the material, it should not be used. The entire sheet should be set aside. Laminate dealers and manufacturers are very good at standing behind their product. Do them the courtesy of not cutting up or damaging the sheets of laminate before you return it to them.

3. The Calibration Program should cover all tools and instruments used to measure or control processing of the product: calipers, micrometers, beta back-scatter and magnetic measuring instruments, heater controllers, rectifiers, numerical control drilling/routing equipment, and multilayer lamination presses. There is absolutely no control over the processes without this very basic requirement. Calibration is merely the process of verifying that the device is correctly indicating the parameter being monitored. The company performing the calibration service must have standards which are accurate and traceable to the National Bureau of Standards.

4. There must be instructions for the use and logging of the inspection stamps. These instructions must include who will have access to the stamps, when the stamps are to be used, and how they are to be stored.

5. The inspection supervisor's duties should be outlined. All of these duties and responsibilities are delegated from the duties of the Quality Assurance Manager. The scope of these duties should be outlined and clear.

6. The section on first article inspection (also called incoming inspection) must explain the basic inspection/acceptability criteria for documentation on all new part numbers; this is a vital function. The basic duties of the first article, or incoming, inspector are to review the same documentation (blue print, procurement specification, artwork, and purchase order) which the Planning Engineer reviewed prior to writing a Job Traveler. (The Job Traveler is a set of instructions which includes every operation the job is to go through in making the product, and the requirements at each of those steps). After doing this, the job traveler must be reviewed for accuracy and completeness. It is good for the first article inspector to look for special notes which should be pointed out to other people who will be processing the job later on—for example, the need to make a small change to the artwork during solder masking, or to perform a special cut-out during fabrication. If there are no discrepancies with the Job Traveler, (1) the artwork must be catalogued and inspected for front-to-back

registration (or layer-to-layer registration for multilayers), (2) the circuitry must be inspected for quality of line definition and usability, (3) minimum air gap must be determined (to determine whether the part must be dry film imaged or screen imaged), and (4) the part number and revision level must be checked for agreement. After the first article inspector "buys off" the Job Traveler, and the customer's artwork, the entire documentation package is sent to the drilling programmer. After programming, a first article (one drilled panel) must be run. This drilling first article immediately goes to the first article inspector. The drill tape for programming the N/C drilling equipment must not be used for production until the first article inspector has approved its accuracy.

7. Inspection procedures include instructions on what is to be inspected during pre-solder mask inspection (first inspection), and during final inspection (after fabrication and prior to shipping). The section must also explain exactly how to process product which is judged to be nonconforming to the customers' requirements. There must be a good discussion on acceptability guidelines. It is almost as bad to have an inspector rejecting perfectly acceptable product as it is to have an inspector letting poor product pass. Too often inspectors receive little training on acceptability. The result is that, after being criticized for letting poor quality product pass, the inspector will start rejecting product which has even the slightest cosmetic defect. There is no reason, for example, to reject a printed circuit for a slight scratch across a contact finger if there is no electrical connection to that contact finger, even though the scratch may cover the entire length of that finger.

8. Instructions for processing reworked printed circuits should spell out what needs to be done to ensure that the records are kept straight, and that shabbily reworked printed circuits are not shipped. Also, with some types of work, specifically military work, no rework is allowed. Military customers tend to be more specific about what is definitely not allowed, and it is more difficult to obtain waivers for non-conforming product (Fig. 6-1).

9. Military printed circuits need to have serial numbers assigned to each panel. Every test coupon and every board on a panel must be identified with the serial number of the panel. This is critical for traceability during Group A and Group B laboratory testing (see the section on military requirements in this chapter).

10. Group A inspection and laboratory testing procedures must detail the requirements on check lists. These forms must be carefully filled in during Group A processing. A defect found on a test coupon during micro-sectioning should have the serial number of that coupon written on the laboratory form. If this is done, then the boards from that panel can be separated from the rest of the boards with coupons that passed the Group A testing.

11. Boards which have hardware installed (standoffs, lugs, connectors, eyelets, etc.) should be inspected to verify that the correct items were installed

where they belong, and that they are made of only the specified metals, and are oriented correctly. A checklist should be used for this purpose.

12. Material Review Board. Sometimes it is necessary to make a careful examination of one or more boards to determine their suitability for shipping to the customer. Perhaps the printed circuit does not conform well to requirements. Before shipping questionable product (even though the customer may be in desperate need of the boards), it is wise to get a panel together and review the problems. This is good practice so that if the customer is to be called, the person doing so will have all the facts at hand; and it may be necessary to build a case as to how the customer can justify accepting questionable products. Before questionable product is shipped, the printed circuit manufacturer owes it to himself and to the customer to review the situation. It may be that the material review board members decide against shipping the product after reviewing all the facts; this too, will have to be explained to the customer.

13. Processing of printed circuits returned from the customer should be carefully documented. Not only must the purchase order numbers and debit memos be kept straight, but the board quantities and reasons for return must be identified. If the boards are to be reworked, they must be re-inspected after the rework has been performed. The Discrepant Material Report (DMR) must have room to record all the information. Everyone who will handle this material and the paperwork must know exactly what they are to do with it when it comes to them.

The Quality Assurance Manual, once it has been written and approved by the powers that be, should be typed on the company's standard procedure form. This form must have blocked in areas for titling, procedure number, revision, date, page number and number of pages, and signature of approval. After this has been completed, and a table of contents drawn up, the quality assurance manual should be printed and bound. The manual should also be prominently displayed in the inspection areas.

It is even good practice to have a page in the front of the manual for signatures and dates of reading by all quality assurance personnel. A record such as this helps to ensure training of personnel; and this is just how the quality assurance manual should be used, as a training manual. It is not enough to leave it for personnel to read; the quality assurance manager should actively train by holding sessions until all aspects of the quality program and procedures have been covered.

The Quality Assurance Manual should not only be complete, it should be used. Customer survey teams will look for compliance with the manual: completed log books/log sheets, signed job travelers, calibration stickers/documents, scrap reports, and isolated areas for storing rejected materials. The manual is a sales tool, and this fact should never be neglected. The manual must be presentable, and so must evidence of its use. Once the manual has

Defect code number	Major defects
12A1	Bonding of conductor to board and peeling of conductor: Any looseness of bond on any conductor length, any peeling of conductor (defect most prevalent at terminals and at ends of conductor contacts).
12A2	Broken eyelet: Part of eyelet missing. Circumferential splits.
12A3	Seating of eyelets: Eyelet not properly seated perpendicular to the board.
12A4	Plated-through hole: (1) In excess of 10 percent void of surface in hole rejectable. (2) Voids at interface of hole rejectable. (3) Circumferential separation rejectable.
12A6	Superfluous conductor: Potential cause of short. Clearance less than that specified in master drawing for electrical spacing.
12A7	Eyelets: Total voids in solder fillet around eyelet exceeding 30 percent of the periphery in funnel flanged eyelets. Cracks in solder around eyelet.
12A8	Reduction in width of conductor at any point exceeding 20 percent of the minimum specified conductor width on the master drawing (see 3.6.7).
12A11	Conductor and lead spacing violating the requirements of 3.6.9.
12A12	Plating on top surface of conductors and in plated-through holes: Lack of plating or less than minimum specified. Plating adhesion defect (see 3.7).
12A13	Cracks, chips, or bulges on board surfaces.
12A14	Unspecified removal of board material: Any visible unspecified removal of board material (such as removal about conductors to increase insulation resistance or indication of contamination). A minor change in surface appearance due to removal of superfluous copper shall not be considered as removal of unspecified surface material.
12A15	Warp or twist of board: Warp or twist in any board exceeds that specified in 3.9 or the master drawing.
12A16	Spacing of holes: Other than that specified on the master drawing.
12A17	Annular ring (at narrow point): a. The projecting flange of eyelet or standoff terminal extending beyond the annular ring. b. Measuring less than 0.015 inch (0.38 mm) beyond the edge of unsupported hole.

Fig. 6-1. List of major and minor defects from MIL-P-55110C.

Defect code number	Major defects

c. Measuring less than 0.002 inch (0.05 mm) for internal terminal areas.
d. Measuring less than 0.005 inch (0.13 mm) beyond the edge of plated-through hole except as allowed in 3.6.1.1.

12A22   Terminal holes: Ragged holes with chipping or cracking in the wall of the holes, and bulging around the holes or reduction of the hole diameters with base laminate material such as fibers. Ragged metal foil edge. Metal foil deformed into the hole, torn or lifted. Size of holes not as specified in the master drawing.

12A23   Delamination: Internal or external separation of layers of base material (paper or glass).

12A24   Spurs or whiskers: Presence of spurs or whiskers.

12A26   Nodules: Cause reduction in hole diameter to less than specified on the master drawing.

Minor defects

12B18   Cuts, cracks, or scratches in conductor: Board visible through copper (copper may be visible through overplating when overplating is specified). Cuts, cracks, or scratches completely across conductor or more than ½-inch along conductor (when this type of defect can be considered a reduction in area of conductor, evaluate in accordance with defect 12A8.

12B19   Size of holes: Size of holes (other than terminal holes) not as specified in master drawing.

12B21   Loose standoff terminal: Any standoff terminal that can be turned or removed by hand. (This condition cannot be determined after soldering as the application of solder will cause the stud to be tight. This condition is unacceptable as it may cause a high-resistance joint to develop which is very troublesome and difficult to locate after soldering.)

Fig. 6-1. (*Continued*)

been incorporated into use by inspection and manufacturing, the quality of the product being inspected should increase, while the rate of product returned from the customer should decrease.

OTHER PROCEDURES

Once the goals, policies, and procedures of the Quality Assurance department have been established, the next goal has got to be carrying out the policy. The next logical area to address is establishing written and agreed upon procedures for the other manufacturing operations. Obviously, this requires working closely with the managers and supervisors of these operations. This affords an excellent opportunity for the quality assurance manager to develop a working relationship with the rest of manufacturing, and to demonstrate that the goal of quality assurance is not to shut down the plant when a problem arises; but to optimize quality, processes, procedures, and training.

There should not be two methods for performing an operation (except as a back-up measure), there should only be one, and this should be the one used by all personnel, on all shifts. Once this is firmly established in the company, then all personnel will be able to tell when the procedure is not being followed (as if variation in quality were not enough of an indicator).

Establishing written procedures will also help bring understanding of those procedures. It may be the first time a department has studied the process enough for it to be documented. Furthermore, the authority of a supervisor can be extended once a procedure has been agreed upon. Too often in the printed circuit industry a process is performed differently by two people working side by side, because too much is left totally to the discretion of the person performing the work. The days of 'you do it your way, and I'll do it my way" must cease if uniformity, consistency, and optimization are to be brought to a manufacturing environment.

The written procedure also furnishes a trainee with a source of information to turn to when in doubt about an operation, or when a dispute exists. The written procedure thus becomes the platform upon which a supervisor or manager can stand. The quality assurance manager should encourage each supervisor to write down the procedures which he/she believes to be the optimum method for performing each special function under their supervision. It may be that the supervisor feels uncomfortable writing, so encouragement and the offer of aid in this area may be needed. The supervisor should be encouraged to discuss the individual procedures with the workers performing the operation, and to solicit their comments. Just like the quality assurance manual, the process procedures must be typed on the company's procedure form with operation name, number, revision, date, and signature of approval. This document, when typed and bound, should also contain a signature sheet at the beginning to document that all operators have read the procedure, and understand its provisions. The document must be conspicuously displayed in the work area, for all to read and refer to.

It must be the goal of manufacturing and quality assurance to build quality into the product. This cannot be done without giving attention to the individual people who actually do the work:

- It must be corporate policy that the individual worker be held accountable for the quality of work which he/she produces.

- It must be corporate policy that the supervisor of each operation be held accountable for the quality of work performed by all personnel under his/her direction.
- It must be corporate policy that all managers be held responsible for the level of quality of the company's product.
- It must be corporate policy that all personnel be responsible for looking at the product before beginning work on it—not to ignore an existing problem, but to stop the product as soon as a problem is noticed.

The day that upper management, middle management, supervisors, and workers lose sight of the fact that quality cannot be inspected into the product, is the day quality will begin to slide. The entire thrust of the manufacturing operation must be in the direction of training, and personal accountability. The greater use management makes of inspection as a tool for controlling quality, the more management will demonstrate its lack of commitment to training and personal accountability.

DEALING WITH CUSTOMERS AND PRODUCT RETURNED BY THE CUSTOMER

Most of the contact between the printed circuit manufacturer and the customer will be through the salesman, the quality assurance manager, the planning engineer, and the production control person of the printed circuit shop, with the buyer, quality assurance manager, and engineering personnel of the customer. It is important that the personnel of the printed circuit shop be aware of these types of contacts and avenues of communication. The salesman should endeavor to obtain information on what personnel of the customer should be contacted in case a problem should arise which would impact quality, delivery, or documentation. The salesman should also inform the customer of what personnel at the printed circuit shop should be contacted to handle similar situations.

The buyer is the main contact at the customer's place of business. Some buyers demand that all contact between the printed circuit shop and the company be made through him/her. This is reasonable, since the buyer is responsible for procuring quality printed circuits, on time. Once a working relationship has been established, together with a good track record for quality and delivery, the buyer may recommend that quality and engineering personnel at the company be contacted directly. It is always wise to establish a good working relationship between the contacts of both companies. Whenever possible, face to face meetings should take place, in lieu of a telephone conversation. Nothing promotes understanding and agreement (and hence prevention and resolution of problems) better than eye-to-eye contact.

The above mentioned personnel at the printed circuit shop must understand that the customer has needs and problems which must be dealt with. These people must understand these needs, and how the printed circuit shop can fullfil them. The needs often include more than the simple buying and making of the printed circuit boards of a given purchase order. Perhaps the printed circuit shop's engineering staff can be of assistance in helping the customer avoid common documentation pitfalls of bringing a new process (auto insertion, for example, or CAD) on line, or perhaps the customer needs information on the practicality of certain material specifications or tolerances. When the printed circuit shop is able to help in these areas, they promote good will. Good will on the part of a customer toward a printed circuit shop drastically affects the quality (or at least the acceptability) of the printed circuits. There are times, in any business relationship, when good will makes the difference between accepting or rejecting a marginally nonconforming product.

There will be times when a customer's incoming inspection department will reject and return printed circuits to the vendor. Some of these times, action may have been taken which was just a little bit too hasty, and not thought out. In other words, sometimes rejects are simply not justified on the basis of IPC, military, or other industry standards, including the published procurement specification of the customer. It even happens that customers return product just because their inventory is higher than they would like; they may or may not then look for a flimsy excuse for returning the boards based on a point of quality. Sometimes a customer may return perfectly usable product, even though it may have a slight cosmetic flaw, just to get the attention of the vendor's Quality Assurance Manager. It also happens that the RTV's (Return To Vendor) may put the manufacturing department of the customer in a bind.

It is in the successful resolution of situations like these that the interpersonal skill of a quality assurance manager and his/her knowledge of standards comes into play. Product which has been returned by a customer should always be looked at very carefully. Do not forget, the reject rate on the vendor's record is at stake also. Examine the product to ascertain what the nature of the defect is. Is the defect merely cosmetic? Is it even a defect? Was the defect caused by an error on the customer's part? All of these questions must be answered.

If there is a slight defect in the product, it is always worth while to call the customer to see if the product can be used, or used with slight rework. In some cases, the product may be cosmetically defective and rejectable; yet the customer may want to use the boards and needs only a little encouragement from the vendor's Quality Assurance Manager to do so. Do not be afraid to document a case in favor of the customer using the product. Search the industry acceptability literature. Perform tests. It may be that the Quality Assurance Manager able to do this is performing a real service to his/her own company as well as for the customer.

No one likes to see product they have manufactured returned from the customer; customer returns are as disconcerting to production and sales personnel as they are to Quality Assurance. Sometimes, less than perfect product does get through inspection to shipping. In fact, almost all the time, less than perfect product goes to shipping. It is rare that a perfect printed circuit is manufactured; no matter how perfect a printed circuit or any other product may appear to be, there is always a way to find something wrong with it. The ability to find defects may only be dependent upon the equipment and techniques available for inspection. So, if it can be accepted that perfect printed circuits are rarely manufactured, then deciding upon the acceptability of a printed circuit becomes a matter of determining whether or not the product conforms to the customer's procurement specification, purchase order, and drawings.

It is important that all inspection personnel understand this fact. If inspectors believe that they are doing their job by accepting or rejecting printed circuits on the basis of whether or not the printed circuit is perfect, then they are wasting time and money. One of the jobs of inspection is to determine whether or not the product meets the specifications required for that printed circuit. This job requires knowledge of the customer's requirements, industry standards, and sometimes of the customer.

The requirements for a quick turn-around prototype and a major production run may be quite different. An engineer who needs a functional board as soon as possible may not have the time or money to obtain the finest quality drawings and artwork. On the other hand, he probably will not reject the board for defects caused by his poor documentation, and maybe not even for defects from poor workmanship. He just needs the board for functional evaluation. This is where knowledge of the customer is required. There is no point in rejecting poor quality artwork, for instance, when it is good enough to meet the customers requirements.

The quality of documentation required for a major production run may be quite different. The printed circuit manufacturer should be more demanding in documentation (drawings, artwork, purchase order), and the level of documentation required (red lined drawings and changes over the telephone may be permissible for prototype/quick turnaround, but not for a $200,000 production run).

There are documents available to help decide whether or not a circuit meets the customer's or the industry standards. IPC-A-600, MIL-P-5511OC, MIL-STD-275D and the customer's procurement specification (spec sheet) are of great help. The customer's specification should be the last word. Sometimes there are honest differences of opinion as to whether or not a circuit meets the requirements; and this may be the case when product has been returned by the customer.

When product is returned by the customer the reason for the return must be looked into as soon as possible. Good customer relations demands that these things not be allowed to fall through a crack and forgotten. Immediate response is a plus; especially since other vendors of that customer may not be very prompt. If the defect is able to be remedied (in the case of, for example, forgetting to add legend, or thin gold on the contact fingers), it should be done so.

Sometimes it is possible to work with a customer to remedy a situation on a one-time-only basis—so the customer can receive the boards as soon as possible, and so that the printed circuit manufacturer does not have to remake the circuits. However, sometimes it is necessary to scrap the product; and the reason for doing so is so obvious that there is no need to call the customer. No matter what course of action, the customer should be called as soon as the reason for the defect is understood, and the possible return date to the customer decided upon.

If a customer is not sure whether or not to use the defective product, perhaps the quality assurance manager of the printed circuit shop can be useful in helping the customer make up his/her mind. The quality assurance manager must not be afraid to build a case, if possible, for using the circuits, or for allowing any needed rework. There are some cases where a defect may only be cosmetic in nature. When this is the case the customer can be encouraged to run samples and tests, or the printed circuit shop quality assurance manager can do this for the customer and present it. Also, if situations involving the specific defect have been dealt with in published literature, this can be presented to help make a decision.

There are some users of printed circuits who have special requirements: color of legend, thickness of nickel on contacts, revision level or other markings, etc. The quality assurance manager must make these peculiarities known to the manufacturing personnel. Even though a note may be written on the work order traveler to call attention to the requirement, there is nothing wrong with posting a note in a work area as a reminder.

UNDERWRITERS LABORATORIES (UL)

The program which a company follows to keep itself in good stead with Underwriters Laboratories usually falls within the domain of the Quality Assurance department. The type and extent of recognition which a company maintains with Underwriters Laboratories is of great importance; and the use of UL recognition as a marketing tool should not be overlooked.

The basic purpose of Underwriters Laboratories is stated in the beginning of the Recognized Component Directory (Footnote 3-1):

TESTING FOR PUBLIC SAFETY

Founded in 1894, Underwriters Laboratories Inc. (UL) is chartered as a not for profit, independent organization testing for public safety. It maintains and operates laboratories for the examination and testing of devices, systems and materials to determine their relation to life, fire, casualty hazards and crime prevention.

ZPMV2   April 30, 1982
Component—Wiring, Printed

SAE CIRCUITS INC   **E44075 (S)**
                                                    (B-cont. from A card)

Type	In.	Cladding Conductor Width Min In.	Min Edge Mils	SS/DS+	Max Area Dia In.	Sold Lts Temp C	Time Sec	Oper Temp C	UL94 Flame Class
Single layer printed wiring boards.									
UL-16	0.013	0.039	1.3	DS	6.0	274	10	105	—
UL-16V	0.013	0.039	1.3	DS	6.0	274	10	105	94V-0
UL-9	0.012	0.025	1.3	DS	6.0	260	10	105	—
UL-9V	0.012	0.025	1.3	DS	6.0	260	10	—	94V-0
94V-0@	—	—	—	—	—	260	10	—	94V-0
Multilayer printed wiring boards.									
M	0.010	0.030	0.34	DS	2.38	260	7	105	—
12M	0.010	0.030	0.34	DS	2.38	260	7	105	—
12MV	0.010	0.030	0.34	DS	2.38	260	7	105	94V-1
UL-9M	0.012	0.025	1.3	DS	4.0	260	10	105	—
UL-9MV	0.012	0.025	1.3	DS	4.0	260	10	105	94V-0

+DS-Double or single sided
@Followed by company identification; investigated for flammability only

Fig. 6-2. UL recognized component card, often referred to as the *yellow card.*

A company which has had products tested successfully to UL requirements is issued Component Recognition Cards (often referred to as Yellow Cards). These cards list some of the more important features and the parameters for which the component is recognized (Fig. 6-2). UL publishes all of the recognized component cards in a multi-volume Recognized Component Directory, available for a small fee from UL.

Explanation of the Component Card

1. The number in the upper right hand corner is the UL file number for the company. (Example: E69171.)

2. *Type.* This refers to the type of printed circuit for which the company has been granted recognition. The designation is assigned by the company, and any code may be used. Type may refer to whether or not the printed circuit is double sided or multilayer; FR-4, polyimide, teflon or some other material; ⅛, ¼, or 1 ounce copper foil; or even a metal core board.

3. *Cladding Conductor.*

- *Min. In.* This refers to the minimum conductor width, looking perpendicular to the surface of the circuit. Testing and approval of, for instance, 0.005 inch conductors, will automatically result in recognition on all thicker conductors.
- *Min. Edge In.* This refers to the minimum distance from the edge of the board to the conductor.

- *Thk. Mils.* This refers to the minimum thickness of metal foil the board type may be built on. Testing on ½ oz (0.72 mil) foil will result in approval on foils up to 2 oz (2.88 mils). Testing on thinner foils, like ⅛ or ¼ oz, will not result in approval on the thicker foils.

4. *SS/DS.* This refers to whether the board is a double sided or a multilayer printed circuit. If the board is a multilayer, it is listed under Type, as multilayer type. Double sided recognition automatically means single sided recognition; however, the reverse is not true.

5. *Max. Area Dia. In.* This refers to the diameter of a circle which can be inscribed in an unbroken land area on the circuit.

6. *Sold. Lts.* This refers to the maximum solder temperature, and the immersion time at that temperature, to which the recognized component may be exposed.

7. *Max. Oper. Temp.* This refers to the maximum operating temperature in which the recognized component may be used. A telecommunications printed circuit built on FR-4 material, with a maximum operating temperature of 105 degrees Fahrenheit, may not safely be used in a 120 degrees Fahrenheit environment for extended periods of time.

8. *UL94 Flame Class.* This refers to the flammability rating assigned to the printed circuit. The 94 rating is generally assigned on the basis of the 94 rating of the material used to build the circuit. 94V-O is the highest classification of fire retardancy.

9. *Marking.* This is the logo, or UL logo of the company. When boards being built are to be UL recognized, they must have the logo, followed by the appropriate board type etched onto the printed circuit's surface. This marking must go only on printed circuits of the appropriate type, for which the company has received UL recognition.

The document UL796 (available from UL, see UL address at end of this section) explains how to go about receiving UL recognition. The basic steps are to submit a letter of request for testing. UL will respond with an application and letter stating the fees for the testing. The fees must be paid at the time sample boards are submitted for testing. A circuit of the appropriate type must be manufactured. This circuit must contain circuitry (and plated through holes) which will result in optimum parameters being listed on the component card. UL796 contains pictures of example testing patterns.

When the application, fees, and sample boards are submitted, a letter briefly stating the processes used to manufacture the printed circuits (chemicals and temperatures) must also be included. The letter from UL, sent in response to the letter stating a desire to submit for testing, will provide information on how many boards must be submitted. After the testing period, recognition will be granted if the samples successfully pass the rigorous testing.

The extent of UL recognition which a company maintains is very important to that company. One of the first acts of a quality assurance engineer or buyer of a prospective customer may be to look up the recognition which a potential vendor has attained. The extent of recognition which a printed circuit manufacturer has attained is a selling point with many companies.

Some manufacturers do not have the capability for manufacturing multilayer printed circuits in-house. This does not mean that the company cannot receive UL recognition for multilayer-type circuits. If this manufacturer wishes to become involved with limited multilayer processing (usually after lamination), all that need be done is to send a letter to UL, stating the source company for the mass lamination service. UL will then grant recognition for multilayer types, with the same parameters as that manufacturer's double sided boards. The only requirement is that the source for the mass lamination service also be UL recognized for multilayers; and that the provider of the mass lamination service have parameters at least as tight as the customer's.

If a UL recognized printed circuit manufacturer wished to buy laminate from a source other than the laminate upon which the testing sample was submitted, then a letter of intent, and the name of the other laminate source must be sent to UL, along with a fee. If the new laminate meets the recognition parameters of the circuit manufacturer, the new laminate will be added to the approved list of laminate sources for that printed circuit manufacturer. This "add on" procedure is available only for double sided circuits, not for multilayers.

UL maintains field offices with a staff of inspectors to follow up on the manufacturer. Regular plant surveys are conducted to see that only recognized materials, and processes submitted on the letter for testing, are being used. Usually, at least once a year samples are collected for UL testing.

Underwriters Laboratories Offices and Testing Stations
333 Pfingsten Road
Northbrook, IL 60062
(312) 272-8800

1285 Walt Whitman Road
Melville, L.I., NY 11747
(516) 271-6200

1655 Scott Blvd.
Santa Clara, CA 95050
(408) 985-2500

Marine Department
602 Tampa East Blvd.
Tampa, FL 33619
(813) 621-9754

MILITARY CERTIFICATION

MIL-P-55110C Scope (Footnote 3-2): "This specification covers the certification and performance requirements for rigid single-sided printed wiring boards, rigid double-sided printed wiring boards, and rigid multilayer printed wiring boards with plated-through holes." This specification calls out the general requirements for many military printed circuits. What sets this specification apart from other specifications are the requirements for extensive testing on an ongoing basis, the requirements for sample boards to be manufactured and sent to a government approved laboratory for evaluation and testing, and the requirement for the Defense Electronic Supply Center (DESC) to approve the final report on that testing.

This specification is enormously important for most printed circuit manufacturers to comply with. Paragraph 6.6 states:

> Contracts and awards will be made only to suppliers which have tested and have passed the certification inspection test . . . and have been certified by the cognizant certification organization. In order to be eligible for award of contract, the attention of the supplier is called to this certification requirement. Supplier certification inspection shall be performed in accordance with the procedures described in this specification and appendix.

What this means is that no printed circuit manufacturer is going to build circuits under a government contract unless they have received DESC certification. Failure to obtain this certification will mean that the manufacturer is cut off from a major source of business. Also, failure to obtain DESC certification under the provisions of MIL-P-55110C may cause a lot of buyers for commercial printed circuits to cast a jaundiced eye upon the printed circuit manufacturer. There is no underestimating DESC certification as a powerful sales tool for any printed circuit shop.

Steps to Certification

1. The printed circuit manufacturer submits a copy of "Application for Supplier Certification Testing" to DESC-EQ (Fig. 6-3).

A. On the form the manufacturer lists the type of printed circuits for which certification testing is desired:

- Type I. Single sided.
- Type II. Double sided. Testing on type II automatically covers type I.
- Type III. Multilayer.
- B. The manufacturer must also list the materials on which testing is desired. It is necessary to be certified on each material for which the manufacturer desires to build circuits: FR-4, polyimide, teflon, etc.

APPLICATION FOR SUPPLIER CERTIFICATION TESTING
DEFENSE ELECTRONICS SUPPLY CENTER DIRECTORATE OF ENGINEERING STANDARDIZATION

CHECK TESTING OPTION DESIRED

☐ AT THE PLANT OF THE MANUFACTURER (IN-PLANT)
(Complete Sections I, II, and IV)

☐ AT A NON-GOVERNMENT TEST LABORATORY (OTHER THAN IN-PLANT)
(Complete Sections I, III, and IV)

☐ AT A COMBINATION CONSISTING OF THE ABOVE OPTIONS
(Complete All Sections)

SECTION I

COMPANY NAME AND MAILING ADDRESS (NAME AND ADDRESS TO APPEAR ON CERTIFICATION)	DATE OF APPLICATION

PLANT NAME & ADDRESS WHERE PRODUCT(S) IS MANUFACTURED

SPECIFICATION NUMBER, TITLE & DATE	AMENDMENT DATE
DETAIL SPECIFICATION(S) OR SHEET(S) & DATE(S)	AMENDMENT DATE

QUANTITY	IDENTIFICATION OF PRODUCT TO BE CERTIFIED

Fig. 6-3. Application for supplier certification testing.   PAGE 1 OF 4

SECTION II
TESTING AT THE LABORATORY OF THE MANUFACTURER (IN-PLANT)

Complete this section ONLY if testing (either complete or partial) is to be conducted at the laboratory of the manufacturer (in-plant). If partial testing is to be conducted in-plant, the information furnished in this section should cover only those tests to be run in-plant. Information covering those tests which will not be run in-plant must be furnished by completing Section III. Under this test option, several different products covered by the same specification may be included in one application; however, a new application must be submitted for products not previously covered by an application.

1. NAME & ADDRESS OF THE MANUFACTURER'S LABORATORY

2. HAS THE MANUFACTURER'S LABORATORY BEEN INSPECTED BY DESC PERSONNEL? ☐ YES ☐ NO
   INSPECTION WAS CONDUCTED ON _____ (DATE) BY _____

3. WERE ALL TEST FACILITIES REQUIRED FOR THIS CERTIFICATION INCLUDED IN THIS INSPECTION OR SUBSEQUENTLY APPROVED BY DESC? ☐ YES ☐ NO IF NOT, INCLUDE LIST OF ALL TESTING FACILITIES ADDED SINCE LAST INSPECTION OR APPROVAL.

4. IF THE MANUFACTURER'S LABORATORY HAS NOT YET BEEN FOUND BY DESC TO BE SUITABLY EQUIPPED AND STAFFED FOR PERFORMING TESTING UNDER THE SPECIFICATION LISTED ON THE FIRST PAGE OF THIS APPLICATION, INDICATE BELOW THE DATE THE EQUIPMENT PROPOSED FOR USE WILL BE AVAILABLE FOR

   INSPECTION BY DESC PERSONNEL _____ (DATE)

5. WILL ALL OF THE TESTS BE PERFORMED AT THE MANUFACTURER'S LABORATORY?
   ☐ YES ☐ NO IF THE ANSWER IS "NO" SECTION III MUST BE COMPLETED

6. IF TESTING IS AUTHORIZED BY DESC, WHEN WILL THE TEST SPECIMENS BE READY AND TESTS STARTED?

6A. WHEN WILL THE TESTS BE COMPLETED?

7. DOES THE MANUFACTURER'S LABORATORY HAVE A GOVERNMENT QUALITY ASSURANCE REPRESENTATIVE?

   ☐ YES   ☐ NO

   ☐ RESIDENT ☐ ROVING   QAR'S NAME _____
                            ADDRESS _____

Fig. 6-3. (*Continued*)

8. NAMES AND ADDRESSES OF CALIBRATION SOURCES

NOTE: The following must also be available at the time of the inspection.

    a. Certified calibration records for the calibration standards and in-plant test equipment.
    b. Calibrations standards.
    c. Test personnel concerned with testing at the laboratory.
    d. All test equipment in operating condition.
    e. MIL-P-15662 Documentation

9. IS MILITARY SECURITY CLEARANCE REQUIRED TO GAIN ACCESS TO ANY OF THE MANUFACTURER'S FACILITIES? ☐ YES ☐ NO   DEGREE OF CLEARANCE REQUIRED ☐ CONFIDENTIAL ☐ SECRET

Fig. 6-3. (*Continued*)

MIL-P-55110C
APPENDIX

SECTION III
TESTING AT A NON-GOVERNMENT TEST LABORATORY (OTHER THAN IN-PLANT)

Complete this section ONLY if testing (either complete or partial) is to be conducted at a Non-Government test laboratory (other than in-plant). If partial testing is to be conducted at a Non-Government test laboratory (other than in-plant), the information furnished in this section should cover only those tests to be performed at that laboratory. Information covering those tests which will be run at the laboratory of the manufacturer (in-plant) must be furnished by completing Section II. Under this test option, several different products covered by the same specification may be included in one application: however, a new application must be submitted for products not previously covered by an application.

1. NAME & ADDRESS OF NON-GOVERNMENT TEST LABORATORY (OTHER THAN IN-PLANT)

Fig. 6-3. (*Continued*)

2. WILL ALL THE TESTS BE PERFORMED AT THE ABOVE LABORATORY?
☐ YES   ☐ NO   IF THE ANSWER IS "NO", LIST BELOW THE TESTS THAT
WILL BE PERFORMED AT THAT LABORATORY

3. HAS THE LABORATORY BEEN INSPECTED BY DESC PERSONNEL? ☐ YES
☐ NO
INSPECTION WAS CONDUCTED ON _____ (DATE) BY _____

4. WERE ALL TEST FACILITIES REQUIRED FOR THIS CERTIFICATION
INCLUDED IN THIS INSPECTION OR SUBSEQUENTLY APPROVED BY
DESC?   ☐ YES   ☐ NO   ☐ IF NOT, INCLUDE LIST OF ALL TESTING
FACILITIES ADDED SINCE LAST INSPECTION OR APPROVAL.

5. IS MILITARY SECURITY CLEARANCE REQUIRED TO GAIN ACCESS TO
THE LABORATORY OR CALIBRATION AREA? ☐ YES   ☐ NO   DEGREE
OF CLEARANCE REQUIRED   ☐ CONFIDENTIAL   ☐ SECRET

6. IF TESTING IS AUTHORIZED BY DESC, WHEN WILL THE TEST
SPECIMENS BE FORWARDED TO THE TEST LABORATORY?
_____ (DATE)
WHAT IS THE ESTIMATED TEST TIME?

7. DOES THE TEST LABORATORY HAVE A GOVERNMENT QUALITY
ASSURANCE REPRESENTATIVE?

☐ YES   ☐ NO

☐ RESIDENT   ☐ ROVING   QAR'S NAME _____
                           ADDRESS _____

Fig. 6-3. (*Continued*)

SECTION IV
CONDITIONS

THE APPLICANT CERTIFIES THE FOLLOWING:

a. That he is the manufacturer of the product.

b. That he has determined from actual tests (within the limits of test equipment commonly available unless otherwise specified) that the product conforms to the applicable specification.

c. That he will supply items for test which are representative of the manufacturer's production.

d. That he will supply for use of the purchaser, products which meet the requirements of the specification in every respect.

e. That he will not apply for retest of the product until satisfactory evidence is furnished that all of the defects which were disclosed by previous tests have been corrected.

f. That he will not state or advertise that the product is the only product so certified or in any way imply that the Department of Defense endorses his product.

g. That he will notify the Defense Electronics Supply Center (DESC-EQ), Dayton, OH 45444 of any change in his product after approval and will state at the same time whether in his belief the change will or will not prejudice the capability of the product to meet the qualification test requirements; whether he intends to submit new samples for testing or desires to have his product removed from certification; and whether the changes will affect the applicants brand designation for the product.

h. That he will notify the certifying activity of the discontinuance of manufacture of a product.

i. That he will notify the certifying activity of plans to move a plant where it is desired to manufacture certified products at the new location.

j. That he will subject all products identified on page one of this application to the Quality Conformance Requirements as specified in 4.7 of MIL-P-55110C.

k. That he will accumulate data for retention reports as specified in 4.5.5 of MIL-P-55110C.

Fig. 6-3. (*Continued*)

THE UNDERSIGNED CERTIFIES THAT THE INFORMATION SUBMITTED TO DESC IN THIS APPLICATION IS TRUE AND ACCURATE TO THE BEST OF HIS KNOWLEDGE AND FURTHER, AGREES TO THE CONDITIONS AS SHOWN IN SECTION IV ABOVE.

TYPED NAME AND TITLE (RESPONSIBLE OFFICIAL OF THE APPLICANT)	DATE
SIGNATURE	

PAGE 4 OF 4

Fig. 6-3. (*Continued*)

- C. The manufacturer must also list the name of the government approved testing laboratory which will do the certification testing. Once a testing laboratory has been located and contacted, they will be very helpful in assisting with the paper work and other steps needed for certification. The testing laboratory will also, generally, supply a copy of MIL-P-55110C. Two of the oldest and largest government approved testing laboratories are:

Delsen Testing Laboratories, Inc.
1031 Flower Street
Glendale, CA 91201
(213) 245-8517

Trace Laboratories, Inc.
P.O. Box 8644
Baltimore/Washington International Airport
Baltimore, MD 21240
(301) 859-8110

2. The manufacturer must also obtain artwork from IPC, as the IPC artwork is used to build the sample boards for testing. Type II circuits require IPC-B-25 artwork; and type III circuits require IPC-B-27 artwork, for a 10 layer multilayer (Figs. 6-4 and 6-5). The artwork is available from:

IPC
1717 Howard Street
Evanston, IL 60202
(312) 491-1660

3. About two weeks after applying for permission to begin testing, the permission will arrive in the mail. As soon as permission has been applied for, the manufacturer should begin building the test boards, but the boards should not be sent for testing until permission has arrived.

Before beginning any work on the test boards, the manufacturer should read MIL-P-55110C repeatedly until all requirements, documentation, and procedures are clear. Any questions can be answered by the testing laboratory.

4. When the sample boards are completed, the manufacturer must perform inspection and laboratory testing per Group A requirements (Fig. 6-6). If the tested boards successfully pass all these requirements, the testing laboratory

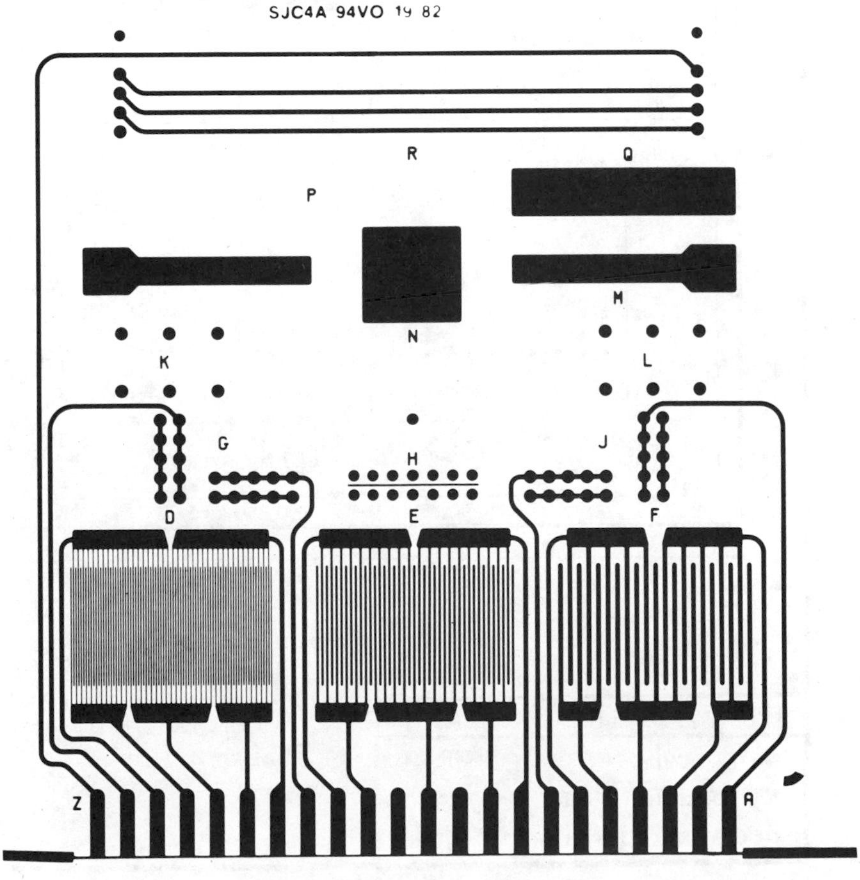

Fig. 6-4. IPC-B-25 artwork.

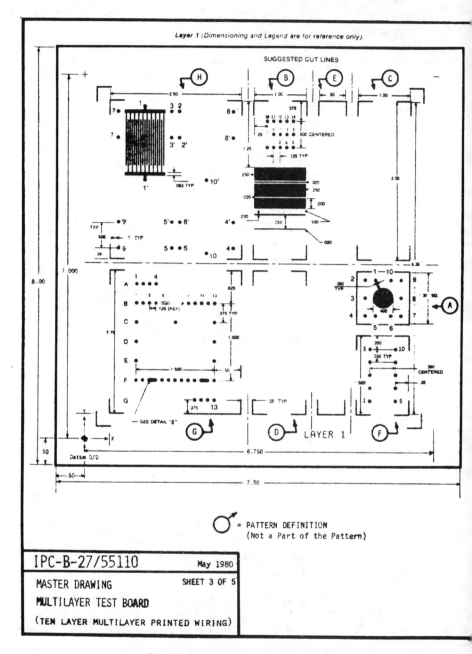

Fig. 6-5. Diagram of IPC-B-27 artwork.

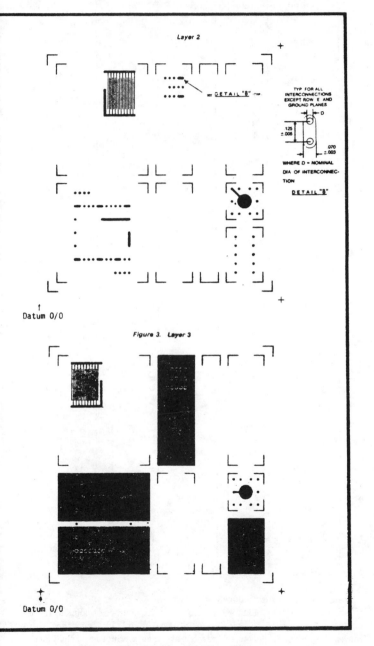

Layer 2

TYP FOR ALL
INTERCONNECTIONS
EXCEPT ROW E AND
GROUND PLANES

SEE DETAIL "B" TYP.

.125
±.005

.070
±.003

WHERE D = NOMINAL
DIA OF INTERCONNEC-
TION

DETAIL "B"

Datum 0/0

Figure 3. Layer 3

Datum 0/0

Fig. 6-5 (*Continued*)

Examination or Test	Requirement Paragraph	Method Paragraph	Production Board	Test coupon by type 1/			AQL (percent defective)	
				1	2	3	Major	Minor
Visual:								
Incoming material	3.4 thru 3.4.7	4.3	–	–	–	–	–	–
Surface Examination	3.4.3, 3.19 3.20	4.8.1 and 4.8.1.1	X 9/	–	–	–	2.5 2/	
Plated-through hole (microsection)	3.5 and 3.5.1	4.8.1 and 4.8.1.2 thru 4.8.1.2.2	–	–	A	A	3/	
Etchback (applicable to etchback process boards only)	3.6.5	4.8.1.2 thru 4.8.1.2.2	–	–	–	A	2.5	2/ 3/
Marking	3.19	4.8.1	X	–	–	–	1.0	4.0
Workmanship	3.20	4.8.1	X	–	–	–	1.0	4.0
Dimensional:								
Plating 8/	3.5 and 3.4.3	4.8.1 and 4.8.1.1	X	–	–	–	1.0	4.0
Annular ring (external)	3.6.1.1	4.8.1.6	X	–	–	–	1.0	4.0
Hole pattern location	3.6.1.2	4.8.1	X	–	–	–	1.0	4.0
Dielectric layer thickness	3.6.2	4.8.1	X	–	–	8/	1.0	4.0
Undercutting 8/	3.6.4	4.8.1	X	–	–	–	1.0	4.0
Conductor pattern	3.6.6 and 3.6.7	4.8.1.4	X 7/	–	–	–	1.0	4.0
Conductor overhang 8/	3.6.8	4.8.1.5	X	–	–	–	1.0	4.0
Conductor spacing	3.6.9	4.8.1	X	–	–	–	1.0	4.0
Bow and Twist	3.9	4.8.4	X	–	–	–	1.0	4.0
Plating adhesion	3.7	4.8.2	X	–	–	–	1.0	4.0
Thermal stress								
(solder float)	3.13	4.8.7	–	–	C	B	6/	
Plated-through hole	3.13	4.8.7	–	–	C	B	3/ 4/	
Layer to layer registration	3.6.1.3	4.8.1.3	–	–	–	B	3/	---
Min. annular ring (internal)	3.6.1.1	4.8.1.6	–	–	–	B	3/	---
Hole solderability	3.11.1	4.8.8.1	–	–	C	B	3/	
Surface solderability	3.11.2	4.8.8.2	–	D	–	–	1.0	4.0
Circuitry	3.16	4.8.11	X (Type 3 only)	–	–	–	100% inspection 5/	

1/ See MIL-STD-275.
2/ AQL is based on the number of boards produced.
3/ 1 coupon per panel shall be microsectioned for type 3 boards; the number of coupons to be microsectioned for type 2 boards shall be based on a statistical sample in accordance with MIL-STD-105 General Inspection level II of the number of panels produced and shall meet an AQL of 2.5 percent defective.

Fig. 6-6. Group A inspection.

Examination or Test	Requirement Paragraph	Method Paragraph	Certification test specimen number (see 4.5.1)	Test coupon by type			Whole Specimen
				1	2	3	
Visual:							
Material	3.4 thru 3.4.7	4.3	---	–	–	–	
Cleanliness	3.17 and 3.17.1	4.8.12	1, 2, 3, 4	–	–	–	X
Bow and twist	3.9	4.8.4	1, 2, 3, 4	–	–	–	X
Plated-through hole	3.5 and 3.5.1	4.8.1	4	–	J	B	–
Etchback (when specified) 1/	3.6.5	4.8.1.2 thru 4.8.1.2.2	4	–	–	B	–
Marking	3.19	4.8.1	4	–	–	–	X
Workmanship	3.20	4.8.1	4	–	–	–	X
Dimensional:							
Plating (surface) 4/	3.4.3	4.8.1 and 4.8.1.1	4	–	–	–	X
Annular ring (external)	3.6.1.1	4.8.1.6	4	–	–	–	X
Hole pattern location	3.6.1.2	4.8.1	4	–	–	–	X
Dielectric layer thickness	3.6.2	4.8.1	4	–	J 4/	B 2/ 4/	X 3/
Undercutting	3.6.4	4.8.1	4	–	J 4/	B 4/	–
Conductor pattern (external layers)	3.6.6 and 3.6.7	4.8.1.4	4	–	–	–	X
Conductor overhang 4/	3.6.8	4.8.1.5	4	–	–	–	X
Conductor spacing	3.6.9	4.8.1	4	–	–	–	X

Fig. 6-7. Supplier certification inspection.

Examination or Test	Requirement Paragraph	Method Paragraph	Certification test specimen number (see 4.5.1)	Test coupon by type			Whole Specimen
				1	2	3	
Plating, adhesion	3.7	4.8.2	1, 2, 3, 4	N	N	B	–
Bond strength (terminal pull)	3.8 and 3.8.1	4.8.3 } a/	1	K	L	F	–
Plated-through hole (microsection)	3.8 and 3.8.1	4.8.3	1	–	L	F	–
Moisture and insulation resistance	3.15	4.8.10	2	B	B&J	A&H	–
Dielectric withstanding voltage	3.10	4.8.5 and 4.8.5.1 } b/	2	B	B&J	A&H	–
Circuitry	3.16	4.8.11	1	–	R	G	–
Thermal shock	3.12	4.8.6 } c/	1	–	R	G	–
Circuitry	3.16	4.8.11	1	B&R	B&R	G 6/	–
Thermal stress (solder float)	3.13	4.8.7	3	–	J	F	–
Plated-through hole (Microsection)	3.13	4.8.7 } d/	3	–	J	F	–
Layer to layer registration	3.6.1.3	4.8.1.3	3	–	–	F	–
Min. annular ring (internal)	3.6.1.1	4.8.1.6	3	–	–	F	–
Hole solderability	3.11.1	4.8.8.1	3	–	J	F	–
Surface solderability	3.11.2	4.8.8.2	1	–	–	–	X (Type 1 only)
Interconnection resistance	3.14	4.8.9	2	–	R	G 5/	–

Fig. 6-7. (*Continued*)

will perform the supplier certification testing (Fig. 6-7) required by MIL-P-55110C.

5. Assuming that all boards pass testing successfully, the laboratory will send a fully documented report to DESC-EQ.

6. If DESC-EQ approves the test report, the printed circuit manufacturer will be placed on DESC's list of approved suppliers, and the manufacturer will be eligible for a Federal Supply Code For Manufacturer's (FSCM).

7. On subsequent military business, at least two test coupons (per MIL-STD-275D) will have to be submitted to the testing laboratory monthly, for Group B testing (Fig. 6-8).

This is a simplified explanation of the work involved. The internal documentation, written procedures and controls needed to meet these requirements are extensive. Generally, if the company's documentation and quality program meets MIL-I-45208, they will be on the way to successfully passing the DESC certification.

Examination or Test	Requirement paragraph	Method paragraph	Test coupon by type 1/			Production board
			1	2	3	
Cleanliness	3.17 and 3.17.1	4.8.12	–	–	–	X
Bond strength (terminal pull)	3.8 and 3.8.1	4.8.3 a/	A	A	F	–
Plated-through hole	3.8 and 3.8.1	4.8.3	–	A	F	–
Interconnection resistance	3.14	4.8.9	–	C	D or G	–
Moisture and insulation resistance	3.15	4.8.10 b/	B	B	E	–
Dielectric withstanding voltage	3.10	4.8.5 and 4.8.5.1	B	B	E	–

Fig. 6-8. Group B inspection.

SUMMARY

The Quality Assurance department should be much more than a tattle-tale or policeman. Functioning of the Quality Assurance department can be likened to the oil which helps machinery to function smoothly. Since just about every aspect of work life and the work environment affect the quality of the product, these aspects come under the domain of the Quality Assurance department. A quality problem in one department affects the quality and profitability of the entire company, as well as the reputation of the entire company. For these reasons, the quality assurance department must help to foster an atmosphere of cooperation, a desire to listen, and a desire to resolve and prevent problems.

7
Quality Assurance Manual

ELEMENTS OF QUALITY ASSURANCE

1. Identifying Production Problems by Customer and Part Number

a. All problems which cause scrap product are to be documented on the job traveler: how many boards were scrapped and the reason.

b. When problems are noted, the Quality Assurance Manager will contact all supervisors whose areas were involved, to address the quality problem.

c. The travelers for past jobs are reviewed prior to issue of travelers for subsequent orders. At this time the planning engineer has the opportunity to review previous difficulties and verify that corrective action has been taken.

2. Customer Returned Material

a. Returned boards are logged in by Shipping and Receiving, and a DMR form (Defective Material Report) is initiated.

b. Quality Assurance Manager reviews incoming customer paperwork and returned boards for quality problem.

1. Q.A. Manager will disposition boards for: Scrap, Rework, or Return to the customer.
2. Q.A. Manager will keep a running inventory of all returned product, by customer and part number.

3. Production Problems

a. If a problem will impact delivery date, sales and production control are to be notified immediately.

b. Problems with customer artwork, prints, or design requirements are to be brought to the attention of the customer by Planning before the job has been released to Production. Every effort should be made to anticipate manufacturing problems caused by customer documentation.

4. Laminate Inspection

a. All material must conform to the type designation of MIL-P-13949 specified on the purchase order.

SHEARING LOGBOOK

DATE	JOB NO.	CUSTOMER	PART NO.	MAT'L TYPE	MAKER	THICKNESS REQUIRED	THICKNESS MEASURED	PANEL SIZE	QTY	LOT NO	INITIALS

Fig. 7-1. Typical heading for shearing logbook.

b. Each sheet should have either a sticker, or an ink stamp which identifies lot number and military type designation.

c. Drilling/Shearing department will keep a log which records all material sheared, manufacturer, and the lot number and type of material sheared (see Fig. 7-1).

1. Panel size and quantity must be recorded, along with
2. Material thickness as measured with a micrometer.

REJECTED MATERIAL

THIS MATERIAL IS NOT TO BE USED
FOR ANY PURPOSE.
_____

MATERIAL:
MANUFACTURER:
LOT NO:
REASON FOR REJECT: _____
_____
_____
_____

DATE:
SIGNATURE _____

ATTENTION: THIS IS REJECTED
MATERIAL AND IS NOT TO BE
USED FOR ANY PURPOSE.

Fig. 7-2. Rejected material tag.

d. 5 sheets per lot shall be inspected by QA prior to shearing for pitting, dents, scratches, creases, bumps, and voids.

1. Sheets of laminate with this condition shall be set aside for return to the manufacturer.
2. When possible, one of the identification stickers is to be affixed to the back of the job traveler (see Fig. 7-2).

5. Process Specifications

a. Quality Assurance Manager will see that procedures are written for every process in the company.

1. Procedures will be under revision control, and posted in each work area.

b. The written procedure will serve as a training manual and contain the following information:

1. Step-by-step procedure for performing the operation.
2. A discussion of what the product is to look like at the end of the process.
3. Maintenance schedule, and detailed instructions on how to perform it.

4. A discussion of how the process works.
5. Safety information.

6. In-Process Inspection

a. There shall be a person with the title "In-Process Inspector;" however

b. Every employee is trained to know what good quality is, in terms of conforming to requirements.

c. It is the duty of the supervisor to see that employees understand the requirements for the job, and receive adequate training.

d. Supervisors are responsible for the workmanship coming out of their area.

1. They must know what is going on in their department and take proper corrective action, if there is a quality problem.

7. Pre-Soldermask Inspection

a. 100% inspection of every job prior to soldermask or fabrication.

b. Severe problems to be brought to attention of Quality Assurance Manager as soon as noted.

c. Production Control to be notified if more than 5% of job is rejected.

8. Final Inspection

a. An A.Q.L. sample of 10% is selected for inspection; this sample is inspected thoroughly.

b. If more than two defects are found in the A.Q.L. sample, a 100% sample is selected and inspected.

9. Inspection Reports

a. Inspection reports are filled out during Pre-Soldermask and Final Inspection. These are forwarded to the Quality Assurance Manager at the end of each day.

b. The report will list each job by Customer. Part Number, Revision, and Job Number. Ordered, Run, and Accepted quantities will be listed, as well as every reason for rejection.

10. Yield Report

a. The Quality Assurance Manager will issue a Yield Report on a weekly basis.

b. Yield Report will discuss quality issues and policy; report will also serve to keep employees alerted to the weekly yield at the company.

CALIBRATION PROGRAM

1. The calibration program is set up to ensure that maintenance and test equipment yield accurate readings; so that our products conform to design and contract requirements.

2. The following measurement and test equipment is subject to calibration:

   a. Calipers.
   b. Micrometers.
   c. Height and thickness measuring gauges.
   d. Beta backscatter, magnetic, and resistivity measuring equipment for measuring thickness of surface and plated-through-hole metals.
   e. Ovens for baking plating resist, soldermask/legend, and laboratory testing.
   f. Heaters for plating tanks, and hot oil reflow.
   g. Rectifiers for plating tanks.
   h. Scribes, as used in Fabrication.

3. The following equipment is excepted:

   a. Quick-check tapered pin hole gauges: are forbidden from use in Manufacturing and Inspection.
   b. Plug pin hole gauges: these are continuously monitored as they are used.

4. Calibration shall be performed by an outside service which maintains reference and transfer standards calibrated and traceable to the National Bureau of Standards.

   a. The calibration service, and their standards, shall meet all provisions of MIL-STD-45662.

5. All equipment covered by this specification shall be logged according to serial or identification number, and department which is responsible for the equipment.

   a. The identifying number shall be on a name plate, or scribed in with an electric pencil.
   b. All equipment shall have a calibration sticker affixed to it.

      1. Calibration sticker shall denote date of last calibration, and date calibration is next due.
      2. At no time shall calibration due date pass before calibration has been performed.

6. Calibration cycle

a. All hand measuring tools, ovens, heaters, and scribes shall be calibrated every six months.
b. Beta backscatter, magnetic, and resistivity measuring equipment for metal thickness shall be calibrated every 12 months.

7. Equipment found to be out of calibration or damaged.

a. Hand tools, ovens, heaters and scribes shall be repaired and calibrated immediately, by the calibration service, when found to be in need of attention.

   1. Any hand measuring tool which cannot be repaired shall be taken from service and replaced with new equipment.
   2. Any piece of equipment which falls from calibration, or is damaged, between calibration dates, shall be returned to a state of good repair and calibration as soon as possible.

b. Beta backscatter, magnetic, and resistivity equipment for measuring metal thickness shall be sent for calibration and repair.

   1. If more than 48 hours will pass before it is returned, and if plating must be performed during that time, a rental unit can be obtained.

INSPECTION STAMP CONTROL

1. Inspection stamps are controlled by the Inspection Supervisor, and are issued only to Quality Control Inspectors.

a. Each Inspector is responsible for the security of the stamp which has been issued to him/her.
b. The stamps are entered into a logbook which is kept by the Inspection Supervisor. Also entered in the logbook are:

   1. An impression of each stamp
   2. To whom the stamp has been assigned.
   3. Dates of issue and return.
   4. List of other inspection paraphernalia issed:

      ● Eye loupe
      ● Burnishing tool
      ● X-acto knife
      ● Hole poker.

2. The presence of an inspection stamp on a part indicates the part has been accepted as conforming to requirements.

3. Inspection stamps shall be applied to the following:

a. Drilling first articles.
b. Drilling and routing tapes: stamp applied to tape box label.
c. Router blocks or templates.
d. Printed circuit boards which have been accepted after Pre-Soldermask Inspection or after Final Inspection.
e. Any other jig or tooling fixture used to manufacture printed circuit boards.

INSPECTION SUPERVISOR DUTIES

The Inspection Supervisor is responsible for the training, performance and maintenance of all personnel and equipment in every area of Inspection. It is the duty of the supervisor to be in charge of Inspection. Personal output is secondary to the proper functioning and accuracy of the Inspection department. A more specific breakdown of duties is as follows:

1. To train inspectors to read and understand the Job Traveler and other documents which accompany it.

2. To train inspectors to read and understand blueprints and other drawings of the part being built:

a. Dimensioning and tolerance requirements.
b. Markings—requirements and location:

1. U.L. logo, Flammability (94V-O), Date Code, Part Number, Revision Level.

c. Plating requirements: type and thickness of all metals.
d. Subsequent screening requirements: soldermask and legend:

1. Type, location, color.

e. Material: laminate type, thickness, and color.
f. Special notes on the traveler, drawings or accompanying documents.

3. Train inspectors to understand the parts of a printed circuit board, and the functions which produce them.

a. Contact fingers, traces, groundplanes, pads, holes -plated and unplated, plating bars, soldermask, legend, chamfer, second drill, etc.

4. Train inspectors to use and understand measuring tools and equipment:

a. Calipers.
b. Micrometer.
c. Hole pin gauges.
d. Magnetic and beta-backscatter thickness measuring equipment.
e. Height gauge for warp and twist testing.
f. Conductivity testers for measuring thickness of copper in holes.
g. Ohmmeter or conductivity tester.

5. To train inspectors in the set up and use of automatic electrical bare board testing equipment.

6. To train inspectors to identify and understand the significance of the major and minor defects in printed circuits; and to understand the acceptability guidelines of customers' specifications, IPC-600, and in house guidelines.

7. To train inspectors in the best procedure to use when beginning work on a job: check contact fingers, then hole sizes, etc.

8. Train inspectors in rework techniques for minor and major defects in printed circuits, and in the tools for that rework: soldering iron, burnishing tool, X-acto knife, tapered pin, etc.

9. To train inspectors to properly fill out travelers, logbooks and other documentation: reject reports, throughput sheets, individual sheets.

10. To train inspectors for First Article inspection. This includes verifying traveler accuracy, acceptability of customer artwork, drill first article, and Photo lay up of new artwork.

11. To see that all equipment, materials and supplies required for inspection are present, maintained and calibrated.

12. To audit the work of other inspectors to ensure that all receive proper training on acceptability, reworkability, and on rework procedures. The results of the audit should be shared with the inspector immediately.

13. To inform supervisors of quality problems arising from their areas.

14. To inform Quality Assurance Manager or production control of inspected jobs which do not have enough acceptable pieces to meet the ordered quantity.

15. To see that all inspectors keep and maintain all equipment which has been passed out to them.

16. To maintain files or customer specifications and electrical test point drawings.

17. To pass along information to the inspectors which is important for them to know: company information and changes in documentation requirements affecting jobs which they may be working on, including communications with other shifts.

18. To expedite jobs through the inspection area according to due date or to a list made up by Production Control.

19. To maintain staffing of personnel, and good attendance and punctuality of inspectors.

FIRST ARTICLE INSPECTION

Duties: First Article Inspector

1. Inspect incoming artwork for suitability to manufacture printed circuits to customer's requirements.
2. Review traveler, drawings, artwork and customer's specification for agreement on part number, revision level, material call out and soldermask/legend requirements.
3. Inspect drilling first articles for correctness and accuracy.
4. Inspect working film used in silk screening, soldermask, legend, and dry film.
5. Review artwork and drill first articles for jobs where a revision change has been made, and Planning has called for the use of existing artwork or drill tape.

Equipment Required

1. Pin hole gauges.
2. Calipers.
3. Light table.
4. 7× to 10× eyepiece.
5. Micrometer.
6. x-y coordinate measuring equipment.

Traveler Inspection

1. Obtain the following:

   a. All drawings for the part number.
   b. Job traveler.
   Customer manufacturing specification.
   d. All artwork supplied by the customer, or working film if repeat job.
   e. Drill tape.

2. Review this documentation for completeness and accuracy in the following areas (see Fig. 7-3.):

   a. Do part number and revision level on all artwork and drawings agree with each other, and with the traveler?

TRAVELER VERIFICATION

1. Manufacturing Specification referenced on drawing

2. Material Type and Thickness: Measured

    : Traveler

    : Drawing

3. Part Number: Artwork

    : Drawing

    : Traveler

4. Revision: Artwork

    : Drawing

    : Traveler

5. Hole Sizes and Tolerance: P-Plated Through   NPT-Non-plated Through

Hole size   Drill Size   Hole Size   Drill Size   Hole Size   Drill Size

6. Hole size and drill size agree with the program tape yellow card?

7. Drill size takes into account whether or not the hole is plated or non-plated?

8. Drill size, for non-plated holes, takes into account whether the hole is to be plugged during plating, or whether it is piloted for second drilling?

9. Soldermask: Component Side   Solder Side

    : Type   Color

10. Legend: Component Side   Solder Side

    Type   Color

11. Special Notes for Soldermask and/or Legend:

12. Pattern Plating: Type and Thickness

    Copper _____ Tin-Nickel _____
    Tin-Lead _____ Tin-Lead on Tin-Nickel _____

13. If Tin-Lead is called for on Tin-Nickel, as special screening operation must be called for after Tin-Nickel.

14. Contact Finger Plating: Nickel   Gold

15. Other Special Notes:

Inspected by: _____
Date: _____

Fig. 7-3. Checklist for Quality Assurance verification of job traveler accuracy.

    1. If not, does the print call out which artwork to use; and is that the artwork being used?

  b. Does part number and revision level on drill tape match that of drawing and traveler?

  c. If revision levels and part numbers do not match, the print must call out which part/revision levels to use. If this is not the case, *do not* release the job. Planning must be notified.

  d. Does the material called out on the drawing match the laminate called out on the traveler?

  e. Do soldermask and legend call outs on traveler match call out on drawings?

    1. Which sides take soldermask or legend?

    2. What color and type of material is to be used?

  f. Are hole plugging and second drilling call outs correct, and documented on the traveler and drawing?

  g. Do hole drilling size call outs on traveler agree with those on drill tape yellow card?

  h. If slots are needed, are they called out on the traveler?

  i. If hardware is needed, is it called out on the traveler?

Artwork Inspection

1. Obtain the following:

  a. All pieces of incoming customer film.

  b. Copies of all drawings for that part number.

  c. Customer's manufacturing specification.

  d. Blank Customer Film Inspection Check List (CFICL).

2. Fill out a customer film inspection check list and document all results on it.

3. Make a photographic reversal (positive to negative or vice versa) of the side which will be used to program to. This is usually the component side or the side detailed on the drawing.

4. Film Inspection Procedure:

  a. Look at part number and revision level on each piece of film, and record them on the CFICL.

    1. Also record whether the film is positive or negative

    2. Notify Planning immediately if there is a discrepancy between part numbers or revision levels which is not already documented on the drawing. Sometimes the part numbers and revision levels differ on the documentation, for the same part number. When this happens, the drawing must call out what the correct part/revision number is.

b. Examine the general quality of the artwork, and record results on the CFICL.

    1. Nicked, broken or ragged circuitry.

    2. Excessive protrusions from circuitry.

    3. Poor trace crossings or trace/pad intersection.

    4. Pinholes, voids or spots.

    5. Readability of letters and numbers on all artwork.

    6. If these defects are present to the extent that more than 10 or 15 minutes of touch up is required (using pen and exacto knife), the artwork is to be rejected, and the reasons documented on the CFICL. But continue with the artwork inspection.

c. Front to back registration:

    1. Obtain a photographic reversal (positive to negative or vice versa) of the artwork which will be used to program for drilling.

       • This is usually the component side or the side detailed on the drawing.
       • Tape this to a light table, emulsion up.

    2. Lay up every other piece of artwork supplied by the customer to this reversal, and register as well as can be done.

       • Component side.
       • Solder side.
       • Inner layers.
       • Soldermask.
       • Padmaster.

    3. Look at the alignment of:

       • Hole to hole.
       • Contact fingers.
       • Targets.

- Fab lines—note if the fab lines are used as a reference in the print.
- Alignment targets on symmetrical patterns.

4. Reject the artwork if the misregistration is .003 inches or more.

   - Document this on the CFICL. Planning should be notified.
   - If the pad diameter is 0.020 inches, or more, greater than the hole to be drilled, document the misregistration on the CFICL, but do not reject the artwork.
   - A digitizing machine, as used for drill program plotting, can be used to accurately measure artwork alignment.

5. Soldermask artwork should have a clearance around pads of sufficient to prevent solder mask from getting on pads.
6. Check airgap, between traces, between pads, and between pads and traces.

   - Jobs with air gap of less than .008 inches should be rejected, and Planning notified.
   - Sometimes, the artwork will show plenty of air gap, but the drilled hole is actually larger than the pad on the art work. This condition can be checked for by holding a pin gauge against the area in question. The pin gauge should be about 0.005 inches larger than the optimum hole size after plating. It is also obvious when the artwork is checked against the drilled first article.

7. Check minimum trace width.

   a. The minimum trace width must not be less than any requirements listed on the drawing or customer specification.
   b. In no case should it be less than .008 inches, or other company standard.

8. Check artwork against the drawing for accuracy and agreement.
9. Special note: Jobs with front to back registration problems should not be flip-flopped in drill programming.
10. At conclusion of artwork inspection:

    a. Document all results on the Customer Film Inspection Check List.
    b. Do not sign off a traveler, log book or CFICL as acceptable if there are unacceptable conditions. The customer must be notified by Planning. They may want to submit alternative documentation

to us, or they may tell us to run as is. If the customer wants us to run with documentation we have rejected, it must be documented on the CFICL, or an engineering form stapled to it stating:

- What the discrepancy is.
- What problem it will cause for us and/or them.
- Whether they agree, and will submit new documentation, or
- They want us to run as is and will buy off any scrap which results from the unacceptable, but, best effort condition.
- The date of customer contact, who was talked to, and the name the person who contacted the customer.

   c. Make a copy of the CFICL, and keep it with the artwork package; the original should be kept in the control file for that part number.
   d. Planning is to be notified immediately of discrepancies noted during the inspection process.

Artwork Definitions

1. Customer Film/Artwork: Artwork submitted by the customer: component side, solder side, padmaster, soldermask, legend.
2. Copy-To-Make-Copies (CTMC): Duplicate of customer film used to generate step and repeat pattern. This piece of film should be touched up before shooting step and repeat exposures.
3. Working Film/Copy: The phototool used in production to shoot screening stencils and expose dry film photoresist. Working Film and Phototool are interchangeable terms.
4. Silk Screen Film: Phototool used to expose circuit pattern screening stencils.
5. Soldermask Film: Phototool used to expose soldermask stencil.
6. Legend Film: Phototool used to expose legend stencil.
7. Legend or Nome: The set of marking designations screened onto the board with epoxy ink to identify areas on the board for the customer.
8. Padmaster: Artwork which has no circuitry, only pads. This may be used to generate soldermask film when the customer so requires.
9. Reading Right Emulsion (RRE): Letters and numbers on artwork read from left to right, when viewing film with emulsion facing up.
10. Reading Right Base (RRB): Letters and numbers on the artwork read from left to right from mylar (base) faces up (emulsion down).
11. Duplicating Film: Positive acting film for transposing RRE to RRB or vice versa.
12. Reproducing Film: Negative acting film for reversing the circuitry from negative to positive or vice versa.

13. Circuit Positive: Traces and other circuitry are black.
14. Circuit Negative: Traces and other circuitry are clear.
15. Logo, or U.L.: The designation which Underwriters' Laboratory allows the manufacturer to put on printed circuit boards which have passed their testing.
16. 94V-0: A flame retardation classification by Underwritters' Laboratory.
17. Fab Lines: Lines on the artwork showing the boarders of the part.
18. Plating Bars: A trace on the artwork which connects the edges of the panel with the contact fingers. This trace must be present for tip plating. It is removed during Fabrication.
19. Hole Centers: Dots on the Customer artwork which show the geometric location of the center of the hole. Hole centers are burned out after step-and-repeat on the working film, to prevent resist from getting into the hole during screening.
20. Dry Film Photoresist: A gelatinous film laminated onto a panel or inner layer. When exposed to ultraviolet radiation, it polymerizes into a hard plating or etch resist. Dry Film is a negative acting photoresist.
21. Negative Acting: When a film reverses a circuit from negative to positive, or positive to negative, it is said to be negative acting.
22. Positive Acting: When the image is not reversed from positive to negative, or negative to positive, the film is said to be positive acting.
23. Flip Flop: When a panel is drilled for imaging with half the panel being component side, and half solder side.
24. Circuit or Solder Side; Side of completed and loaded board that will be exposed to the molten solder during wave soldering. It is the side opposite the component side.
25. Component Side: Side of completed board from which components will be loaded. It is the side opposite the solder side.
26. Film Package: The large envelope which contains smaller, separate envelopes for screening film, soldermask and legend film, inner layer film, and copy to make copies.
27. Customer Film Inspection Check List (CFICL): Artwork inspection form. (See Fig. 7-4.)

Drilling First Article Inspection

This procedure is extremely important; it is the last opportunity to identify and rectify errors which could cost thousands of dollars, before production work begins.

1. Equipment needed:

a. Pin hole gauges.
b. 7× to 10× eyepiece.

CUSTOMER FILM INSPECTION CHECK LIST

CUSTOMER _____
PART NO. _____ REVISION _____

1. Inventory of film   Positive   Negative

	Positive	Negative
Component		
Solder		
Soldermask: Component		
: Solder		
Legend: Component		
: Solder		
Inner Layer:		
:		
:		
:		
:		
:		

2. Front-to-back or layer-to-layer registration

3. Circuit Quality
   a. Nicks, cuts, protrusions

   b. Raggedness

   c. Pin holes

   d. Air gap minimum

   e. Minimum trace width

   f. Other Notes:

4. Is revision level on the artwork? Does it agree with drawing and traveler?

5. Artwork Accepted   Artwork Rejected

6. Inspected by _____ Date _____

Fig. 7-4. Inspection checklist and inventory for incoming customer artwork.

c. Micrometer.
d. Calipers.

2. Inspection Procedure

a. Obtain the following:

1. Traveler.
2. All drawings for that part number.
3. Drill tape.
4. Drilling first article.
5. Drilled mylar.
6. All customer artwork for that part number.
7. Blank First Article Inspection Check List.

b. Panel size and boards per panel (number up) must agree with traveler.
c. Panel thickness and type of material must agree with drawing.

1. Material type, manufacturer, and lot number must be written on back of traveler.

d. Lay artwork over first article drilled panel, register and check for:

1. Registration:

● No break out is allowed, all pads must allow for at least 0.005 inch annular ring around holes.
● Missing or extra holes:

   ● Look for second drilling pilot holes, or no drilling at all for holes which will be second drilled.
   ● Check the traveler and drawing to see that second drill and plugging call outs are correct.

● Misplaced special tooling holes.
● Screening holes correct.
● Holes which will cause an airgap problem due to closeness to traces or other circuitry.
● Check first article with soldermask artwork for missing/extra holes also. Clearance around holes must be at least 0.015 inches.
● Use mylar to lay against panel to check registration and missing/ extra holes also.

e. Hole sizes.

- Check coding on all holes.
- Use pin gauges to check every for correctness on each size of hole.
- Check traveler and drawing to verify correctness of second drill and plugging call outs.
- Drilled holes must match call out on drill tape yellow card and on traveler.

f. Document all results of first article inspection on First Article Inspection Check List.

- Do not sign off a first article which has discrepancies.
- Discrepancies must be brought to the attention of Planning as soon as possible.
- When discrepancies have been resolved, that resolution must be documented on either an engineering form, or on the First Article Inspection Check List.
- A new first article will have to be run and inspected if a change needs to be made to drill tape.
- When the first article is acceptable:

  - Q.C. stamp the drill tape.
  - Sign the traveler and First Article Inspection Check List. (See Fig. 7-5.)
  - First article and traveler go to Photo.

INSPECTION PROCEDURES

First inspection should be performed on all panels after the tin-lead fusing operation. After inspection the next operations are soldermasking, legend, and fabrication. Once they have been performed it is difficult to identify defects or rework boards. One of the purposes of first inspection is to identify these problems and resolve them while product is in the panel form.

There is a definite sequence to follow which will prevent wasting time from working on boards which will be scrapped later, or worse, which have problems that go undetected until the customer finds them.

Another purpose of first inspection, along with any inspection operation, is to determine when we have a process running out of control or continuously putting out substandard quality—if the problem has not already been noticed by the supervisor of that department. The information on defects listed for every job will be used to issue yield reports and inspection disposition reports.

DRILLING FIRST ARTICLE INSPECTION CHECK LIST

1. Drilling Registration With Artwork: Component Side   Solder Side   Mylar

2. Missing or Extra Holes:

3. Air Gap Problems

4. Drill Sizes For Holes: Traveler and Yellow Card Agree:

5. Were hole sizes correctly drilled for plugging and second drill requirements?

6. Were correct drill sizes called out for plated and non-plated through holes?

7.	Drill Size	Measured Size	Drill Size	Measured Size	Drill Size	Measured Size

8. Extra Tooling Holes?

9. Panel Size and Thickness: Traveler   Measured

10. Number Up

11. Other Observations and Special Notes:

Inspected By   Date

Fig. 7-5. First article drilling inspection.

These reports will be circulated throughout the shop. Because of this, it is critical that the reasons for defects be listed for each board

FIRST INSPECTION

A. Contact Finger Plating

1. Review traveler, drawing, and part to make sure that part number and revision level agree.

   a. Check requirements for gold thickness on contact fingers.
   b. Make sure traveler and drawing agree.

2. Pull a 10% A.Q.L. sample. (A.Q.L. stands for acceptable quality level.) Take the boards from a random sampling.
3. Use the beta backscatter equipment to measure thickness (Micro-derm).

4. If the readings are satisfactory per requirements, do not check any more panels.

5. If two panels read low, but within 10% of required value, check another 10% sample.

  a. If there are no low readings on the next 10%, do not check any more panels.
  b. If there are more low readings, check 100% of the panels. Those with low readings must be replated and should be returned to Plating.

6. If any boards read more than 10% below required thickness, check another 10% sample.

  a. If anymore low readings are found, check 100% ofthe panels.
  b. All panels with low readings must be returned for more Plating.
  c. If no more low readings are found in the second sample, do not check more panels.

7. Tape testing should be done to all panels checked for gold thickness. If metal peelers are found, the entire job must be tape tested.

8. Do not continue with panel inspection until it is determined that A.Q.L. panels meet gold thickness requirements and pass tape testing.

B. Hole Sizes

1. After reviewing traveler and drawing for hole size and through hole plating requirements, check a 10% A.Q.L. sample for hole sizes.

2. Use only pin gauges to check sizes: do not use quick check tapered pins.

3. For each hole size:

  a. Take the pin which corresponds to the smallest size at the lowest tolerance allowed for each hole size.
  b. Check at least four holes of each size; check corners, edges, and the center of the board.
  c. If there are no undersize holes, do not check more panels.
  d. If any of the holes are undersized:

    • If 50% of A.Q.L. sample is no more than 0.001 inch under tolerance, pull another 10% A.Q.L. sample and check the hole sizes.
    • If 50% of the second sample is more than .001 under tolerance, do not check more panels. Notify Quality Assurance Manager.

● If any of the holes in an A.Q.L. sample are 0.002 inch or more under tolerance, notify Quality Assurance Manager.

e. Do not continue with panel inspection until the Quality Assurance Manager has made a disposition on the undersized holes.

C. Hole Quality

1. Check every board on every panel in the job for:

a. Plugged or noduled holes.
b. Voided holes.
d. Dark or dull holes.

2. All holes should be shiny, bright, and smooth.
3. If a hole is plugged, carefully press the obstruction to the side of the hole using a tapered pin, as long as hole is not damaged.
4. If the holes have nodules which do not reduce hole diameter, neglect them.

a. If the nodules are large enough to reduce hole diameter, press them to the hole side using a tapered pin, without damaging hole.
b. If there are too many nodules obstructing holes, notify the Quality Assurance Manager.

5. If there are a few dark holes, fill them with solder, then remove the solder. If this has failed to brighten the hole, reject the board.
6. Dull holes will have to be returned to the Plating department for rework.
7. Stains or spots on the tin-lead must be cleaned off by the Plating department.
8. Solder slivers are caused by overhang of tin-lead on the etched conductor. It can be tested for using tape; boards with slivers should be sent back to the Plating department.

E. Burned Plating

1. This is important because it affects the condition of the circuitry and holes. The discussion on conductors will deal with more specific examples.
2. Light burned (noduled) plating along the edges of circuitry will not be cause for rejection.

a. Burned/noduled plating along edges of contact fingers is rejectable if the burned plating is higher than the surface of the contact fingers. Burned plating here will create reliability problems when it comes to making good electrical contact in the electronic equipment in which the board will eventually be used.

3. Heavy burning along traces, or pads or in holes is rejectable.

a. Burned plating like this will be difficult to soldermask and will create soldering problems.

4. Burning on numbers, letters, and logos only are generally acceptable, as long as they are legible.
5. Burning which reduces air gap or which extends more than 0.005 inch from a trace is generally rejectable.

F. Conductors

1. Line widths should be within 0.002 inch of artwork. If they are not, oversized traces present no problem as long as there is sufficient air gap between circuitry.
2. Reduced traces can be caused by a number of reasons:

a. Nicks from imaging.
b. Raggedness due to poor artwork, silk screening or over-etching.
c. Some degree of raggedness, for whatever reason, is acceptable, if the raggedness does not exceed 20% of the thickness of the trace.
d. Protrusions from the circuitry are usually acceptable, as long as they do not cause a short circuit or air gap problem.
e. Over exposure in dry film.

3. Air gaps below 0.008 inch are rejectable for most part numbers.

a. Air gaps down to 0.005 inch are acceptable on prototype boards, and boards where the air gap is on the artwork.
b. It is often necessary to obtain the artwork when an air gap problem is noticed. Photo and the Quality Assurance Manager should be notified when there is a major air gap problem.

4. Pinholes, pits, dents, and scratches:

a. Contact fingers:

- Pitting is acceptable within the guidelines of IPC-A-600.
- Pin holes and dents must not be more than 0.015 inches across.

- Scratches are allowed, as long as they are in the gold only, and no copper is showing.

b. Traces and ground planes:

- Pin holes must not decrease trace width more than 20%.
- Dents and pits are acceptable.
- Scratches are acceptable as long as the copper is not exposed.
- Pin holes in ground planes or bus traces (wide traces) do not present a problem and are acceptable.

5. Conductor Peeling:

a. Tape testing panels will determine whether or not the metal will peel.
b. If solder slivers are detected, the boards should go back to plating.
c. Gold peeling off nickel on contact fingers: The boards must be tape tested 100%, and returned to plating to have the gold stripped off and replated.
d. Nickel peeling off copper, on the contact fingers: This will result in the board being rejected and scrapped out.
e. Copper to copper or tin/lead to copper peeling will result in the board being rejected and scrapped out.
f. Lifting of the conductor off the base laminate will result in the board being rejected and scrapped out.
g. Logo, lettering, and numbers: Peeling here will not be cause for rejection; but the circuitry must be carefully checked by digging at it with a knife blade, for peeling.

6. Laminate Defects:

a. Delamination, separation, blisters, and broken fibers are rejectable.
b. Measling:

- Accept if it is only on 5% of the board surface.
- Reject if the measle connects circuitry.

c. Crazing—connected measles: Reject, unless located on less than 2% of board surface and not connected to circuitry.
d. Weave Exposure—butter coat removed: Reject, unless that side of the board which has weave exposure gets soldermask.
e. Scratches:

- Accept normally.
- Reject if the weave is exposed, or the scratch gouges below the surface.

f. Included Particles:

- Accept if particle does not lie between conductors.
- Reject, if particle bridges conductors, or lies beneath conductors.

7. Registration:

a. Registration is acceptable as long as there is 0.005 inch of annular ring around the holes, for plated through holes; less, if customer allows.
b. IPC-A-600 acceptability guidelines should be followed for most cases.

8. After Inspection:

a. The Inspection Supervisor should take examples of rejected or problem boards to the Quality Assurance Manager, if the defects are wide spread and causing a major scrap problem.
b. Defects in circuitry which appear to be repeating should be brought to attention of Photo and Screening supervisors.
c. All defects for every job must be tabulated and logged into scrap report.

9. Every job which does not have enough acceptable parts to meet shipping quantity should be brought to the attention of the Quality Assurance Manager, before that job leaves Inspection area.

10. All boards which are determined to be acceptable are to be stamped with a registered Inspection Stamp.

FINAL INSPECTION PROCEDURE

The actual inspection is similar to first inspection, and acceptability guidelines remain the same. The only difference is that the boards have already been inspected once, and they have had soldermask and legend applied, and they have been routed from panel form into discrete printed circuit boards.

Since they have been inspected, there should be no need to repeat inspection on the entire job. Only soldermask, legend, fabrication and the contact finger chamfering have been performed to the boards. The inspection should concentrate on these areas. Since work has been done to the contact fingers, a very easily damaged item, they should be looked at for scratches and nicks.

1. Pull a 10% A.Q.L. sample, selected at random.

2. Perform inspection on soldermask, legend, fabrication and contact fingers, and reinspect for everything checked during First Inspection.

3. Review traveler and drawing for all manufacturing requirements.

4. Use calipers to measure dimensions of the board.

a. If the boards are oversized and out of tolerance, take them back to Fabrication.

b. If there is a problem with boards being undersized, notify Quality Assurance Manager.

5. If second drilling was performed, check hole size, location, and correctness per drawing and traveler instructions.

6. Check location and size of slots. If slotting was done between contact fingers, look for lifting of the electrical contacts.

7. Check for complete removal of the fab lines.

8. Check board thickness with micrometer.

9. Use a flat stone, or surface, to measure warpage.

10. Soldermask:

a. Soldermask must be of the type and color specified by customer.

b. Registration:

- Soldermask should be around holes, and not on pads—minor bleeding onto the pads presents no problem.
- Clearing around pads should not be so large that adjacent circuitry is exposed.
- Check soldermask instructions on drawing, many parts require clear areas on the board.
- Check to see that correct side(s) have been soldermasked.

11. Legend

a. Check to see that correct color was used.

b. Check to see that correct side(s) were legended.

c. Check legibility of legend, and registration. Legend ink on pads of component side is acceptable, if that is what artwork called for.

12. If defects are found in the A.Q.L. sample, check the rest of the job.

13. All defects must be logged into the scrap log.

14. Notify Qualify Assurance Manager every time the acceptable quantity is not enough to meet shipping requirements.

PROCESSING NON-CONFORMING PRODUCT

1. Non-conforming product is material which does not conform to requirements by reason of poor workmanship, material or plating defects, poor artwork or unauthorized engineering changes.

2. Nonconforming material is to be set aside and identified as follows:

a. A red arrow marker is to be affixed to the board, pointing out the defect.
b. The nonconforming pieces are to be taped together and a NONCON-FORMING tag applied with the following information:

- Customer.
- Part Number.
- Job Number.
- Date.
- List of discrepancies.
- Number of pieces involved.

3. Nonconforming material is to be logged out on the Daily Reject Report.
4. The nonconforming material is also to be written off the traveler.
5. Requirements for work being processed under MIL-P-5511OC:

a. All nonconforming material will be stored for a period of three months, in a location specified by the Quality Assurance Manager.

PROCESSING RE-WORKS

All boards which have reworkable defects shall be processed according to this document. The purpose of establishing a written procedure, and logbook, is to meet the following objectives:

- To prevent improperly reworked boards from getting back into production.
- To accurately track the movement of reworked panels.
- To account for who performed the rework, and subsequent re-inspection.

1. All boards which are to be reworked shall be logged into the Rework Logbook. (See Fig. 7-6.)
2. When boards are logged into the logbook, a note shall be placed in the traveler for those boards, which notes the quantity of boards that are being sent for rework.
3. When boards are returned from the rework facility, they must be logged back into the Rework Logbook, and be reinspected.
4. No boards shall be sent for rework without being properly logged out.
5. No boards shall be placed back into production without being properly logged back in, and without being re-inspected by Quality Control.
6. If a board which has previously been inspected and stamped by Quality Control requires rework, that board must be reinspected and stamped by Quality Control when it is determined that the board conforms to requirements.

RE-WORK LOGBOOK   FOR ALL RE-WORKED BOARDS

DATE SENT	RETURNED	JOB NO.	CUSTOMER	PART NO.	REV.	BOARDS QTY.	SENT TO	DEFECT	RE-INSPECTED BY	DATE

Fig. 7-6. Log sheet for tracking rework.

MILITARY SPECIFICATION PRINTED CIRCUITS

Serializing Military Specification Printed Circuits

MIL-P-5511OC is the specification which lists the requirements for manufacturing printed circuit boards for the military. This specification requires that

test coupons be run on every panel. These test coupons are eventually removed from the panel and are used for other testing.

The test results of the coupons are considered by the military to be indicative of the rest of the boards built on the same panel as the test coupons. Because of this, every panel must have its own individual serial number—and that number must be stamped onto each board and each coupon on the panel. The serial number must be applied during first inspection.

Panels should be serialized as follows:

1. A hand held printer with movable numbers, and epoxy legend ink or other indelible ink (such as CAT-L-INK) should be used to apply the serial numbers.
2. For 1 to 99 panels in a job, use serial numbers 01, 02, 03 through 99.
3. For 100 to 999 panels in a job, use serial numbers 001, 002, 003 through 999.
4. Each panel should be serialized with consecutive numbers; and only one number is to be used per panel.

   - This number should be stamped onto every board and every test coupon, near the part number or manufacturer's logo.
   - Note: Care must be used to keep the numbers from smearing.
   - After the panels have been serialized, they must be baked for 15 minutes minimum in the appropriate oven at 275°F.

5. After serialization, the traveler must be marked with the serial numbers applied. *Example:* For a 25 panel job, "Serial numbers 01 to 25" should be written in the comments section.
6. Whenever a board is lost or scrapped out, or sent for rework, the serial number becomes part of the identification of that board, just like the part number and revision level.
7. During First Inspection and Final Inspection, when boards are rejected, the serial number must be recorded on a special inspection form used only for military jobs. This Military Inspection Report is to be kept on file in the Inspection area. (See Fig 7-7.)
8. This Military Inspection Report shall detail board quantity, serial numbers, and reasons for nonacceptance for every job.

Group A Inspection and Laboratory Forms

The Group A Inspection and Laboratory Forms shall be used for all printed circuits being built under MIL-P-5511OC. There are two parts of the Group A inspection, visual and laboratory:

MILITARY REJECTION REPORT		REPORT NO.	DATE
CUSTOMER	PART NO.	REVISION	

QUANTITY ORDERERED   QUANTITY RUN

SERIAL NO.'S	QUANTITIES	DISCREPANCY	REWORK	SCRAP	MRB

OTHER NOTES:

SIGNATURE   DATE

Fig. 7-7. Log sheet for inspection of military jobs.

- Group A Inspection Form is to be used for all visual inspection documentation.
- Group A Laboratory Form is to be used for all laboratory testing and microsection documentation.

No military job is to leave Final Inspection until the Group A Laboratory form has been completed and forwarded to Final Inspection for evaluation and disposition:

- All unacceptable conditions noticed during laboratory testing shall be listed on the Group A Inspection Form, section 17.
- These forms become part of the permanent documentation for the particular job and part number.

Normal inspection procedures shall be followed, except for what is written on this form. Boards which have defects—rejectable or reworkable—shall have their serial number written under the appropriate section on the Group A Inspection Form. A defect noted during laboratory testing of a coupon shall be considered a defect on every board with that serial number. All boards with that serial number shall be set aside for rework, reject, or Material Review Board consideration.

Group A Inspection

1. All panels, boards and test coupons shall be serialized per Mil Spec Serialization procedure, during First Inspection. All serial numbers are to be written on Group A Inspection Form.

2. All test coupons shall be forwarded to the Laboratory, during Final Inspection. The serial numbers forwarded to the Laboratory shall be written on the Group A Inspection Form, section 2.

3. Markings present on boards and coupons. The following shall be present on both the individual boards and the test coupons; and shall agree with the traveler and drawing:

a. Part Number.
b. Revision Level.
c. Date Code.
d. UL Logo.
e. Federal Manufacturer's Supply Code Number.
f. Serial number: the Inspector shall affix this number.

4. Contact Fingers. Both the nickel and gold shall be measured.

5. Tape Test. Both the contact fingers and plated circuitry shall be tested with ½ inch wide cellophane tape.

a. Overhang (slivers) which may be removed by the tape from the edges of conductors, are acceptable.

b. Lifting of any plated conductor shall be cause for rejection.

6. Registration and annular ring (internal or external pads) shall be as specified on the drawing.

7. Conductors shall be inspected for the conditions listed under section 7.

8. Holes shall be inspected for conditions listed under section 8.

9. Contact fingers shall be inspected for the conditions listed under section 9.

10. Board surface shall be inspected for conditions listed under section 10. All panels which are to be soldermasked must be cleaned in a 75% isopropyl alcohol and water mixture, and tested for cleanliness by the Solvent Extract Resistivity test; results of the tests are to be logged into section 16 of the Group A Inspection Form.

11. Soldermask shall be inspected for the conditions listed under section 11.

12. Legend shall be inspected for the conditions listed under section 12.

13. Fabrication shall be inspected for the requirements of the drawing.

14. Warp and twist shall be checked during Final Inspection, and corrected prior to bagging.

15. Hardware shall be installed after Final Inspection; and checked prior to final cleaning and bagging. The Hardware Inspection Check List shall be used for this purpose.

16. Section 17 shall be reserved for any other pertinent notes and observations.

17. Section 18 shall list any Group A Laboratory problems noted on the Group A Laboratory Form. The Group A Laboratory Form shall be completed and evaluated before any work is to be released from Final Inspection. (See Fig. 7-8.)

Group A Laboratory Form

1. All coupons must have the serial numbers listed on them; and those serial numbers shall be written in section 1 of the Group A Laboratory Form (Fig 7-9).

2. The plated through holes of coupon A shall be microsectioned and inspected for the conditions listed under section 2, view at 50× to 100×.

a. Figures 7-10(a–c) shall be referred to for double sided boards.

b. Figures 7-10(a–e) shall be referred to for multilayer boards.

c. Dielectric thickness for multilayers is the actual measured thickness, minus the thickness of laminate voids.

● This shall be measured for all inner layers.

	GROUP A INSPECTION FORM		DATE
CUSTOMER	PART NUMBER	REVISION	JOB NO.

1. Serialize Panels

2. Coupons To Laboratory

3. Markings Present On Boards And Coupons
   a. Part No.   b. Revision   c. Date Code   d. U.L. Logo
   e. Federal Supply Code   f. Serial No.

4. Contact Fingers: Nickel

                : Gold

5. Tape Test: Circuitry

           : Contact Fingers

6. Registration/Annular Ring

7. Conductors
   a. Plating Peelers or Conductor Lifting

   b. Protrusions: Spurs, Spikes, Nodules

             : Un-etched Copper

             : Resist Breakdown

             : Metal Bridging

   c. Width Reduction: Minimum Airgap

             : Broken/Cut Traces

             : Nicks

             : Over-etched

             : Munchies

             : Raggedness Not From Over-etching

   d. Solder: Dewetting

           : Bumps

           : Exposed Copper

8. Holes
   a. Voids

   b. Nodules

   c. Dark/Dull

   d. Hole Sizes: Plated

             : Non-plated

9. Contact Fingers
   a. Pitting/Scratches

   b. Plating Bubbles

   c. Nicks

Fig. 7-8. Group A inspection checklist.

CUSTOMER	JOB NO.	GROUP A INSPECTION FORM

10. Board Surface
    a. Foreign Particles

    b. Measles/Delamination/Scratches

    c. Contamination

11. Solder Mask
    a. Registration

    b. S/M In Nonplated Holes

    c. Uncovered Traces

    d. Coverage Between Holes And Traces

    e. Color/Types

12. Legend
    a. Reading Right

    b. Legible

    c. Color

13. Fabrication
    a. Chamfer

    b. Slotting

    c. Dimensional Tolerance

14. Warp And Twist

15. Hardware—USE HARDWARE INSPECTION CHECK LIST

16. SOLVENT EXTRACT RESISTIVITY (FOR INSPECTED AND SCRUBBED BOARDS PRIOR TO SOLDERMASK)

    RESISTIVITY _____
    SERIAL NO. _____

17. Other Notes And Comments

18. Micro-section/Laboratory Results

Inspector _____   Date _____

Fig. 7-8. (*Continued*)

GROUP A LABORATORY FORM

CUSTOMER	JOB NO.	DATE

1. SERIAL NO.'S

2. PLATED THROUGH HOLE: COUPON A
   A. PLATING THICKNESS: COPPER

   : TIN-LEAD

   B. DIELECTRIC THICKNESS

   C. UNDERCUTTING

   D. ETCHBACK

   E. SMEAR REMOVAL

   F. CRACKS

   G. NODULES

   H. SEPARATION

   I. VOIDS: COPPER

   : LAMINATE

   J. NAIL HEADING

3. THERMAL STRESS: COUPON C—DOUBLE SIDED, COUPON D—
   MULTILAYER
   A. CRACKS

   B. SEPARATION

   C. MEASLING

   D. DELAMINATION

   E. LAMINATE VOIDS

   F. RESIN RECESSION

4. SOLDERABILITY (USE THERMAL STRESS COUPONS)
   A. HOLE

   B. SURFA CE

Inspector ——————————————   Date ——————————

Fig. 7-9. Group A laboratory form.

- Two sheets of prepreg, for a minimum thickness of 0.0035 inches, shall be the minimum thickness allowed. All measurements must fall within the tolerance specified on the drawing.

d. Undercutting of copper conductors shall be measured from the micro-section of the pad of a plated through hole. The undercut shall not exceed the thickness of the clad and plated copper.

e. Etchback and Smear Removal [Figs. 7-10 (a and d)]:

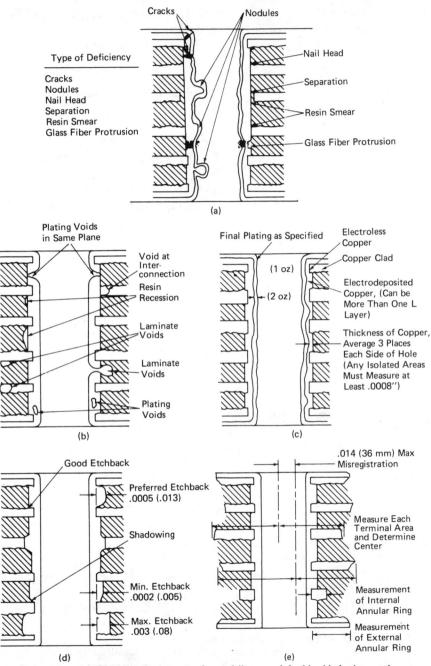

Fig. 7-10. Group A inspection requirements for multilayer and double sided microsection examination. (a) Plated through hole deficiencies. (b) Voids. (c) Plating thickness. (d) Forms of etchback. (e) Layer-to-layer registration and annular ring measurement.

- Etchback shall be 0.0002 to 0.0003 inches back from the edge of the inner layer when required on blueprint.
- No smear shall be present at the inner layer connection.

f. Layer-To-Layer Registration—microsection coupon B, view at 50× to 100×. Perform two microsections, one parallel to × direction, and one parallel to Y direction (length and width).

- Determine where the center line of the hole is, on each microsection and plot them on a graph per Fig. 7-10(e). This must be done for each layer.
- The misregistration is the distance between the center lines of the layers, and shall not exceed 0.014 inches, unless otherwise noted on the drawing.

g. Annular Ring.

- Inner layer: 0.002 inch minimum.
- External: Plated Through Hole 0.005 inch minimm.
  Non-Plated Hole 0.015 inch minimum.
- A 20% reduction, not to exceed .004 inches, is permissible due to nicks and pinholes.

h. Workmanship [Figs. 7-10 (a,b)].

- Cracks: none allowed.
- Separation of conductor interfaces: none allowed.
- Resin smear on multilayers: none allowed.
- Glass Protrusion: none allowed.
- Nailheading on multilayer: 50% of foil thickness maximum.
- Nodules: shall not reduce hole diameter below tolerance.
- Voids: Total length 5% of barrel length.
  Total area maximum of 10% of barrel area.
  3 voids maximum.
  None allowed at conductor interface.
  None allowed on both sides of a hole, in the same plane.

i. Plating Thickness, measure at 200× to 400× [Fig. 7-10(c)].

- Copper: 0.001 inch minimum average.
  0.0008 inch in isolated locations, except for voids.
- Tin-Lead: 0.0003 inch minimum at fused crest.

j. Other Comments.

- Laminate voids: maximum 0.003 if dielectric spacing minimum is not violated.
- No measles, delamination, or blistering.
- Solder shall completely fill plated through holes; wetting all areas of hole wall and extending outside hole onto pads.

3. Thermal Stress: Use Coupon C for double sided, and Coupon B for multilayer.

a. Saw or shear the test coupon, flux it, and emerse in molten solder at 550° for 10^{+1}_{-0} seconds, maximum depth of 1 inch.
b. Clean sample, microsection and view at 50–100× magnification.

- Cracks: none allowed.
- Conductor separation: none allowed.
- Blisters, delamination: none allowed.
- Resin recession: maximum 40% of hole wall length.
- Solderability: Hole—all areas filled and wetted.
  Surface—no dewetting on pads.

4. Cleanliness/ionic contamination testing: prior to soldermasking.

a. Use a 75% isopropyl alcohol in water mixture.
b. Spray 100 ml of above mixture over both sides of the production board, from a wash bottle.
c. Solvent must measure 6 megohms when tested with a specific resistivity meter.

Hardware Inspection

1. Prior to installing hardware into printed circuit boards, those boards must have been bought off by Final Inspection and contain an Inspection stamp.
   2. Boards which are sent outside for hardware insertion must be shipped in slotted carrying boxes, slip sheets, or in slotted racks with tape running over the boards to hold them motionless.
   3. After hardware insertion, and prior to shipping, the boards must be inspected by Quality Control.

a. Commercial boards—a 10% sample is drawn for inspection.
b. Military boards—the entire job must be inspected.

4. Hardware shall be defined as any eyelet, spacer, stand-off, pin, post, lug, or connector of any kind which is inserted through any of the holes in the board or is clamped onto the board.

HARDWARE INSPECTION CHECK LIST

1. HARDWARE	PART NO.	SIDE	ORIENTATION	METAL	QUANTITY

2. ARE THREADS PRESENT ON ALL PARTS WHICH REQUIRE THREADS?

3. ARE THERE THE CORRECT NUMBER OF PARTS ON EACH SIDE OF THE BOARD?

4. ARE PARTS FASTENED INTO THE CORRECT HOLES OR OTHER LOCATIONS?

5. ARE PARTS CORRECTLY ORIENTED?

6. DO PARTS WIGGLE OR ROTATE?

7. ARE PARTS MADE OF THE CORRECT METALS?

8. ARE PARTS SECURED WITH PROPER TYPE OF FIT: FLARED OR ROLLED ENDS?
CRACKS EXTENDING INTO HOLE BARRELS?

9. DOES DRAWING CALL FOR ANY PARTS TO BE SOLDERED?
IF SO, HAVE THEY BEEN?

DATE _____   INSPECTED BY _____

Fig. 7-11. Hardware Inspection checklist.

5. Inspection shall be as follows:

a. Obtain the job traveler, part drawings, customer manufacturing specification, and Hardware Inspection Check List (See Fig. 7-11).
b. Separate the inspection sample from the rest of the job: pick the sample at random.
c. Read the hardware call out on the drawing; and list the parts on the Hardware Inspection Check List as follows:

- Part name and number.
- Side of board part belongs on.
- Orientation part is to have. *Example:* Forked Lug—flat side parallel to short side of board.

- Quantity for each part and orientation.
- Acceptable metals.
- Forbidden metals.

d. Check each board in the sample for the required part, side, orientation, quantity, and metal. *Note:* To distinguish aluminum from silver, place a drop of 1 Molar Sodium Sulfite solution of the hardware part. If that part turns dark, it is made of silver.)

e. If any of the hardware parts have threads (either internal or external) check each part for the presence of threads.

f. Check the type of fastening: flared end or rolled end. If there are any cracks which extend into the barrel of the hole, reject that part.

g. Check the parts for tightness; they should not wiggle or rotate.

h. Check drawing for any requirements that the part also be soldered into place.

i. If any discrepancies are found in the sample, the entire job must be inspected.

Material Review Board

1. A Material Review Board shall be called to disposition printed circuits which do not conform to customer's manufacturing specification; or, in the case of jobs being processed for the military, which do not conform to MIL-P-5511OC.

2. A formal Material Review Board shall be composed of the Production Manager, General Manager, Quality Assurance Manager, and Planning Engineer.

3. The Quality Assurance Manager shall present the problem, together with the manufacturing requirement. For military work, the Quality Assurance Manager shall cite all relevant sections of MIL-P-5511OC.

4. If the nonconformance will not adversely affect the function, reliability, or safety of the printed circuit, the MRB may issue approval to ship the boards: subject only to final approval of the customer, for nonmilitary work.

Nonconforming boards being built under provisions of MIL-P-5511OC shall be processed and documented under the provisions of this specification; except that written approval from the customer must be obtained prior to shipping the MRB approved and documented boards.

5. The disposition of the Material Review Board will usually call for one of the following types of action:

a. *Use As Is.* This type of disposition may result in the need for customer approval. It shall be left to the discretion of the Quality Assurance Manager to decide whether or not to contact the customer and seek their approval, prior to shipping.

b. *Rework.* Most rework dispositions can proceed without contacting the customer; the only exceptions being when the customer has required, in writing, that they be notified when rework is needed.

c. *Scrap.* Scrap dispositions will not usually require notifying of the customer. If the scrap will adversely impact the delivery date, the customer shall be notified immediately. All scrapped boards shall be kept in a secure location by the Quality Assurance Manager for a period of 30 days after the job has been shipped to the customer.

6. All proceeding and activities of the Material Review Board shall be documented on the Material Review Form (MRF, Fig. 7-12).

a. The MRF becomes a part of the history for each part number which has been subjected to a Material Review Board.

b. In the event that it becomes necessary to seek approval of the customer, prior to shipping nonconforming material, a copy of the Material Review Form shall be forwarded to the customer.

c. A copy of the Material Review Form shall be kept in the control file for that part number, and another copy shall be kept by the Quality Assurance Manager, in the Material Review File.

7. Documenting Material Review Board action, on the Material Review Form

MATERIAL REVIEW FORM

CUSTOMER:		PART NO:	REV: DATE:
JOB NO:		P.O.NO:	CONTRACT NO:
QUANTITY	DEFECT		SIGNATURE

Fig. 7-12. Form for material review board.

QUANTITY   DISPOSITION

QUANTITY   FOLLOW UP ACTION AND DISPOSITION

Fig. 7-12. (*Continued*)

a. The following information must be listed on the form:

   ● Customer.
   ● Part Number.
   ● Revision Level.
   ● Job Number.
   ● Date of Material Review Board.

b. If boards being run for military work under MIL-P-55110C are under consideration, the serial number must be listed under Quantity, along with the quantity.

c. The defects and quantity of each defect must be listed. There is ample room for writing in comments about requirements, and other useful and pertinent information.

d. The disposition section is used to document the action which was taken by the Material Review Board. There is ample room for a full discussion of the action taken.

e. The Follow Up Action and Disposition section is to document the results of the action which was decided upon by the Material Review Board. Examples of what may be included in this section are: rework which is unacceptable, customer disapproval of rework, and a customer's decision to accept substandard boards, and perform extensive repair, in order to make their own shipping requirements.

PROCESSING OF RETURNED PRODUCT

1. Customer will be assigned a number from DMR log.

a. DMR stands for: discrepant material report.
b. The number assigned is the next available in the log.
c. If a customer should call and request a return authorization number, the DMR number will be the number assigned.
d. The DMR log is kept in the Shipping/Receiving area.

2. All returned material is to be processed by the Shipping/Receiving department. No material is to be brought in otherwise.

a. When discrepant material arrives Shipping and Receiving will fill out the DMR form (Fig. 7-13) as follows:

- Customer.
- Part number.
- Revision level.
- Returned quantity.
- Customer return number.
- Original P.O.
- Return P.O.
- Packing slip number.
- Date of return.

DISCREPANT MATERIAL REPORT NO _____

RETURN MATERIAL AUTHORIZATION NO _____

DATE ___/___/_____

№ 501

MARKETING:

CUSTOMER _____ PART NO _____ REV_____

ORIGINAL P.O. _____ REPLACEMENT P.O. _____ QUANTITY_____

UNIT PRICE _____ CREDIT & REPLACE _____ PACKING SLIP NO._____

CUSTOMER D.M.R. NO. _____ ROUTING _____
    NO. CARTONS_____

PICK-UP DATE ___/___/_____ AUTHORIZED BY _____
    BUYER_____

RECEIVING:

WORK ORDER NO. _____ QUANTITY RETURNED _____

DISCREPANCY: _____

_____

_____

LOCATION _____ QUANTITY REPAIRED _____
    QUANTITY SCRAPPED_____

ACCOUNTING:

CREDIT MEMO AUTHORIZATION_____ DATE ___/___/_____
AMOUNT OF CREDIT _____ CREDIT MEMO NO _____
Form No. 1020   6-81

Fig. 7-13. Form for tracking customer returns.

b. Shipping/Receiving will make one copy of paperwork from customer which accompanied the returned material.

c. Shipping/Receiving will then obtain the traveler on which the materials were built and do the following:

- Paper clip the traveler and copy of return paperwork to the back of the DMR form; this should all be placed in a plastic folder.

- Place the returned boards in the Inspection area.
- Place DMR package on the desk of the Quality Assurance Manager.
- Place the original return paperwork in Production Control office.

   d. It will now be up to the Quality Assurance department to examine and make a disposition on the boards.

3. Quality Assurance dispositioning of returned material will be done within 48 hours of the receipt of the DMR form from Shipping/Receiving. The Quality Assurance department will make the decision to rework or scrap the returned product.

   a. Scrap all, issue credit. This should be done only if the customer requests. The Quality Assurance manager will write this on the DMR form.
   b. Scrap all, replace.

- The quality assurance manager will write this on the DMR form.
- Production Control notify the Production Manager, or the General Manager of the need to replace boards.

  - The replacements will come from stock or from an existing run, or a remake will be authorized by the Production Manager or General Manager.
  - If boards are pulled from an existing run, Production Control will log them into the Open Order Log, and make a note on the job traveler for quantity and shipping P.O.
  - If a remake is authorized, it will be logged into the Open Order Log and tracked by Production Control.

   c. Scrap, partial return, replace. Follow the same procedure as above, except than those boards which are good, or have been reworked, will be returned immediately to the customer.
   d. In all cases where a remake is needed, the quality assurance department will issue an Inspection Disposition Report. This I.D.R. will detail the reasons for the lost boards.
   e. If the boards are to be reworked, the DMR form will have the rework instructions written on it, and it will serve as a traveler to accompany the boards until they ship.

4. Shipping/Receiving department will ship the reworked boards back to the customer.

   a. Boards will be returned on a return P.O. provided by the customer.

5. The completed DMR form will have the copies distributed as follows:

White	Quality Assurance Manager
Green	Accounting
Yellow	P.O. File
Pink	Production Manager
Gold	Sales

SECTION THREE
IMAGING AND ARTWORK PROCESSING

8
Artwork Processing

Photo is one of the areas in printed circuit manufacturing where the product is never actually handled. Yet artwork is critical to manufacturing. Artwork, as supplied by the customer, is part of the documentation package and should never be taken for granted. It is just as critical to study the artwork supplied for a job as it is to study the drawing, manufacturing specification, and purchase order.

During the job planning stages, the planning engineer should consider the impact which the artwork will have on the product. Those people involved with using and making changes in the printed circuit artwork should also consider that impact.

Among any list of considerations should be the following:

1. Pad to hole size. Many circuit designs call for a pad (annular ring) of about 0.010 inches around the hole. This requires the pad on the artwork to be about .020 inch greater in diameter than the hole. If the minimum annular ring acceptable is 0.005 inch, this will allow 0.0025 inch of registration tolerance.

If a new revision is issued, where the purpose of that revision is to increase hole size, care should be taken to measure the pad on the artwork. Printed circuit manufacturers often find customers requiring increasing hole sizes which result in the hole breaking out of the pad. From looking at the drawing or artwork, it may appear that clearance at a hole is more than adequate for proper airgap. However, once the hole is drilled, a different view is evident. The photo technician must never assume the customer's artwork will not result in problems like this.

2. Front-to-back registration, or layer-to-layer registration. Usually, the job will be drill programmed to either the component side, or the side detailed in the drawing. If it is a multilayer, the component side is commonly used, unless the drawings call out otherwise. A diazo (transparent) copy or a photographic reversal (negative to positive) should be made of this side. All other pieces of customer artwork should be layed up and registered to this side. If the registration is more than 0.002 inch off, watch out for possible hole break out and air gap problems. Also, jobs where front-to-back registration is poor should not be programmed flip-flop. When a job is not run flip-flop, an extra piece of work-

ing film will be required for every operation performed to both sides of the board: imaging, soldermask, legend.

3. Artwork acuteness and definition. Some artwork which is supplied to printed circuit manufacturers is of such poor quality, that hours of touch up for raggedness, pin holes, nicks and broken circuitry barely suffice to turn that film into artwork which can be successfully used to make a printed circuit. Unless the customer is informed of the low quality, and given the opportunity to replace it, the manufacturer should proceed with the job only after obtaining the customer's agreement to buy any scrap produced as a result of that artwork being used.

4. Soldermask may be added to one or both sides of the board. If required for both sides, it is a good idea to lay the soldermask artwork against the artwork for the circuit patterns and ask the following questions:

- Is clearance in soldermask artwork sufficient to avoid getting soldermask on the pads?
- Is clearance so great that adjacent circuitry is uncovered? This may result in solder bridges forming during wave soldering.
- Is there a clearance provided for every hole, and are there clearances where no hole is indicated? Every printed circuit manufacturer has at least one customer who wants soldermask to flow through electrical feed through holes; but no customers who want soldermask in component or tooling holes.

5. If the job is a revision change, find out what that change is; make a determination as to whether or not the existing artwork can be used merely by changing the revision number. Typically, if the only change is that of hole sizes, there is little problem in using the existing artwork. A circuitry change, obviously, requires new artwork.

The artwork is so important that users and generators cannot afford to make assumptions on its quality or suitability for use: it must be examined expertly.

A printed circuit manufacturer should review the technology available to assist in the processing of artwork from customers film into working film. Some basic functions of any printed circuit photo area are:

1. Artwork inspection and touch up.
2. Exposures for duplicating, reversing, step-and-repeat, and drilling inspection and programming.
3. Developing.
4. Assuring registration during its use in manufacturing: especially during multilayer manufacturing.

The equipment for artwork inspection and touch up should include:

1. Light tables; generally, the larger, the better.
2. X-acto razor blade type knives, with various styles of blades.
3. Graphic pens, together with touch up brushes (personal preference is up to the photo technician), and opaquing ink.
4. Magnifying devices, from a stereo zoom mincroscope or magnifying projector to $7\times$ or $10\times$ eyepieces. There are even eyepieces mounted on a measuring track which can be used in place of a drill tape digitizer to take measurements off the film, for comparison with the drawing.
5. An assortment of clear and red litho tapes; double sided tape; press on letters, numbers, and logos; and an assortment of red litho pads.

EXPOSING

Step-and-repeat, reversing, and duplicating are all accomplished by exposing. The basic exposure equipment is a contact frame and a light source. There are several refinements of this equipment which will improve the quality of exposure. A well equipped darkroom and photo area should be able to process silver and diazo films. Each of these has its own strong points, which the printed circuit manufacturer should be able to utilize.

To take full advantage of the relatively inexpensive technology available, the photo area should contain the following:

1. Darkroom, equipped with red safety lights, and also yellow safe lights for use with diazo or diazo-containing films.
2. White light exposure source. This should include a contact frame equipped with a vacuum drawdown, and a point source light bulb several feet above it. If a light integrater photo cell is used to control exposure, the film will receive exactly the correct amount of light energy. This frees the technician from differing exposures which result when timers are used, and the light bulb begins to age.
3. Ultraviolet (UV) exposure source. This is required to expose diazo and diazo-containing silver films.
4. Developing equipment trays, faucet, sink.

Silver film has been around the longest and is most widely used in printed circuit graphics. There is a great deal of prejudice in printed circuit manufacturing against trying anything new, or trying it and actually trying to make it work. People fall into a rut over the years, and are reluctant to learn or try anything new. Whenever the mask of prejudice is lifted, ignorance and superstition—also known as lack of training and lack of knowledge—are uncovered. Those involved with manufacturing printed circuits should actively seek to lift this mask. In almost all cases, the vendors who manufacture and supply materials, equipment, and processes to the printed circuit industry are willing and able to train people in the use of new products, equipment, and processes. This type of support is abundant in the industry and should be taken advantage of. The films available, other than traditional silver film, are among those products which would be more widely used if printed circuit photo technicians and managers would evaluate them.

Diazo film offers several advantages over traditional duplicating film:

1. No darkroom is required. Diazo can be used under yellow safe light.
2. No wet processing chemicals are required for developing. A simple ammonia vapor developer is used. This type of developing is quick, clean, and takes little room; it can even be mounted on a wall.
3. Diazo has a transparent emulsion. The fact of being able to see through it aids greatly in registration for step-and-repeat, drilling inspection, artwork front-to-back inspection, and dry film photoresist.
4. No pin holes. Pin holes are a major drawback to silver films, and require much time for touch up.
5. Tougher emulsion, which is less sensitive to scratches from dirt. It also stands up better to scrapes received during handling.
6. Excellent emulsion stability and resolution capability. The imaging comes from molecules of diazo material which absorb ultraviolet light. This permits visible light to pass through. It also means resolution is on a molecular scale, and not limited to large crystals of silver halide.

Another type of film available is a silver film which can be processed under yellow lights, like diazo film, but develops in exactly the same way as traditional silver film. The obvious advantage is that there is no need to be confined to a darkroom for processing under poor lighting conditions. Exposure is by ultraviolet light, but exposure times are short. Multiple exposures can be made on one piece of film, by masking out all areas except the exposure being made. This eliminates the need for shooting several pieces of film, only to tape them together to shoot the working film. The working film can be stepped-and-repeated without an intermediary lay-up; and this can be done under excellent lighting conditions of a yellow light room. The diazo-containing silver film offers increased reliability of registration with a drastic reduction in stepping-and-repeating time. There is also a savings in film cost, since the stepped and repeated film is not a lay-up, it is the working film. This type of silver film is available in both duplicating and reprographic type emulsions, and can be used for any application served by traditional silver film.

DEVELOPING

Developing of silver film is accomplished by immersing the exposed film in three chemicals: developer, stop bath, and fixer, followed by water rinsing and drying. The chemicals are in open trays which should set in a sink equipped to provide a temperature controlled water bath. A timer should also be present for proper developing times. After developing and rinse, the film may be clipped to a suspended line or, preferably, run through a film drier to quickly and reliably complete the developing process.

Also available today are film processors which develop, rinse, and dry the film. These cost several thousand dollars; however, reduced developing times and more consistent results may be worth the money. The cost is minor when compared with the cost of an etcher, a drilling machine, or a plating line.

Of great importance is artwork registration. This can be a real problem for inner layer artwork on multilayer boards. Registration can be accomplished with pins inserted through the artwork into tooling holes. The holes may be either punched or drilled. Punching offers some advantages over drilling. The punch can quickly be set up to punch holes in artwork or inner layer laminate which will register ±0.001 inch with each other. The punch can be stored in the photo area, and another in the dry film area for the thin clad inner layer laminate. The punch can be built to accommodate several panel sizes. This eliminates the need to bundle up the unexposed film for drilling.

DEFINITIONS

1. Customer Film/Artwork: Artwork submitted by the customer: component side, solder side, padmaster, soldermask, legend.
2. Copy-To-Make-Copies (CTMC): Duplicate of customer film used to generate step and repeat pattern. This piece of film should be touched up before shooting step and repeat exposures.
3. Working Film/Copy: The phototool used in production to shoot screening stencils and expose dry film photoresist. Working Film and Phototool are interchangeable terms.
4. Silk Screen Film: Phototool used to expose circuit pattern screening stencils.
5. Soldermask Film: Phototool used to expose soldermask stencil.
6. Legend Film: Phototool used to expose legend stencil.
7. Legend or Nome: The set of marking designations screened onto the board with epoxy ink to identify areas on the board for the customer.
8. Padmaster: Artwork which has no circuitry, only pads. This may be used to generate soldermask film, when the customer so requires.
9. Reading Right Emulsion: Letters and numbers on artwork read from left to right, when viewing film with emulsion facing up.
10. Reading Right Base (RRB): Letters and numbers on the artwork read from left to right when mylar (base) faces up (emulsion down).
11. Duplicating Film: Positive acting film for transposing RRE to RRB or vice versa.

12. Reproducing Film: Negative acting film for reversing the circuitry from negative to positive or vice versa.
13. Circuit Positive: Traces and other circuitry are black.
14. Circuit Negative: Traces and other circuitry are clear.
15. Logo, or UL: The designation which Underwriters' Laboratory allows to be put on printed circuit boards which have passed their testing. Do not put logos on the following types of boards: Any multilayer, double sided, or fine line board for which the company has not yet received UL recognition.
16. 94V-O: A flame retardation classification by Underwriters' Laboratory.
17. Fab Lines: Lines on the artwork showing the borders of the part.
18. Plating Bars: A trace on the artwork which connects the edges of the panel with the contact fingers. This trace must be present for tip plating; it is removed during Fabrication.
19. Hole Centers: Dots on the Customer artwork which show the geometric location of the center of the hole. Hole centers are burned out after step-and-repeat on the working film, to prevent resist from getting into the hole during screening.
20. Dry Film Photoresist: A gelatinous film laminated onto a panel or inner layer. When exposed to ultraviolet radiation, it polymerizes into a hard plating or etch resist. Dry Film is a negative acting photoresist.
21. Negative Acting: When a film reverses a circuit from negative to positive, or positive to negative, it is said to be negative acting.
22. Positive Acting: When the image is not reversed from positive to negative, or negative to positive, that film is said to be positive acting.
23. Flip Flop: When a panel is drilled for imaging with half the panel being component side, and half solder side.
24. Circuit or Solder Side: Side of completed and loaded board that will be exposed to the molten solder during wave soldering. It is the side opposite the component side.
25. Component Side: Side of completed board from which components will be loaded. It is the side opposite the solder side.
26. Film Package: The large envelope which contains smaller, separate envelopes for screening film, soldermask and legend film, inner layer film, and copy-to-make-copies.
27. Customer Film Inspection Check List (CRICL): Artwork inspection form.

CUSTOMER ARTWORK INSPECTION

1. Obtain and carefully review:

   a. All drawings for that part number.
   b. Manufacturing specification for that customer.

c. Sales Quote Summary.
d. All pieces of customer artwork:

- Component side.
- Circuit side/solder side.
- Soldermask—one or two sides.
- Legend—one or two sides.
- Padmaster.
- Inner layers.

e. Copy of the traveler.
f. Customer Film Inspection Check List (CFICL).

2. Check part number and revision level on artwork for agreement with Sales Quote Summary and drawings. Notify Planning of any discrepancy.

3. Examine image definition: artwork requiring extensive touch-up is not acceptable:

a. Nicked, broken or cut ircuitry.
b. Excessive protrusions from circuitry.
c. Pinholes, voids and incompleteness, or spots which would have to be removed to render artwork usable.
d. If any of these defects are present to the extent that more than a few minutes of touch up with an X-acto knife and pen are required, the artwork is to be rejected at the end of the inspection process. The reasons for rejection are to be written on the CFICL.

4. Front-To-Back Registration:

a. Shoot the reverse image of the component side artwork, if the reversal is not already supplied by the customer. This is a positive to negative type of reversal.
b. Use the reversed component side to lay up to the other pieces of artwork to check registration, unless the drawing or customer specification requires using a side other than component. Do not use the padmaster, as it has no contact fingers.
c. Lay the reversed component side to each piece except for the legend:

- Solder side.
- Inner layers
- Soldermask.

d. Check:

- Hole to hole.
- Contact fingers.
- Ground areas.
- Alignment targets.

e. If registration is not perfect, using 7× or greater magnification:

- Measure the misalignment: Calibrated disk microscope/eyepiece.
  digitizer.
- If the measured value is 0.003 inch or greater, reject the artwork and notify Planning.
- If the customer wants to run with the film, it must not be flip flopped; notify Planning.

5. Check artwork against drawings for agreement; notify Planning if noted otherwise.

6. Measure minimum trace width and air gap, to ensure compliance with customer's specification or drawing requirements.

a. Do not accept artwork with an airgap of 0.008 inch or less.
b. It is possible to use artwork with airgaps of 0.007 if the customer gives permission to photographically choke the circuitry.
c. Out of tolerance artwork is to be rejected and documented on the CFICL, and Planning should be notified.

7. At the completion of Customer Artwork Inspection:

a. Thoroughly document all findings on the CFICL.
b. Return all artwork, drawings, Sales Quote Summary, customer specification, film envelopes, and CFICL to Planning.

COPY TO MAKE COPIES

1. Assemble the following:

a. Customer Film.
b. Customer Film Inspection Check List.
c. First Article Panel.
d. All drawings for that part number.
e. Customer specification.
f. Copy of traveler.

2. Carefully read the traveler and drawings.

a. Look for special notes and requirements, especially for markings and their location requirements.
b. When adding markings:

- Do not place them in areas which will be cut out during Fabrication. Study the drawings and note these areas.
- Notice whether the requirement is for markings to go on the circuit, soldermask, or legend film—or whether it is to be hand stamped on.

3. Shoot the Copy-To-Make-Copies (CTMC) for each piece of film supplied; use 7 mil duplication film.

a. Never perform touch up to the customer film, only to the CTMC and working film.
b. All step and repeat is done on duplication film.

4. Touch up the CTMC:

a. Nicks, breaks, pinholes, etc. High density, fine line circuitry can be touched up on a circuit positive, then reversed to a negative for step and repeat.
b. Where there are tight airgaps, do the touch up on the base side. Excess ink may then be scraped away with a No. 16 X-acto knife blade.
c. Except for the above, it is probably easier to do touch up on the emulsion side.

5. Fab Lines should be removed by taping over with ⅟₃₂ inch red litho tape.
6. Plating Bars: run ⅟₃₂ inch red litho tape all the way across the CTMC at the contact fingers, just outside the fab line area.
7. Markings: Add logo, 94V-O, revision letter, and date code as required. For Date Code use the digital 8888's, and mark out where required to form the date code.

STEP AND REPEAT (USING DIAZO CONTAINING SILVER FILM)

1. Put on white cotton or polyester gloves to handle virgin film.
2. Take a sheet of yellow room light safe 7 mil duplicating film.

a. Use Kodak RL or Dupont Highlight film.
b. This film shows finger prints if bare fingers touch the emulsion.
c. Cut the film to panel size and tape it to the drilled panel using double sided tape on all four sides—emulsion up.
   An alternative procedure is to cut the virgin film 1 inch shorter than the

panel in both directions. When this is layed on the panel, a ½ inch border of bare copper is showing. The virgin film must be taped to the panel, using ½ inch clear litho tape, on all four sides.

3. For a panel that has up to four circuits, use Method A; for more than four exposures, use Method B.

Method A:

1. Use ½ inch red litho tape to make a border around the virgin film.
2. Register the touched up CTMC to one of the drilled circuit patterns. Tape this down on two sides.
3. Place entire panel in the UV contact printer, and block out all areas except for the CTMC with black cardboard.
4. After vacuum drawdown, inspect registration. If it has changed, open frame, remove the panel and correct.
5. Use button labeled Duplication, for exposure.
6. After exposure, carefully lift taped CTMC from virgin film and reregister. Continue in this manner until all exposures for step and repeat have been made.
7. After last exposure, turn panel over with virgin film still in place, and expose. This will burn out the hole centers. It will also make pads for holes which are to be plugged during plating. Do this step even if there are no hole centers to be burned out.
8. Remove red litho tape border and develop the film.

Method B: This method is to be used when there is a two-dimensional matrix of more than 4 circuits. Example:

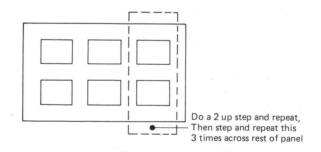

Do a 2 up step and repeat,
Then step and repeat this
3 times across rest of panel

Fig. 8-1. Performing Step-and-Repeat.

Do a 2 up step and repeat; then step and repeat this 3 times across rest of panel.

This is basically the same procedure as method A, except that an intermediate CTMC is generated. The intermediate CTMC made of one complete row in the matrix. This intermediate CTMC will reduce the number of steps and exposures, and increase the accuracy of registration.

4. If registration is off after using method A or B, do the following:

a. Cut the stepped and repeated film into individual circuits.
b. Cut a piece of clear mylar to panel size and pin/tape it to the drilled panel.
c. Register and tape all individual circuits together *emulsion* up on the clear mylar.
d. Use ½ inch red litho tape to add a border to the mylar.
e. Remove the mylar with the taped circuits and place on a fresh sheet of cut to size virgin film, so that the circuit emulsion is down and in contact with the emulsion of the virgin film.
f. Expose and develop. Check for proper registration.

5. Flip Flop. If the panel has been programmed flip flop you must use both component and solder side for generating step and repeat pattern. However, only one piece of working film need be made, since it will be used for both sides during imaging.

6. Once you have a well registered phototool:

a. Add 0.125 red litho pads to idiot and screening holes, 5 per panel.
b. Use ½₂ inch red litho tape to extend Plating Bars to panel edges, if necessary.
c. Inspect for proper marking, registration, touch up and hole centers.
d. Check for proper emulsion and reading orientation:

Silk Screen Film	RRE	Negative pattern
Soldermask Film	RRE	Negative pattern
Legend Film	RRE	Negative pattern
Inner Layers—dry film	RRB	Negative pattern
Outer Layers—dry film	RRB	Positive pattern

e. Place 2 inch white labels on the emulsion side of the film, in the border area. Write in Customer, Control Number, operator initials.

7. Place the completed phototool in the properly labeled envelope for the type of phototool it is:

a. Silk Screen Film.
b. Soldermask Film.

  c. Legend Film.
  d. Inner Layer Film.

8. Special note: When the back side of a panel is to be solid metal, do not generate artwork for it—even if the customer has supplied it. This side of the panel will be left blank after screening. If there are questions about how to handle a job, Planning should be consulted.

SOLDERMASK

1. Soldermask, when supplied by the customer, must be laid up to the drilled first article, even if it is only one up.

2. Check drawings and traveler for holes which are to be plugged during plating.

  a. All holes to be plugged must have pads added to soldermask film. The pads must be larger than the hole. This will prevent soldermask from getting into the hole. Use a pad about 0.010 inches larger than hole.
  b. If there are other holes which do not have pads, leave the film that way. Do not add pads to keep soldermask from flowing into feed through holes.

3. If customer supplies two pieces of artwork, one will be for the circuit side, and one for component side.

  a. Read prints to find which is which. Keep them separate during step-and-repeat.
  b. Two pieces of artwork for soldermask usually indicates there are holes which have differently shaped pads, depending on which side of the board they are on.

4. If only one piece of artwork was supplied by customer for soldermask, read prints and traveler to see whether one or both sides is to be soldermasked. If both sides require soldermask, the same working film may be used for both sides.

5. When making any working film, but especially soldermask, watch for:

  a. Pads on the film where there are no holes in the board.
  b. Poor drill registration.
  c. Holes in the board where there are no pads on the soldermask.
  d. Notify Planning or Quality Assurance manager when any of these conditions are noticed.

6. When customer requires soldermask, but does not supply the artwork:

a. Unless the print or traveler specifically calls for vendor to generate soldermask film, notify Planning that the soldermask artwork was not in the customer film package.

b. Otherwise, Photo must generate the soldermask working film, use padmaster when supplied.

7. Drilling may generate a stencil for soldermask using entry material:

a. Measure the pad size on the artwork for each hole size.

b. Drill entry material out 0.010 inch over pad size.

c. For holes with two or more pad sizes for the same size hole:

- Drill all holes at the smaller size.
- Selectively redrill larger pad sizes out 0.010 inch over pad size for the larger pad.
- Holes with no pads should be drilled out 0.010 inch over hole size.

d. Expose and develop duplicating film, using drilled entry material as phototool.

- Check pad size, registration and completeness against the plated panel.
- If pads are too small, do a photographic spread on the soldermask film.
- If pads are so big that circuitry is exposed, do a photographic choke to the soldermask film.

8. Photo may use a padmaster, if one has been supplied, to generate the soldermask film:

a. Perform a photographic spread on the padmaster, until pad size has been spread 0.010 inch in diameter. Measure airgap on outer layer circuitry; solder mask must cover all circuitry.

b. Add special pads for holes that have been plugged for plating. The pad must be larger than the hole diameter.

c. Add pads for screening and idiot holes, five per panel.

MULTILAYER

1. Obtain and read:

a. Customer specification.

b. All drawings for the part number.

c. Customer Film Inspection Check List (CFICL).

    d. Traveler.
    e. Drilled first article.

2. *Special Note:* Ground Planes on Inner and Outer Layers. The layers may be virtually *all* copper, or *no* copper. It is important that the prints be studied and compared against the artwork. It is very easy to make an error where the result is a layer with no copper, when in fact the layer is supposed to be a ground plane.

Power and Ground layers must be registered against each other to varify that no common electrical holes exist.

3. All inner layer artwork is to be Reading Right Base, when completed.

4. Outer layer artwork is generated the same for multilayers as it is for double sided boards.

5. On double sided artwork, the front to back registration is checked. For multilayers, all layers must be checked for registration against the side used for programming.

6. When the artwork for all the layers has been completed, it must be pinned together and checked for registration accuracy.

7. To assure good layer registration during the fabrication process, tooling holes are punched into the artwork and the separate layers of thin clad material.

    a. The tooling holes (six of them per panel) are arranged in two rows, with one hole designated as the common hole. This results in there being only one way that artwork or panel may be pinned together.
    b. During programming, drilling and Photo, keep the common hole in the upper left hand corner, with the panel being viewed lengthwise.

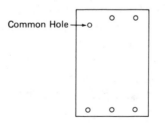

Fig. 8-2. Tooling hole layout to prevent misorientation.

8. Determining emulsion orientation on artwork, with reference to the common hole.

    a. Look at the back side of the traveler, to see how the layers are arranged.
    b. Take one sheet of yellow light safe silver film for each layer of the mul-

tilayer panel; stack the film up as if it were being laid emulsion down against the copper; use duplicating film.

c. Take a sharp-pointed marking pen and write on one corner the number of the layer to which the film corresponds.

d. Punch each sheet of virgin film in the punch (preset for the size of panel to be run). Punch film as stacked.

e. Restack the film in the sequence which corresponds to the layers.

f. All the layers should have the common holes properly aligned.

g. *Example:* typical 6 layer board using 3 part construction:

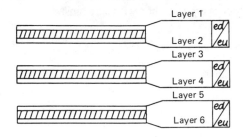

Fig. 8-3. Layer orientation for multilayer construction.

h. If the virgin film is not labeled and punched all at once, it is easy to see how the emulsion should be oriented, by looking at the layer sequence on the back side of the traveler. This diagram assumes the viewer to be looking at the panel with the common hole oriented in the upper left hand corner. The *eu* and *ed* in the blocks on the right stand for emulsion up and emulsion down.

9. Step and repeat is performed much the same as it is for double sided, with a few changes:

a. The virgin film is pinned to the drilled first article, not taped. The pins (hat pins) are taped to the panel to prevent them from falling out.

b. Inner layer border:

- Use red litho tape to block off the boarders, as for double sided boards.
- After step and repeat, the film is developed. When it has been determined that registration is acceptable, use vented or striped litho tape to make the border. A honey comb dot pattern also works very well.
- Add a ½ × ½ inch strip of red litho tape over the tooling holes. This will provide copper metal around the tooling holes on the completed inner layer.

   c. Markings, such as date code, UL logo, 94V-O, need not be added to inner layer artwork, only the outer layers.

   d. Microsection coupons are drilled into every multilayer panel along one edge, outside the circuit area. Place a strip of red litho tape over these holes, on every layer.

   e. Outer layers are clear, not vented.

10. Just as for double sided board, the step and repeat registration must be checked after the film has been developed. However, for multilayers, each completed layer should also be checked for registration to the layer used for programming. Pin the layers together with hat pins.

REPEATS

1. Assemble the following:

   a. Traveler.
   b. Copy of all drawings.
   c. CFICL.
   d. Customer Specification.
   e. Drilled F/A: labeled with Customer, Part Number, Revision.
   f. Working Film.

2. Look for special notes written on any of the documents or attached to film packages.

3. Do not assume that the film package which has been prepared for past runs has been prepared correctly.

4. Review instructions on the traveler and customer's requirements for markings:

   a. UL markings: Logo, 94V-O.
   b. Revision level—if none, place a dash (–) after Rev: Rev–.
   c. Date Code.
   d. Method to be used for applying the markings:

      • Etched on via the artwork.
      • Screened on during legend process with epoxy ink.
      • Stamped on by hand.

   e. Location for the markings.

5. Inspect the artwork to ensure that it complies with the customer's requirements.

6. Change the Date Code:

a. Remove old date code.
b. Add fresh Digital 8's.
c. Ink out the 8's to form correct date code.

7. Revision Level normally goes on after the part number or in a specially designated location. Be sure that revision level is not accidentally added after the assembly number.
8. Reinspect for airgap, broken or shorted traces. pinholes, nicks and protrusions.
9. Check registration of artwork against the drilled First Article which is stored in the Photo area. It may be necessary to re-layup the artwork.

a. Touch up
10. Check labels:

a. Located only on emulsion side of film.
b. Customer name.
c. Artwork preparer's initials.
d. Control Number or Part Number.
e. Revision Level.

11. It is also important that artwork be checked for proper orientation.

Silk screen film	RRE	Negative pattern
Soldermask film	RRE	Negative pattern
Legend film	RRE	Negative pattern
Inner layers—dry film	RRB	Negative pattern
Outer layers—dry film	RRB	Positive pattern
Outer layers—silk screen	RRE	Negative pattern

SPREADS AND CHOKES

1. Assemble the following:

a. Artwork which is to be modified.
b. 4 mil and 7 mil clear polyester sheets.
c. Light diffusion sheet.
d. Black, nonreflective cardboard.
e. Duplicating contact film.

2. A *choke* is performed to:

a. Decrease trace width.
b. Decrease pad diameter.
c. Increase air gap.

3. To perform a choke:

a. Obtain a positive copy of the artwork.
b. Lay black cardboard in the bottom of the contact frame.
c. Lay a sheet of duplicating film, emulsion up, on top of the cardboard.
d. Lay a sheet of clear polyester on top of the duplicating film:

  - 4 mil for spreads of 0.002 inch or less.
  - 7 mil for spreads of 0.003 or more.

e. Lay the artwork, emulsion up, on top of the polyester.
f. Close the contact frame and draw a vacuum.
g. Place the light diffusion sheet on top of vacuum frame.
h. Expose film. Exposure time will vary, depending on the degree of choke required, but in no case will it be less than the setting used to expose duplicating film under normal conditions. 50 to 100% increase is typical.
i. Develop and dry the exposed film.
j. Measure the degree of choke. If choke is insufficient, repeat using either more exposure time, or using 7 mil polyester, if 4 mil had been used.

4. A spread is performed to:

a. Increase trace width.
b. Increase pad diameter.
c. Decrease air gap.

5. To perform a spread:

a. Obtain a negative copy of the artwork.
b. Lay the cardboard on the bottom of the contact frame.
c. Lay a sheet of duplicating film, emulsion up, on top of the cardboard.
d. Lay a sheet of clear polyester on top of the duplicating film:

  - 4 mil for spreads of 0.002 or less,
  - 7 mil for spreads of 0.003 or more.

e. Lay the artwork, emulsion up, on top of the polyester.
f. Close the contact frame and draw a vacuum.
g. Place the light diffusion sheet on top of the contact frame.
h. Expose film. Exposure time will vary, depending on the degree of spread required, but in no case will the time be less than that required to expose duplicating film under normal conditions.
i. Develop and dry the exposed film.

j. Measure the degree of spread: If spread is insufficient, repeat using either more exposure time, or using 7 mil polyester, if 4 mil had been used.

Note: a machine called **MICRO MODIFIER** is now available which will perform very accurate chokes and spreads.

Experiments

There are two controls on the degree of choke or spread achieved: thickness of the clear polyester spacing, and length of exposure time. Since performing spreads and chokes is such a basic procedure for the printed circuit photo technician, it is worthwhile to develop a familiarity with just what can be done with exposure time and spacer sheets between artwork and film. This can be done by performing two experiments, one with 4 mil polyester, and another with 7 mil.

1. Obtain the following:

a. Resolution test pattern (a Dupont Spread Check Pattern will do).
b. Black, nonreflective cardboard.
c. 4 mil and 7 mil clear polyester sheets.
d. Duplicating film.

2. Lay the black cardboard on the bottom of the contact frome.
3. Lay the duplicating film emulsion up on the cardboard.
4. Lay a sheet of 4 mil polyester on top of the duplicating film.
5. Lay the resolution test pattern emulsion up on top of the polyester sheet.
6. Close the contact frame, draw a vacuum.
7. Lay the light diffusion sheet on top of the contact frame.
8. Expose film, using normal setting for duplicating film.
9. Develop film, and dry.
10. Measure the degree of line width change, and note the quality of line acuteness or definition.
11. Record this information. Repeat the process using longer exposure times until line acuteness beging to disintegrate.
12. Construct the following table:

| | 4 mil polyester | |
Time	Line width change	Line acuteness
T_1	?	Acceptable
T_2	?	Acceptable
⋮	⋮	⋮
T_x	?	Unacceptable

13. Repeat the experiment using 7 mil polyester spacing sheet.

PHOTO SUPERVISOR

1. Check all incoming customer artwork.

   a. This may be delegated this on a job-by-job basis.
   b. Drill and programming film is to be shot at this time by the Photo Department.

   2. Review all film packages coming out of Photo.

   a. All silk screening, soldermask, legend, and inner layer artwork is to be complete.
   b. Check for the presence and location—against the print—of UL, 94V-O, and date code.
   c. Check for the correctness of date code.
   d. Check silk screening, solder mask, and inner layer artwork for any needed touch up.
   e. Pads on holes with pads smaller than hole diameter are to be about 0.010 inches smaller than the hole.
   f. During lay up, artwork is to have a border added approximately ½ inch wide. No part of the border is to extend inside the fab lines.
   g. Use the stereo zoom microscope to check fine line and high density artwork for nicks, breaks, and pinholes. The microscope can also be used to touch up.
   h. When discrepancies are noticed, take them to the photo technician who prepared the job.
   i. Check registration against a drilled panel.

   3. Get around to talk with the screeners, dry film and touch up people on both shifts. Listen to any complaints and suggestions they have.
   4. Every night:

   a. All work areas should be left clean and straight—including the darkroom.
   b. All tools should be locked in the cabinet.

   5. In the morning, check to make sure that the second shift left the area straight and with the tools locked in the cabinet.

Part of this material has appeared in *Printed Circuit Fabrication* magazine.

9

Artwork Registration Systems for Dry Film Imaging

Phototool registration for double sided and outer layer multilayer applications is either visually registered or pin registered; phototool registration for inner layer multilayer applications is always accomplished by pin registration. Where registration is critical or difficult to achieve, only pin registration should be used.

Pin registration systems depend upon the dry film technician only to orient the phototool to the tooling holes; the pins register and hold the phototool in place. Visual registration depends on the technician to correctly orient and register the phototool repeadedly; with tape to hold registration once the phototool is in place.

PIN REGISTRATION

Pin registration is accomplished by taping "hat" pins (Fig. 9-1) into tooling holes which have been punched into the phototool (see Chapter 8). The tape holds the pins in the phototool. When the phototool is laid over the panel or inner layer, so that all tooling holes are aligned, the pins are pushed into the matching tooling holes in the panel or inner layer. This locks the phototool into a position of correct registration. The tooling holes and pins have performed all the work and care of registering to tight tolerance, and assure consistently good results for multiple exposures.

The artwork can be set up for exposing both sides of a panel at one time. To accomplish this, tape pins into the tooling holes of the phototool; pins are taped into the artwork for both sides of the panel. One of the phototools is taped onto

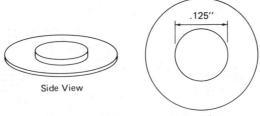

Side View

.125″

Top View

Fig. 9-1. The hat pin.

the glass of the vacuum frame with emulsion up. The panel is laid against this, with the hat pins inserted into the tooling holes of the panel. This will securely register one side of the panel. The phototool for the other side is laid up and pinned, emulsion down, to the panel. Phototools are always laid emulsion down against the cover sheet of the photoresist. When the vacuum frame lid is closed and a vacuum pulled, the correctly registered artwork is firmly and reliably held in place for the exposure.

VISUAL REGISTRATION

Visual registration, also called eyeball registration or eyeballing, uses tape, instead of pins, to hold registration. The registration is achieved with manual positioning of the phototool. Holes are punched into the four corners of each phototool, in the red border area. An alternative to hole punching is to cut squares (about ½ × ¼ inch) in the same area of the red bordered corners. These are referred to as *tape windows* (Fig. 9-2). A strip of tape, preferably red litho tape, is laid across these tape windows. The tape must be stretched tightly over the windows, with no creases or folds; slack, creases, and folds result in loose holding ability of the tape, and poor registration. Clear cellophane tape may be used, but it must be covered over with red litho tape, or photoresist will be exposed where the windows are located. This will not affect image quality, but resist in the border area of the panel may interfere with a good electrical connection to the plating racks.

Registration is achieved by placing the printed circuit panel on a light table. The light table must be equipped with yellow safety lights, no white lights. The phototool is placed over the panel until the panel holes and artwork pads are approximately aligned. Visability is enhanced by placing red mylar over light table. The artwork is held by the dry film technician at two diagonal corners.

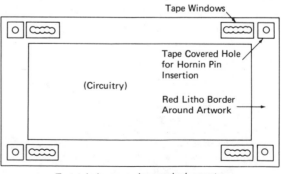

Tape windows may be punched, or cut
with an exacto type razor knife

Fig. 9-2. Artwork set up for Horning pin registration.

The artwork is then registered as finely as possible near one of the diagonal corners; and the tape over the tape window is pressed firmly to lock that corner in place. Artwork should always be registered with the aid of an eyepiece. The corner on the other end of the diagonal is next positioned in a like manner. When two corners have been positioned, the remaining corners are registered. It may be necessary to lift one or more of the corners for slight re-positioning to achieve the best possible registration. Eyeball registration may require several minutes per side; and may entail near endless repositioning of the artwork. This can lead to technician fatigue and frustration.

The drawbacks to visual registration are obvious: (1) it is, highly dependent on the operator; (2) unreliable tape/tape windows are used to hold artwork during positioning for registration and for vacuum drawdown; (3) it is a time consuming operation which must be duplicated for each exposure. Clearly, visual registration does not lend itself well to high volume productivity or consistent quality. What, then, are the problems associated with pin registration? As is turns out, there are a few, and most of them are related to the traditional hat pin.

Figure 9-1 shows the typical hat pin, which may be made of brass or steel. It is designed to be inserted into a tooling hole which is about $0.125 + 0.003/ -0.000$ inches diameter. This presents a problem. During dry film lamination the tooling holes are covered with dry film and the protective cover sheet. To open the holes, dry film technicians use a hot soldering iron, with a pointed tip, to poke at and burn out the holes. Aside from releasing toxic monomer-bearing fumes to the work area, and slowing production, this hole burn-out technique leaves a cauterized bead of resist and polymer around and inside the hole. This bead can make hat pin insertion quite difficult at times.

The burn-out problem is avoided for inner layer registration by punching tooling holes in the thin clad laminate *after* dry film lamination. Not all printed circuit manufacturers use punches for their thin clad, preferring instead to drill the tooling holes after blanking laminate to panel size. It should be pointed out that the main reason some shops still drill multilayer tooling holes, instead of punching them, is that a good set of matched punches for artwork and thin clad laminate may cost over $20,000. Drilling of the tooling holes, instead of using precision punching jigs, adds another potential registration problem— tooling hole location. The inherent accuracy of drilling machines in locating holes may be no better than ± 0.002 inch; a little better for some machines, and a little worse for others. A good punching jig will consistently and quickly place tooling holes with better than ± 0.0005 inch accuracy. Drilled hole inaccuracy is further compounded by drill bit splay, the tendency of a drill bit, when drilling through layers of material, to drill at an angle from true perpendicular. Only by punching tooling holes after dry film lamination can hole burn-out or hole location problems be avoided for inner layer imaging. For double sided panels being hat pin registered, hole burn-out remains a problem.

There is one characteristic of multilayer panels which should not go unmentioned, as it may contribute to a severe artwork registration problem: multilayer panels may change dimensionally during lamination and post-press baking. The thermal processing associated with lamination pressing and related baking operations may cause the panel to change slightly. MIL-P-13949F allows 0.0005 inches of dimensional change per inch of panel dimension; this means that a multilayer panel which is 20 inches across may display change of 0.010 inch. If hat pins are inserted into the phototools of the outer layers, and an attempt is made to use the multilayer tooling holes to register the outer layer artwork, misregistration will occur. The degree of change can be minimized greatly by baking epoxy FR4 laminate for several hours at 300 degrees Fahrenheit; but those involved with imaging multilayer printed circuit boards should always be aware of the potential problem.

Panel change will present itself as a problem when one is trying to register artwork using the punched tooling holes; there will be a bubble in the artwork once it has been pinned to the panel. It may not be possible for the technician to register the artwork and have it lie flat against the panel at the same time. If the circuitry does not require extremely accurate registration parameters, and the degree of panel change is small, there will probably be little problem; indeed, the bubble and panel change may go totally unnoticed.

However, in cases of high circuit density, fine lines, tight air gaps, and/or little annular ring, any deficiency in registration must be minimized. These problems can be dealt with by reverting to visual registration, or pinning to tooling holes other than those which resulted from punching of the thin clad laminate: tooling holes which have been programmed and drilled along with the other holes in the panel after lamination.

Using visual registration and pinning to alternate tooling holes may alleviate some registration problems and allow "splitting-the-difference" of the misregistration across the panel, but there are other registration problems which must be faced from time to time. The artwork itself may have changed dimensionally. This can happen from its exposure to temperature and humidity beyond tolerance levels recommended by the film manufacturer. Dimensional instability of the film can also result from poor quality of polyester base on which the film has been coated.

If the artwork has changed dimensionally, it is best simply to obtain freshly prepared working film from the Photo department. However, there may be instances where the artwork cannot be replaced, because of time factors or unavailability of master artwork. The dry film technician must accommodate the distorted artwork to the greatest extent possible.

Note: The condition previously discussed, of multilayer panel dimensional change, is commonly misdiagnosed as dimensional instability of the phototool; measuring tooling hole positions in the panel and/or artwork will quickly resolve differing opinions on etiology of panel/artwork misregistration.

To visually align such artwork would be to compound all the difficulties associated with this method already. Splitting the difference where the margin for error has been further reduced is almost certain to result in unsatisfactory results. Recent experiments have proven another pin registration system to be effective in overcoming the limitations of both the visual and the hat pin registration methods.

THE HORNING PIN REGISTRATION METHOD

The tight design requirements for modern printed circuits and the constraints of manufacturing realities often leave no option but to use a combination of visual and hat pin registration methods. There is one alternative to both of these methods which utilizes the most desireable features of each method; with none of the drawbacks. This method uses the Horning pin.

The Horning pin is shown in Fig. 9-3. (See the appendix for full engineering details on the Horning pin.) The Horning pin is only remotely similar to the hat pin in appearance, design, use, and function. A discussion of how it is used will best explain its benefit. The Horning pin is used to position and hold registration of phototools to double sided, and multilayer panels not to inner layers during layer imaging. With it, artwork can be registered using any available tooling holes, and it makes no difference whether or not the tooling holes have had the photoresist/cover sheet burned out after dry film lamination.

Registration with the Horning Pin

1. Visually register the artwork to the panel which will be exposed with that artwork.

   a. Either a production panel or a drilled first article may be used successfully.

   b. The panel should be set on a light table for optimum registration, and also in order that artwork can be taped to hold registration with the panel.

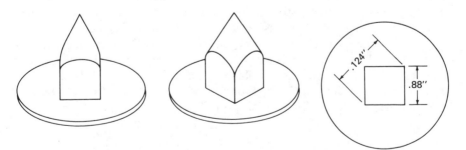

Fig. 9-3. The Horning pin.

2. Use magnification (an eyepiece or a magnifying glass) to check that registration is optimum. The registration obtained here will be the same registration which will be held for all subsequent panel/layer exposures.

3. A tape window must be cut, or punched, over the tooling holes which will be used for registration.

    a. The holes can be marked (with a pen or pin), then punched, or they can be cut with an X-acto type razor knife.

    b. The tape windows should be only slightly larger than the 0.125 inch tooling holes in the panel.

4. Place the Horning pins directly over the tooling holes and press them firmly into the holes. Although the pins go into the holes correctly whether or not the holes in the panel have been burned out with a soldering iron, there is no reason to waste time burning out the holes during dry film lamination.

5. Carefully stretch tape over the heads of the Horning pins, and firmly press the tape over the pin heads. Red litho tape may be used for this; if it has not been used, red litho tape should be applied over the tape which was used, to protect the integrity of the panel border.

6. The panel should be turned over, and the pin insertion instructions repeated on artwork for the other side.

The artwork can now be correctly and consistently positioned for multiple exposures and for exposures requiring great precision for tight registration requirements. The registration is achieved simply by laying the artwork, pin side down, over the dry film coated panel and pressing the pins into the tooling holes. The uniquely shaped Horning pin will pull the artwork into precise registration as it is pressed through the photoresist cover sheet; and it will position itself in the center of the tooling hole for consistent and reliable registration; which is unachievable by any other method of registration. The Horning pin system can be used for any circumstances which require splitting-the-difference for misregistration.

Since the Horning Pin is pointed, it punctures the photoresist cover sheet. The square shape of the pin shaft helps to center it in the tooling hole, even if it must cut through a burned-out bead of photoresist, or the cover sheet. The Horning Pin registration system is superior to all other visual and pin registration systems, including glass masters and image plane technology.

The inventor, David Horning, was a Dupont, Riston Division, technical representative who entered another business before bringing the Horning pin to commercial fruition. It is also available through Cal-Micro, 431 Perrymont Avenue, San Jose, CA 95123, Tel. (408)297-6470, and Eaton Colby BOX 511, Gloversville, NY 12072, Tel. (518)725-7157.

Part of this material has appeared in *Printed Circuit Fabrication* magazine.

10
Dry Film Imaging

The dry film photoresist method of imaging offers numerous advantages over screen printing:

1. Fine line definition.
2. Plated circuitry has straight sidewalls.
3. Excellent conformance to artwork dimensions.
4. Fast set up and turnaround time on small jobs.
5. Superior registration for multilayer inner layer circuitry.
6. Hole "tenting" eliminates the need for hole plugging. Tenting provides a photoresist seal over a hole, which in turn eliminates the need for second drilling operations, in most instances.

There are disadvantages associated with dry film imaging; chief among them are:

1. Dry film photoresist is expensive to buy.
2. A large capital investment is required to purchase processing equipment needed for dry film processing.
3. It leaves a monomolecular layer of adhesion promoter on the copper surface from which it has been developed, and this layer must be chemically etched off, or removed by pumice scrubbing. Failure to adequately remove this layer is a chief cause of plating peelers, where dry film photoresist has been used for imaging.

Dry film photoresist is generally easy to work with, yields excellent results, and operator training time is short, when compared to training time for a screen printing technician. The basic operating steps are:

1. Electroless copper deposition (through hole plating).
2. Acid copper flash (to add extra copper through the holes).
3. Bake to dryness.
4. Laminate panels with dry film photoresist.
5. Expose dry film photoresist with artwork and ultraviolet light.
6. Develop exposed panel to remove unexposed resist.
7. Touch up image.
8. Pattern plating/etching.

The three key steps directly related to dry film photoresist imaging are:

1. Lamination.
2. Exposure.
3. Developing.

The better the understanding and control of these operations, the better will be the image quality, and the overall satisfactory performance of dry film photoresist in meeting the imaging goals of the printed circuit shop. The peculiarities of each operation, as related to achieving optimum performance from photoresist imaging, must be discussed separately. There are a number of considerations throughout the entire printed circuit manufacturing process which are affected by the imaging choice: screening or dry film photoresist.

Photoresist contains an adhesion promoter which forms a bond between the copper surface and the photoresist. After developing of the unexposed photoresist the copper, which has been bared and is available for plating, appears to be a clean surface. In reality, it is covered by a monomolecular layer of chemically bound adhesion promoter. The adhesion promotor must be removed prior to pattern plating; otherwise the metal subsequently plated (copper) will peel off. There is no way to avoid this problem of adhesion promoter. (*Note:* There is one manufacturer of aqueous dry film photoresist which claims their dry film photoresist to be free of adhesion promoter, and its associated problems.) Adhesion promoter affects the electroless copper/through hole plating process and the cleaning line for pattern plating because of the need to etch copper to remove it.

There are three methods used for removing the adhesion promoter layer:

1. Pumice scrubbing of the patterned copper surface.
2. Chemical etching.
3. Electrochemical etching.

Pumice scrubbing is an effective method for cleaning copper surfaces, but it does have limitations. Scrubbing any imaged surface introduces the possibility of damaging the image. Pumice scrubbing is very labor intensive and operator dependent. It lends itself neither to high productivity, nor to consistently high quality. Great care must be ued to ensure even and uniform scrubbing in all areas—including fine traces. Nothing is accomplished by scrubbing 99% of the panel, only to have a peeler on the other 1%. Pumice is a very fine and difficult to rinse substance. Failure to thoroughly remove pumice from the panel surface, and from the panel contact areas of plating racks, may result in other problems—problems of surface plating quality, and problems related to dragging this material into the plating baths.

Still, there are some reasons for using pumice: it is easy to use, and cheap to insert into a pre-pattern plate cleaning operation. There is no need to buy chemicals, holding tanks, and heaters. Another point favoring pumice is that only the surface is scrubbed and cleaned: no copper is removed from the holes. This fact alone is why most shops which use pumice do so. Pumice scrubbing allows the printed circuit manufacturer to use dry film photoresist imaging without having to: (1) set up additional tanks for microetching, (2) acid copper flash electroplate to put more copper in the holes for microetching, and (3) resort to using a thick deposition electroless copper.

Chemical methods are the most widely used means for removing adhesion promoter. Chemical etching usually requires that more copper be deposited in the hole during or after electroless copper than does pumice scrubbing. The 20 microinches of copper deposited in the holes by room temperature electroless copper baths does not provide an adequate margin of copper thickness for reliable copper etching, without having to be concerned about voiding all or part of the hole. For this reason, panels which are to be imaged by dry film photoresist are typically "flashed" (panel plated) briefly to add an extra 0.0001 to 0.0002 inch of copper in the hole.

Another common method used to put extra copper inside the plated hole is the use of high speed electroless copper, which will deposit up to 0.0001 inch of copper in the hole. High speed electroless coppers deposit copper at a linear deposition rate, and operate at higher temperatures (100+ degrees Fahrenheit). Since they deposit about five times as much copper as room temperature baths, they require about five times more attention to their analyses and additions.

Common chemicals used for etching the surface of dry film photoresist imaged panels prior to pattern plating are ammonium persulfate, sodium persulfate, and sulfuric acid/hydrogen peroxide. This type of etching is called microetching because only a few microinches of copper need to be removed.

Anodic cleaning, also called reverse current or electrocleaning, is effective and fairly common. Here the panels are immersed in an electrolytic tank, with a positive polarity applied to the printed circuit panel and a negative polarity applied to stainless steel electrodes which are suspended opposite both sides of the panel (Fig. 10-1). Electrocleaners "de-plate" to remove copper from the panel surface. Electrocleaning is an effective method of removing the adhesion promoter and cleaning the copper surface. However, it requires setting up a large electrocleaning tank, with a large rectifier, to move large panel throughputs. Even though the etching rate is more uniform and controllable than with chemical microetching methods, the expense of setting up for high volume production has kept electrocleaning from becoming more widely used, especially in job shops.

Electrocleaning also requires that an immersion cleaning tank be used prior to electrocleaning; the requirement for this tank, and for a large electrocleaning tank, means a fairly large amount of floor area in the plating shop will be taken

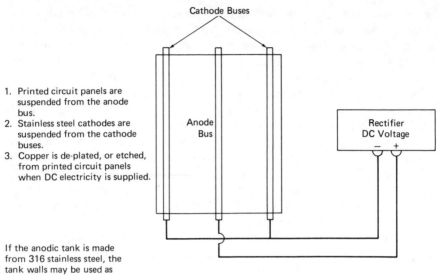

1. Printed circuit panels are suspended from the anode bus.
2. Stainless steel cathodes are suspended from the cathode buses.
3. Copper is de-plated, or etched, from printed circuit panels when DC electricity is supplied.

If the anodic tank is made from 316 stainless steel, the tank walls may be used as the cathodes.

Fig. 10-1. Bus bar layout for anodic electrocleaning.

up. The need for the soak tank, as well as the anodic tank, means that the panels must be handled in an extra operation. All this places a heavier demand upon the plater, increases the time for cleaning, and increases the opportunity for panels to become damaged.

The basic operating sequence for the three common cleaning methods used prior to plating are:

A. Pumice scrubbing (Fig. 10-2):

1. Wet panels.
2. Pumice scrub.
3. Spray to rinse pumice.
4. Soak cleaner (acid or neutral).
5. Spray rinse.
6. Immersion rinse.
7. Sulfuric acid soak.
   (On to plating).

B. Chemical microetch (Fig. 10-3):

1. Hot soak cleaner.
2. Spray rinse.
3. Immersion rinse.

4. Microetch.
5. Rinse (spray or immersion).
6. Sulfuric acid.
   (On to plating).

C. Electro-clean microetch (Fig. 10-4):

1. Hot soak cleaner.
2. Electrocleaner.
3. Spray rinse.
4. Immersion rinse.
5. Sulfuric acid.
   (On to plating).

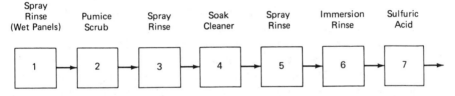

All three spray rinse operations may be performed using the same spray rinse tank.

Fig. 10-2. Steps for pumice cleaning of dry film imaged panels.

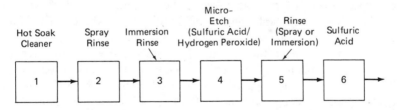

Fig. 10-3. Panels should have about 0.0001 inch of copper in the plated through holes to prevent voids.

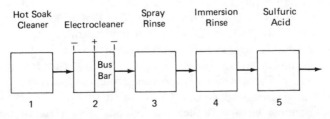

Fig. 10-4. Tank sequence for anodic electrocleaning.

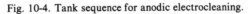

There is another alternative to these cleaning methods. Most of the companies which manufacture dry film photoresist also manufacture soak cleaners for removing fingerprints and oxide due to handling, as well as for etching the layer of adhesion promoter (Dupont's AC-500, and Thiokol/Dynachem's LAC-41). These cleaners must be made up in high concentration, 30–50%, for successful operation. When heated to 130–155 degrees Fahrenheit, they become sufficiently active to etch the copper surface of the dry film imaged panel. The etch rate is controlled by (1) immersion time, (2) temperature, and (3) concentration of cleaner. Each shop must experiment and arrive at their own satisfactory parameters for cleaner operation. Successful determination of etch rate parameters, coupled with proper training of plating personnel, allows the printed circuit shop to use dry film without having to use extraordinary through hole plating or cleaning techniques. This means that no special cleaning tanks, or hazardous etching chemicals, or extra handling steps are needed either.

The drawbacks encountered in relying upon these types of cleaners as the sole source of cleaning and copper etching are few, but important. Of relatively minor consideration is the fact that these cleaners are viscous and tenacious—in other words, they are extremely difficult to rinse. Some care must be used when rinsing them. The panels should be spray rinsed, one at a time, until sudsing has stopped, then immersion rinsed. Of course, this is only what should be done when rinsing any soak cleaner.

Of greater importance is the general care and attention which must be shown by the platers when operating with these cleaners. When a separate microetchant is being used, or when screen printing is being used for imaging, the acid soak cleaner serves to remove light oxide and oils, such as fingerprints. This alone is an important function being fulfilled by the cleaner. However, when these special cleaners are being used as a combination soak cleaner and microetchant to remove the dry film photoresist adhesion promoter, a new dimension of care is needed to reliably operate without voids or peelers.

The goal of the cleaner operation is to remove a consistent amount of copper: enough to prevent copper-copper peeling, but not enough to cause a voided hole. This may not be easy, and all personnel involved must understand these goals and never forget them. The factors of cleaner *concentration, temperature,* and *time* are critical for this operation, whereas they are only important for a normal cleaning operation. There is an operating window for these cleaners, and the challenge is to find and operate inside of that window. The reward for successfully operating within the window is that it will not be necessary to go to the trouble and expense of: (1) copper flashing with electroplated copper after through hole plating; (2) setting up the special operating line and analytical procedures for running high speed electroless copper, and having to deal with all the associated headaches; or (3) having to set up separate microetching tanks and chemical baths, with their required analysis and/or continual dumping and make-up.

The best way to go about determining the optimum parameters is to experiment. Set up the cleaner tank with a 50% make-up (or whatever the maximum recommended make up is of the manufacturer), at the medium to high side of operating temperature, and try cleaning test panels starting with the maximum recommended immersion time. After cleaning and subsequent plating, the holes must be inspected for voids, and the plating must be tape tested and/or dug at with the blade of an X-acto type razor knife. Once a successful set of operating conditions is arrived at, the boundaries of those conditions must be determined. Also, the cleaning bath should not be run as long as a traditional soak cleaning bath: it serves not only to clean, but also to etch; and it not only becomes depleted, it becomes loaded with copper. The optimum square footage of panel area should be determined, and square footage tracked via plating logbook, in very much the same manner as the logbook is used to track electroless copper throughput. The square footage log can easily be used to keep track of when it is time for the plater to make up a fresh bath.

It will be necessary to continually monitor the holes for voids and the plating for peelers—but this is a good idea for any plating operation which uses dry film photoresist. When the cleaner and etchant are combined into one cleaning bath, the need to watch for voids and test for peelers is absolutely critical; not to do so is very unwise.

MICRO-ETCHANTS

When separate micro-etchants are used, the panels are generally "flashed" with copper before imaging.

Ammonium Persulfate

Ammonium persulfate will do an excellent job of etching a copper surface. After a soak cleaner to remove light oils and oxide, the panel is rinsed then etched for one to three minutes by immersion in ammonium persulfate, followed by a rinse and immersion in sulfuric acid.

The ammonium persulfate is made up at 1 to 2 pounds per gallon. The printed circuit panel is immersed until the copper has become uniformly dull matte pink. Failure to achieve the uniform matte surface may result in peeling of the copper which will be plated. There is no set time for the immersion; the panel is simply immersed until the surface has become dull matte pink. The time will vary with freshness of the ammonium persulfate, concentration, temperature, and perhaps other factors more difficult to get a handle on, such as whether or not the panel had been stripped of dry film photoresist during the imaging process, whether or not the copper surface had been scrubbed prior to dry film lamination, and others.

Immersing the panel in sulfuric acid is required after ammonium persulfate etching; there is a copper/persulfate complex formed which should be removed from the surface for adequate cleaning of the panel. The acid is made up 10% to 20% by volume. Immersion time of a minute or two will prove sufficient; however, longer immersion time will not be detrimental.

Ammonium persulfate is a rapid and aggressive etchant. It should be made up with hot water to aid dissolving (hydration of ammonium persulfate is a very endothermic reaction which will drop the bath temperature into the 30's degrees Fahrenheit). The addition of 1% by volume sulfuric acid will aid the etching process by removing and preventing formation of oxide on the copper. Etching should not be attempted until the ammonium persulfate is completely dissolved; as there will be a tendency for pits to develop on the copper surface, and this must be avoided.

Another point to consider when using ammonium persulfate is that it forms a complex with the copper ions, thus posing a waste control problem. It may be cheapest to simply pump the spent bath into plastic or plastic lined drums and have it hauled away; or avoid using ammonium compounds altogether.

Sulfuric Acid/Hydrogen Peroxide

Sulfuric acid/hydrogen peroxide performs well as a microetch. This system offers several advantages over ammonium persulfate. The sulfuric/peroxide etchant operates continuously in a batch operation. The bath only requires dumping and make up if it becomes contaminated. Acid and peroxide are analyzed for and added on a daily basis. It operates with high copper loading capability. To reduce copper, the bath need only be cooled down overnight and cross filtered into a separate tank. Copper drops out as copper sulfate pentahydrate crystals—which can be saved and sold for copper/copper sulfate reclaiming. The etch rate is a little slower than ammonium persulfate, and a little easier to control. The steady state operation with virtually no generation of hazardous waste, or generation of complexed and difficult to remove copper, is a real plus many printed circuit manufacturers take advantage of. Also, since the bath is operated steady state, there is no need for down time while a new bath is made up.

The important things to remember when using dry film photoresist are that a thin layer of adhesion promotor must be removed prior to pattern plating. The options used to remove that layer affect how much copper must be deposited in the holes before imaging. Pumice scrubbing, and some etching soak cleaners allow the use of room temperature electroless coppers without copper flashing by electroplating. Anodic cleaning, ammonium persulfate, and sulfuric acid/hydrogen peroxide microetches require the use of an electroplated copper flash or high speed electroless copper capable of depositing about 50 to 100 microinches of copper in the holes prior to dry film lamination.

COMPATABILITY WITH OTHER PROCESSES

Some consideration must be given to the compatability of the photoresist to the processes in which it will be used: plating and etching. The aqueous and semi-aqueous resists are, to some extent, sensitive to pH, temperature and some processing chemicals. A resist which holds up in acid copper may not do so in pyro copper (pH 9.0+, temperature of 125 degrees Fahrenheit); and a resist which functions perfectly in pyro copper, may not hold up well in cyanide gold, tin-nickel (160 degrees Fahrenheit), or tin-lead with fluoboric acid over 60 ounces per gallon. High-throw tin-lead can present a problem to some aqueous resists if the fluoboric acid content rinses too much above 60 ounces per gallon; the resist becomes porous and will lift or break down. Cyanide gold is rough on several fully aqueous resists, which may break down. Cyanide silver, with a pH of 10 to 11, is another bath which is difficult for many resists to withstand. Fully aqueous resists generally break down quickly and will even strip from the panel surface; while semi-aqueous resists will hold up, but leach into the bath. In cases like this, the resist may be able to withstand the bath better than the bath can withstand the resist. It is common for high pH baths and baths which operate at elevated temperatures to leach components of the resist; the baths will even become discolored from leaching resists with their associated dyes. Some plating baths, especially those operating without organic brighteners, may tolerate the leaching and be capable of continuous operation with a carbon pack filter cartridge in the filter chamber. A simple carbon pack cartridge will remove leached resist with sufficient speed and completeness that even plating baths which are sensitive to the presence of organic agents, such as plating baths used to deposit metal for die attach bonding, are unaffected.

Almost any resist will, given sufficient time, leach appreciably into virtually all plating baths. It must be remembered that all plating bath manufacturers recommend periodic carbon treatment and/or filtration.

Generally the heartiest of all resists is the solvent developing dry film photoresist. Since pH and alkalinity are not factors in its developing or stripping operation, it is totally immune to their effects. The solvent developing resists are also the least susceptible to leaching in strong chemicals and elevated temperatures—but leaching does occur and will eventually be noticed in the plating quality of any plating bath. Although solvent developing resists hold up the best for cleaning, plating, and etching operations, they do have one chief drawback—the need for solvent for developing and stripping. Solvents (the chlorinated hydrocarbon varieties) are expensive, difficult to work with, and have a tendency to form scum during developing and stripping if these operations are not carried out in a precisely correct fashion. The requirement for solvent adds greatly to the expense of processing equipment and chemicals, as well as adding to the burden for air and water pollution control.

Most dry films stand up well to alkaline etchants. These etchants are operated at elevated temperatures and have a fairly high pH of 8 to 9, and sometimes greater if the chemistry is off slightly. As with all chemicals the order of greatest resistance to alkaline etchants will be solvent, semi-aqueous, and fully aqueous developing resists. If fully aqueous resist is being considered for an etching operation, it is advisable to obtain samples of all the resists brands for evaluation, and to discuss the application with the resist salespeople: not all aqueous dry films are equivalent in this application.

DRY FILM LAMINATION

As already mentioned, prior to dry film lamination (or coating) the decision must have been made whether or not to electroplate the panel after electroless copper. If no flashing is to be performed, the panels may be given an optional gentle scrub. A grit of 320 or finer should be used for rotating brushes of compressed fiber or bristle. Oscillating scrubbers, such as the Tru-Scrub, should use either the brass or stainless steel bristles. The scrub is optional as long as the electroless copper deposit is not powdery, and is adherent after a tape test. If the panels have been flashed with electroplated copper, a scrub with 320 grit for rotating brushes, and 240 grit for oscillating brushes is very much recommended for good mechanical bonding of the laminated resist.

Panels should always be baked shortly before resist lamination. Moisture will prevent good bonding and contribute to resist lifting or break down (plating beneath the resist and resist lifting at the edges). Aside from moisture removal, there is another reason for baking; a positive correlation exists between good resist adhesion and copper surface temperture at lamination. Photoresist bonds best to a warm copper surface.

The heat may be supplied by a hot shoe laminator (Fig. 10-5) or by a hot roller laminator (Fig. 10-6). The hot roller is a more efficient machine and does a good job of supplying heat to both the resist film and to the panel on which it is being laminated. The hot shoe is much less efficient at heating the panel, since it does not make contact with it. The hot roller laminator can be successfully used to laminate room temperature panels, while panels to be laminated with a hot shoe machine must be baked (175–195 degrees Fahrenheit) and laminated while they are hot. Only 6–12 panels at a time should be removed from the baking oven. The oven used for baking should be located near the hot shoe laminator to avoid cooling while being transported. If a delay of more than a minute or two occurs between lamination and the time panels are removed from the oven, they should be rebaked.

The conveyor speed on the laminator is also important. Panels being laminated while hot can always be laminated reliably at a faster rate than panels being laminated at room temperature. Inner layers of multilayers (thin clad laminate) can be laminated at a faster conveyor speed than should be used for thick, double sided panels.

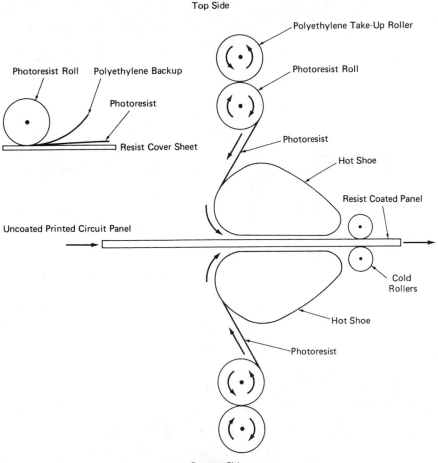

Fig. 10-5. Hot shoe heats the photoresist. Firm and intimate contact of resist and preheated panel is provided by the cold rollers.

As the panels or layers come out of the laminator they should be trimmed on all four edges with a razor knife, to remove excess and overhanging resist. Neat trimming of the four edges will prevent flakes of resist from breaking off and clinging to the panel surface, the artwork, or the contact printer vacuum frame. Lay the panels flat on a sturdy surface for trimming.

Laminated panels should be stacked no more than two inches high, and then only for short periods of time. As soon as possible, they should be stored vertically, along the longest edge of the panel. Photoresist is, after all, a soft, gelatinous material which can be deformed by pressure.

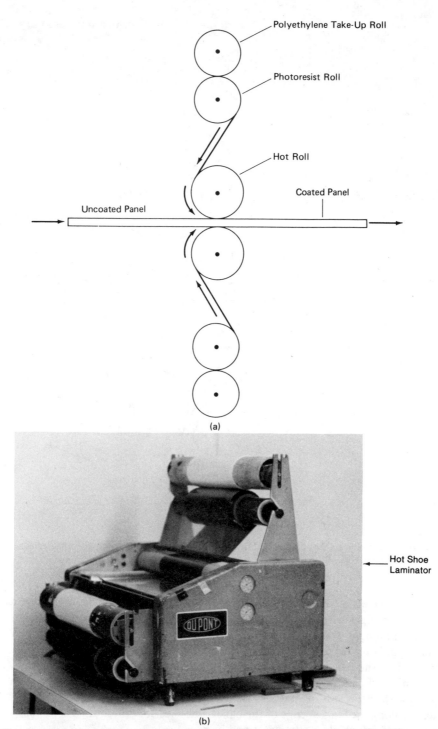

Polyethylene Take-Up Roll

Photoresist Roll

Hot Roll

Coated Panel

Uncoated Panel

(a)

Hot Shoe
Laminator

(b)

Fig. 10-6. (a) Hot rollers assure firm contact of photoresist and panel. (b and c) Hot roll laminators.

Hot Roll
Laminator

(c)
Fig. 10-6. (*Continued*)

If inner layers are being laminated, the tooling holes will have to be punched. The layers should already have been stamped to identify the ounces of copper. After lamination the layer drawing (multilayer construction diagram) should be consulted and layers stacked according to copper orientation: 1/1, 2/1, 2/2, 1/0, and 2/0. Once this has been done the punch can be set up for punching the inner layers. Do not punch the inner layers before they have been laminated. The dry film and cover sheet will interfere with insertion of the artwork tooling pins.

EXPOSURE

A lot of factors go into generating quality images with dry film photoresist. After the panels have been laminated with dry film, the next concern is to obtain a good exposure of that dry film using the artwork. Important elements of a good exposure are:

1. Cleanliness: of the artwork, the exposure source, the panels, and the working environment.

2. Determination of the proper exposure setting, and routine monitoring of the setting using a step tablet, as well as examination of the developed product.
3. Phototool quality: proper ultraviolet density, good dimensional stability and registration with the drilled panels, and good touch up of the circuit image.
4. Good control of the developing process: clean developing, proper rinsing, and control of developing time.

CLEANLINESS

The need for cleanliness is important in all areas of printed circuit imaging. The entire imaging process must be conducted in areas free of dirt, dust, lint, oils, and moisture. A dry film imaging area should have the following routine performed in it every day:

1. Vacuum-clean floors, tables, light tables, vacuum frame, and all other horizontal surfaces; it is a good idea to turn on an anti-static fan, also.
2. Wipe down the glass and polyester vacuum frame with glass cleaner and a lint-free rag, to remove dust and dry film flakes.
3. Wipe the reflectors of the ultraviolet exposure source, with glass cleaner and a lint-free rag.
4. Check the humidity and temperature. They should be within film manufacturer's recommended limits: around 70 degrees Fahrenheit, and 50% humidity.

It is a good idea to keep the anti-static fan running at all times during operation of the dry film area. The fan should be blowing over the vacuum frame and the light tables. The anti-static fan is a small fan which draws air over a radioactive source (shielded in plastic); then blows it over the work area. The radioactive source ionizes the air, placing an electrical charge on the air molecules. Dust and germs float around the room suspended in the air by an electrical charge. When the ionized air contacts the floating particles, which are charged opposite to the ionized air, their charge is neutralized and they fall from the air. Once their charge has been eliminated and they have been removed from the air the particles can be vacuumed up. The air ionizer will help to keep the air clean and the work environment dust free.

Once the vacuuming has been done, and the air ionizer is on, it is possible to begin working with a degree of confidence that the artwork and panels will not be contaminated.

THE PHOTOTOOL

The phototool is artwork which has been prepared for use exposing dry film photoresist. The term "phototool" can be used interchangeably with "artwork" and "working film." The importance of a quality phototool cannot be over-stated. With all else being optimum, the dry film image obtained on the panel can be no better than the phototool provided to expose the dry film. (For more information on artwork, see Chapter 8.)

The phototool must be kept clean, in a state of good touch-up, and stored in a temperature/humidity controlled environment. When used, it must be registered correctly to the panel, with good contact made between artwork and dry film coated panel during vacuum drawdown and exposure. The emulsion must be placed directly against the photoresist cover sheet for maximum resolution and acuteness. Some common problems related to phototools are:

1. Poor state of touch up, due to:

- Scratched emulsion.
- Dirt and foreign particles (photoresist flakes).
- Old tape, or tape residues, left on from date code, UL logo, or other markings.
- Registration system, and problems related to them.
- Red touch-up arrows left on.

2. Poor line acuteness, due to poor duplication of customer artwork because of:

- Failure to duplicate emulsion-to-emulsion.
- Improper diazo developing:

  - Developer not sufficiently heated.
  - Insufficient ammonia.

- Partial exposure of virgin film, accidently, from stray light.

3. Poor contact achieved during drawdown, due to air bleed line not being located near artwork.
4. Poor film dimensional stability.
5. Poor customer artwork:

- Tight air gaps, ragged traces, nicks, pinholes, protrusions.
- Poor front-to-back or layer-to-layer registration.
- Image ultraviolet density not consistent, and poor.

The correctness and suitability of artwork (from the customer) and the phototool should never be taken for granted: correctness and suitability must always and continuously be questioned.

Touch-Up

Touch-up is one of the ongoing considerations which should be considered on every exposure of a panel or inner layer. The first time the phototool is pulled from its envelope the dry film technician must examine it for:

- Part number.
- Revision level.
- UL logo and board type: multilayer, double sided, etc.
- UL flammability rating: 94V-0, 94V-1, etc.
- Date code.
- Federal supply code for manufacturers (for military printed circuits).

The presence of all these and other markings must be checked during the touch-up stage. The job traveler should be read, as well as the part drawing supplied by the customer. The date code will have to be changed every time the part number is run as a job. It is sometimes critical that markings be applied in a certain location; the dry film technician must verify that the markings have been applied in a specific area on the board, if called out by the customer on the drawing.

After correctness of markings has been checked the phototool must be examined for touch-up needs. It is poor practice to assume that the Photo Department, which generated the working film from the customer's artwork, made no mistakes or oversights. Process engineers, managers, supervisors, and technicians must be encouraged to double check everything before work begins on a job; and they must be held accountable for what work comes out of their departments. These people in turn have to train and instill vigilance and pride of workmanship into the personnel working in their areas.

The artwork should be spread over a light table for viewing and touch-up. Every square inch of the phototool must be examined for pinholes, scratches, and other items requiring touch-up. In checking for the presence of dry film photoresist flecks, the finger tips of an ungloved hand can be run over the film surface; this is a quick and reliable method to aid visual inspection. When resist flecks are removed, pieces of the emulsion may come off the artwork; this is especially the case where solvent developing dry film photoresist is being used. The dry film technician must be aware of the need for touch-up when removing the photoresist.

Touch-up of phototools is an art which must not be taken lightly. There are differences between the surfaces of silver and diazo films; generally, the silver is easier to touch up. The gelatinous silver emulsion can be removed easier than the harder, more tenaceous diazo. Because of this, the polyester base of the diazo phototool tends to become slightly gouged out when scraped with an X-acto type razor knife. When touch-up ink is applied, the ink flows into the gouges and may cause ink to be placed where none is desired. Many technicians will scrape unwanted emulsion from the emulsion side, where that emulsion is not desired; but apply ink only to the base side of the phototool. This prevents having ink flow into the gouged out areas.

The gelatinous silver emulsion is so easily removed with a razor knife, that there is no need to scrape the polyester base. Touch-up ink can usually be applied to the emulsion side without the problem of ink wicking into gouges.

When fine lines or tight spaces are being touched up it may be difficult to find a pen or brush fine enough to accurately apply ink. It is common for the pen tip to be wider than either the trace or the spaces between the traces. When this occurs, apply the least amount of ink available equipment will allow; then lightly and carefully scrape excess ink from unwanted areas with the razor knife blade. Since it is necessary to scrape away ink, there is always the possibility of accidentally removing emulsion where not intended. This is another reason to apply ink only to the base side of the artwork. Figure 10-7 graphically demonstrates how to touch up situations like this, and how to touch up fine line, tightly spaced circuitry.

There are no short cuts to proper touch-up; it demands patience. Although only a tiny area requires touch-up, technicians must be trained to realize that lack of touch-up, or improper or sloppy touch-up, will result in scrap circuits being manufactured. The ink pen and razor blade knife can be used to touch up nicked or broken circuitry, protrusions, pinholes, unwanted spots, and scratches.

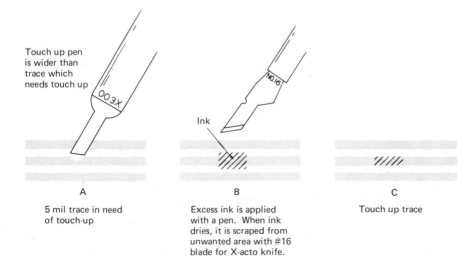

Touch up pen is wider than trace which needs touch up

Ink

A — 5 mil trace in need of touch-up

B — Excess ink is applied with a pen. When ink dries, it is scraped from unwanted area with #16 blade for X-acto knife.

C — Touch up trace

Fig. 10-7. (a) Touch-up pen is wider than circuit trace. (b) Excess ink is applied. When dry, excess ink removed with a No. 16 X-acto blade. (c) Touched up trace.

Poor Circuit Quality Due to Poor Quality of Customer's Artwork

There is little point in trying to make a good exposure from artwork of quality which is so poor that it will result in scrap. Ragged, nicked, and broken traces with protrusions and unreasonably tight air gaps are sometimes built into the artwork which is provided by the customer. The Planning and Photo departments should inspect incoming artwork for unacceptable conditions, and decide whether or not the artwork can be salvaged without an unreasonable amount of touch-up and photo manipulation. These departments should get back in touch with the customer as soon as they have determined that the artwork is unacceptably poor. Although this is what is supposed to happen, every dry film technician must be trained to understand acceptability criteria for artwork, and to be continuously alert for artwork which is going to cause problems. The dry film technician who takes for granted that all artwork which passes through Photo is suitable artwork, is making a mistake—a mistake which will, sooner or later, lead to processing and quality problems. Once artwork is accepted, the printed circuit manufacturer is liable for all scrap produced as a result of that artwork.

Artwork Registration

Poor artwork registration with the drilled panel is another problem common to phototools which result in an unacceptable exposure. Poor registration may be due to several factors, one or more of which may be in play in any given situation (Fig. 10-8):

1. Poor dimensional stability of the polyester base.
2. Circuitry growth (expansion) due to poor duplication procedure or equipment. This may be especially common if it was necessary to perform a choke

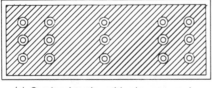

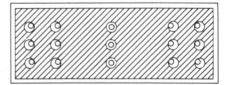

(a) Good registration with adequate annular ring (pad) around all drilled holes.

(b) Poor registration with inadequate annular ring. Registration demonstrated in b is called "splitting - the - difference," where misregistration is divided as equally as possible.

Fig. 10-8. Phototool registration with a drilled panel. (a) Good registration with adequate annular ring (pad) around all drilled holes. (b) Poor registration with inadequate annular ring. This type of registration is referred to as "splitting the difference" in misregistration across the entire panel.

or spread of the circuitry to eliminate an air gap problem with the customer's film.

3. Poor drilling or drill programming:

● The drilling equipment may have moved out of calibration or away from the zero point.
● Drill splay may have been excessive; this is the tendency of a drill bit to drill at an angle.
● The programming may have been inaccurate; or, more commonly, the customer may have sent in artwork produced from a hand tape-up, and then required the printed circuit manufacturer to program and drill to grid. If the hole pads are not located on grid, then the drilled panel and the artwork will automatically misregister.
● The front-to-back registration of the customer's artwork may have been off. Even though the job was programmed per the artwork, only one side is going to register perfectly, the side to which programming was performed.

4. Poor photo-reduction from designer's tape-up.

There are many reasons for misregistration of artwork and the drilled panel. The dry film technician must not proceed with panel dry film exposure until either the cause for the misregistration has been determined and corrected, or the customer notified and given a chance to accept or reject the registration, or the quality department has been notified and buys off the condition.

Photo Density of the Artwork

The dry film technician must watch for yellow or light edges on the circuitry of diazo phototools; and watch for grayness or low opacity of the circuitry of silver phototools. Circuitry of diazo phototools must be of the same uniform color and density, whether on the edge of a trace or in the center of a land area (Fig. 10-9). If silver phototools have noticeably low opacity, or if diazo circuitry is not uniformly colored and dense, scumming of the photoresist will occur; that portion of the photoresist which is to remain unexposed, and be cleanly removed during development, will become partially polymerized.

After developing, chunks of gelatinous photoresist will remain clinging to sidewalls and copper surfaces (Fig. 10-10). This may not be noticed by the naked eye or even under magnification; however, it will be apparent after plating or etching. There are other causes of scumming, but this is a common reason which is often overlooked. One technique for determining the presence of scum, which will not damage the panel or resist for subsequent use, is to apply an etchant such as ferric chloride with a cotton swab or Q-tip. The etchant

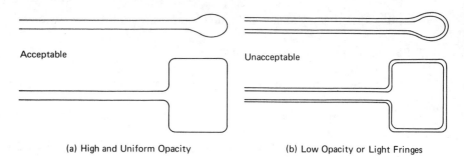

(a) High and Uniform Opacity   (b) Low Opacity or Light Fringes

Fig. 10-9. (a) High quality phototool image. Uniformly high opacity. (b) Poor quality phototool image. Low opacity or light fringes.

darkens the copper wherever it comes into contact with the copper. The slightly etched surface is then viewed under magnification of $25\times$ to $100\times$. Bright, unetched copper along walls of photoresist indicates the presence of scum, which prevented etching or staining of the copper by the etchant (Fig. 10-11).

Scumming along sidewalls is not to be confused with a normal "foot" of photoresist which occurs at the copper surface along the photoresist walls (Fig. 10-12). If gross scumming is detected, do not try to redevelop the panel; it must be stripped of all photoresist, cleaned, and reprocessed like a fresh panel. A phototool which causes scumming, or which has uneven color density (diazo) or is gray, translucent (silver) must be discarded and replaced by a new phototool. The new one should be shot from the best available copy of the artwork. If the new phototool exhibits similar undesirable characteristics, the photo technician should examine the exposure setting on the equipment used to shoot the artwork, and examine the quality of the virgin film stock.

If scumming is detected, and the phototool appears to be of high quality, the scumming may have been caused by either accidental exposure of the exposed panel to stray light prior to developing, or by poor contact of the phototool with

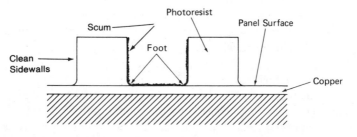

Fig. 10-10. Cross section of photoresist sidewalls: excessive foot and scumming due to poor phototool density.

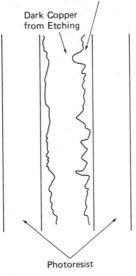

Partially polymerized photoresist
forming excessive resist foot and
scum. The copper here is bright
and unetched.

Dark Copper
from Etching

Photoresist

Fig. 10-11. Etch is applied with a cotton swab to a developed panel (ferric chloride works well). Circuitry is then viewed under 100× magnification. Bright (unetched) copper extending from sidewall indicates presence of resist scum.

the photoresist cover sheet during vacuum drawdown and exposure. Another possibility is that the exposure was made by positioning the phototool emulsion up, away from the photoresist, with the polyester base next to the cover sheet (Fig. 10-13).

The polyester base of a silver or diazo phototool is 0.007 inches (7 mils) thick. By orienting emulsion incorrectly, as shown in Fig. 10-14, an extra 7 mils of distance has been added between the artwork image and the dry film photoresist. This will decrease the ability of the dry film imaging technique to hold close tolerances and maintain sharp circuit acuteness.

EXPOSURE

The process of determining correct exposure setting, and of maintaining the exposure within acceptable limits, is achieved only by following all the information presented thus far, as well as all the information presented in the section entitled Determining Correct Exposure Setting. Almost every aspect of dry film imaging has an effect on exposure. Dry film technicians must be trained to understand why this is, and to process panels and inner layers

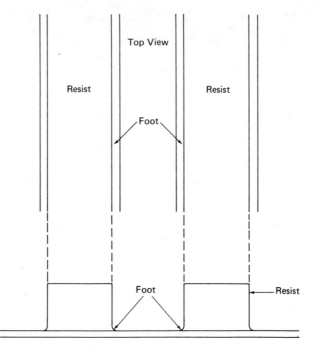

Top View

Resist   Resist

Foot

Foot

Resist

Cross-Sectional View

Fig. 10-12. An image with accurate exposure and good development will exhibit a slight, well defined resist foot at the copper/sidewall junction.

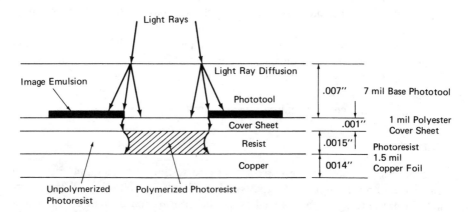

Light Rays

Image Emulsion

Light Ray Diffusion

Phototool   .007"   7 mil Base Phototool

Cover Sheet   .001"   1 mil Polyester
                                          Cover Sheet
Resist   .0015"   Photoresist
                                          1.5 mil
Copper   0014"   Copper Foil

Unpolymerized   Polymerized Photoresist
Photoresist

Fig. 10-13. Light travels through (1) phototool, (2) cover sheet, and (3) photoresist. Sharpest image is obtained when phototool emulsion is directly against photoresist cover sheet: light diffusion through resist is minimized.

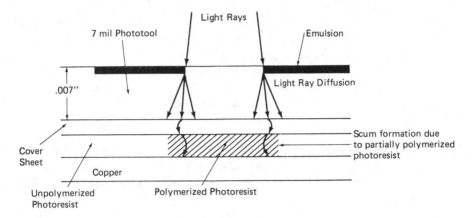

Fig. 10-14. When phototool is laid with emulsion up (and away from cover sheet), light rays are diffused before reaching photoresist. Diffused light causes polymerization to occur in a more random manner: the result may be scum formation due to partial polymerization.

through imaging with this in mind at all times. However, this section is concerned with determining the correct degree of photoresist polymerization needed to achieve accurate reproduction of the artwork on the panel surface. The first step in determining the proper exposure setting is to turn on the ultraviolet source, and let the mercury-halide lamp warm up for 10–15 minutes.

How Dry Film Photoresist Works

Photoresist is a combination of one or more monomers, free radical photoinitiators, plasticizers, dye, adhesion promoters, and a methacrylic (or other) binder which holds it all together. When a layer of photoresist (0.0015 inch thick) is applied to the surface of a copper panel, the ingredients are uniformly dispersed throughout the binder polymer, the binder serving as a framework to form the film. When this resist film is exposed to ultraviolet radiation (330–400 nanometers wavelength) the photoinitiators generate free radicals. A free radical chain reaction (polymerization reaction) is set off. The free radicals join with the monomers to form a larger free radical. This new free radical joins with still other monomers to form still larger free radicals. These larger free radicals join with each other to form a crosslinking network of photopolymerizing free radicals, all wound around the binder polymer.

While the photoresist is exposed to ultraviolet light, the photoinitiators are being ignited. The free radical crosslinking induced by the ultraviolet light continues for several minutes after the photoresist has been removed from the presence of the ultraviolet. When the free radicals have extinguished themselves, most of the available monomer in the light path will have become polymerized into the tightly bound network dispersed throughout the light path and the binder matrix. What had formerly been a soft, gelatinous film, easily washed away by the developing solvent, has now become a hard, durable, and chemically resistant surface, suitable as a plating or etch resist.

The quality of the developed image depends on holding the panel for a sufficient amount of time, to allow the polymerization reaction to complete itself. Also, the free radical polymerization reaction is extremely heat sensitive. This can be demonstrated by performing the following experiments:

1. Expose three panels under identical circumstances (a feat not likely to occur in a manufacturing environment because of thermal considerations) with a Riston 17 or Stouffer 21 step tablet and a resolution pattern.
2. Develop one panel immediately after exposure.
3. Place another in an oven at 150 degrees Fahrenheit for 15 minutes, then develop.
4. Let the third panel sit for 30 minutes at the 70 degrees Fahrenheit room temperature of the dry film imaging room; then develop.

The three panels will exhibit great differences in circuit reproduction acuteness and degree of polymerization. Panel number one will exhibit the least degree of polymerization, since the free radical reaction had not been allowed to complete itself. Panel number two, the one baked in the oven, will exhibit the greatest degree of polymerization, with possible loss of acuity due to apparent overexposure. Panel three, the one held for 30 minutes at room temperature, will exhibit the highest quality side walls, circuit definition, and overall acuity with the artwork.

The Step Tablet

The easiest method for tracking degree of polymerization is to use the step tablet, or step wedge. The Stouffer 21 step and the Dupont Riston 17 step tablets are the two commonly used tablets for exposure control (Fig. 10-15). The step tablet contains a series of consecutively denser steps of silver emulsion. The phototool is placed over a photoresist coated panel, while the step tablet is placed between the phototool and the dry film cover sheet in a clear area of the phototool outside the circuit pattern (Fig. 10-16). When the panel is exposed to a given amount of ultraviolet energy the photoresist beneath each step of the tablet receives a different amount of energy, and achieves a different degree of polymerization.

After exposure, the panels should be set aside for 15–30 minutes before developing, to complete the free radical polymerization process. The holding time before developing will yield more reliable exposure information. Once development has been completed, the area of the step tablets is examined. On

Fig. 10-15. The two common step wedges are the Stouffer 21 and the Riston 17.

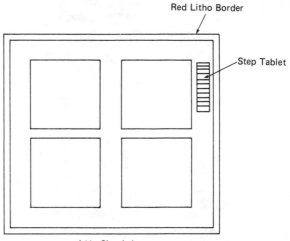

Fig. 10-16. The step tablet should be inserted between the phototool and the resist cover sheet.

the step tablets, the clearer the step, the more polymerization will occur; the denser the step, the less polymerization will occur. Each step is numbered, 1 being the clearest step for both the Stouffer and the Riston tablets. Thus, after developing, the number of each polymerized step is recorded on the test panel. The higher the step number polymerized, the greater the energy used to expose the panel. If one panel shows a 6 step, and another a 9, the 9 step panel achieved a greater degree of polymerization.

The next task is to correlate degree of polymerization, as measured with the step tablet, to optimum circuit acuteness and desirable sidewall characteristics. This task has been performed by the manufacturers of dry film photoresist already. Each manufacturer cites an optimum step range for a given step tablet. For instance, a given film may yield excellent results with steps 6–9 (Stouffer 21) being displayed as the highest polymerized number. (*Note:* Any time 50% or more of a step is left after development, that step is considered present and polymerized. See Fig. 10-17). This is true even if that step is little more than a gummy residue.

The step tablet must always be placed beneath the phototool, since ultraviolet light must also pass through the phototool on its way to polymerizing the photoresist. The D_{min} (phototool density in the clear area) frequently are not the same from one phototool to another; and there are several reasons for this: age of virgin film stock, age of phototool, amount of stray light impinging on film prior to development, degree of exposure, degree of development, thickness of base polyester, type of emulsion, and other reasons. Because of these D_{min} variations, the step tablet must be located beneath the phototool for exposure testing to have any degree of validity. There is also the possibility of varying

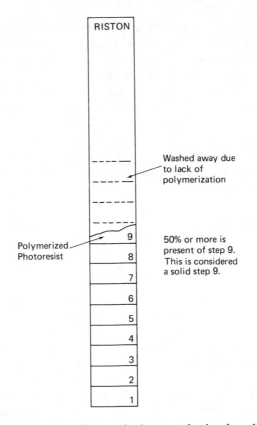

Fig. 10-17. Appearance of step wedge image on developed panel.

amount of light being reflected, depending upon whether or not the step tablet is located on, or beneath, the phototool.

For a dry film photoresist with an exposure range of steps 6–9 (Stouffer 21), what is the effect of obtaining a step 5 or a step 10? In other words, what is the effect on image quality of under- and overexposing the dry film photoresist? Generally, the width of a trace (positive image phototool) will be wider for underexposure than it will be for a fuller exposure. Conversely, a trace on an overexposed panel will be narrower than that on a panel which received less exposure. Underexposing will cause poorer sidewall straightness, and finer traces may not be sufficiently polymerized to withstand the developing operation. Overexposed circuits will tend to lose the finer resolution ability. The exposure step ranges chosen by dry film photoresist manufacturers reflects the balance between achieving sufficient polymerization, and not receiving enough. The ability of the dry film to resolve and hold fine circuitry, and to withstand the chemicals which the resist will see, are the guiding factors.

A resolution target (such as the Stouffer pattern shown in Fig. 10-22 later in this chapter) will enable the dry film technician to perform his/her own resolution testing. When a step tablet is "shot" and developed, the step achieved indicates whether more exposure time is required, or less time to be within the range recommended by the dry film photoresist manufacturer, or whether any change in exposure setting is required at all.

If both sides of a panel are to be exposed simultaneously, two step tablets should be used, one for each side. The tablet used for the underside is usually notched in a corner; this will enable the technician to tell which side of the panel is top and which bottom after development. Depending on the step range visible in the developed panel, the lamps·for either side can be adjusted for more or less time.

Tracking Exposure

As more panels are exposed, the vacuum frame of the contact printer becomes warmer. With continuous exposures (a typical production situation) the glass will become very warm. Heating the vacuum frame glass causes increased polymerization, with no increase in time setting. To maintain a given step reading, it may be necessary to reduce the exposure time. The need to change exposure setting can easily be determined by "shooting" a step tablet at regular intervals of either time or panel quantity.

All that is necessary to easily mold this routine into production is to shoot steps on a production panel, and set it aside for 15 to 30 minutes, then develop it, and look at the result. If upper and lower sides are within recommended range, no adjustment is required. If one or both sides is out of range, simply increase or decrease exposure setting for the side, and shoot another step on a production panel. If no panels have been shot for a long time, like an hour or more, it is always a good idea to shoot step tablets to note any change in the degree of polymerization as the vacuum frame has cooled down.

When the exposure source has just been turned on, let it warm up for 10–15 minutes. After the warmup period, run several exposures with no panels in the frame. This will remove the chill from the cool, unused frame.

Note: a light integrator or millijoule meter is not sufficient for determining and tracking exposure; since they do not account for heat.

Exposure Log

It is always good practice to keep a log, such as that of Figure 10-18. As steps are shot they can be recorded and tracked. This type of recordkeeping encourages dry film technicians to monitor exposure continuously; this is critical, because the time setting will almost always need to be changed throughout the day. Use the log to set exposure for the first step tablets shot each day.

Date	Time	Customer	Part No.	Job No.	Time Upper	Lower	Steps Upper	Lower	Lamp Hours
6-14	0800	XYZ	5220-1	7401	14	15	7	7	115
6·14	0930	DBH	3742-B	7414	14	15	6	7	116
6·14	10:00	DBH	3742-B	7414	16	15	7	7	117

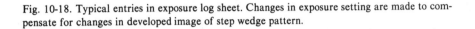

Fig. 10-18. Typical entries in exposure log sheet. Changes in exposure setting are made to compensate for changes in developed image of step wedge pattern.

When the machine is first turned on, and the dry film technician is about to shoot a panel to check exposure setting, the log is invaluable. Since there are so many factors which affect the degree of polymerization, the log can help get a handle on them. The log will help point to the first exposure setting of the day, and allow the technician to keep track of polymerization throughout the day. But there is also one more critical bit of information which can be tracked quantitatively with a log: the aging of the ultraviolet lamps. The lamp hours can be tracked and correlated with exposure setting.

Why Track Lamp Hours?

The spectral output of mercury halide lamps (wavelength of emitted light) changes as the lamp ages. Only the spectral light emitted in the 330–400 nanometer range is useful for effective dry film photoresist exposure. As the lamp ages and the wavelength shifts out of this range it will be necessary to increase exposure setting to compensate. This will be reflected as a gradual upward creeping in exposure times as the days and weeks pass. The lamp which yields a step 8 after 14 seconds when the lamp is new may require 45 seconds to achieve a step 8 when the lamp has aged for 400 hours. The 45 second exposure is going to heat the vacuum frame, and the panels being exposed in it. The heat will cause the resist polymerization to be more random, and less likely to proceed along the carefully defined paths outlined by the artwork.

As the hours mount on the ultraviolet lamps, it is good practice to monitor more critically the quality of the dry film photoresist image; scum formation, growth of the photoresist foot, and loss of fine resolving power of the resist are some of the areas to pay special attention to. It is possible to have excellent control over all other factors in dry film photoresist imaging, only to lose image quality because of an over-aged ultraviolet lamp. Once the number of hours of lamp life recommended by the lamp manufacturer has been reached, it is good practice to replace the lamp with a fresh one. Whenever the lamp is changed, a note should be made in the log, together with the lamp hours and date.

Note: lamp aging will also cause poor quality diazo phototools to be produced; if not monitored.

The Vacuum Drawdown

In order that a good exposure be made, intimate contact must be achieved between the phototool and the photoresist cover sheet. As discussed earlier, photoresist must always be exposed with artwork emulsion as near to the cover sheet as possible. Actually, an even better contact could be made by removing the cover sheet just prior to exposure, and placing the phototool emulsion directly against the photoresist. There are two reasons for not doing this:

1. The photoresist would adhere at least partially to the phototool emulsion; the touch-up required for multiple exposures would be prohibitive. For a special case, where resolution happened to be absolutely critical, this technique would be feasible.

2. With the cover sheet removed, oxygen would diffuse into the photoresist. Oxygen is a free radical inhibitor and would retard photopolymerization; potentially, the exposure time could change and become unpredictable and erratic in localized areas, such as tight air gaps and fine traces.

The vacuum frame is commonly used to obtain the close contact required. The "drawdown" is the process of pulling a vacuum, eliminating air and bubbles between the mylar cover and the glass of the frame, panel and phototool (Fig. 10-19). The vacuum frame is a glass drawer with a clear Mylar or polyester cover for a lid. The glass frame has a rubber liner which mates with the lid when it is closed. When the lid closes, a vacuum pump turns on and draws the air out through evacuation holes. One end of a cord (called the bleeder cord, bleed line, or air bleeder) is placed next to the phototool and panel while the other end is placed next to the air evacuation holes. As the vacuum pulls the cover tight against the panel and phototool, air is bled from beneath the phototool and intimate contact is achieved.

Drawdown is considered complete when the vacuum meter reads 26+ inches of mercury, and Newton rings (swirling, rainbowlike color patterns) form over the entire surface of the phototool and panel. At this time, the entire vacuum frame drawer is shoved into the exposure source. The drawer is pushed in with steady, even pressure. When the EXPOSE button is pressed, the shutters of the lamps are opened for the pre-set duration of time. There are several points which the dry film technician must always keep in mind:

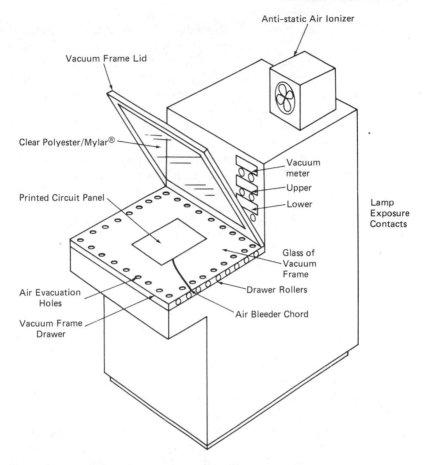

Fig. 10-19. (1) The lid is closed and a vacuum drawn. (2) Polyester sheet is drawn tight around panel, as air is withdrawn. Intimate contact forms between phototool and resist cover sheet. (3) Drawer of the vacuum frame is shoved into the UV source for exposure. (4) Exposure is provided by shuttered lamps in upper and lower reaches of machine.

1. Artwork registration must be checked prior to, and after drawdown. If the phototool has shifted, break the vacuum, open the frame, and reposition the phototool.

2. Glass and polyester must be clean for every exposure.

3. Bleeder cord must always be positioned next to the phototool and panel, with the other end at the evacuation holes.

4. When pulling the vacuum:

- The polyester lid must be capable of remaining closed by vacuum pressure alone. If this is not the case, no intimate phototool/panel contact is possible.
- If vacuum fails to reach its normal drawdown value, check for air leaks.

- Newton rings must be present and uniformly distributed across the phototool surface before the drawer is shoved in for exposure. Insufficient Newton rings indicate lack of intimate phototool/panel contact.
- Vacuum must remain intact while the exposure is underway. If it is broken, even for a second, the resist must be stripped from the panel, and the panel cleaned for reprocessing.
- There must be no obstructions between the circuit pattern portion of the phototool and the panel surface. Anything which will raise the artwork off the panel surface will cause unacceptably poor reproduction of the circuit image. (*Note:* The step tablet and border tape, which are outside the imaging pattern on the phototool, will not cause a problem.)

Just about any discrepancy in the drawdown will cause an unacceptable exposure.

DEVELOPING

The developing operation entails washing away unexposed resist, immediately followed by removal of the developing solution by water rinse. Each of the three types of photoresist (solvent developing, semi-aqueous developing, and aqueous developing) has different considerations for the two steps of developing. Resist developing must be carried out with the same attention to detail presented in each of the previous sections. It is the completion of the exposure process; and none of the other criteria discussed so far can be properly evaluated without excellent quality in the developing process.

Developing may be carried out in immersion/batch type equipment, non-conveyorized spray chambers, or in conveyorized spray equipment. For consistancy, ease of control, and maximum output, the conveyorized spray type developer is generally best. All of these types of developers can be successfully used, however; and all can be used for solvent developing, semi-aqueous developing, and totally aqueous developing types of photoresists.

SOLVENT DEVELOPING PHOTORESIST

This type of resist was the first developed for printed circuit imaging. The solvent used may be either 1,1,1-trichloroethylene, or chlorothene. These solvents are quite volatile, and this presents a number of problems which will be discussed.

All viable solvent developing systems (batch immersion, non-conveyorized spray, and conveyorized spray) utilize multiple developing chambers, through which the panels progress; each chamber being less contaminated with dissolved resist than the preceeding. The final step is always a water spray rinse. The goal of multiple solvent chambers is to obtain a panel for final rinsing which has as little resist residue as possible. Should resist residue bearing solvent evaporate prior to water rinse, a thin resist scum will be left on the panel surface. The scum will inevitably cause plating peelers and unacceptable circuit quality. Once the scum has been deposited on the copper surface, no amount of pre-plate cleaning or etching will remove it.

It is critical that the final solvent chamber be visually pure; no resist coloration, no cloudy or particulate matter, and no water floating on the surface. It is just as critical that each panel be water rinsed (by high pressure spray) immediately after it leaves the last solvent chamber. The resist scum, or residue, will not dissolve in the rinse water, which is why the final solvent must be clean and removed immediately.

Non-Conveyorized Developing

Fig. 10-20 shows the chamber layout for a typical non-conveyorized developing unit. The diagram shows what the developer would look like if the front were glass, and therefore transparent.

Chamber 1. This is the primary developing chamber, it is also the only solvent spray chanber. 10–15 gallons of solvent lie in the sump. There is a lid which fits over the top of the chamber. The lid has clips which hang from it, and into which the printed circuit panel is clipped. When the lid is in place, the panel hangs between the two banks of spray nozzles. The spray pump is actuated by a timer/controller which can be set to any desired time interval. At the conclusion of the preset spray time interval, the pump is shut off. This

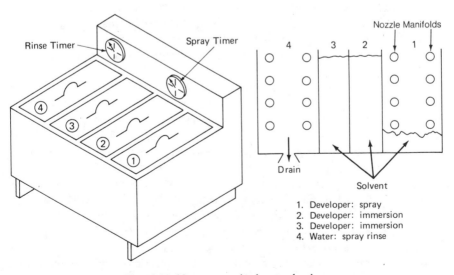

1. Developer: spray
2. Developer: immersion
3. Developer: immersion
4. Water: spray rinse

Fig. 10-20. Non-conveyorized spray developer.

is where the actual developing-off of the photoresist takes place (removal of unexposed photoresist). The next two solvent immersion chambers function to remove the contaminated solvent from the panel to eliminate the possibility of scum formation.

Chamber 2. After the panel has been allowed to drain briefly, it is removed from the spray chamber and immersed in chamber 2. This is a chamber which is filled almost to the brim with developing solvent. The panel, which is still attached to the lid of chamber 1, should be stroked up and down four or five times. The stroke length must be almost the same length as the panel. At the completion of the last stroke, the panel should be moved immediately to chamber 3. Do not bother to drain the panel.

Chamber 3. In this chamber the immersion rinse/stroking procedure is repeated exactly as it was in Chamber 2. This chamber is also filled almost to the brim with developer solvent. The purpose of these two solvent rinsing chambers and the up-and-down stroking motion is to flush the contaminated solvent from the surface and from between the finest traces, of the developed circuitry. Failure in this flushing operation can be devastating to the developing process. It is important that the panel not be drained after this last immersion rinse: but, go immediately to chamber 4.

Chamber 4. This chamber is similar to chamber 1, as far as the layout of spray nozzles. However, no solvent is sprayed. High pressure tap water is sprayed onto both sides of the panel to blow off the solvent. The critical step is to get the panel into this chamber fast enough to prevent solvent from evaporating, in order to prevent the possible deposition of photoresist scum. There is no sump of water; the spray rinse goes immediately into the drain. 15–30 seconds is adequate water spray rinse time. At the completion of this spray rinse, the panel is removed from the clips in the lid. The panels may be set in a rack to dry, or passed through an air knife type of drier.

There are four points of good practice for solvent developing by this method which should never be neglected:

1. After the initial developing operation performed in solvent spray chamber 1, the panel must be allowed to drain. The solvent in the spray chamber is heavily loaded with dry film photoresist, even after developing only one panel.

2. Vigorous stroking in each of the immersion solvent chambers is critical. The stroke length must be at least 12 inches—longer for panels which are more than 12 inches long. There is simply no other way to guarantee removal of resist residue from the small developed-out reaches of the panel. A shorter stroke length, 2–4 inches for example, will not prove sufficient in most cases, especially if fine lines and tight spaces are present in the circuitry.

3. Rinsing is an integral part of the developing process. Its purpose is to remove resist bearing solvent from the copper and circuitry before that solvent has a chance to dry. The rinsing operation for solvent developing photoresist is very different from most rinsing operations involved with printed circuit man-

ufacturing; neither the solvent nor the resist residue in that solvent are soluble in water. This is a significant fact. If the water is not (1) sprayed on, (2) immediately, and (3) with high pressure and fine water particle size, the developing operation may not be successful; the likelihood of scum forming is great. The scum may not be near great enough to see, even with magnification, but it may be enough to cause a copper peeler, or loss of circuit definition in a critical location.

4. Solvent purity is very important to good developing. Part of the developing includes panel progression through a series of succeedingly cleaner solvent chambers. This system of multiple chambers is necessary because neither the resist, not the solvent, are soluble in water. Trichloroethylene does not have the high loading capacity typical of aqueous and semi-aqueous developers. The 10–15 gallons of solvent in the spray sump of chamber 1 must be changed every 20–50 panels (for a plating pattern) depending on the thickness of photoresist being developed and the size of the panels. An etching pattern for print-and-etch applications such as inner layers of multilayer panels may require the solvent to be changed more frequently, perhaps twice as often.

There is another type of solvent contaminant which must not be overlooked during processing: water. Water forms on the solvent surface by moisture condensation. As the solvent evaporates, it cools everything which it contacts, including itself. Moisture from the air condenses on the spray pipes, chamber walls, and even on the solvent itself. This condensed moisture forms a puddle of water on the surface of the solvent. The solvent, pipes, and chamber wall will actually become very cold during processing. The water which forms should be skimmed from the surface, and wiped from the inside of the developing equipment periodically. This water forms in the immersion tanks as well, and must be removed from them too. The reason that water removal is so important is that water causes scumming, which greatly interferes with photoresist development.

When it becomes necessary to change the solvent in chamber 1, the spent solvent is pumped out of the chamber. It can be pumped into a spent solvent drum, to be carted off for solvent recovery, or it can be pumped directly into a solvent still for immediate reclamation. The economics of operating with solvent developing photoresist will probably dictate that a solvent still be installed along with the other resist processing equipment.

There is a filter cartridge in series with the spray pump; this filter must be emptied, and the filter cartridge replace every time the solvent is changed. The filter contains solvent which is contaminated with resist and with moisture. The cartridge will actually clog up with gelatinous resist scum from loading and moisture if it is neglected. Also, as the filter cartridge clogs, the spray pressure will drop, all of which contribute to a really poor developing operation. Once the tank has been emptied it must be wiped dry of solvent and water with a dry rag. This is the only way to be assured of removing all scum/resist residues.

When chamber 1 is ready to refill, the solvent should be taken from chamber 2, while chamber 2 should be replenished from chamber 3. Fresh solvent, or freshly distilled solvent, should only be added to chamber 3, which is the final solvent chamber and must be the cleanest. The levels of chambers 2 and 3 will be low, due to replenishment of chamber 1. While they are low, a ring of scum may be noticed just above the solvent line. This ring must be wiped from the tank walls. Changing the solvent, and replenishing in this manner will result in optimally pure solvent and economy.

Prior to beginning operation after changing the spray solvent in chamber 1, the spray nozzles must be checked for obstructions which will impede optimum spray volume and pattern. With the tank lid off, simply jog the spray pump. It need only operate long enough to demonstrate that all nozzles are free and functioning properly. From time to time, particles of dirt and resist and tiny pieces of cover sheet get into the tank and the spray nozzles. They are a regular nuisance, and it is rare that an entire batch of solvent can be operated without having to unclog one or more nozzles. They must be removed and the swirl vane taken out. It may be helpful to poke debris from the nozzle with a paper clip point, or other sharply pointed instrument. Failure to keep the nozzles clean will result in spotty and unacceptable developing.

Periodically throughout the day, check the level of solvent in each of the three solvent tanks. The solvent evaporates rapidly, and is dragged out of the tank with every developed panel. If, for example, one-third of the solvent in a tank has evaporated, the contamination level in that tank has increased 50%, and the lower operating level may prevent correct stroke length during rinse operations in chambers 2 and 3. The solvent levels must be raised to optimum every time the dry film technician begins resist developing.

The solvent used for developing can cause health problems, especially after prolonged exposure. It is a chlorinated hydrocarbon and can cause headache and nausea upon even brief exposure. Long-term exposure may have even more serious complications. The minimum precaution must be to provide ventilation to each of the chambers. The air volume for ventilation must be sufficient to prevent fumes from escaping to the work environment. It might also be advisable to provide and insist upon the use of gas masks fitted with the correct cannisters for removing chlorinatd hydrocarbons. As long as the lids for each solvent chamber are on, and sealed tight, it will only be necessary to turn on vent fans during operation of the developing equipment. If the chamber covers do not fit air tight, it will be necessary to keep the vent fan on continuously, which will contribute greatly to solvent evaporation.

Conveyorized Developing Equipment

Fig. 10-21 shows the chamber layout for a typical conveyorized developer. There are three solvent spray chambers, and one spray rinse chamber. Addi-

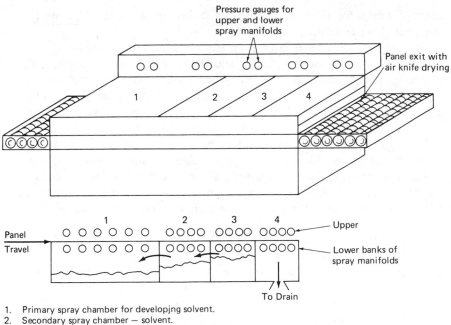

1. Primary spray chamber for developing solvent.
2. Secondary spray chamber — solvent.
3. Tertiary spray chamber — solvent (cleanest solvent).
4. Water rinse chamber.

Fig. 10-21. Conveyorized spray developer.

tionally, mounted across the conveyor after the water rinse chamber may or may not be an air knife for drying the panels after they exit the developing process. The reason for partitioning solvent spray into three chambers remains the same: to provide successively cleaner solvent for spraying on the panel surface, leading up to the water rinse and subsequent removal of all solvent.

Non-conveyorized spray developing equipment requires regular halts to production while solvent is changed after achieving maximum loading with washed off resist; conveyorized equipment requires no such production halt. There is a continual bleed-and-feed of the solvent; and except for water buildup on the solvent surface, contamination level (resist loading) remains constant. The most heavily loaded solvent is removed from chamber 1 (bleed), while the developing system is replenished (feed) through solvent addition to chamber 3. As solvent is added to chamber 3, it cascades into chamber 2; and from chamber 2 it cascades into chamber 1. The tendency for water to build up in any solvent developing operation is not quite the problem it is in non-conveyorized equipment. The panel never has to be submerged in the solvent; consequently, it does not have to pass through the layer of water which has condensed on the solvent surface. The possibility of scumming from contact with water has been reduced, except for water which has been sucked up by the spray pumps from the layer of water floating on that surface.

Conveyorized developing equipment will yield greater production throughput, with less down time and greater consistency than any other method. This is true no matter which resist system is being used: solvent, semi-aqueous, or aqueous. It is easier to ventilate and prevent fumes from escaping conveyorized equipment, since the chambers are covered at all times and there is no need to remove the covers during normal developing operation.

Conveyorized equipment does require periodic filter changes, and this can be monitored by observing the spray pressure gauges for significant decreases in manifold pressure. There are usually three sets of manifolds, one for each chamber. However, since each spray chamber may require more than one set of manifolds, each chamber may contain more than one pump, and more than one filter.

Since more spraying time elapses per solvent change and shut down, there is increased opportunity for: (1) formation of water condensation, (2) solvent evaporation, and (3) nozzle obstructions and clogging. Even though solvent cascades and eventually ends up in chamber 1, each chamber will develop a layer of water on the solvent surface. The water layer will build to an unacceptable level, however, only in chamber 1. This chamber must be drained before the water level reaches the point where it can be sucked up by the pumps and sprayed onto the developing panels. Dry film technicians must monitor water level in chamber 1, and be aware of the danger of water displacing solvent. The chamber may look to be full of solvent, but most of that volume can actually be water.

The problems of nozzle partial or complete clogging are accentuated in conveyorized equipment. There are more manifolds, more nozzles, and more chambers. The fact that spray is directed at the same time upward from the lower manifolds and downward from the upper manifolds makes nozzle clogging more difficult to see. A poor nozzle spray pattern is especially difficult to see, since the spray pattern is viewed through a glass top which has solvent impinging on it.

Prevent problems, do not wait for them to occur. The ease and continuous operating mode of conveyorized equipment require special vigilance on the part of the dry film technicians, since there is no regular required shutdown which would normally allow access to the insides of the developing equipment. It is advisable to let all panels run out of the equipment and perform a periodic check list.

1. Spray nozzles for each manifold in each chamber must be checked. The top manifolds can be checked by closing the solvent supply valves to the lower manifold. While the lower manifold is shut off, the spray pattern on the upper spray nozzles can easily be viewed. Nozzles which do not present a full spray pattern must be removed and taken apart for cleaning, then reinstalled. It is

common to find resist scum and small particles of dirt or cover sheet in the nozzles. After all the nozzles in the upper manifold have been viewed, their solvent supply valve can be closed and that of the lower manifold opened and the spray pattern of the lower manifold nozzles checked.

2. Pressure gauges for each manifold must be watched; gradual, or more abrupt, pressure reduction indicates the need for filter changes.

3. The buildup of moisture, as a layer of water, must be carefully monitored. This can only be done by either viewing the solvent depth from the side, if the chamber has glass walls, or by submerging a beaker and lifting it from the solvent. The depth of the water layer in the beaker will be the same as the depth of the water layer in the chamber. When that layer is deep enough that it can be sucked by the spray pumps, it must be drained. Do not forget that the level in any chamber is going to decrease as soon as the pumps are turned on.

4. Check the length of travel through spray chamber 1 required to clear a panel of resist. (See Determining Proper Development.)

5. Check the level of solvent inside the chambers.

6. Check the level of solvent available for feeding to chamber 3, in the storage tank or barrel.

7. Check the level of spent solvent, in the spent solvent storage tank or barrel, which has been bled from chamber 1.

8. Check solvent, in all chambers, for coloration. Any color other than clear or light yellow indicates contamination and possible incorrect processing: check bleed-and-feed hookups for connection to the proper barrels or tanks. Do not forget that the developer is located in a room which contains yellow safe lights; it may be a good idea to take a solvent sample and look at it under white lights. A yellow solvent when viewed under yellow light appears to be clear white.

9. Always check both sides of a panel prior to placing it on the conveyor for development. It is easy to pull off a cover sheet and have a corner of that sheet tear off and remain on the panel. If that cover sheet covers part of the circuitry, the panel may have to be stripped and reprocessed.

Also, both sides of each panel should be checked when removed from the conveyor after development. Not only must panels be checked for incomplete removal of cover sheet, but the quality of development should be inspected continuously.

Determining Proper Development: Conveyorized and Non-Conveyorized Equipment

This section is common information for all dry film photoresist systems: solvent, semi-aqueous, and aqueous. The bulk of the developing information will be presented here; during later discussions on resists other than solvent, information peculiar to those systems will be presented.

One point to keep in mind, now and during resist processing, is that developing times and setting must be determined for each type, brand, and thickness of dry film photoresist. The surface upon which the resist is coated will also affect the developing rate (as well as the exposure rate). If resist is coated on double treated or black oxide coated copper, the development rate used for bare copper may not be valid. Only experimentation and evaluation of the results can adequately determine development settings.

Non-Conveyorized Equipment.   The developer spray time setting is determined by developing a panel; a full sized panel, coated but unexposed can be used for determining setting on a new type of resist. The actual development time for an exposed panel will generally be somewhere between two and two-and-one-half times the time required to clear the unexposed resist from the panel surface.

1. The resist cover sheet is peeled from both sides of the unexposed panel; which is then clipped to the lid of chamber 1. The timer, which controls the length of time the spray pump is on, should be set to an arbitrary time, say 2 minutes (since it is unlikely that developing off the resist will require more time than this).

2. The panel is loaded into the spray chamber and the pump timer activated. At regular intervals, say 10 or 15 seconds, the panel should be quickly lifted from the spray chamber and inspected for resist removal. Should resist still remain on the surface, simply lower the panel into the spray chamber for more developing.

3. When all the resist has been removed, simply note the elapsed time. This time is the Test Panel Clear Time (TPCT).

4. The next step is to develop a test panel which has been exposed using artwork.

The exposure should be made with a special phototool containing a number of test patterns, such as Fig. 10-22. This pattern contains plating and etching type test patterns: dots and holes, lines and spaces. The patterns should converge, so that the resolving power of the imaging process, including development, can be determined. The pattern should also contain a step tablet. Ideally, a series of three panels should be run, each exposed the same time. The panels should be developed as follows:

Panel A: use spray time of $2 \times$ TPCT
Panel B: use spray time of $2.5 \times$ TPCT
Panel C: use spray time of $3 \times$ TPCT

After each panel has been developed and dried, inspect the image quality and resolution (tightest spacing on test panel to be cleanly developed and resolved). If no elaborate test image is available, artwork which has the finest traces and closest spacing should be used for exposing panels A, B, and C.

Fig. 10-22. Stouffer test pattern.

When these panels are exposed, a step tablet should be used, to verify that exposure was made within the manufacturer's recommended range. Also, developing and exposure are integral parts of the dry film imaging process; one aspect cannot effectively be considered without giving consideration to the other. An image produced on the high end of the exposure range will withstand the developing process more effectively than will an image produced on the low end of the exposure range. Poor attention to developing details may result in resist scumming, resist breakdown, ragged traces, and other problems associated with poor attention to cleanliness, touch-up, phototool quality, and exposure.

The A, B, and C test panels should be evaluated for optimum acuteness, resolution, and steps of exposure. A 4 mil (4 thousandth) line on an etch pattern may look good, while the 4 mil line on the plating pattern may not be cleanly developed. Or, the 4 mil plating line may look well developed, with straight side walls and sharp traces, while the 4 mil line on the etching pattern may look ragged and overdeveloped. The optimum spray developing time is the time which yields the highest quality circuitry and optimum resolution for the type of pattern which needs to be imaged; an etching pattern may or may not be developed for the same length of time as a plating pattern. This should be borne in mind by all dry film technicians.

When experimenting to find optimum exposure or developing settings, the technician must not hesitate to vary either of these, as a change is made to the other. Such experimentation will only broaden understanding of the dry film photoresist imaging process, and the interrelation of each aspect.

Just as there is a range in which ultraviolet exposure can be made, so is there a range in which the film can be developed and yield excellent results. The dry film technician does not have to be concerned with pinpointing the developing time to within one second; rather, the technician should discover and understand the width of the operating windows of the processes and materials being used. Not all equipment and materials will have the same parameters and windows.

Monitor Developing. Once the range of developing times has been determined for optimum conditions, the technician should realize that the actual time for developing in production situations will vary as conditions of solvent temperature, solvent loading, evaporation, and even the quality of the solvent vary (see the section on Solvent Purity later in this chapter). The quality of the developing being performed must be looked at critically at all times.

One of the parameters most critical and easily monitored is the time to clear resist from the panel borders. This time, on fresh solvent, will be about the same as the test panel clear time. Usually the border and even the circuitry clear time can be checked merely by lifting the panel from the spray chamber for an instant. If the panel is not clear of resist in the border areas, the developing time will have to be increased When the developing time has increased 50–100% of initial time using fresh solvent, it is time to change solvent in the spray chamber. The solvent must be kept at optimum volume inside the spray chamber, as well as the immersion chambers. If too much life is sought from the solvent, the image quality will deteriorate, and scumming, poor resolution, raggedness will result. Also, the purity of the immersion tank solvents will deteriorate. Another reason for changing developing solvent with increasing developing time is that too much time can be wasted in the spray chamber, and this can present a serious drawback to productivity.

Conveyorized Developing Equipment

Determining development settings for conveyorized equipment is much the same process as for non-conveyorized equipment; only the setting to be determined is a conveyor speed, not a timer setting for the spray pump controller. The cover sheets from a coated but unexposed panel are removed, and the panel placed on the conveyor. As the panel progresses through the developer's first spray chamber, the point at which the panel is cleared is noted. Initially, just about any conveyor speed setting may be chosen. If necessary, the conveyor speed can be adjusted and another panel run. The correct conveyor setting is the one which will clear a panel by the time it passes 40–50% of the distance through the first spray chamber (Fig. 10-23). Several panels may be run to determine this setting.

After this conveyor setting has been determined, several test panels should be exposed using the test pattern phototool. The conveyor speeds tested should be centered around 40–50% clear points. After developing through all chambers, and drying, the test image must be evaluated for acuteness, resolution, and other aspects already discussed.

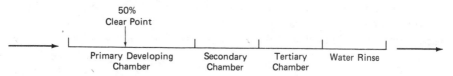

Fig. 10-23. Clear point for conveyorized developing equipment. As panel passes the clear point, it should be clear of unexposed photoresist.

Fig. 10-24. The fan nozzle: spray pattern is flat, and fans out from nozzle opening.

SPRAY NOZZLES

One important consideration to look for in developing equipment is the type of nozzles used in the spray manifolds: Fan or Conical (Figs. 10-24 and 10-25). Each of these nozzle types is available for use by manufacturers of spray developing equipment.

It is the author's opinion that fan pattern nozzles perform a superior job of dry film photoresist image developing and rinsing than do conical nozzles. The area covered by a manifold which contains fan pattern nozzles is more uniform across the width of the panel which passes beneath them. The fan pattern impinges upon the panel in a line across the panel (Fig. 10-26), and can be set at an angle to force the developing, or rinsing, solution to flow across the panel

Fig. 10-25. The conical nozzle: spray pattern is conical.

Fan Type Spray Nozzles

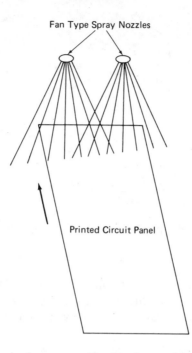

Printed Circuit Panel

Fig. 10-26. Spray impinges on panel in virtually a straight line during developing.

if desired. The conical spray pattern is more diffuse (Fig. 10-27) and covers a wider area of impingement. It cannot be set to impinge at an angle—without drastically increasing the spray area and reducing the force of impingement so necessary to good developing and rinsing. When purchasing spray equipment for photoresist processing, look at and consider the type of spray nozzles used; they will have an effect on functionality.

SOLVENT, DISTILLATION, RECLAMATION, AND STORAGE

A solvent still is an economic necessity for operating with solvent developing photoresist; usually two stills are required, since solvents are required for stripping as well as developing. The solvents are volatile and lend themselves to easy reclamation by distillation. The ease of distillation is fortunate, since great quantities of solvents are used in developing and stripping operations. Failure to distill, in favor of placing spent solvent in barrels to be carted away, would lead to incredible added waste storage problems, and to added expense of operation.

Figure 10-28 shows the layout for solvent developing or stripping areas which include distillation recovery systems. Fresh solvent is best delivered in bulk. The savings in time, safety, and cost are worth the effort and expense for

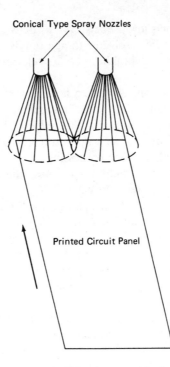

Fig. 10-27. Spray pattern is diffused over a wide circular area.

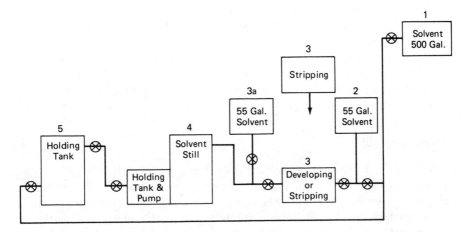

Fig. 10-28. Solvent processing layout.

the type of installation shown. The solvent is delivered in bulk and pumped into the storage tank. The fresh solvent storage tank should be at least 500 gallons, and located outside the building on an elevated platform, supported by concrete.

The fresh solvent is transferred to a holding tank of about 100 gallon capacity, located in the vicinity of the still. This tank is used to feed directly into the developing or stripping equipment. The solvent should flow by gravity force; little pumping should be necessary up to the processing equipment. The solvent may flow either directly into the processing equipment (for a non-conveyorized manual operation) or into a holding area near the equipment (for a conveyorized automatic bleed-and-feed operation.)

A conveyorized bleed-and-feed operation will pump spent and fresh solvent simultaneously. If the solvent flows into a 55 gallon holding drum, and another 55 gallon drum is used to hold spent solvent being bled from the machine, the filling or depletion of any one of these drums will serve as a convenient opportunity to temporarily halt the developing operation for a few minutes and perform the preventative checks (solvent coloration, spray nozzles, water buildup, etc.,) already discussed as good operating practice. It may need to be that solvent is pumped into the developing equipment faster than the spent drum is filling with solvent. This is due to solvent evaporation.

The spent solvent can be pumped directly into the still, as capacity will allow, or it can be stored temporarily in drums, until the still can process it. Many solvent stills are not designed for maximum efficiency. In fact, often they are quite rudimentary. The distilled solvent may collect in a small compartment in the still. This temporary holding compartment must be pumped out into a larger holding tank about once an hour. It is from the larger holding tank that solvent will flow back to the processing area. The periodic, or hourly, pumping of this temporary holding tank must be worked into the daily routine of the dry film technicians, unless the transfer pump is float actuated. Also the still must be checked regularly throughout the day for the need to add solvent. The still must not be allowed to run dry. If this should happen, the coils will burn out. This represents not only a cost but also a fire hazard. The task of checking the still should be assigned to all personnel in the dry film department, and double checked by supervisors.

Cleaning the Still

There is a limit to how much solvent the still can reclaim before it needs to be cleaned. Both water and photoresist contamination concentrate in the still. It should be cleaned either on a regular basis, such as once every two or three days, or whenever "fogging" is noticed during the hourly checks. When the lid is removed from the top of the still and only a dense "pea soup" type of fog is visable, the presence of excess water and little solvent is indicated. Further distillation yields little or no solvent because of the buildup of water in the still. If more spent solvent is added, distillation will continue, but a point will eventually be reached where the still is completely filled with water.

The still must be shut off and allowed to cool. It should then be pumped out with a plastic barrel syphon pump, the kind that costs about $20. An expensive pump should not be used because this type of still waste tends to form sludge. The contents must be dumped into a drum labeled "still sludge" and the appropriate E.P.A. waste labels attached. When all that can be pumped out has been, a scoop must be used to remove the rest from the bottom and coil area of the still. When this is completed, several gallons of spent solvent should be pumped into the still, and the still swabbed to remove caked-on sludge from the coils; a mop, broom, or brush works well for this. This solvent can now be pumped into the sludge drum, which should not be sent for reclaiming. It is strictly a waste product, unlike regular spent solvent, which can be sold for a few cents on the dollar for reclaim.

The still is now clean and ready for recharge. Before solvent is added, about 25 pounds of flocculent (recommended by the still manufacturer) should be added per 50 gallons of still capacity. If the flocculent is not added, the mother liquor in the still will foam, instead of boiling and distilling, when spent solvent is added and operation commenced. Once this material is added, spent solvent can be filled to the recommended level, and cooling water and power turned back on. A log should be posted on the still, and an entry made, with the date, by the person who last cleaned the still.

Solvent Purity

One of the potential hazards of working with photoresist is the presence of monomers. Monomers are a volatile and toxic component of unexposed photoresist, and can cause allergic reactions to develop in people who have become sensitized to them. Because of their volatility, monomers will distill and carry over into the reclaimed solvent. The developing operation is accomplished using reclaimed solvent as much as possible, for economic reasons. Fresh solvent is added to the developing system only when required to replenish adequate operating volume. This means that eventually monomer content will build to an unacceptable level. Either the solvent will begin to reek of monomer, or it just will not do the job effectively, through long developing times or residue being left on the panel surface. The solvent will also take on a deep yellow color and may even leave heavy residues inside the holding tank. Solvent which has become polluted with heavy monomer residues must be pumped into 55 gallon drums and sold to a company which has the facility to reclaim such material. There is another possibility, and that is to set up the equipment necessary to recirculate this type of solvent through activated carbon.

The solvent used for stripping dry film photoresist (methylene chloride) can be reclaimed by distillation with little residue build up. There is little monomer present in exposed photoresist, the components of which are considerably less volatile.

Safety

It is always good to avoid breathing solvents or letting them contact bare skin. They remove skin oil and may cause irritation. A solvent which has monomers in it must be handled with the same consideration as is given to handling or breathing vapors from the unexposed photoresist: use ventilation, wear gloves when handling the solvent, and wash hands as soon as possible after any of this material has contacted skin. Safety glasses should always be worn when processing any solvents. It is also a good idea for people to wear smocks, to keep fumes and residues off their clothes.

Cost of Solvents

A very important consideration in using the solvent developing types of photoresists are the costs of solvent and associated equipment. These cost considerations apply not only to developing, but also to stripping the photoresist.

Chlorinated solvents cost several dollars per gallon. For a panel size of 12 × 18 inches and 1.5 mil thick resist, the following are approximate values based on $4.00 per gallon of solvent for both developing and stripping. These numbers do not include solvent losses from evaporation (which are considerable) and from dragout, which goes down the drain.

    Developing   30 panels, double sided:
                 15 gallons × $4 = $60 for developing
    Stripping: 30 panels, double sided:
                 25 gallons × $4 = $100   for stripping
                                        $160   solvent cost for processing
                                               30 panels

    Average cost per panel = $5.33 for solvent developing and stripping 30
                                               panels

The solvent cost for a dry film imaging system is probably the single most expensive consideration on the panel (including the cost of FR4 laminate and gold). This is a major drawback of solvent developing resists. Other costs are those of buying airtight processing equipment for developing and stripping, waste disposal considerations, health considerations, and the need to buy distillation apparatus for the developing and stripping operations.

SEMI-AQUEOUS DEVELOPING RESIST

The semi-aqueous developing photoresists are very similar to the solvent developing resists as far as capability in accomplishing imaging tasks. The resist can be processed in either conveyorized or non-conveyorized equipment. The biggest differences are in the ease of handling and processing semi-aqueous photoresist. The ease is mostly due to the fact that both the resist and the developer are soluble in aqueous solutions. There is no need for a series of progressively cleaner chambers of developer; the resist dissolves into the developer with no scumming problems being caused by contact with water. The developer is not volatile, and is only about 10–12% solvent, and that solvent is soluble in water. It is free rinsing, with no scum formation during water rinse.

The developer is a mixture of 1 part butyl Cellosolve or butyl Carbitol, 8 parts water, and 1% by weight sodium borate. It must be heated to 80–85 degrees Fahrenheit for best results. In mixing, the sodium borate should be completely dissolved in 5 gallons of hot tap water before it is dumped into the spray chamber.

Non-Conveyorized Developing Equipment

The same equipment used for solvent developing may also be used for semi-aqueous developing resist. This equipment is depicted in Fig. 10-19. Chamber 1 contains the developing solution, chambers 2 and 3 are empty and not used for semi-aqueous developing resists. The only chambers utilized in semi-aqueous developing are the developer spray chamber and the water spray rinse chamber.

Developing

The developer must be heated to operating temperature of 80 to 90 degrees Fahrenheit. After carefully and completely peeling the resist cover sheet from both sides of the panel, it is clipped into the lid of chamber 1 and placed into the chamber. After the spray time has elapsed, the panel may be drained briefly, then placed into the water spray rinse chamber. There is little chance of the developer evaporating so quickly that scumming will occur.

There is a major difference between the purposes of the water rinses in solvent and semi-aqueous processing. The semi-aqueous developer and resist residue actually dissolve into the water, and are diluted and washed away by it. The solvent developing system depends on multiple chambers of solvent to dilute the residue bearing solvent, and depends on the water spray rinse to blow the solvent from the panel before it can evaporate. The semi-aqueous water rinse joins hands with the solvent to promote cleanly developed copper and resist surfaces and sidewalls. The solvent in the developer aids the rinse process.

Developer Purity and Developing Characteristics

Semi-aqueous developer has much higher resist loading capacity, than does the solvent system. Water does not condense into the developer forming scum. The developer is not volatile. All these factors which promote scum in the solvent system are inoperative in the semi-aqueous system. It is possible to adjust surface tension, and add emulsifiers, to increase desirable developing characteristics. This does not eliminate the need to change the developer and monitor the developing quality. The time to clear panel edges (monitor developer loading) must still be watched, and developing time increased as loading occurs. When developing time has been increased 50–100%, the developing solution and filter must be changed. Depending on the water pollution control regulations of the area, and the pollution control equipment in a given printed circuit manufacturing plant, the spent developer may go directly to the sewer; or be acidified to precipitate dissolved resist, then dumped to the sewer.

Conveyorized Developing Equipment

Most dry film developing is done with conveyorized equipment. There are many equipment manufacturers, who make excellent equipment for conveyorized spray purposes; from developing to stripping, etching, black oxide application, and smear removal.

The equipment required for conveyorized developing of semi-aqueous resist is simpler and cheaper than that required for solvent systems. Only two chambers are required, developing and rinsing. The equipment does not have to meet quite the stringent ventilation requirements of solvent equipment. The conveyorized developer is depicted in Fig. 10-29.

The method for determining conveyor speed remains the same. During pro-

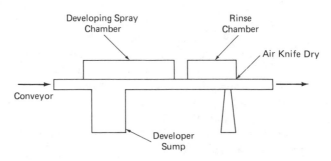

Fig. 10-29. Aqueous/semi-aqueous developing equipment.

duction operation, the dry film technician has only to keep a watchful eye on the panel borders, which should be clear of resist 40–50% of the way through the developer spray chamber. As developer loading increases, the conveyor speed should be decreased until it has reached about one-half to two-thirds of the initial value of the fresh developer. When it is so loaded, the developer and filters should be changed. The developer changing procedure is the same as for non-conveyorized equipment. There is no bleed-and-feed operation for developer, nor do the developing chambers cascade into one another. There is only one developing sump, and one spray chamber for developer.

FULLY AQUEOUS RESIST

The fully aqueous developing photoresists are developed almost exactly the same as semi-aqueous resist. Some fully aqueous resists can even be developed quite satisfactorily in the semi-aqueous developing solution (better to test first, before trying production boards). The fully aqueous developer is 0.5–1% by weight sodium carbonate monohydrate dissolved in tap water; there are no solvents. A slight amount of anti-foaming agent may be added to the sump, however. The developer is heated to about 80–90 degrees Fahrenheit.

Besides the type of developing solution used, the only major consideration to be aware of is the need for a slightly longer rinse chamber. The solvent in the semi-aqueous system aids rinsing. Since no such rinse aid is present, more water rinse time is required for the fully aqueous system. If water rinse pressure is a problem, the printed circuit manufacturer would do well to install a pump to provide continuously high pressure for the water rinse. When considering what processing equipment to buy for processing dry film photoresist, it would be wise to consider the rinsing requirements of the aqueous system.

SPECIAL CONSIDERATIONS

1. Any time dry film photoresist is being worked with, long-sleeve smocks should be worn. When through working, wash hands with soap and water.

2. The monomers in the resist are toxic; it is possible for employees to develop allergic reactions to the resist and fumes. Avoid exposure.

3. Always work with ventilated equipment, and use the ventilation in developers, strippers, and ultraviolet exposure sources.

4. Stripping of the photoresist is discussed in the chapter on resist stripping.

11
Screen Printing

SCREEN PRINTING VERSUS DRY FILM IMAGING

The two most common imaging techniques currently used in printed circuit manufacturing are screen printing and dry film photoresist. A major drawback of photoresist is the need to etch the copper surface of dry film imaged boards prior to pattern plating; this can only be done reliably if the panels have been copper plated after electroless copper deposition, or if a high deposition/high temperature electroless copper has been used to place extra copper inside the plated through holes. The only occasion where copper etching prior to plating is of no concern, is when the image is for print-and-etch, as with multilayer inner layers. This requirement for added copper in the holes is a real disadvantage of photoresist imaging. There is no such requirement for screen printing.

Dry film photoresist requires a very high capital investment. A hot roll laminator may cost $15,000, and a well collimated ultraviolet exposure source may cost an equal amount. A modern conveyorized developer may cost $20,000. Thus, it is fairly common for a manufacturer to spend $50,000 just to purchasing the major pieces of equipment needed for dry film imaging. The cost of any one of these pieces of equipment would be more than enough to set up an entire multi-station screen printing area, including actinic exposure source, developing equipment and chemicals, rinse-out sink with temperature controlled water source, separate hot water heater for the stencil making area, high pressure water pump for screen cleaning, screen drying and storage racks, frames, screening tables, stainless steel and polyester screening mesh, and all other chemicals and stencil films. There is a need for ovens to cure the screened-on resist; however, there is also the need for ovens for other operations in printed circuit manufacturing, including backing for dry film photoresist lamination.

Screen printing generally requires operators of a higher level of skill than does dry film imaging. The training time for a skilled person is longer for screen printing than for dry film imaging, and the salary is necessarily higher as well. Where dry film is capital intensive, screen printing is labor intensive. There is an exception to be made: automatic screen printers enable less skilled operators to maintain high quality, productivity, and consistency.

Because manual screen printing is labor intensive, there is a good deal of inconsistency from one operator to the next, and even from one print to the next. These inconsistencies become more pronounced the larger and more difficult the panel and image being screened, and the longer the production run. Operator attention, squeegee pressure, angle of contact, and other factors enter into the quality picture. These are overcome to a large degree by the use of automatic screen printers.

Screened-on resist generally has better adhesion, and better tolerance for unfriendly environments, than does photoresist. Photoresist often breaks down in plating baths like tin-nickel (165+ degrees Fahrenheit), and will not last at all in immersion tin (165+ degrees Fahrenheit), electroless nickel (200+ degrees Fahrenheit), or in black oxide (200+ degrees Fahrenheit). See the section Soldermask over Bare Copper: Alternatives, Advantages, and Manufacturing Techniques.

Screened-on resist generally is easier to strip, strips cleaner, and strips faster than does photoresist. Dry film photoresist tends to soften and then strip off in strands and/or particles. As the resist strips, it may leave photoresist particles along the edges of circuitry, especially between traces. The tighter the air gap, or spacing, the worse is the problem of photoresist particles left on the surface of the panels. This situation may be misdiagnosed as resist breakdown if magnification is not used to examine the panels. These particles may not be visible to the naked eye (usually they are visible with $7\times$ to $10\times$ magnification). However, after the panel has been etched, there remains unetched copper along the traces. This situation can lead directly to a major scrap problem if the manufacturing procedure calls for etched panels to be immersed in tin prior to reflow. The immersion tin covers the unetched copper along the traces, and will prevent any amount of restripping and re-etching from correcting the problem. Screened on resist does not have the complex binder polymer/photopolymer structure. The net result is clean and fast stripping with no particles or strands.

It is very difficult to get away from screen printing for soldermask application. The basic classifications of soldermask are those which are ultraviolet (UV) curable, including the screened-on and the dry film varieties, and those which are not. The most common, widely accepted, and trouble free varieties of soldermask are the oven cured, screened-on varieties. There are dry film soldermasks currently available, so that technically speaking it is possible to manufacture printed circuits with soldermask, and without using screen printing. However, dry film soldermask is UV curable, and is subject to the drawbacks of UV curable materials, e.g., narrow exposure window for determining degree of cure. Dry film solder mask is also extremely capital intensive, labor intensive and time consuming. Should the exposure be insufficient, the soldermask will be damaged during wave soldering. When all is said and done, it is the customer who decides which soldermask will be used. Right now, most commercial and military printed circuit users call out for PC-401 (or equivalent) and SR1000 soldermasks, which are screened-on varieties.

The realm of fine line (less than 0.008 inch) traces and spaces is dominated by dry film, which is unsurpassed for fine line definition, especially for plated circuitry. However, screen printing is very adequate for traces as fine as 0.008 inches. The main problem with screen printing circuitry much finer than 0.010 inches is not resolving ability; it is line growth during plating. Once the surface plated metal exceeds about 0.5 mil, the metal begins to mushroom over the surface of the resist. The net result is that the spacing between traces decreases, as the width of traces increases. Screen printing cannot compete with dry film imaging for holding tight tolerances in plating, unless the thickness requirement for the plated metal is in the area of 0.0 to 0.5 mil.

Dry film photoresist offers another advantage over screen printing where there is a need for numerous non-plated through holes. Typically, non-plated through holes are either plugged (when there are only a few), or they are second drilled after plating. However, when photoresist is used for imaging the holes may be drilled during first drilling, then tented over during dry film imaging. The artwork, or phototool, simply allows the photoresist over the holes to be polymerized. The hole is covered with photoresist during plating, so that no plating occurs. After resist stripping and copper etching, the hole has no metal inside. This cannot be done reliably using screen printing.

WHAT IS SCREEN PRINTING?

Screen printing is the art of forcing ink through a stencil which is mounted on a tightly stretched screen. A rubber or plastic squeegee is used to push a small puddle of ink (plating resist) across the stencil area on the screen, which forces the ink through the open areas on the screen to print the circuit pattern on the copper clad panel.

Screen printing is an old process; however, with the need for imaging electronic circuits, screen printing has taken giant technological steps forward. The screen printing requirements for printed circuits are the most challenging and demanding of all screen printing currently being done in a production mode. Much research effort and money has been expended to develop the components of printed circuit screen printing.

BASIC COMPONENTS OF A SCREEN PRINTING OPERATION

The following list is basic to any printed circuit screen printing operation. Good screen printing practice demands that the importance of each component be understood and respected at all times.

1. Frame, also called a chase.
2. Screen material.
3. Stencil processing area: developing, washout, drying, and storage.
4. Ultraviolet exposure source.

5. Screen printing station, and squeegee.
6. Plating resists.
7. Drying oven.
8. Solder mask.
9. Legend ink.
10. Screen stripping requirements.

FRAMES AND SCREEN MATERIALS

The article by Robert Nersesian in the appendix entitled "Screen Printing Is the Answer," presents an excellent discussion of requirements and considerations for the frame, screen material, and the squeegee. This information will not be duplicated here. The engineer or supervisor responsible for the screen printing area would do well to experiment with the various screening materials available, to gain an understanding of what types of circuitry can be screened with the less costly materials, and which demand stainless steel mesh. Also, too often the requirements for tensioning are neglected, the result being rapid stencil breakdown, excessive touch-up, and the need for shooting additional stencils to complete a production run. Tensioning of screens, both for new screens and for existing screens prior to stencil application, should not be left to the discretion of "how tight it feels" to the stencil maker. A device for measuring tension should always be used, and the tension requirements of the screen manufacturer followed. A tensionmeter, or serimeter, for measuring Newstons/centimeter can be purchased from any screen printing supply house.

CLEANING NEW SCREENS

One item which should never be neglected is the cleaning of new screens prior to stencil application. Stainless steel must be scrubbed with a highly caustic solution, such as 1–3% trisodium phosphate (TSP), or sodium hydroxide of similar strength. Heating the caustic solution aids cleaning. Solvent cleaners should not be used because of their tendency to leave a residue. The cleaning solution can be brushed on with a lint free rag or a nonmetal polyester/nylon brush. The screen should be rinsed well with water. However, water alone will not remove a caustic substance. The screen must also be rinsed with a mild acid, such as vinegar, and then water rinsed again. The vinegar will neutralize the caustic for freer removal by water.

Polyester screens are cleaned in a different manner. The surface of these screens will not allow stencil materials to adhere well. The surface of the screen fiber must be abraded with an abrasive grit, such as silicon carbide powder. The screen should be wetted with water, and the powdery silicon carbide liberally sprinkled on. Scrubbing should be done using a lint-free rag and circular motions, until all areas of the screen have been abraded. The silicon carbide will easily wash away. The high pressure water used to clean screens during the stripping process can also be used to rinse newly abraded polyester screens. This process will not have to be repeated for the life of the screen.

ULTRAVIOLET EXPOSURE SOURCE

The actual source of ultraviolet light may be a carbon arc, mercury halide, or pulse xenon. The carbon arc is used less today than in the past because it tends to be slightly inconvenient and emit fumes; also, the newer mercury halide and pulse xenon are easier to work with.

The basic requirement for the exposure source is determined by the type of stencil media to be used: indirect, direct or direct/indirect.

1. Indirect stencils are made from presensitized stencil film which comes in a roll, and is exposed, developed, and washed out before being applied to the screen. Indirect stencils are presensitized to ultraviolet light.

2. Direct stencils are coated, as an emulsion, directly onto the screen itself. There is no intermediate step of cutting film and exposing. The emulsion is not photosensitive at this stage. After the emulsion has dried, it is coated with a sensitizing solution and dried again. Since the stencil is photosensitive, and already on the screen, the exposure source must be capable of accommodating the entire screening frame, and will have a plastic or rubber sheet on the back of the contact table to conform to the screening frame.

3. Direct/indirect stencils are cut from a roll of nonphotosensitive stencil film and applied directly to the screen. They are processed like direct stencils from this point on, and have the same requirements for the exposure source.

The exposure source required for direct and direct/indirect can also be used for exposing indirect stencils. If the equipment is just being purchased, or specified, it would be wise to purchase the type of source which will accommodate the entire frame, since this will allow greater process versatility (Fig. 11-1).

The best exposure source to use is one which most closely approximates a point light source with well collimated light rays. If a unit has a reflector and a light source which is fairly close to the contact frame, it is good practice to evaluate the collimation by performing a test. Simply measure the dimensions of a piece of artwork, then expose a piece of stencil film. After developing the film, measure the dimensions of the reproduced artwork. An Accu-Scale ruler works well for this. After development has taken place, wait several minutes to allow stencil film to achieve temperature equilibrium.

The exposure source must not be neglected during the course of the work day. The glass of the contact frame must be inspected for stencil residues, tape residue, dirt, fingerprints and any other type of contamination. Tape residues typically come from the phototool. Almost any kind of contaminant can find its way onto the contact frame by way of the stencil film, if the film is cut on a dirty surface. It is not enough simply to wipe the contact frame; it must be

Fig. 11-1. Flip-top platemaker for exposing screening stencils.

carefully inspected for contamination. When the contact frame is wiped, a liquid glass cleaner and lintless rag or tissue (such as Kay Dries) should be used; newsprint or other material may leave fibers which will cause image quality problems.

EXPOSURE CONTROL

Every photosensitive medium has an exposure window through which proper exposure may be obtained. Too long an exposure will result in a brittle stencil with unacceptable line growth. Too short an exposure will yield a stencil that is too thin and has poor image quality. A host of other stencil and image problems are also associated with over- and underexposure.

Optimum exposure setting can be determined and monitored with the use of

THE AUTOTYPE
EXPOSURE CALCULATOR
is designed so that you can quickly and simply determine the best exposure time for your Autotype photostencil materials. With it you can also check definition and, in certain cases, light geometry effects (size and placing of the light source).

The Autotype Exposure Calculator consists of five definition targets and columns of type. The first column has no filter behind it, and has an exposure factor of 1.0. Behind each of the other four target columns there is a neutral density filter which reduces the amount of light reaching the stencil by a factor of 0.7, 0.5, 0.33 and 0.25. It is therefore possible to make five exposures at one time.

To use the exposure calculator, estimate what you think should be the correct exposure time and then double it. Expose the calculator to the photostencil film in the normal way, emulsion side against the stencil. Afterwards process the stencil as usual and select one of the five target columns for best stencil thickness, colour and definition. If your original exposure estimate was correct, the preferred target column should be the centre one (0.5). To establish the exposure required for the job, multiply the factor at the top of the selected column by the exposure time given for the test.

Example
You may consider 5 minutes as the likely exposure time required. Double it and expose the stencil, i.e. for 10 minutes.

If you find after processing the stencil that the column with factor 0.33 gives the best result then multiply the exposure time by this factor.
10 × 0.33 = 3.3 minutes—say, 3¼ minutes for convenience.

Should you find that you cannot produce a suitable stencil from the first test, select the end column which gave the best result.

a) If it is the first column (Factor 1) then re-test at double the exposure time.

b) If it is the last column (Factor 0.25) then re-test at half the exposure time.

Always use a test exposure time twice what you think it should be so that you are likely to fall in the centre of the calculator's range.

DEFINITION
Definition targets are made up of four concentric rings of tapered lines. The approximate thickness of these lines at various points is given below in thousandths of an inch (0.001″—25 microns).

The change in definition caused by altering the exposure conditions can be observed in the target, where the image spread (filling in) gradually increases outwards from the centre to the edge of the radial pattern. Accurate comparisons are facilitated by noting the degree of spread in relation to the concentric rings in the target.

Fig. 11-2. Autotype exposure calculator.

either a step tablet (such as a Stouffer 21) or an Autotype calculator. The step tablet has a series of progressively denser steps. It is placed between the phototool and the stencil film during exposure. After developing the film, the number of steps which remain solid provides an indication of degree of exposure. If the stencil film data sheet provided by the manufacturer specifies a range of steps, or an optimum step left solid, then the exposure setting must be adjusted until the desired step is achieved.

The Autotype calculator is slightly different, and can be used for determining more than exposure (Fig. 11-2 and 11-3). This tool contains five resolution targets. Each of these targets has a differing degree of optical density. When one exposure is made, the stencil film records the targets as if each had received a different exposure. The exposure time used for the test should be twice what is guessed to be in the vicinity of a realistic time; it can actually be off by quite a bit, however. The sample of stencil film is measured with a micrometer at each of the targets; the target which measures nearest to 0.5 mil for indirect

Factor   **1**   **0.7**   **0.5**   **0.33**   **0.25**

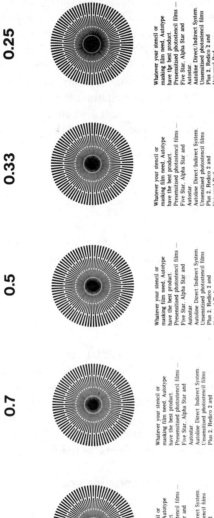

Whatever your stencil or masking film need, Autotype have the best product. Presensitised photostencil films — Five Star, Alpha Star and Autostar.
Autoline Direct/Indirect System.
Unsensitised photostencil films Plus 2, Redico 2 and Universal Red.
Hand-cut stencil films Autocut and Solvent Green.
Automask the complete masking film.

Whatever your stencil or masking film need, Autotype have the best product. Presensitised photostencil films — Five Star, Alpha Star and Autostar.
Autoline Direct/Indirect System.
Unsensitised photostencil films Plus 2, Redico 2 and Universal Red.
Hand-cut stencil films Autocut and Solvent Green.
Automask the complete masking film.

Whatever your stencil or masking film need, Autotype have the best product. Presensitised photostencil films — Five Star, Alpha Star and Autostar.
Autoline Direct Indirect System.
Unsensitised photostencil films Plus 2, Redico 2 and Universal Red.
Hand-cut stencil films Autocut and Solvent Green.
Automask the complete masking film.

Whatever your stencil or masking film need, Autotype have the best product. Presensitised photostencil films — Five Star, Alpha Star and Autostar.
Autoline Direct/Indirect System.
Unsensitised photostencil films Plus 2, Redico 2 and Universal Red.
Hand-cut stencil films Autocut and Solvent Green.
Automask the complete masking film.

Whatever your stencil or masking film need, Autotype have the best product. Presensitised photostencil films — Five Star, Alpha Star and Autostar.
Autoline Direct/Indirect System.
Unsensitised photostencil films Plus 2, Redico 2 and Universal Red.
Hand-cut stencil films Autocut and Solvent Green.
Automask the complete masking film.

Autotype International Limited
Grove Road
Wantage, Oxon OX12 7BZ
Telephone 02357 66251
Telex 837275

Fig. 11-3. Autotype exposure calculator.

223

stencils is best.* For direct and direct/indirect stencils, the image quality and line width must be evaluated to determine optimum exposure setting. Once the optimum target has been decided upon, there is a multiplying factor associated with that target. This factor is multiplied by the exposure time used for the test; the result is an exposure setting which will yield the desired degree of stencil exposure. In addition, the image quality of the converging lines may be examined, and the line widths measured on the stencil film for comparison with the Autotype calculator. The Autotype calculator can also be used for evaluating developers, washout temperature, and just about anything else which affects the image quality of the stencil.

STENCIL AND SCREEN PREPARATION

Indirect Stencil

This is perhaps the most common stenciling system currently being used; it is cheap, it is easy, and requires the least amount of training. The stencil is pre-sensitized to UV light, so that there is no extra step involved in sensitizing the film.

The basic steps are: (1) tensioning the screen, (2) cutting stencil film to size, (3) exposure of stencil, (4) development of stencil, (5) lay-up of stencil to screen, (6) drying of stencil.

1. A screen should be chosen that will allow at least six inches of space between the edges of the circuitry and the sides, front, and back of the frame. This is critical for avoiding unnecessarily high off-contact distance as well as screen and squeegee distortion. The screen must also be inspected for cleanliness; no dirt, ghost image from a previously used stencil (as occurs when the screen is not properly cleaned with bleach), or residue of any kind can be present. Also, there should be no tears or broken mesh fibers, as these will adversely affect the image quality.

A screen mesh of 325–270 should be used for printing fine lines with tight spacing (0.010 inch traces and spaces or finer).

The screen should be retensioned after every use. The best time to do this is after it has been cleaned from the previous use. The bolts should be tightened in a sequence such as that shown in Fig. 11-4. A tensiometer or serimeter should always be used. Do not tension the screen after the stencil has been applied, as this procedure will distort the image and make registration extremely difficult or impossible.

2. Much care is needed to cut the stencil to size. The film is very sensitive to fingerprints, and only the edges of the film should be touched, not that portion which will carry the circuit image. It is best to roll it out on a wide, smooth,

---

How To Get The Most Out of A Stencil. Peter Brown, Autotype USA, 1978 Fall Symposium, Calif. Circuits Association

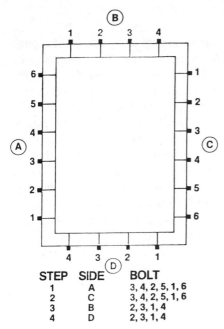

STEP	SIDE	BOLT
1	A	3, 4, 2, 5, 1, 6
2	C	3, 4, 2, 5, 1, 6
3	B	2, 3, 1, 4
4	D	2, 3, 1, 4

AUTOROLL AUTOCHASE DIRECTIONS

Read Entire Sheet Before Proceeding

Mounting New Mesh

1. Move the floating bars so that they are straight and parallel to, and approximately ¼-inch from the inside rails. Then tighten the end bolts, smoothly bowing the floating bar outward at the ends an additional ¼-inch. NEVER TIGHTEN ANY BOLT MORE THAN 1-TURN AT A TIME. NEVER PERMIT MORE THAN 2-TURNS DIFFERENCE BETWEEN NEIGHBORING BOLTS.

2. Tear or cut the mesh approximately 4-inches larger in length and width than the chase inside dimensions. Be certain that the mesh edges are parallel to the weave, and that the mesh is smooth and unwrinkled.

3. Lay the mesh on the chase, with 2-inches of overlap on each floating bar. Using a smooth plastic tool such as a piece of Plexiglas, board material, or a plastic ruler, gently push the mesh into the floating bar groove along one long edge of the chase. Be certain not to wrinkle the mesh.

4. Using the plastic tool as a leader and guide, slide a nylon rod through the hole at the chase corner, into the floating bar groove, and over the mesh, trapping the mesh in the groove. Slide the nylon rod completely into the groove. The rod must not extend beyond the ends of the floating bar.

5. Pull the mesh cut edge upward gently, along the floating bar, pinching the mesh in the groove. Repeat the above operations for the opposite long edge, then the two short edges.

Tensioning

1. NEVER TIGHTEN ANY BOLT MORE THAN 1-TURN AT A TIME. NEVER PERMIT MORE THAN 2-TURNS DIFFERENCE BETWEEN NEIGHBORING BOLTS. Proper tensioning technique dictates that tensioning be performed on alternate sides of the chase, rather than consecutively around the chase. In addition, you should tension bolts starting at the rail center and work alternately left and right toward the corners, rather than consecutively from corner to corner. This technique is shown in the attached diagram. If this technique is followed, you will obtain a chase with unsurpassed tension level, uniformity and stability.

2. Following the proper sequence, tension the center bolts about 1-turn and the outer bolts about ½-turn, until the floating bars are straight and parallel to the inside rail.

3. Continue tensioning each bolt about 1-turn until the desired tension level is reached. It is often desirable, at the end of the tensioning process, for the floating bar to be bowed slightly so that the bar ends are about ¼-½-inch closer to the inside rail than the bar center.

4. Trim excess mesh with a razor blade by cutting along the outer edge of the floating bar. Do not trim by tearing the mesh. Apply tape to the mesh, covering the floating bar and inside rail to prevent emulsion from entering the precision floating bar groove.

Mesh Removal

1. Return the floating bars to starting position, press the nylon rods downward with the plastic tool used for meshing, and slide the rods outward through the holes at the chase corners.

2. Before re-meshing, be certain that the floating bar grooves and the nylon rods are clean and free of emulsion.

Fig. 11-4. Correct tensioning sequence–autochase.

and very clean surface, as this will help avoid formation of crinkles. Cut enough stencil film so that it will form at least a one inch border around the phototool. After the stencil has been cut, carefully roll up the remainder of the stock and place it back into its storage tube; replace the light proof cap on the tube.

3. The exposure source should already have been checked for cleanliness: the glass looked at for visible amounts of tape, dirt, or oily residue, and wiped down with glass cleaner and lintless rags or paper tissue. Sometimes it is best to feel the surface of the glass with the tips of the fingers, and to scrape any noticeable bumps with a razor blade. Also, glass cleaner may not be adequate for cleaning the glass. Opaquer and tape residue may not come off until the glass is cleaned with a squeegee cleaner, such as NS Thinner. The bottom of the contact frame should be black cardboard (which should also be checked for dirt or contamination). The stencil film is laid emulsion up on the black cardboard bottom of the contact frame. The phototool, which is oriented reading right emulsion, is laid emulsion up against the stencil film.

The top of the contact frame is closed and a vacuum pulled. As the vacuum is pulled the films should be inspected for any bumps of debris which might now be visible. Vacuum is complete when Newton rings (swirling rainbow patterns) are visible on the contact frame cover, and vacuum gauge reads at least 25 inches, or 15 pounds of pressure. At this time the contact frame is turned over, or on its side, to face the exposure source. The stencil film is then exposed for the preset amount of time. When the exposure is over, the contact frame is turned back to the initial position, the vacuum is turned off, and the top opened. (*Note:* never turn vacuum off before the exposure has been completed. The stencil film is now ready for developing.

4. Development is accomplished by immersing the stencil film in a tray of either hydrogen peroxide (diluted to film manufacture's recommendations) or in two-part powder in aqueous solution. The manufacturer's recommended immersion time should be followed; one minute of immersion time is a typical value for hydrogen peroxide development. The film should be immersed emulsion up, with the entire surface immersed at all times. It is good practice to rock the tray to agitate developer over the film. The film will turn cloudy white as it hardens.

When the immersion time is up, the stencil should be lifted from the developer, and placed on a flat surface, emulsion up, at an angle in a large sink. At this time, a gentle stream of temperature controlled (105 $\pm$ 5 degrees Fahrenheit) water is directed at the stencil to wash off the unexposed stencil gelatin. The water hose should be connected to a hot/cold water mixing valve for controlling and indicating the water temperature. This too, should be done for the time recommended by the stencil film manufacturer; two minutes is a typical value. A good rule of thumb is to rinse about one minute past the time of clean rinse out. This warm water rinse is followed by a cold (70 degrees Fahrenheit) rinse to set the gel on the stencil. The stencil is now ready to be pressed on to the screen.

5. The emulsion side of the stencil is laid on the bottom side of the screen (the side which faces downward during the printing operation). Care must be used to avoid wrinkling the stencil, or entrapping air. A roller should be used to very gently press the stencil emulsion onto the mesh of the screen, although many screen printers use the palm of their hand for this. Every stencil-making operation should have a raised platform about ½ inch high, with the other dimensions slightly smaller than the inside of the screen frame. Upon this platform lay a sheet or two of newsprint. Turn the screen frame over and lay it upon this newsprint-covered platform. Gently blot the stencil with a roller, or the palm of the hand. Lift the frame, place fresh newsprint on the platform, and repeat the blotting operation. This blotting should be performed three or four times to remove excess water. Sometimes isopropyl alcohol can be sprayed onto the stencil before blotting, as an aid to removing water. The stencil and screen are now ready for drying.

6. The newly prepared screen is dried by placing it before a fan. The entire stencil-making area must be located away from thoroughfares, so as to keep dust and dirt at a minimum. It is critical that the air in the screen drying area be as free as possible of airborne particles, which would adhere to the stencil. After about 15–20 minutes, the stencil will probably be dry. At this time it can be set in the screen storage rack until it is needed for printing.

Direct/Indirect Stencil

The procedure for preparing a screening stencil using direct/indirect film is slightly different and more involved than that for preparing indirect stencils. This is the second most widely used stencil system. One of the chief advantages of this system over indirect stencil is the longevity of the stencil and its ability to hold sharp lines. Since it is a stronger stencil, it needs less support from the screen mesh.

There are several things which all screen printing operators and engineers should keep in mind at all times: the skill of the stencil maker; the care which he/she displays on the job; the cleanliness of the stencil making and screen printing area; and the quality of the exposure source, washout sink, and other stencil-making equipment. These are all much more potent factors in determining the complexity of circuitry which can reliably be screen printed than is the stencil-making system being used.

The basic steps in preparing screens for printing with the direct/indirect stencil system are: (1) tensioning of the screen, (2) cutting the stencil and sensitizing it to the screen, (3) exposing the stencil, (4) developing the stencil, and (5) drying the stencil.

1. All the information presented on tensioning the screen for indirect stencils applies equally to any stencil system. This is simply good screen printing practice.

2. When cutting the direct/indirect stencil film, the same requirements for

cleanliness apply as for indirect. The film must be handled only at the edges. The flat surface upon which it is rolled out must be free of dirt, dust, oils, and moisture. The film must not be crinkled.

The stencil film should be cut 1–2 inches larger than the phototool on all sides, so as to form a border around the circuitry. After the stencil film has been cut to the desired size, the remainder of the stencil stock must be rolled up and returned to its protective storage area for safe and clean keeping. The stencil film is not, however, photosensitive at this point.

All stencil-making areas must have a platform which is slightly smaller than the inside dimensions of the frame. It should be very flat, and about ½–1 inch thick. The newly cut piece of stencil film is laid emulsion up on this platform. The underside of the screen is laid upon this piece of film. Great care may be needed to ensure that the screen and stencil film make wrinkle free, intimate contact. The screen and stencil are now ready for the sensitizing step.

Sensitizing Emulsions. There are two types of sensitizing emulsions available: bichromate and diazo. There are advantages and disadvantages to each, and a printed circuit manufacturer may want to use both types. The advantage of bichromate is that it dries faster than does diazo, and requires considerably shorter exposure time. Diazo sensitizer requires several hours to dry. The longer exposure time also places greater emphasis upon the quality and ultra-violet density of the phototool, to prevent bleed-through and partial exposure in undesired areas. Longer exposure also means that the contact frame will tend to heat up, especially with repeated exposures; a heated contact frame will lead to changes in photospeed, and more random exposure less defined by the phototool.

In most job shop printed circuit manufacturing environments, it is important that there be the capability of setting up screens for printing fairly rapidly. However, it is also desirable to have sensitized screens ready in advance.

Bichromate sensitized screen stencils should ideally be used within a few hours; where as diazo may be stored (in the dark) for longer periods of time. The printed circuit manufacturer must evaluate the needs, and capabilities of the shop before locking onto one system over the other. Bichromate will be used for all examples in this section.

Sensitizing the Stencil. Sensitizing emulsion can be poured along the screen just outside one edge of the stencil, or it may also be used to block out the entire screen. The sensitizer makes an excellent screen block-out. Excess sensitizer should be saved for use as a block-out agent, instead of discarded. The sensitizer is then squeegeed across the entire surface of the stencil film. The squeegee should be soft (50 durometer) with smooth, even edges. Make only one pass with the squeegee. There is no advantage to multiple passes, and the stencil quality will deteriorate by being forced further into the screen mesh.*

---

*Chromaline Direct-Indirect Stencil System. The Chromaline Company, 4832 Grand Ave., Duluth, Mn 55807 (218) 628-2217

The screen must now be set in a darkened room before a fan for drying. The dried and sensitized screen is now ready for exposure.

3. The same considerations for proper cleaning of the contact frame, and for exposure control discussed under the indirect stencil system apply equally for direct/indirect.

The polyester base must be peeled from the stencil when the screen is to be exposed; this can be done as soon as the screen is dry, or immediately before exposure. The screen is placed on top of the contact frame, with the underside (stencil side) of the screen up. The correct phototool to use is negative reading right emulsion. The phototool is placed emulsion to emulsion with the stencil film. The contact frame is closed, and a vacuum drawn. The vacuum is adequate when Newton rings appear at the surface of the contact frame, and vacuum has reached 15 pounds or 25 inches. When this occurs, turn the contact frame to face the exposure source; and begin the exposure for the predetermined amount of time. At the conclusion of the exposure, turn the contact frame back to its rest position, disengage the vacuum, and open the contact frame. Never disengage the vacuum while the contact frame is facing the exposure source. The screen is now ready for developing.

4. There is no requirement for hydrogen peroxide or any other two-part proprietary developer. The screen is set against the back of the wash-out sink and developed by gently spraying warm water onto the stencil. Cold water can be used, but requires a longer rinse for a clean screen. A temperature of 85–90 degrees will wash out the stencil rapidly. The same water control valve for mixing hot and cold may also be used, but temperature is not critical. The spray should be continued for about one minute after the screen appears to be clear of stencil emulsion. The underside of the screen, the side with the stencil film on it, is the side which must be sprayed during development. Some manufacturers of screening stencils sell a stencil hardener. Generally, the stencil will be hard enough. Unless there are weather conditions contributing to rapid stencil breakdown, no additional hardener will be required. The screen is now ready for drying.

5. Drying is accomplished in the same manner as is used for indirect stencils; and the cleanliness requirements already discussed also apply. Newsprint is placed on the screen platform. The underside of the screen is placed against the newsprint, and gently blotted. This operation is repeated three or four times to remove excess water. After blotting, the screen is set before a fan for drying, and should take no more than about 20 minutes. Dried screens must be stored in a clean storage area.

Direct Stencil Systems

Direct stencils are coated directly onto a screen using a roller. The advantage of this type of system is that the stencil thickness can be controlled as required. The sensitizing, exposing, and developing procedures used for direct/indirect are the same for direct stencils.

Other Considerations

The stencil requirements for soldermask are slightly different than for printing on a copper surface. Some stencil manufacturers make a tougher stencil for soldermasking, such as Dorn's SM-1000, in place of EG-1000. It is common for printed circuit manufacturers to use the same stencil for both operations, but to increase exposure time about 25% for solder masking. There is no such increase in exposure for legend stencils, as resolution is critical.

When setting up stencils for each of the requirements (plating resist, soldermask, and legend) the proper screen fabric must not be forgotten. Polyester and nylon are almost always used for soldermasking. Approximate meshes are 110–150 for soldermasking, and 150–230 for legend. It is good practice to color code screens used for different purposes, e.g., pink screening material for soldermasking and white for legend.

SET-UP FOR SCREEN PRINTING

The basic operations are: (1) screen and stencil inspection, (2) image registration and frame set-up, (3) pre-printing considerations, (4) first article, and (5) production printing.

Screen and Stencil Inspection

Once the screen printer has chosen a job and obtained the corresponding screen with stencil, he/she must inspect the screen prior to use. If the stencil is of the indirect variety, the operator must pull the polyester backup sheet from the stencil, if this has not already been done. It should be inspected for debris on the screen and beneath the stencil. A shiny film between circuit traces indicates inadequate washout of the screen during developing; screens which exhibit this shininess must not be used. The screen should be returned to the stencil maker for stripping, and another stencil prepared.

The circuitry of the stencil must be inspected for pinholes, raggedness, broken and nicked traces, and any other defect or condition which will adversely affect the image quality. The stencil can be touched up with opaquer to a certain extent. All touch-up must be applied to the underside (the stencil side) of the screen. There is a limit to what can be done to touch up a stencil. Excessive touch up generally means that either the film supplied by the customer and processed by the photo department was not good to begin with, or the stencil preparation was poorly executed. It may be that as much time can be saved by preparing another stencil as can be saved trying to salvage a poorly prepared one. The screen printing operator must also consider the amount of touch up that will be needed on each panel after the job has been screened.

Just as the stencil maker must be trained to read the job traveler, and to study the blueprints and read the notes on them, so must the screen printing operator be trained. The number of panels should be counted, the stencil checked for UL logo, date code, and any other required markings. The screen printing operator must always make sure there is a trace connecting the ends of the contact fingers with the panel borders. This is called a gold plating bar, and is necessary for plating nickel and gold on the contact fingers.

Once the screen operator has decided that the stencil is acceptable for printing, he or she should sign their initials in one corner of the stencil so that it will show up on a panel outside of the fab lines. The frame is now ready to be fastened on to the screening table and set up for registration.

Frame Set-Up and Image Registration

The next task is to block out the rest of the screen, which is not already covered by stencil. (This step is not necessary for direct/indirect stencils already blocked out with sensitizing emulsion.) The screen can be completely blocked out around the stencil with clear cellophane tape. However, this is time consuming and uses excessive amounts of tape. Another method is to add a tape border to the existing 1–2 inch border already around the circuitry, which is formed by the stencil overhang. After the tape border has been added, a resinous screen block-out liquid is squeegeed over the rest of the available screen. The block-out resin dries in about 5 minutes when the screen is set in front of a fan. There is another reason to use a block-out resin over tape. Ink gets beneath the tape and dissolves the tape's emulsion. This emulsion may get into the screen and make spots in the imaged area of the printed circuit panels. The tape emulsion cannot readily be removed from either the panel surface or the stencil, as ragged traces are likely to result from touching the stencil with any cleaning substance. The frame is set into the hinged clamps of the screening station, and the clamps tightened securely.

At each screen printing station is a platform similar to the one used to blot the screens dry during developing. There are some differences, however. This platform is larger than the frames, and is secured to the screening table by a precision bolting device. This device allows the platform to be moved small increments in an X-Y coordinate scheme, for precise registration of the panel to the stencil. (Fig. 11-5).

The screen is now ready for registering to the panel. This can be accomplished by inserting a panel beneath the stencil on the screen. The screen operator can use the palm of a white gloved hand, or a round cylinder, such as from a tape roll, to gently press the screen to the panel. The panel can then be shifted manually to approximate registration, within about 0.100 inch. The frame is then lifted out of the way, and Carlson pins inserted beneath the panel and in the screen tooling holes (Fig. 11-6). The Carlson pin has double sided tape to hold it to the table. The operator should place an arm across the panel, and lift

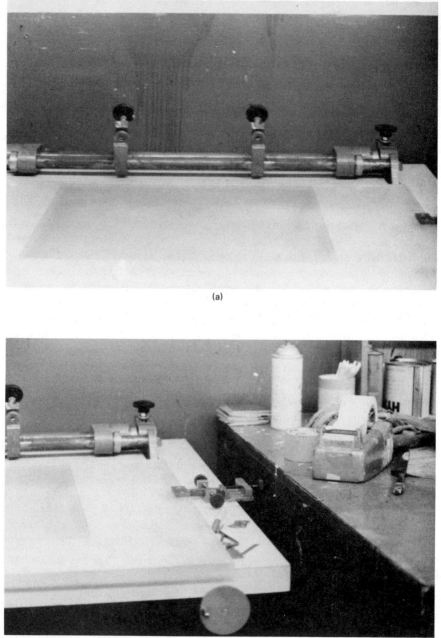

(a)

(b)

Fig. 11-5. (a) Chase (screen frame) is held by these two clamps. (b) These controls are used for making fine adjustments in panel position to aid registration.

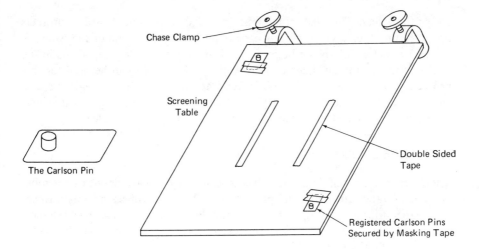

Fig. 11-6. Screening table set up with Carlsen pins for panel registration. Pin nipple fits into screen tooling hole of panel.

one end at a time to keep it from slipping while the Carlson pins are being inserted. After the pins have been inserted they are to be covered with masking tape, to protect the screen.

The panel should be removed from the platform, and two rows of double sided tape laid down. The double sided tape will hold the board during the printing stroke, and will help overcome some of the difficulties of screen printing warped panels.

The panel should be replaced on the platform, and the frame brought down. The off-contact distance (OCD, the height of the screen above the substrate being printed) should next be set. The screen height above the panel to be screen printed should be around 1/16 to 1/8 inch. But the screen fabric being used for printing, the size of the image being printed, the looseness of the screen, and the type of resist being printed all come into consideration. The off-contact distance should be as shallow as possible for reliable printing. Too high an off-contact distance and the image becomes distorted.

Once the off-contact distance has been set, the adjustment bolts for the platform, and the adjustment bolts on the screen frame can be used to finely register the image to the panel. Generally, the stencil at one end of the panel is registered by platform movement; the rest of the panel is registered by wrenching the screen. After registration is achieved, the platform is securely taped to the printing station using masking tape.

The underside of the screen should be taped all around the stencil. This will prevent creases from forming in the mesh as a result of printing. There should be at least six inches from the end of the panel to the end of the frame. There should be a piece of laminate inserted between the end of the panel and the end of the screen. This piece of laminate is called an extender board. It protects the screening fabric when the squeegee is pushed over the end of the panel.

The last act which should be performed to the frame is to tape off the inside of the frame on the top side of the screen. Masking tape can be used for this procedure. This tape is a great aid in keeping ink off the frame, and an aid in cleaning up after the run is complete. The edge of the squeegee where it is fastened into its wooden handle can also be taped off for easy cleaning.

Pre-Printing Considerations

There are a number of details which should not be overlooked, but rather must be built into the screen printing procedures. The screen operator must inspect each panel prior to printing. Both sides must be checked for cleanliness. Panels with excessive oxide, spots, fingerprints, or other contamination should be returned to Plating for cleaning.

Prior to printing, and usually during the panel inspection step, the panels must be uniformly oriented. They must be stacked up with the same side facing upwards, and the tooling holes all aligned. When this has been done, there is less chance that the printed image will be on the wrong side, or backwards.

First Article

Another panel should be pinned to the platform and the registration double checked visually. Then ink may be applied to the screen, usually the back end. A squeegee is then drawn lightly over the stencil to flood it. It is good practice to wait a minute or two after flooding the stencil, for ink penetration. A piece of clear polyester should be placed over the panel, and the panel printed. When this has been done, the polyester covered panel is checked for registration and image quality. Any adjustments to either the image or the registration can be made at this time.

The job traveler and blueprint should be consulted to identify which tooling holes are going to be remain unplated, and which are to be plated. The stencil should contain pads to prevent ink from getting into plated through tooling holes. If it does not, the screen operator should add the pads before beginning production screen printing.

After the clear polyester image has been checked and adjustments made; the first print on a copper panel can be made. Generally, every panel should be wiped with a lintless rag or tissue (tack rag), after it has been pinned on to the platform. Several panels should be printed, and the image looked at after each one. If there are no defects noticed, the third or fourth panel should be baked for about 15 minutes (at the ink manufacturer's recommended temperature), and sent to the touch-up department for inspection. No more printing should be done until Touch-up has bought off the first article. If there is no problem, the job is ready for production printing.

Production Printing

It is good practice to have Touch-up inspect about every twenty-fifth panel; this does not mean that the screen operator should not be inspecting every panel as it comes off the screening table.

The screen printer should not put any more ink on the screen than is going to be used. Excess ink wastes ink, and makes it more difficult to keep the screen clean.

Double flooding, which is two flooding strokes being used before the print stroke, is a good technique to increase the speed with which the print stroke can reliably be made. Without double flooding of the stencil, a rapid printing stroke will tend to cause dry areas on the edges of the screen.

If the stencil has a tendency to bleed ink into areas where none is desired, the screen operator must be aware of this. It might be necessary to blot the screen every one or two prints. Blotting paper should be used for plating resist; newsprint is fine for blotting soldermask.

Warm panels should not be wiped with cheese cloth as a tack rag, as the wax will contaminate the panel surface; another lintless rag or tissue will have to be used.

The thicker and more difficult a resist is to print, the harder the squeegee blade should be. If the squeegee blade is too soft, it will buckle and cause a printing void. A squeegee hardness of 70–80 on the durometer will be adequate for most printing requirements. Only squeegees with clean, sharp, and nick free blades should be used for any screen imaging operation. The squeegee should not extend more than about one inch from either side of the circuit pattern.

The squeegee should be held at about 30 degrees from the front of the frame (Fig. 11-7), and at an angle of about 60 degrees from the surface of the panel being printed (Fig. 11-8). It should be held securely with both hands, and pressure applied firmly and evenly across the width of the squeegee; and for the entire length of the printing stroke.

The oven used to cure screen printed plating resist must not be used for soldermask curing; separate ovens must always be used. The ovens must provide rapid air circulation. If a thinned out resist is being printed, good practice suggests that only every other slot in the rack be filled. The added spacing will allow increased air circulation. Inadequate air circulation may result in ragged traces.

A lot of problems with screen printed panels are due to the fact that the screening operator did not read all the notes on the job traveler, or did not read the notes on the blueprint. Other problems are caused by the screen operator

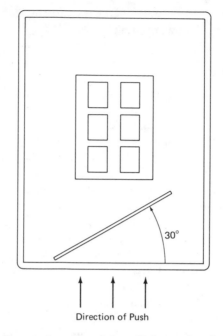

Direction of Push

Fig. 11-7. Squeegee is positioned about 30 degrees from front of chase.

not looking at what is being done. Sometimes the phototool used to expose the stencil was not prepared correctly. Sometimes panels which appear to be symmetrical really are not. The stakes are too high to allow production personnel to work on jobs without training them to read and think before they act.

STRIPPING THE SCREEN AFTER USE

When all printing is completed for a given screen, it must be stripped of all resist and stencils. Screen cleaning compounds made of trichlorothylene, or of

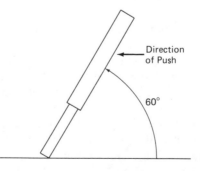

Fig. 11-8. Squeegee is tilted about 60 degrees from panel surface.

toluene base, work well for removing all resists and most of the stencil. After these have been stripped, there is a ghost image of the stencil left on the screen which can be removed with household bleach. The bleach is just poured right over the screen's surface, which is then scrubbed lightly with a nylon brush, then rinsed off with high pressure water. The disadvantages of toluene and trichloroethylene solvents are that they are dangerous and expensive.

Alkaline strippable plating resists can be removed in a caustic solution, then rinsed with water and bleach. The big disadvantage to caustic strippers is that they attack the frame, shortening its working life, and making it difficult to adjust the tensioning bolts. There is one type of noncaustic, nonflammable or chlorinated screen stripper which will not attack the screen, even with prolonged exposure (Liquistrip 10, Magna-flux Co.). The use of dangerous strippers should be avoided.

AUTOMATIC SCREEN PRINTERS

There are numerous manufacturers of automatic screen printing equipment (Fig. 11-9). Advantages to using this type of equipment are:

1. The pressure used during the printing stroke is uniform and consistent.
2. The squeegee angles with respect to the surface of the panel, and the direction of push, are consistent.
3. Throughput rates are dramatically higher.
4. Operator training time is shorter, and the demand for high priced skilled labor is greatly reduced.

It often happens that a company will invest the $25,000 or so needed to purchase an automatic printer, only to have it sit unused. Manual screen printing operators often do not like to see this equipment come into their work area. If a company does invest money in one, management should assign one person the task of getting the printer set up and running successfully.

PLATING RESISTS

Plating resists can be divided into three main categories: vinyl, alkaline soluble, and ultraviolet (UV) curable.

Vinyl Plating Resist

Vinyl plating resists have been used successfully for decades. These resists have excellent adhesion and screening characteristics, and they can be used with all plating, etching, and other immersion processing baths. Their resistance to cyanide, alkaline etchants, and high pH plating baths is well established. The vinyl

Fig. 11-9. Automatic screen printing machine.

resists are the only resists capable of withstanding black oxides. (See the article "Soldermask over Bare Copper: Alternatives, Advantages, and Manufacturing Techniques," in the appendix.)

Vinyl resists have been supplanted since the 1970s by aqueous stripping resists. Vinyl resists give off noxious fumes during screen printing and oven curing, which makes them unpleasant to work with. Also, they require chlorinated hydrocarbon (trichloroethylene, methylene chloride) or flammable hydrocarbon (toluene base) strippers. These solvents are expensive and unpleasant to work with, and pose a substantial health and fire hazard. Spent solvent must be distilled, and eventually hauled away for disposal at substantial cost. Solvent strippable resists will form resist scum on panel surfaces if the stripper becomes too loaded with resist.

Alkaline Soluble Plating Resists

Alkaline soluble resists have found wide acceptance, and are used almost exclusively throughout the printed circuit industry where screen printing is used.

Alkaline resists give off virtually no noxious fumes during printing or drying. They withstand all plating and etching baths which are neutral to acidic. These resists also withstand the rigors of high temperature processing for such baths as tin-nickel and immersion tin (both at 165 degrees Fahrenheit).

These resists strip extremely well in alkaline/caustic solutions, even at room temperatures. At 125 degrees Fahrenheit their stripping solutions have an extremely high loading capacity, and can be water rinsed without scum formation. The stripper is cheap and easily disposed of, often with little more than acid neutralization before going to the sanitary sewer.

Quality performance is another reason alkaline strippable resists are so widely accepted. They are formulated for ease of printing, good adhesion, ability to resolve fine lines and spaces, and resistance to bleeding (flowing into tiny crevices in the panel surface). The screen printing operator can use the resist most suited to his/her purpose. If speed of printing with minimum operator fatigue is desired, low viscosity, high solvent resists are available. If a double screen printing operation is needed (screen printing on top of tin-nickel for selective solder plating), a more tenacious, higher viscosity, higher molecular weight resist is also available.

For the finest resolution, a 325 mesh stainless steel screen with a high viscosity resist is needed. The high viscosity resists are a little more difficult to print in a manual operation, but there are good reasons for staying clear of low viscosity, high solvent inks. The lowest viscosity resists are truly easy to print with, but they tend to load up the drying oven with more solvent fumes than can be carried away. The net result is ragged traces. When low viscosity inks are used, it is advisable to double space the panels on their racks for maximum air circulation.

Ultraviolet Curable Resists

The UV curable resists may also be alkaline soluble. The advantage to printing with UV curable ink is that there is no solvent to be dried off. The resist is simply screened on, and run through a conveyorized ultraviolet light source for curing.

The UV curable resists have several disadvantages over oven curables. The UV curables are far more expensive than traditional resist. Since the oven drying resists do cure fairly rapidly (15–20 minutes), little is gained over them by UV curing. These resists tend to strip like photoresists. They do not completely dissolve; and since they do not, one of the advantages of using screened on resist is lost. Any UV curable leaves tiny particles of resist which can cause other problems, as already discussed. Another problem with UV curables is their toxicity. They contain photoinitiators and monomers which can cause mild irritation, or even severe allergic reactions. It is difficult to avoid breathing their vapors, or eliminating all skin contact. Good practice requires frequent washing of hands, always and in particular before eating food or touching the skin anywhere on the body. Wearing protective clothing, such as a smock, helps prevent contamination of clothes and skin.

SOLDERMASK

Soldermask is a coating applied to printed circuits which reduces the ability of molten solder to adhere to the board's surface. The soldermask is typically applied by screen printing using artwork which enables the entire board surface to be covered, except for the holes, pads, and contact fingers.

Ideally, after any soldering operation solder adheres only to the solder plated hole, and the component lead in the hole. In reality, without soldermask, solder will not only stick to the dielectric substate (the fiberglass and plastic material which the board is fabricated from), it will also bridge across that substrate and short circuit the conductors. With the application of the hard, resistant surfaces of soldermasks, the ability of molten solder to adhere to the board surface is greatly reduced.

There are other reasons for applying soldermask. In actuality, soldermask need only be applied to the solder side (that side which will be in contact with the solder wave); however, it is common for printed circuit users to require soldermask on both the component and solder sides. Soldermask also acts as a barrier which prevents damage to the circuitry due to scratches. It also prevents short circuits from forming when stray wires or dirt contaminate the surface. Printed circuits which must function in environments of high humidity are better protected from corrosion with the aid of soldermask.

Types of Soldermasks

The four common types of soldermasks are: (1) two part epoxy, (2) one part epoxy, (3) ultraviolet curable, and (4) dry film soldermask. All of these are screen printed onto the board, except for the dry film soldermask. The dry film soldermask is laminated onto the board surface, exposed with a phototool and UV light, and developed like photoresist, but using more complicated equipment and procedures. (More information is available from Dupont, Dynachem, and Hercules).

Two part epoxy soldermasks must be weighed and mixed thoroughly before they are screen printed onto the board. Failure to weigh the components properly, or to mix them completely, may result in streaky discoloration of the soldermask, and poor adhesion during wave soldering (Fig. 11-10). The applied soldermask is cured to hardness, heat resistance and chemical resistance by baking in an oven with hot air circulation.

The one part epoxy soldermask requires no mixing. It is slightly more difficult to print, due to high viscosity. The material also cures by oven baking; however, it requires about 50% longer baking time than does the two part sold-

Fig. 11-10. Wave soldering machine.

ermask. With both masks, failure to cure sufficiently will cause white spots or a white film to appear on their surfaces after wave soldering.

UV curable soldermasks offer two potential advantages over oven curing solder masks: much shorter curing times, and low volatiles content. These masks are screen printed, then run through a conveyorized ultraviolet light source for curing. For best results they should also be baked a little to complete the cure. Because of the low volatiles content, the ovens are not burdened with having to remove large amounts of solvent. At the same time, there is much less contamination of the pads and holes due to soldermask fumes.

The main drawbacks with UV curables, outside of their cost, are the need for excellent exposure control, and the need for excellent cleanliness and dryness of the panel about to be soldermasked. Failure to obtain adequate exposure may result in blistering of the mask during wave soldering, just as failure to print on a sufficiently clean and dry surface may result in poor adhesion as well as blistering.

Dry film solder masks are currently available which draw upon acrylic photopolymer technology, already discussed under the chapter on dry film photoresist imaging. The chief advantages of dry film solder masks are:

1. Ability to apply soldermask between tightly spaced conductors.
2. Ability to cover large panels with high density circuitry, without distortion of the soldermask pattern; this leads to finer registration.
3. Total absence of soldermask bleeding onto pads and other conductor terminal areas.
4. The thickness of the soldermask (3–4 mils) provides enhanced mechanical and electrical protection for the printed circuit.

These are all valid advantages over screen printed photoresist. However, this type of soldermask does have its disadvantages:

1. It is acrylic based, not epoxy based; surface cleanliness and dryness are critical. Acrylics lack the inherent adhesion of epoxies.
2. Dry film soldermask is expensive and complicated to use.
3. The vacuum lamination equipment needed for soldermask application is expensive to buy and to maintain (Fig. 11-11).
4. As with any UV curable resist, exposure control is critical for soldermask resistance to processing chemicals and temperatures.

The disadvantages and added costs have to be passed on to the customer. However, even though there are disadvanges and expenses which must be tolerated, perhaps the one overriding argument in favor of a printed circuit manufacturer investing in dry film soldermask is the competitive advantage that manufacturer has, in being able to offer the customer a process not available elsewhere.

Fig. 11-11. DuPont's vacuum laminator for dry film solder mask.

The manufacturers of these masks have spent much research on developing easy exposure control methods for use by screen printing operators. Exposure control must not be neglected, even for a few hours. The UV curables are manufactured from modified acrylic resins, which lend themselves well to photopolymerization. Polymerization is by free radical reaction (chain reaction), and is heat sensitive. Heat applied to such a reaction will hasten it to completion. For this reason the panels should be baked to provide a final soldermask which is sufficiently resistant to molten solder and cleaning solvents.

Photopolymers chemically bond to copper or other metal surfaces; and the adhesion is due in part to the chemical bond. For this reason, the surfaces must be very clean and dry. Where epoxy solder masks which are oven cured will bond well even over fingerprints, oils, and oxidation (oven curing removes moisture), photopolymers will not. Screen printers must understand this need and not neglect it.

Screen Printing Considerations

Soldermasks are more difficult to print than are plating resist, for several reasons. The material has a high viscosity and is printed over a raised surface: the plated circuitry. While any screen printing operation should be done with as little pressure as is needed to get the job done, soldermask requires an extra firm grip on the squeegee, and extra firm pressure on the stencil. The printing stroke should be slow enough to flood between the circuit traces. A coarse screening mesh, for both polyester or stainless steel, should be used. Many screen printers expose the stencil about 25% longer than is required for printing plating resist. This extra exposure time chokes the holes (makes them smaller) around pads, and makes the stencil a little thicker and stronger.

Sometimes bleeding (printed substance flowing where it is not desired) becomes a problem, and the soldermask flows onto pads. When this happens, the screen operator must blot the stencil by periodically printing over blotting paper. It may be that a soldermask with more thixotropic properties is required. Thixotropic resists are resists which have a lower surface tension while they are being printed, than they do immediately after the printing stroke. These types of resists flow less than do other materials of virtually equivalent properties.

Each panel must be inspected before it is printed, and also as it comes off the printing table. Once soldermask is applied and cured, it is on for good. Soldermask cannot be stripped without damaging the surface of the printed circuit. The panels must be checked for contamination, scrubbed lightly to roughen the surface of the solder, and stacked with the same orientation, to minimize printing the wrong side of the panel.

Since soldermask application is such an irrevocable step, it is especially critical that the phototool be checked for compliance with the blueprint. It is a fairly common oversight by photo technicians to miss notes on the blueprints which require that there be no soldermask in certain areas on the board surface. It should always be the responsiblity of the screen printing operator to read the blueprint, and understand the requirements of each job before beginning work on it.

Uncured soldermasks can be stripped with chlorinated hydrocarbons or with flammable hydrocarbons like toluene based screen strippers.

LEGEND RESISTS

The legend, or nomenclature, is screen printed onto many printed circuits to identify areas of the circuit for loading and testing purposes. The legend is often referred to as "Silk Screen" on blueprints. The ink used to apply the resist is usually a two part epoxy, which is oven cured. When the inks are mixed, they are usually left to stand for 30–60 minutes; this gives the catalyst and resin time to mix for a proper curing reaction. The manufacturer's instructions must be followed to the letter, if proper printing, curing, and adhesion results are to be obtained. Printing legend can be difficult; it is not only being printed over raised circuitry, but it must be legible. Since legend is permanent, once it is cured, legend printed panels must also be checked for legibility and for completeness of the printed image.

SECTION FOUR
N/C PROCESSING

12

Numerical Controlled Drilling And Routing

Louis T. Verdugo
Hewlett Packard
Corvalis, Oregon

INTRODUCTION

Printed circuit board fabrication today consists of many factors. The two most important factors are *drilling* and *routing*. It should also be remembered that the design of the printed circuit board is also a variable that greatly affects drilling and routing. This chapter focuses on some design considerations and deals with drilling and router programming.

Printed circuit (PC) design begins with the schematic. The schematic should be drawn to its simplest and relative form at the outset. This means holding crossovers to a minimum and putting it in a general format pertinent to the finished product.

It is often easier to use layout templates and/or the physical components themselves to make many trial and error wiring diagrams. This is done until the interrelationships of all components and wiring are compatible with good design practice, with respect to mutual coupling and inductive and capacitive effects. This process, with many rearrangements of components and conductors between each step, will evolve into more refined designs that will be production ready.

In designing PC boards, the designer/engineer interface is a must. Prototypes should always consider the production process capabilities. The age old statement of "We made the prototypes, why can't you make production?" can be avoided if these philosophies are kept in mind.

Design and layout have a serious effect on basic circuit performance and stability. The design and layout are major factors in controlling deterioration by effects such as temperature, vibration, shock, vapor, and moisture. Also, it should *always* be remembered that *design* will influence every stage of production from prototype to the finished printed circuit board. These factors also affect production, test repair, maintenance, and factors other than the initial design costs.

DESIGN CONSIDERATIONS

1. *Material Selection.* For high frequencies, insulation resistance and dielectric properties of the insulating base material of the PC board must be given high consideration. This becomes critical when stability is required for circuit performance under environmental extremes.

2. *Trace or Conductor Width.* The voltage drop and temperature rise of the conductors should be given special thought. With the new electronic devices, such as integrated circuitry, the proximity of a trace, or the width of a trace can cause an integrated circuit's performance to falter. Trace width can generate noise, capacitance, and temperature transfer.

3. *Location of Components.* The position of each component should lend itself to the overall function of the board. Switches, trimmers, and test jacks should be readily accessible to the user. It is also desirable to have every component in a position that is as close to its schematic order as possible to avoid coupling. Also, with the advent of automatic insertion equipment, it is most desirable to have the components in the same X and Y axis. Components should be located so that physical interference can never occur. When components touch each other, thermal or electrical problems can occur.

4. *Trace or Conductor Routing.* As to this consideration, input from the electrical engineer is a must. Many of the paths taken by a trace can affect circuit performance. For instance, to eliminate crosstalk requires adequate space or ground between traces. The end result of examples such as this can mean longer and wider traces.

5. *Dimensions.* This topic alone is sufficient for an entire book. Dimensioning of a printed circuit board can run the cost of the board far beyond original cost estimates. Obviously, tight dimensions and tolerances are attainable (if you are willing to pay for them!), but why incur extra costs when the following careful design considerations can be followed?

- Reference point. All dimensions should be taken from a common point. This point, for example, could be a tooling hole or the board edge. This must be done to eliminate the accumulation of tolerances.
- Most PC boards today are dense and require small conductor widths. If so, they should be kept to a minimum of 0.015 inch wide, with a space of 0.013 inch.
- If a dimensioned hole is required, try to leave as much tolerance in the dimension as possible to eliminate the need for accurate artwork.
- Keep identifying letters or figures as far away from traces as possible. Small lettering or figures can be a problem during the etching process. Consequently, they should be as large as possible.
- A space of 0.050 inch should be allowed around the overall circuit pattern.

This is to accommodate any accumulation of tolerances of drilling or screening.

ARTWORK GENERATION

The next step in PC design is making a photographic replica of the design sketch. Here the PC board actually becomes phototooling. There are two basic ways to generate phototooling:

- Computer Aided Design System (CADS)
- Tape Master

1. *Computer Aided Design* (CAD). The CAD system initially starts with an X-Y coordinate device known as a digitizer. A digitizer is a device which converts line drawings and other graphic representations into digital values which can be processed directly by a computer for analysis, storage, or computation. The digital values are recorded by the digitizer onto punched cards, paper tape, or magnetic tape for immediate storage and handling between the digitizer and the computer.

Next, a plotter quickly and efficiently generates graphics, accepting digital values from a computer, producing line drawings, cut masks or exposed film. (Exposed film now becomes the photo tooling for PC boards.)

Typically, plotters are used to produce engineering drawings of all kinds: electrical schematics and logic diagrams, artwork (photo tooling) for printed circuits, and drill drawings for PC boards.

More recently, digitizers and plotters are being combined with small computers to perform the complete cycle from input through output more economically at one location independent of large outside batch processing computers.

An excellent example in relation to PC boards is the computer design aided drafting system which combines the digitizer and plotter with a minicomputer and specialized software to allow the draftsman or engineer to go from the sketch to the final drawing.

2. *Tape Masters*. The tape master is the most widely used method of generating photo tooling for PC boards. Tapemasters are generated by placing pieces of tape over a clear mylar film which represents the various circuit paths. Unlike the CAD system, the tape is laid upon the film manually. Obviously, a shaky hand or a bad day for the layout person can mean lesser quality artwork. Another consideration while doing tapemasters is that the artwork must be kept clean (no dust or other foreign material). What is produced on the film down to the smallest detail will be transmitted throughout the production operation.

ARTWORK INSPECTION

Once the artwork has been generated, it must be inspected for quality in relation to the production process. This can only be brought about by a visual inspection. The inspection department should be the focal point from where a decision must be made to run a job through the production area or not. It is *critical* in the PC board manufacturing process that *all* artwork meet certain criteria for production. In order to accomplish this, the following should be considered:

1. Check spacing, pad to pad, pad to trace, trace to trace, and lettering or symbols to pads. Do they meet the design standards?
2. Check opaqueness of pads and traces.
3. Check correct lettering height and width.
4. Check correct logo information.
5. Check for traces not terminating at a pad.
6. Check for traces terminating past a pad.
7. If multilayer, check inner layer registration of pads to outer layer.
8. If the board has finger connectors, check that they are centered on the connector tab and that the component and circuit sides are aligned.
9. Check fabrication dimensions against the artwork. Are pads centered on the dimension?
10. Are internal tooling holes present? If so, are those dimensions correct?
11. Do you have all the necessary artwork?
12. If a soldermask is required, are *all* the holes relieved?
13. Do you have *all* the necessary artwork for the job?

Though these considerations are brief, much detail must be spent on artwork inspection. Each consideration listed above must meet the design standards of your particular production process. *Remember,* good artwork will mean a good product.

DRILLING

The Good Old Days

In the beginning, there was the hand drill. Numerical Control (N/C) is a relatively recent innovation in the printed circuit industry. Before N/C machines got to be so popular, drillers used a variety of manual and semi-automatic drilling machines to drill single stacks of boards. The artwork for any given board guided the drillers directly, and dimensioned holes, finger alignment, etc., were of no concern whatever to drilling.

With such "primitive" drilling methods, tolerances were loose and pads were large. Drillers got pretty tired of drilling the same pattern over and over. It did not really matter how many times an order was done, every time the board came through the shop the same amount of work went into each board. To put it simply, every stack drilled was like having to reprogram the board. With the advent of NC machines, changes began in drilling.

It all started as a way to make large quantities of the same board without having to put so much work into each board. Program it once, and just run the tape for each stack of boards you want drilled. For a PC house doing any volume at all, the economics were immediately clear. The capital investment for N/C machines is large, but the process is cheaper and more accurate and faster than manual drilling.

It did not take long for drillers to become programmers, and everyone was happy . . . for a while. It soon became clear, however, that tolerances for N/C drilled boards could tighten up, and pads could shrink because these drilling machines were (and are) more accurate than manual drilling. From this point, it was only a small step to realizing that drill programmers could "adjust" the drill tape to correct for sloppy artwork. It has become commonplace to expect minor errors in artwork to be compensated for in drill programming. This approach makes good economic sense, but it requires more and more technical ability from the drill programmer.

Drill programmers need to know something about every operation in the shop. The more closely related the operation is to drilling, the more they need to know about that operation. The more complex it gets, the more important it becomes that one asks questions rather than making assumptions.

Most of us learn to draw before we learn to write, and the famous line, "A picture is worth a thousand words," holds true in printed circuits. Our version of the picture is the print. With every production order, you will get at least one print. There are many names for various prints:

- Drill drawings.
- Blue lines.
- Sepias.
- Fab drawings.
- Blank drawings.
- Drillmaster.
- Subpanel drawing.
- Board blank.

Often you will get part of the information on one print and the rest of it on another. Believe it or not, there are several good reasons for splitting up the information onto more than one print. But, you must have all the relevant information before programming a board.

The first step in programming any board is to get all the information together. Along with all the pictures, lines, and numbers will be several notes. At least one of these notes will be the title (name, number, etc.) of the print. Also, the various pieces of artwork and tapemasters which were used to photographically make the artwork will be listed and numbered as if they were prints as well. You only need one piece of artwork so ignore "component," "circuit," "soldermask," and "tapemaster;" these are merely names and numbers for artwork. The documents you need are all the other prints. Be sure you have them all before going any further.

In addition to the above prints, you should also have a copy of the step sheet. This sheet is a simple drawing of the panel to be made indicating how many images per panel, dimensions of the images, and spacing between those images. There will also be some dimensions which indicate the theoretical distance from the panel edge to the images. These dimensions are for the photo lab to guide them in aligning the stepping machine. However, these dimensions *do not* accurately reflect the actual distances from the reference hole to the images on the final artwork. There is too much potential error in the process of aligning single image artwork on the stepping machine for drill programming to use the dimensions from the panel edge to the reference hole, images, etc. The dimensions on the step sheet which you should be concerned with are:

1. The overall panel dimensions.
2. The single image length.
3. The single image width.
4. The distances between images, in both axes.
5. The distances between tooling holes.

The first thing to be sure of is that you have the correct revisions of all documents. Usually, the piece of artwork you will be programming will have a date on it. This date represents the day, month, and year when the master artwork was generated by the customer. Be sure that this date corresponds to the date on the drill drawing (not necessarily the date on which the drill drawing was drawn, but rather the date under the picture of the boards to be drilled).

Each print will have a revision letter code somewhere on the board. This letter should reflect any changes which have been made in the print and should be accompanied by notes of explanation. These notes can be very useful in revising programs, recognizing possible problems, and keeping track of what has been done to a certain printed circuit board.

Both the date of the artwork and the revision code are used by the expediters in determining whether you are working on the correct version of a certain job.

The next thing to check is the location of any dimensioned holes on both the drill and blank drawings. At this point, just establish that all your information is consistent. Assuming that all your documents are consistent with one another, and they are all the right ones for the job, you can now begin to use the prints. Somewhere on every print will be a series of notes which explain any special features of the board in question. Read all the notes and be sure you understand exactly what is being said. The understanding part may be more difficult than you anticipate; as mentioned before, most of us draw better than we write. It is also not clear that note writers all really understand how you make PC boards.

The Drill Drawing

Drill sizes, the number of holes of each size, whether the holes are to be plated through or not, all should be indicated, as well as a picture (and perhaps some dimensions) to indicate the locations of the holes on the drill drawing.

Different customers use different methods to indicate drill size and finish (plated or unplated). Plated-through holes must be drilled before sensitizing (a light copper deposit covering the whole panel) and before plating. Since the plating process will naturally slightly decrease the diameter of a hole, production sometimes needs to compensate for plating. If the drill drawing only tells you the finished hole size for a series of plated-through holes, production will have to drill these holes slightly larger than the finished size. Generally, production will add (0.006 inch) to the finished diameter of a plated-through hole to arrive at the drill diameter. Often, the customer indicates both the finished hole size and the drill size which will ultimately produce that finished hole. When the customer indicates both, the difference between the finished hole size and the drill size should be checked to see if it is 0.006 inch.

Drill sizes are usually indicated by number, not by decimal fractions of an inch. The numbering system for drills is simply a simple way of talking about drills of different sizes. Each drill size number or fractional drill size corresponds to a decimal fraction (expressed in inches) which is the actual measurement of the drill diameter.

Unplated holes are less confusing than plated-through holes. Unplated holes are drilled after plating, and because production does not have to worry about buildup in the hole, they are drilled at the same size indicated as the finished hole size. Occassionally, customers will make a mistake in indicating the drill size and finished hole size of an unplated hole. They might show it as unplated and yet indicate the usual 0.006 inch difference between finished hole size and drill size (as if it were a plated-through hole). *Do not* assume that they mean this or that. Any mistakes or ambiguities on prints should be straightened out if for no other reason than to bring the mistake to the customer's attention. Never second-guess the print!

Sometimes, the customer does not care whether a hole is plated or not. Then they will indicate that the hole can be "E" (either plated or not). Generally, production will plate such a hole because it is easier to drill all the holes at once, unless, naturally, all other holes on the board are unplated. Any holes with a diameter greater than 0.156 inch (5⁄32 inch and above) can easily be routed instead of drilled. Production can rout holes greater than 0.125 inch (1⁄8 inch and above), but diameters between 0.125 and 0.156 inch are hard to rout because of speed limitations of the routers.

Generally, production cannot drill holes greater than 0.25 inch (1⁄4 inch) on the drills. This is because the spindles on the drilling machines do not generate enough torque (power) to drill large holes. Any plated-through holes which need drill sizes over 1⁄4 inch should be done on the routers. Since most of the boards go on the routers for final fabrication (cutting the boards out of the panel), and since production does have the ability to rout holes, unplated holes over 0.156 inch should be routed when the board is fabricated. The only times this is not true are when:

1. Drilling unplated holes over 0.156 inch gives production internal router tooling which would otherwise not be possible.
2. Production has to after-plate drill some small holes on the board.

Dot to Dot: Color Coding. Virtually every job done in drill programming has artwork and photographic negatives and positive (actual size) of both outside layers of the board. In addition to these pieces of artwork, most jobs also have a positive and negative padmaster. The padmaster is a piece of artwork without circuitry, only pads and fingers. Ideally, this is the artwork to use when programming. On jobs without a padmaster, production generally uses the component side artwork. If the component side has holes which are obscured by traces or a ground plane, production will have to use the circuit side. Whatever set of artwork is used (padmaster, component, or circuit side), production should always use the positive.

Once you have selected the appropriate positive, you must be sure that it is right-side-up before color coding it. There are several ways to determine the right side. One of the easiest is to see if you can read the written notes on the piece of artwork. If you can, that is the right side. If the writing seems to be mirror-image, turn the artwork over. Second, check the pattern on the artwork against the pattern on the drill drawing. Generally, the drill drawing will show the right side if you are using either the padmaster or the component side. If you are using the circuit side, and the drill drawing represents the component or pad side, then the right side of the circuit artwork will show the reverse pattern of the drill drawing.

Once you have the right piece of artwork and have it right-side-up, you are ready to code it. On a multiple image piece of artwork you need only code one image, so the process of coding is about the same for single or multiple images. The idea is to connect all the pads for holes of the same size with a colored line. The line should be as straight as possible (without cutting through other size hole pads); it should be as efficient as possible (the total length of the line should be as short as possible); and the line of any single hole size should never cross itself. Use a different color pen for each color size. These rules make the actual programming of the board as easy as possible while also keeping to a minimum the amount of table travel which the drill will ultimately do.

It is usually easiest to start coding the less numerous of the hole sizes first and leave the greatest number until last. Some programmers like to write the drill size they are coding in the same color pen at the start of the coding line, others don't. It is up to the programmer. It is also a good idea to use the more transparent (but easily visible) colors like red or orange for the smaller, and usually more numerous pads. This way you won't tend to obliterate the holes which should be programmed by dimension. Generally, one of the easiest ways to indicate dimensioned holes is just to circle them with the pen you're using to link holes of that size. Holes that will be routed, not drilled, can simply be crossed out with an "x" through the pad. Don't obliterate these pads—remember you're coding these holes for router, so that whoever programs the board doesn't think you simply missed a hole—you're not trying to wipe the pads out.

Stepped artwork (artwork with more than one image) needs on more thing done after you've finished coding the single image. Stepped artwork has exactly the same image "stepped" several times in either the X or Y (horizontal or vertical) axis (or both). Stepped artwork can be divided into two categories: simple and "flip-flop". Simple stepped artwork merely consists of several images, all oriented the same way. Flip-flop artwork consists of ½ the images in one orientation, the other half of the images in reverse (mirror image) orientation.

In color coding simple stepped artwork, select a single hole from the image you've coded (usually one of the smaller, circular pads), and circle it with a colored pen which differs from the color with which you coded that drill size. Then circle the identical pad on all the other images on the piece of artwork.

In color coding flip-flop stepped artwork, select and circle a pad in the coded image. Then circle the identical pad on all the images which are oriented the same way as the image you coded first. Next, select the image among the reverse orientation images which has the same relative location as the image you first color coded (if you coded the "upper left" image, select the "upper left" image of the flipped images). Then circle the identical pad as you circled in the first image.

Programming

Aligning the Artwork. Once a job has been color coded, all the prints assembled, and all questions answered, it gets programmed. The artwork goes onto the work table right-side up so that the writing is readable (unless there is a specific reason for turning it over).

Place the artwork on the work table and place the glass cover-plate over the artwork. Try to align the artwork parallel to the Y-axis as well as you can as follows:

1. Lift the glass.
2. Place the artwork on the work table so that the single obround hole punched in the artwork is furthest away from you.
3. The row of three holes should be close to you so that the left obround is 4 inches from the center round hole, and the right obround is 3½ inches from the center round hole.
4. The far obround and the near round hole should be in line with the Y-axis edges of the work table.
5. Carefully lower the glass onto the artwork so that the artwork does not shift as the glass lays down.

Now to fine tune the alignment using the scope and the work table adjusting knob. The adjusting knob pivots the work table to align the artwork with the "true" Y-axis of the programming machine. First, though, a word about the scope.

The scope should be set up once during a job and never changed during the programming process. If necessary, adjust the focus so the artwork appears crisp and clear. Adjust the light fiber bundles so that the image in the scope is well lit, without shadows, but also without glare. On the surface of the scope image is a pair of intersecting cross-hairs. A plastic overlay is clamped to the scope screen surface to provide a series of concentric circles for digitizing. Be sure that the center of the overlay aligns with the intersecting cross-hairs. The lines on the overlay will not align with the cross-hairs; they should align with the true X- and Y-axes of travel. This alignment can be done by moving the table in one axis and rotating the overlay until the lines on the overlay seem to align with the direction of travel. Once the scope and overlay are focused and aligned, and the light adjusted, *never touch them until you are done with the job.* Otherwise, the image will shift in the scope. Also, *never turn the digitizer off in the middle of a job.* In the event of a power failure during a job, start over; there is no way to accurately do half a job at a time. For the same reasons, once you start a job, do not let someone else finish it; everyone's eye is a little different. It is simple: do not start unless you can finish.

Back to alignment of the artwork. Run the Y-axis down to the far obround hole (or to the far target on a single image, nonstandard piece of artwork) and center the scope circles on the obround hole. Now run the Y-axis down to the round hole. Don't move the X-axis to center it; rather, turn the adjustment knob until the round hole is centered. Go back up to the obround and center it using the X-axis control. Back down to the round hole and center it using the adjustment knob. Repeat this process until you no longer need to adjust at either end of the artwork. At this point, the artwork will be in line with the true *Y*- and *X*-axes of the programming work table. *This alignment is critical.*

Establishing Working Reference. Now it is *absolutely* necessary to establish the zero reference for the image you will program. Every print has a reference point indicated from which most of the dimensions are measured. We need to establish this point on the work table so that any dimensioned holes can be programmed from that point. However, do not simply go to a location and set the display registers to zero. The correct way to establish the reference point for any board is to work back from the critical areas of the board to the point which theoretically corresponds to the print reference. This is a zero-tolerance operation, so be sure that the display reads *exactly* as it should:

1. Go out to the critical area or areas of the board and establish a point which is dimensioned on the print.
2. Preset both display registers to zero using the red buttons below the display.
3. Move both axis controls back to the negative (opposite) direction from the indicated point to the print reference point and reset both axes to zero.

This will give a base reference. From this point, the operator does the fine tuning on the established reference. Essentially, this means going out to several points and returning to the appropriate zero point, adding or subtracting a few thousandths from the base reference until you get the best possible theoretical reference point which corresponds to the print's reference. By basing the zero reference on the critical areas of the board (rather than on arbitrary areas like corner cutoff lines) production can ensure that these critical areas will be manufactured correctly with a minimum amount of trial-and-error within the various fabrication areas of the shop.

The general set of priorities for critical areas is as follows:

- *Most critical areas:*
  Special areas on print (specified).
  Connectors centered in tabs.
  Dimensioned holes.
  Dimensioned tooling holes.
- *Least critical areas:*
  Cutoff lines.

Once the operator establishes the best possible zero reference for the single image, check the critical areas to be sure that production actually can make this board.

In the following examples, print theoretical dimensions are indicated by enclosure in quotation marks; actual dimensions are unadorned, and corrected balanced dimensions are indicated with an asterik (*).

Example 1. Fingers off center within connector tab.

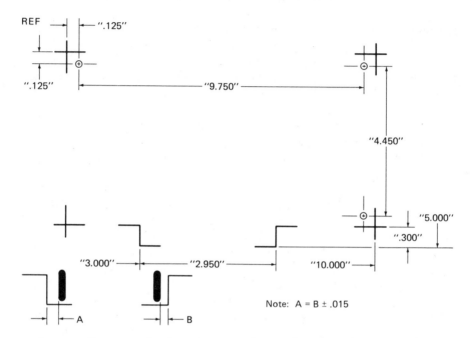

Note: A = B ± .015

Assume, for example, that the artwork matches the print except for one minor detail: the fingers are not centered within the connector tab. The tab cutoff lines may be dimensioned correctly, but A = 0.125, and B = 0.085 (as indicated below).

Drill programming has the capability to correct this error at this step in production. This is accomplished by shifting the base reference point by 0.020 (twenty thousandths) inch to even up the distances between the fingers and the tab edge (dashed lines), then establishing the zero reference at the shifted point, *and programming all dimensioned holes from the new reference.*

The ironic thing is that the good board won't look as good as the impossible board. In the good board, the dimensioned holes will be shifted 0.020 inch off the center of the pad, as shown below:

but when production cuts the board out, it should be right on in the critical area of connector spacing. Without the shift, the connectors would be off. It might be argued that the router operator could accomplish the same thing by offsetting the board to the right. There are two problems to this approach:

1. It takes a lot more time, and is a lot less precise to correct dimensional inaccuracies on the router.
2. If the operator were to offset the board 0.020 inch, the drilled dimension holes would no longer be the correct distance from the edge of the board. Instead of 0.125 inch from the left edge to the hole, the distance would be 0.105 inch. This hole may be almost as critical to the next operation (component insertion) as the connector finger spacing is to the finished board. (See Figs. 12-1, 12-2.)

NUMERICALLY CONTROLLED ROUTING PARAMETERS

Parameters

Numerical control (NC) contouring machines (routers) give production much more flexibility and repeatable accuracy at far greater speed than any former method of blanking printed circuit boards. The cost effectiveness of NC routing stems from three basic elements of the system:

1. "Soft" tooling. Programmable operations and positioning allow quick, inexpensive changes in blank configurations. This eliminates hard tooling cost and inflexibility.
2. Positioning and repeatability accuracy. Both positioning and repeatability accuracy are measurable at magnitude of 10^{-3} inch.
3. Multiple spindles and stations. A single pass on one routing machine can blank as many as twenty standard (0.06 inch thick) PC boards.

Along with these features, however, production finds that NC routing has some limitations in terms of both possible and desirable results. As with any

Fig. 12-1. N/C programming equipment.

system, there are specific results which are impossible to obtain on the routers, and certain results are much easier (and consequently cheaper) to obtain than others. What follows is an outline of specific limitations, operational parameters, normal procedures; and the advisable tolerances, tooling configurations, and design criterion for routing.

Fig. 12-2. N/C drill.

"Impossible" Operations

Square Inside Corners. Inside corners are those in which material surrounds the cutout (see illustration below):

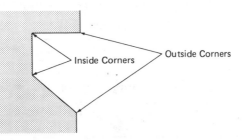

The cutters production uses in the routers are round and generally have a theoretical diameter of 0.125 inch. Consequently, any inside cuts will have a corner radius of about 0.06 inch. Outside corners, on the other hand, are more easily and quickly cut without any radius. A few methods of designing around inside corner radii are illustrated on the following page:

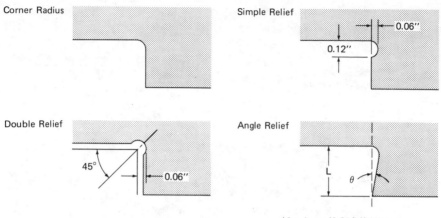

$(\theta = A \tan (0.06/L))$ Minimum

Whenever corner radii are permissible on inside corners, a note to that affect on the blank drawing helps avoid ambiguity. When it is necessary to design around inside corner radii, the simple relief method illustrated above is easier to program than either the double relief or the angle relief methods. Production prefers the simple relief. Whatever method of relieving the corner you see, naturally, land area on the board must reflect that relief, or we will end up cutting into traces or connector fingers.

Size Limitations. The maximum cutting surface under each spindle measures 24 inches × 24 inches. Boards exceeding this limit in either length or width could conceivably be cut in a multi-step operation. However, the board would need internal tooling holes in each 24 inch section, and the probability of holding tight tolerances in this kind of operation (less than ±0.015″) would be very low. See illustration below:

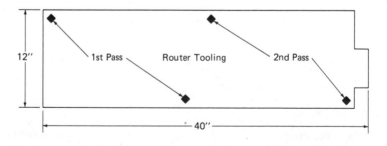

Although it would be theoretically possible to rout such oversize boards in a multistep operation, it would also be necessary to drill and silkscreen in multistep operations. Feasible tolerances would multiply in each operation, and by the time the boards were to be blanked, the results would be unsatisfactory.

Non-Circular Arcs. The routers which we use have the ability to contour arcs or portions of arcs if the portion of the arc being cut can be fit to a circle with radius of 99.999 inches or less. Any other conic sections (parabolic, eliptical, hyperbolic) or other second (or greater) order equations can only be cut by approximation of portions of the equation in straight line or circular segments. The cost of this kind of approximation, in small enough sections to result in a "fit" of any appreciable acceptability, would be *extremely* high.

"Unrealistic" Tolerances. For tolerances within the range of magnitudes between 10^{-3} and 10^{-2} inch it is clear that yield is inversely proportional to tolerance. (The tighter the tolerance, the lower the yield). The factors involved in limiting tolerances and the types of tolerance which we recognize as distinct areas of concern will be discussed at length later. For now, specifying the diameter of a routed hole at ± 0.001 inch or specifying the blank dimensions of a board at ± 0.003 inch create "impossible" situations for NC routing. (*Note:* The above tolerances are indicated as examples only, and do not represent limits for possible versus impossible tolerances.

Z-Axis Control. Fabrication requirements which include cutting partially through the thickness of the board (e.g., countersink holes, engraving) with tolerances of magnitude tighter than 10^{-2} inch are not possible on the routers. Z-axis control is only programmable to the extent that the spindles are either down or up (cutting or not). All settings of depth of cut are entirely dependent on operator set-up. These situations are essentially manual operations in fabrication.

Tolerances in N/C Routing

In N/C routing, production is faced with two basic types of tolerance limitations. For discussion, these are divided here into "absolute" and "relative" limitations. After defining these two types of limitation, we will go into the various implications of specific limitations upon our end results.

"Absolute" Tolerance Limitations. These limitations are simply the limits which a single operation, such as drilling or routing, can hold within that operation. The absolute limits of routing depend only on the routing process itself. Once the material to be routed goes onto the machine, the aspects of the finished part which are fabricated at this time are related to one another by the tolerances production is able to hold under a particular router spindle. If we rout a hole in the board and cut the board out in the same operation, the relationship between the hole's location and the perimeter of the finished board will depend on the repeatable accuracy of the NC routing machine (assuming that programming is correct). Similarly, the relationship between the length, width, and connector tabs (for example) will only vary within the limits of the machine. The set-up, programming, and operation of the machine are related

to these absolute limitations. However, lead-screw drift, spindle shift, table squareness, and other machine variables are more directly related to absolute tolerance limits.

"Relative" Tolerance Limitations. Since the manufacture of printed circuit boards involves a series of operations, the entire process can be seen as if we only made multilayer boards. In the same way that inner layers must align to drilled patterns and outer layers, the various serial operations in making any PC board must align to one another. The relationship between any aspects of a finished part shows the relationship between the processes used to create the aspects in question. The relationship between the location of a drilled hole and the perimeter of the board depends on the relationship between drilling and routing. The relationship between drilling and plating determines the resultant finished hole size of a plated-through hole.

Whenever more than one operation is necessary to obtain a desired result, the reliable limitations on the possible tolerances are functions of the accumulated tolerance limitations of the individual operations. This is just a complex way of saying that the sum of the parts equals the whole tolerance limitation. Therefore, the relative tolerance limitations for NC routing depend entirely on the absolute limits of both routing and whatever other operations are necessary to achieve the end result. The implications for routing are clear. As the last step in PC manufacture, routing depends on virtually every preceding operation in order to rout a "good" board. An error or an excessive amount of shift in drilling, image transfer, or even artwork generation may not show up until it is too late. Naturally, each operation controls and monitors the degree to which it deviates from the theoretical "true" board, which represents the nominal relationships between all of its various aspects. However, as the tolerances of the finished board which are acceptable to the customer approach the limits of the serial relative tolerance limitations, the average yield of any specific order goes down. The chances of manufacturing a maximum number of good boards in an order goes up as the finished board's acceptable tolerances loosen up.

Currently, when we speak of limitations, we assume that M. R. Murphy's law is in full control: Anything that can go wrong will: From that point on it becomes a case of addressing the *probability* of getting a good board.

NC routing generates programming methods which give essentially two options:

1. Contour the board once and hold overall blank dimensions within a certain tolerance (preferably ± 0.015 inch, for reasons which may become clear).
2. Contour the board twice and hold overall blank dimensions within a tighter tolerance.

These two options demand different programming, so it is not a case of simply repeating the cut to get a second pass around the board. When correctly programmed, the single pass method takes about one-half the time to rout a board. Whenever a board can be cut to the looser tolerances, it saves time and dollars to do so. The only drawback to the single pass method is that the finished parts come out a little dustier than those routed with two passes. Consequently, boards with rotary switches may need additional cleaning to ensure that dust lodged on the edge of a board not wear loose and possibly cause contact problems later.

Current programming methods include compensation for reasonable variations in plating thicknesses of the holes which are used for tooling.

Absolute limitations of the routing process (assumption: machine is within operating specs):

Table Positions:
1. Absolute X,Y positioning: ± 0.002 inch.
2. Repeatable X,Y positioning: ± 0.001 inch.
3. Squareness of travel: ± 0.002 inch over 2 foot run.

Alignment:
4. Spindle to work table reference: ± 0.002 inch.
5. Work table's tooling squareness: ± 0.002 inch over 2 feet.

Operating Stability:
6. Variability in true spindle position as it heats: ± 0.003 inch.

All the above limitations are based either upon actual measurement or on the limitations implicit in methods currently used in one operation to align the machines. They are not based on advertized specs which the manufacturers offer. They are given only to show the possible problems implicit in the single operation of routing.

Let us say, for example, that we are going to rout a hole in a piece of material. It takes a minimum of five programmed positioning moves to route any hole. Each of these moves has, at the very least, a possible deviation of ± 0.001 inch in repeatable positioning accuracy. This fact alone accounts for a certain degree of deviation in roundness of routed holes, as well as limiting the hole size tolerance for routed holes to ± 0.004 inch. The position of this hole on parts routed under separate spindles at the same time may vary as much as 0.005 inch due to alignment and spindle position variations.

Absolute Tolerances: Productivity versus Accuracy. There are several methods for controlling the operation of routing which result in minimized potential error. There are also methods for increasing the time available to actually cut out boards. Unfortunately, these two types of method are generally inversely related. To minimize the work table to spindle alignment and squareness tolerances, production could tool each job every time it is run under the spindle by which it will be routed. This would mean placing a drill in each spindle, placing new or partially used tooling subplates under each spindle, running the tooling tape to drill the tooling holes, and mounting pins in the tooling holes. This process would usually take an average of 17 minutes per part number every time the job gets run. Add to this set-up time the time it takes to change cutters and run a first article. Then you must carefully measure the first article and make any offset adjustments to center connector fingers, etc. Now it is finally OK to run the part.

There is really no way to trim down the actual cutting time of the board. Production typically limits table travel as much as possible by maximizing the efficiency of the programming language. To push machines any faster would result in broken cutters or shattered (rather than cut) edges on the finished parts. There may be a minor reduction in the average time to rout the average board as production learns more, but the length of time needed to rout a board is primarily dependent on the designed complexity of the board's perimeter. It takes much less time to cut out a simple square than it does to cut a board which has several areas of complex detail. Boards are frequently designed to fit into irregular areas; this is made possible by the flexibility of the routing process. However, some boards finally look more like an abstract representation of a contour map than like a PC board. Cosmetic detail, such as calling out a radius on all external corners, should be avoided. When specific details are necessary, production should be more than happy to oblige; however, a straight

Fig. 12-3. N/C router.

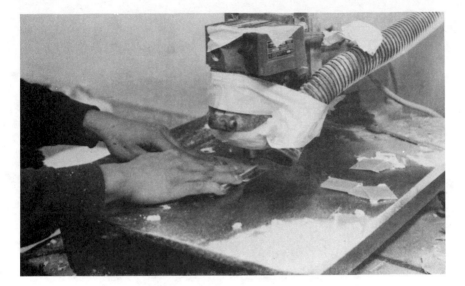

Fig. 12-4. Using a pin router for manual fabrication.

line is still the fastest distance to cut between two points. (See Figs. 12-3 to 12-7.)

The one place where it is possible to save a great deal of time is in tooling. For high volume boards, a set of tooled subplates can sometimes be made and saved to run again whenever the job comes through. This requires safeguards

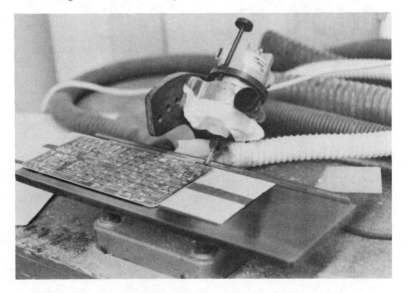

Fig. 12-5. Chamfering (beveling) the leading edges of contact fingers.

Fig. 12-6. Air powered shears for cutting laminate to panel size.

Fig. 12-7. Machine for drilling holes and inserting stack-up pins prior to drilling.

against random spindle-to-spindle error by a system of regular alignment of each spindle with its corresponding tooling plate. Saving tooling subplates can actually eliminate the initial tooling and set-up time from some high-volume, frequently ordered parts. Due to the potential for absolute tooling positioning error, however, first article inspection becomes even more critical in such cases.

REFRENCES

1. Coombs, C. C., *"Printed Circuit Handbook"*, pp 13–16.
2. Haya, G. G., *"Computer-Aided Design"*, IEEE TRANS Comp. Vol. C-18, No. 1, pp 1–10, 1969.
3. Yule, Gordon, *"Drilling & Routing"*, Technical Paper, Vol. 1, No. 1, pp 1–10, 1976.

ACKNOWLEDGEMENTS

My thanks to Gaylen Grover for giving me the opportunity, and to Gordon Yule for his daily help in the drill and routing program area. Most of all, my thanks to the people of Bldg. 15 who make all this happen year after year.

12A
Drilling Procedures

Raymond H. Clark and Chuck Candelaria

SHEARING

Shearing, or blanking, is generally under the authority of the drilling supervisor. Too often, shearing practices are left solely to the discretion of the operator. The more types of laminate used at a shop, the more critical is the need for shearing operator training and rigidly enforced procedures. Some typical problems which arise from a poorly controlled shearing area are:

1. Wrong type of material issued by warehouse:
   A. Polyimide instead of FR4.
   B. FR4 instead of teflon.
   C. Natural color instead of blue.
   D. Wrong thickness issued.
   E. Wrong cladding issued.
2. Unapproved or rejected laminate issued.
3. Cross-warped (cross-grained) multilayer laminate issued.
4. Incorrect panel size cut.
5. Panels cut from the incorrect sheet size, resulting in excessive drop-off being created.
6. Incorrect number of panels being cut.
   A. This can result from insufficient laminate being available and from an error being written on the traveler by Engineering.
7. Material panels cut from are insufficient for the job, due to lack of laminate, and production control not being notified of this fact.
8. Operator failing to verify that he/she has been issued the proper material, or that when cut, the material is being assigned to the correct job.
9. Warp direction not being marked on panels and travelers for the information of multilayer and dry film personnel.
10. Copper cladding (2/1, 2/2, 3/3, 3/2) not being stamped into the material for the information of multilayer and dry film personnel.
11. Laminate manufacturer and lot number information not being recorded on traveler and in logbooks.
12. Panels being pinned together incorrectly for drilling:
    A. Stack height.
    B. None, or wrong, back-up material.
13. Unbaked inner layer laminate getting into manufacturing.
14. Unused material not being identified properly and returned to the warehouse.
15. Drop-off material not being properly identified.
16. Laminate scrapped out during processing being replaced without properly accounting for it.

The shearing department can be a major source of heartache when the operator is untrained and not supervised. All work put into a job is based on the material assigned to the job by the operator.

Because of inner layer registration requirements of multilayer construction, it is good practice to improve the dimensional stability of the core laminate by baking just after shearing. This operation must be built into the process by procedure.

The planning engineering department sometimes writes incorrect information on the traveler. The shear operator must be trained to look at such basic things as the number of boards required to meet shipments, the number of panels called out to be sheared, and the number of circuits per panel. Should a grossly disproportionate number be listed for cutting, the operator must bring this to the attention of his/her supervisor.

To address the problems of material control the shearing operator must not be left out of the training loop.

DRILLING

Part of the reason for establishing written procedures is to determine how best to run a process. There are some very basic considerations which must be recognized before beginning work on a job. Among them are verifying the correctness of the tools being used, and making sure that all instructions have been read and understood. No operator of any process should proceed without doing the necessary thinking and reading needed to prevent problems. The operator must also recognize mistakes which may already have been committed.

Every shop should establish the number of hits to be used on a drill bit, and develop a method for tracking how often a bit has been resharpened. Color coding provides control of the number of resharpenings. The guidelines provided for the number of hits must be based on drill bit inspection before and after drilling a given number of hits, and on examination of microsectioned holes after plating: bit quality and hole quality.

Second drilling is the operation of loading panels, which have already been drilled and plated, back onto the drilling machine to drill holes requiring no plating. This is an operation which should be avoided when the nonplated holes are dimensined to tolerances tighter than ± 0.003. Also, second drilling holes in metal areas can result in burrs and lifting of hole pads from the laminate surface. The Photo department should provide relieved areas in metal surfaces when second drilling is to take place.

All drilling stations should be provided with a microscope for inspecting drill bits. Encouraging drill operators to become aware of drill bit quality will help make them aware of some of the situations which damage drill bits. Even the simple act of setting the ring can chip a bit, if the anvil face of the ring setter is worn. Unfortunately, many drill operators feel confident of drilling good hole quality just because they may be using a new bit, not being aware of the numerous conditions which will damage the bit.

Drill operators should be trained (and the training enforced) to sort panels into stacks according to whether or not the panel is the bottom panel from the drilling stack. These bottom panels must be checked for missing or extra holes, registration, broken drill bits, and wrong hole size. If no defects are found in the bottom panels, there is little chance of defects in the panels drilled above; defects in the bottom panel are indicative of defects in the panels further up in the drilling stack.

Drill operators must sign travelers for the actual number of panels drilled. Should this number differ from traveler requirements, the drilling supervisor must be notified. In fact, supervisors must be notified during any operation where the number of panels being processed is not sufficient to meet shipping requirements, and supervisors must in turn notify production control.

When setting up a machine to drill multilayer panels, it is good practice to first check the zero point against an etched panel (etched detail). Initially, the multilayer tooling holes are used as the reference for registration. It is common for multilayer panels to become slightly distorted. Because of this the etched detail should have only a few holes drilled across the panel. The panel can then be removed from the machine and viewed for registration. It may be necessary to adjust the zero in the X or Y coordinates. Once this has been done, the rest of the panels can be drilled with greater confidence of accurate registration of holes to inner layer patterns. An etched inner layer is sometimes used for this purpose. The etched detail, a laminated panel which has had all copper etched from the outer layers, is still better, since it has been subjected to the stresses of lamination.

First articles are drilled to verify correctness of programming, as an aid to inspection of incoming artwork, and as the pattern used to set up working film. The first article (F/A) must be drilled as though it were a production panel: same drill sizes and feed and speeds. Also, second drilled holes must be drilled into it; and, for multilayer panels, holes representing lamination tooling holes must be drilled. A mylar is drilled at the same time and is submitted to Inspection with the F/A for approval. This mylar will be used by the drill operators on production runs to check for missing and extra holes, and shifts in the zero. F/A's must always be dated and identified with Customer, part number, revision level and company control number.

PROCEDURE FOR RIGID LAMINATE (DOUBLE SIDED BOARDS/WHITE TRAVELER)

1. Only laminate which has been accepted by Quality Assurance is to be issued by the Stock Room for manufacturing.
   A. Shear operator shall use only material issued by the Stock Room.
   B. No laminate shall be issued by the Stock Room unless it has been accepted by QA.
2. Shear operator shall read the traveler to verify that the material which has been issued matches the requirements listed on the Shop Traveler.
   A. Type
      1. FR4 = GFN = G10/FR for most applications.
      2. Thickness: 0.031, 0.059, 0.062, 0.093, or 0.125 are most common.
      3. Copper cladding 1/1, 1/0, 2/2 or 2/0 for most applications. (1/1 means 1 oz copper on both sides, etc.)
      4. The shear operator shall consult with the drilling supervisor for any materials other than those listed above.
3. Laminate is to be sheared according to the panel size and quantity listed on the Shop Traveler, and according to the posted shearing guide for each sheet and panel size.
   A. This posted guide has diagrams which show the layout for cutting panels from any given sheet size, for maximum material utilization.
   B. Leftover laminate (drop-off) is to be identified with the material type, the manufacturers name, and the material lot number.
   C. Panel quantity to be sheared is listed on the traveler just above the PROCESS STAGES heading.
      1. Shear operator shall verify that panel quantity listed is sufficient for approximately 10% overage above what is required to meet ordered board quantity.
         a. Operator shall notify drilling supervisor if too few, or grossly too many panels are listed for shearing on the Shop Traveler.
      2. Sometimes a note has been written that there are parts available in Stock; in which case a reduced panel quantity is being run.
4. Shearing
   A. Shear one sheet at a time.
   B. When the shear has been set up and the first cut made, the operator shall measure the first cut to verify the dimension as correct before proceeding with further cuts.
      1. Readjust the shear if needed.
      2. Proceed with all cuts in that dimension.
   C. When the shear is readjusted for cutting in the cross direction, the first cut must also be measured and verified as correct.
   D. For longer runs, it is good practice to measure a piece approximately every 10 cuts.
   E. At completion of shearing:
      1. Count panels sheared, sort out any which are obviously damaged.

2. Make proper entries on Shop Traveler for:
   a. Material type, lot number, and manufacturer, and number of sheets used.
   b. Shear operator initials, date, and quantity.
3. Make the proper entries in the Shearing Logbook.
4. Bundle drop-off material and identify it by type, lot number, and manufacturer.
5. Bake laminate in stacks which are no more than 3 inches high between steel plates. Bake at 300 degrees for 4 hours.
6. It is good practice to attach a material sticker, if there were any on the laminate, to the back of the Shop Traveler.

PROCEDURE FOR THIN CLAD LAMINATE (MULTILAYER/BLUE TRAVELER)

1. The instructions are similar to shearing rigid, with additional considerations.
   A. Multilayer laminate must have the warp (or grain) direction marked on each sheet prior to shearing.
      1. If the warp direction is not marked on the sheet, the shear operator must reject it back to the Stock Room and notify the drilling supervisor.
      2. All panels cut from each sheet must have the warp running in the same direction.
      3. All travelers must have the warp direction denoted on them.
         a. The diagram on the Shop Traveler labeled "panel" shall have "warp" written in, with an arrow showing the direction of warp.
   B. Multilayer laminate frequently has copper cladding which is uneven (such as 2/1, or 3/2), and each sheet of uncut laminate must be marked with the heaviest side clearly shown.
      1. Unmarked laminate shall be rejected by the shearing operator, back to the Stock Room, and the drilling supervisor notified.
      2. Each panel must have the heaviest cladding stamped in the corner: the number must be imprinted into the foil cladding, not applied by ink.
      3. All laminate clad with other than 1/1 shall be stamped on both sides.
   C. All multilayer laminate must be baked between steel plates according to the instructions on the traveler (4 hours at 300 degrees F).
      1. The panels must cool under pressure for at least 1 hour, before being unloaded from the oven.
   D. The material issued from the Stock Room must be of the exact core and copper cladding specified on the Shop Traveler.
      1. No substitutions are to be made unless Engineering is notified, agrees, and writes the substitution on the Shop Traveler.
   E. Drop-off must be bundled together and identified with:
      1. Type: Thickness
             Cladding
             Warp (grain) direction
      2. Lot number
      3. Manufacturer
      4. Drop-off less than 6 × 6 inches shall be placed in the scrap bin.
2. The shear operator shall never use laminate that does not have all the identifying information on it.

DRILLING PROCEDURE

1. Before starting any job the drill operator shall carefully review the instructions on the Shop Traveler form, paying special attention to comments written by Engineering.
   A. The traveler is to be properly signed and dated on all jobs which come into drilling.

B. All jobs are to have a blue print before the drilling operator begins work on the job.

2. Once the drill tape box has been pulled for the job, it must be reviewed against the Shop Traveler and the drill instruction sheet which is inside of the traveler:

    A. Customer, part number and revision level must agree.

    B. Drilling order and drill sizes must match.

    C. Second drilling instructions on the front of and inside the Shop Traveler must match those on the file card inside the drill tape box.

3. Drill bit control and color coding:

    A. New drill bits shall have either no color coding at all, or may have a yellow dot on their box.

    B. All repointed drill bits shall be color coded:

        1. The butt of the bit shall have been marked with a Sharpie brand marking pen.

        2. Each box shall have a colored dot placed on it.

    C. The colors to be used shall be:

        1. No color, or yellow: New bits only.

        2. Green: First repointing.

        3. Red: Second repointing.

        4. Orange: Third repointing.

    D. Used drill bits shall be inserted point down into their carrier box, and shall not be used again until repointed.

    E. Discarding drill bits:

        1. Drill bits shall not be used after their third repointing.

        2. Drill bits which are not to be re-pointed shall be inserted back into their carrier box point down, and the box turned over to Purchasing for disposal.

    F. All drill bits shall be kept under lock and key by the drilling supervisor; repointed bits shall not be allowed into manufacturing until they have been inspected and color coded by the drilling supervisor.

4. Obtaining drill bits:

    A. Multilayer boards: Only new drill bits shall be used on multilayer printed circuits.

        1. When new drill bits are not available, the multilayer job shall be set aside and the drilling supervisor notified.

    B. Drill operator shall determine the number of bits for each size required to drill the job as follows:

        1. Hits per drill:

            a. Multilayers: 2000 hits (new drill bits only)

            b. Double-sided: 2000 hits (repointed drill bits only)

        2. Number of holes, for each size, per individual circuit pattern.

            a. This information is on the file card inside the drill tape box.

        3. Number of circuit patterns per panel (number up).

            a. This information is on the file card and the Shop Traveler.

        4. Number of panels to be drilled.

            a. The number of panels to be started for the job is recorded on the Shop Traveler in the area above the heading "PROCESS STAGES."

        5. Number of drill bits per size:

$$\text{No. of bits} = \frac{\text{No. Panels} \times \text{No. Up} \times \text{No. Holes Per Size}}{\text{No. Hits Per Drill Bit}}$$

    C. The Shop Traveler and file card shall be taken to the drill supervisor, along with the calculations from B above; all drill bits shall be issued by the drill supervisor from a locked cabinet.

5. Setting up drill bits:
   A. Select enough bits of each size to drill one load of panels.
   B. Use the ring setter to locate the plastic ring on each of these drill bits.
      1. Bits 0.120 inch or less are set on the left side of the ring setter.
      2. Apply soft pressure to avoid chipping drill bits.
      3. Ring must be in location which matches the posted diagram for the drill size.
      4. Use bit microscope to check drill bit for chipping.
      5. If chipping occurs, notify drilling supervisor immediately.
      6. Chipped drill bits shall not be used under any circumstances.
   C. Set drill bits in drill pods from Left to Right, beginning with tool number 1.
      1. Use micrometer to measure every drill bit before it is set into the tool pods.
6. Setting up drill machine (Double Sided):
   A. Load the drill tape into machine for programming.
      1. There shall be a maximum of two tapes per box, one marked Primary, and one marked Second Drill (if needed).
      2. Whenever there are other tapes, or unmarked tapes, in the box the drill operator shall not proceed, but shall notify drill supervisor immediately.
   B. Feeds and speed programmed into controller by the tape shall be checked against the posted feeds and speeds chart for each drill size.
      1. If programmed feeds and speeds do not match posted, the posted values shall be keyed into the machine by the drill operator for each tool number.
   C. Number of hits per tool programmed into controller by the tape shall be checked by the drill operator for agreement with this specification; correct values must be keyed in by operator if not in agreement with below:
      1. Multilayer boards: 2000 hits
      2. Double-sided boards: 2000 hits
      3. Teflon/polyimide: 500 hits
   D. Check zero point programmed into the controller.
      1. Coordinates must match those on the file card.
      2. If programmed zero point coordinates do not match those of file card, drill operator shall key in the correct numbers.
7. Loading panels (double-sided):
   A. Panels pinned with back-up material are set on table according to pinning holes.
   B. Top panels shall be brushed to remove debris.
   C. Tape panels to table along all four sides, using 2-inch-wide masking tape.
   D. Tape precut entry material to top panel; tape on two sides.
8. Adjusting stroke heights:
   A. Adjust lower stroke height according to the value posted on each machine for the type of board being drilled.
   B. Lower the maximum stroke height to provide enough clearance to avoid drills being dragged across panel surfaces.
      1. 0.060 inches is usually adequate.
      2. When drilling screening or other tooling holes (such as pinning holes in multilayer First Articles), it is best to keep maximum stroke height at the highest value until these holes have been drilled.
         a. The intension here is to avoid striking the pins which hold panels to the drill table or carrier plate.
9. Running the job.
   A. Place machine in Auto Run and Cycle Start mode.
   B. Operator shall give attention to spindles at all tool changes by the machine.
      1. Operator shall verify that machine has dropped used bit, and picked up the new bit correctly.

2. Operator shall stop the drill if it failed to drop or pick up bits correctly.

C. Operator shall open spindle window to check drill bits, and inspect panel surfaces for broken drill bits.

    a. Broken bits shall be replaced immediately, and the missing holes picked up.

D. Operator shall monitor drill bit quality with the microscope, looking for chips, excessive debris, and darkly colored flutes.

    1. Bits which are wearing excessively shall be brought to attention of drilling supervisor.

10. Completing a load:

A. Check each stack with the drill mylar prior to deloading machine, looking for missing or extra holes.

B. Use the nail puller to pry stack from the table.

C. Use the arbor press to de-pin each stack (double-sided boards).

D. Examine back-up material for countersinking appearance of drilled holes, which indicates a shallow drilled hole in the bottom panel of the stack.

    1. Shallow drilled holes must be drilled out as if they were missing holes.

E. (Multilayer boards) Panels to be stamped to identify operator, drill head, and stack height:

    1. Each operator is assigned a stamp with his/her number.

    2. The stamp is to be applied as shown in Fig. 12A-1.

F. Reload table with more panels and continue the drilling process.

11. Completing a job:

A. Drill operator shall count all panels to ensure quantity matches that which is to be run on the job, from the Shop Traveler.

    1. Notify drilling supervisor any time fewer panels have been drilled than is specified on the traveler.

    2. Drilling supervisor shall notify Production Control whenever there are insufficient panels to make the job complete.

    3. The job shall not be signed out of Drilling until all panels for the job have been drilled.

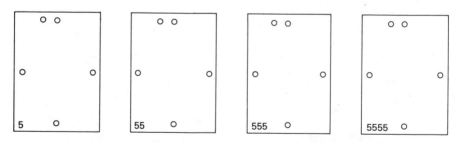

Fig. 12A-1. Multilayer panels drilled in one-high stacks. (Example: stamp orientation for Operator #5.)

    A. 5   Head no.1

       55   Head no.2

       555   Head no.3

       5555 Head no.4

    B. Left Corner: Drill stack is one panel high

       Center: Drill stack is two panels high

       Right Corner: Drill stack is three panels high

B. All scrap shall be noted on the traveler; scrap panels are to be taken to Inspection for log-out and disposal.
C. Traveler is to be signed and dated by the drill operator when the job is completed.
D. The job is to be logged into the drilling logbook also.
E. The daily throughput logsheet for the drilling department is also to be filled out.
F. The entire job will proceed to drilling inspection, which is part of the drilling department, to be inspected for:
   1. Missing or extra holes.
   2. Shallow drilled holes.
   3. Correctness of hole size.
   4. Broken drill bits.
   5. Burrs.
   Any of these conditions shall be reported by the drilling inspector to the drilling supervisor, and QC department, for resolution of the problem.
2. Special instructions for multilayer boards:
   A. Setting drill table up for carrier plates:
      1. Carrier plates are used to pin multilyers to the drill table.
      2. Panels are not stacked and pinned to back-up like double-sided boards are.
      3. Carrier plates are cut from FR4 0.125 1/1 laminate, and are taped and pinned to the drilling table along all four edges.
   B. Drilling tooling holes in carrier plates.
      1. Each size of multilayer panels will have a tooling tape used to drill the 5 multilayer tooling holes.
      2. Choose the correct tooling tape for the panel size.
      3. Set the zeros of the tooling tape, and the feeds and speeds for the 5 multilayer tooling holes.
      4. Tooling holes are drilled with a #31 drill bit (0.120 inch), set in the number 1 tool pod.
      5. Set the lower stroke height to the value posted on each machine.
      6. Load the tooling drill tape to program the drill.
         a. Hit program reset.
         b. Place machine in Auto.
         c. Hit Cycle Start to drill the 5 holes.
   C. Follow the instructions for setting up drill bits.
      1. Remember, only new drill bits are ever to be used on a multilayer job.
   D. Sort all multilayer panels so that all 5 tooling holes are oriented the same, and check for stamped job numbers on each panel.
      1. When cap layers have been used during multilayer construction, there may only be 3 tooling holes to align.
      2. Do not proceed with jobs when they do not have a job number stamped on them; notify drilling supervisor.
   E. Take one multilayer panel and have all the copper etched from both sides; this is called the etched detail.
      1. The etched detail shall be used to set the drill up for adjusting the zero point.
      2. The etched detail shall always be the first panel drilled in any multilayer job.
   F. Loading panels on the drill table:
      1. Insert tooling pins into the carrier plates:
         a. Place a pin in the Pin Insertion Jig.
         b. Plate jig and pin over the drilled tooling hole in the carrier plate.
         c. Tap pin gently with hammer, until it is inserted flush with the surface of the jig.
      2. Lay back-up material on the carrier plate.

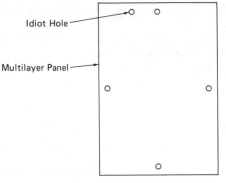

(Front of Machine)

Fig. 12A-2. Multilayer tooling hole orientation for loading panels on drilling machine.

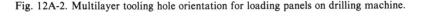

    a. Back-up must be predrilled to match multilayer tooling holes.
    b. Brush debris from carrier plates and back-ups.
  3. Brush debris from multilayer panels and lay them over the tooling pins.
    a. All panels shall have the idiot hole oriented in the upper left hand corner, as shown in Fig. 12A-2
    b. Stack height of panels shall be as follows:
      1. Etched detail: 1 high
      2. 0.062 multilayer boards: 2 high (maximum)
      3. 0.093 or thicker: 1 high
      4. Hole sizes 0.033 inch or smaller: 1 high
      5. Copper foil heavier than 2 oz.: 1 high
      6. When specified on the traveler: 1 high
        NOTE: When panels must be drilled 1 high because of requirements 2, 4, and 5 above, the drilling supervisor shall be notified before drilling begins.
  4. Tape entry and panels to carrier plates on along all 4 sides.
G. Drilling the etched detail:
  1. Use drill tape to program machine, and set up drilling instructions in the same manner as for double-sided boards.
  2. Drill several uncoded holes along the left, center, and right areas of the detail.
  3. Remove detail and examine for any shift required in either X or Y axes to center the drilled hole on inner layer metal pads, or in clearance areas of power and ground planes.
  4. If an X or Y coordinate shift is needed, key in the shift and drill several more holes in the indicated areas to varify.
  5. Drilling supervisor shall approve the etched detail of each multilayer job, by signing and dating it, before the job is run.
    a. Should the drilling supervisor not be available, he may designate another individual to approve the etched detail.
  6. The shift in X and/or Y coordinates of the approved etched detail shall be written on a note taped onto the drilling machine.
H. When the etched detail has been approved, follow the instructions for setting up drill bits as per double-sided boards.
 I. Unloading panels
  1. Check panels with the Mylar before unloading, as for double-sided boards.

2. Stamp all panels for operator, head number, and stack height.
3. Check panels from each head for the presence of shiny copper rings in the inner layers.
   a. Dull copper indicates smear.
J. Run the job following the guidelines given here under special instructions.

SECOND DRILLING

Purpose

To provide instructions for use by all drilling personnel for performing second drilling operations on multilayer and double side panels which have already been through the plating operation.

Procedure

1. The Shop Traveler shall be read for all notes important to the drill operator.
   A. The drill operator shall not work on the job if the traveler has not been accurately and fully signed off coming into the second drilling operation.
      1. Drill operator shall verify that each traveler has a blueprint inside, before beginning work on the job.
2. The drill tape box pulled for second drilling shall be the same box used for primary drilling: Customer, part number, and revision levels of box and traveler shall match.
3. File card inside tape box shall have the same drilling instructions as the drill instruction sheet inside the traveler.
   A. The holes called out for second drilling on the Shop Traveler shall be the same holes as are listed on the file card.
4. Pin FR4 0.125 1/1 carrier plates to the table, and secure with 2-inch-wide masking tape along all 4 sides.
5. Use the primary drill tape (production drilling tape), and a No. 31 drill bit (0.120 inch) to drill screening holes into the carrier plates.
   A. The screening holes will be the tooling holes used to pin the panels to the carrier plates.
6. Use Pin Insertion Jig and hammer to insert 0.120 inch pins into the carrier plates.
7. Align all panels to be second drilled in racks, with the screening holes oriented exactly the same.
8. Brush carrier plates and back-up material to remove debris, and place back-up on the carrier plates.
   A. The back-up is cut to fit inside the tooling holes.
9. Obtain drill bits, measure with micrometer, and set up in drill pods in a manner similar to primary drilling.
10. Read second drill tape into controller for programming.
    A. Check feed and speed, and number of hits per drill tool, to verify correctness with posted values.
    B. Use same number of hits when second drilling, as for primary drilling.
11. Stack height:
    A. Load to the same height as for primary drilling, except for when second drilling is through metal plated areas.
    B. Use a one-high stack height when drilling through metal.
12. Completing a job:
    A. Count all panels to verify sufficient quantity to make shipping requirements.
       1. Drill operator shall notify drilling supervisor whenever there are insufficient panel quantities to make the job.

2. Drilling supervisor shall notify Production Control whenever there is insufficient quantity of panels to make the shipping requirement.
13. Sign all trvelers, and log entries with operator initials, date, and panel counts.

FIRST ARTICLE DRILLING

Purpose

To provide instructions for use by all drilling personnel for performing first drilling articles to verify drill programming.

Procedure

1. Check the drill file card provided by Programming against the drilling instructions inside the traveler.
   A. They must match perfectly.
2. Pin the F/A panel to back-up material and load on drilling table.
   A. The F/A panel is always FR4 062 1/1.
3. Tape a piece of white or red Mylar and entry material to the brushed and debris-free surface of the panel.
4. For multilayer F/A's:
   A. Choose the multilayer tooling hole tape which corresponds to the panel size being drilled.
   B. Use this tape and a 0.250 drill bit to drill the 5 multilayer tooling holes into the F/A panel.
      1. Use the feed and speed instructions posted on the machine.
5. Obtain drills and prepare them for use as per regular drilling instructions.
   A. All F/A's must be drilled like a production panel.
6. Read the primary drilling tape into controller.
   A. Check feed, speed, and hit information as per normal drilling procedures.
7. Read in second drill tape, and perform drilling operations on the F/A.
8. At conclusion of drilling:
   A. Write "LRC" in the lower right-hand corner of the panel as it sets on the drilling table.
   B. Remove panel and write date, customer, part number, and job control number on panel and the Mylar.
   C. Write the number of hits for each tool on the file card which is kept inside the drill tape box; this information is taken from the hit page.
   D. The Mylar will be stored, after buy-off from Inspection, in the drilling room; and will be used to check future runs.
   E. The FR4 first article will be stored in Photo.
9. Fill out traveler, and all drilling log books correctly.
10. Submit F/A, mylar, traveler and tape box to Programming.
    A. Programming will turn all of this over to First Article Inspection.

SECTION FIVE
PLATING AND OTHER WET PROCESSES

13
Metal Thickness Determination

The printed circuit shop must be concerned with determining the thicknesses of the metals it is plating, for a variety of reasons.

1. The thickness of copper and the overcoating of metallic etch resist (tin-lead, tin-nickel, tin, nickel) affect the diameter of the plated through hole, and this diameter is of primary concern in manufacturing salable printed circuits.

2. The thickness of plated metal may be critical and called out by the customer: nickel barrier plate between copper and gold, gold thickness of contact fingers, bondable metals for die attach bonding, reflowed tin-lead to protect solderability and prevent oxidation.

3. Circuit conformance to artwork and design requirements in microwave and controlled impedance multilayer circuitry is absolutely necessary for the proper functioning of the printed circuit board in the equipment for which it is designed.

4. Some metals are quite expensive and are a substantial factor in the manufacturing cost of the printed circuit. When 30 microinches of gold is called out on the contact fingers, nothing is gained by plating an additional 50 micro inches; but the additional gold may have been enough to wipe out the profit margin in the printed circuit.

5. Some metals are plated as an etch resist to protect the copper, and are to be stripped from the copper after etching. This is true of tin-lead in soldermask over bare copper type of circuits. If tin-lead is plated in great excess of what is required to protect the copper during etching, additional time and chemicals must be spent stripping the tin-lead from the circuitry.

There is a new awareness on the part of printed circuit users of just how much metal is needed on the contact fingers (both nickel and gold), and in the holes and on top of the circuitry. Board users are becoming more sophisticated in the approaches they take to measure this metal. It is now fairly common for board users to have beta backscatter, magnetic, and electrical resistivity measuring equipment for measuring metals on the surface and in the holes. This new awareness on the part of users often means the printed circuit manufacturer must use the same equipment to inspect and plate with.

COPPER

Typical requirements for copper are 1.0–1.5 mils (0.001–0.0015 inch) of plating in the plated through hole. This affects the drill sizes used to drill the plated through holes. Once the holes have been drilled copper is chemically deposited inside. This deposited copper serves as an electrical connection between the sides of the board and as a substrate for electroplated copper. The plater tracks the thickness of copper, as it is being plated, by measuring the diameter of the through plated holes (Fig. 13-1).

For every 0.001 inch of copper plated in the hole, the diameter reduces by 0.002 inch; or, for every 0.0005 inch of copper plated in the hole, the diameter reduces 0.001 inch. The plater uses a hole diameter measuring gauge to track the hole diameter. The hole size is measured before plating is begun; the value obtained is the starting reference point. Every 10–20 minutes panel(s) are removed from the plating tanks to have the hole diameters measured by the plater.

There are a number of considerations which the plater, planning engineer, and process engineer must keep in mind regarding hole sizes:

1. The deposition rate of plated metal is a function of current density and current distribution. As was discussed in the chapter on amperage determination, current density is not evenly distributed across the surface of the panel. It is greatest at corners and edges of the panel or circuitry, and at isolated holes; and current density is least in the center of panels and land areas. This fact makes it necessary that holes be checked in numerous locations on the panel and within the individual circuits of the panel. The plater must always be aware of the areas where plating will be least and most, and not neglect these areas.

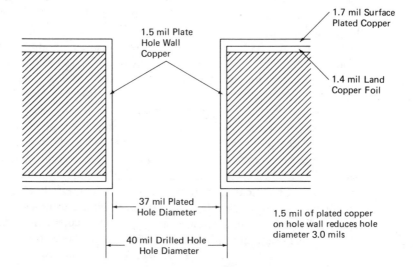

Fig. 13-1. (1) Metals generally plate slightly faster on the surface than in the holes. (2) Each 1.0 mil (0.001 inch) of copper plated on the hole wall, reduces hole diameter by 2.0 mils (0.002 inch).

2. The size of the pad around a hole may affect how fast the metal is deposited in that hole. Many manufacturers of copper plating baths for printed circuit plating advertise in their data sheets that their bath throws 1 : 1 copper in the hole and on the surface. The printed circuit manufacturer must realize the limitations of this information. The copper will plate about 1.0–1.5 times faster on the surface of the pad than it will plate in the hole, more often than not. However, the smaller the pad around the hole, the faster metal will plate inside that hole.

3. Not all holes have the same tolerance of diameter. Some holes have a diameter tolerance of ±0.005 inch, some of ±0.003 inch, and some even less. Obviously, it does little good to meet the tolerance of the ±0.005 inch, only to overplate the holes with lesser tolerance. With the advent of automatic component insertion equipment and press fit connectors, it is common to see plated through hole diameter tolerances of ±0.002 inch and even ±0.001 inch. Tolerances such as these pose a real challenge to the printed circuit manufacturer. In such cases, the planning engineer should make note of the critical holes (such as ±0.002 or less) and designate these as the holes to be plated to: only these holes should be used to measure copper deposition. The press fit connector must be shoved into the board and remain in when pulled with a designated force (such as 8 psi). This can only be accomplished when the hole has been correctly plated within the tight tolerance called out on the drawing. Platers and engineers must be aware of these holes and their criticality to the end user.

The type of hole measuring gauge is important. The two types in general use are the pin plug and the tapered shaft. Pin plug gauges are individual pins of constant diameter, one pin for each hole diameter. They usually come in a box with an assortment of sizes varying in 0.001 inch increments. The tapered shaft type of gauge is used to measure all hole sizes capable of being measured with that shaft. The taper gauge is available in a variety of sizes.

Each of these gauge types has its own advantages and disadvantages. The pin guage has the potential of being the most accurate. However, it is easy for platers to place the pins in the wrong hole when returning them to the pin box. Also, these pins are not usually available in stainless steel. Since they are used to measure wet, acid laden panels, they are exposed to a lot of moisture and acid. Even if they do get placed in the correct slot after use, these pins are subject to rapid corrosion and must be replaced continuously, to maintain accuracy. A supply of pins should be kept on hand and out of the plating area.

The taper shaft is easy to use and very quick. The shafts are usually made of stainless steel, or of a corrosion resistant metal overplate; corrosion is not the main problem. These shafts are not accurate for measuring holes in printed circuits. This is partially due to the plater's technique. The measured result is heavily dependent upon the amount of force used to shove the shaft into the holes, copper being a soft and easily distorted metal. It is common for different people to obtain considerably different results for the same hole. This fact alone makes the use of tapered shafts unreliable for printed circuit plating. Also, they tend to come out of calibration quickly when exposed to the rigors of the plating shop. They are banged, dropped, used as darts, shoved into and yanked out of tight holes. It is possible, in theory, to use these instruments successfully; however, to expect this is to have the utmost confidence in the restraint and diligence of those using them.

There are some instances where there are no plated through holes, but copper of specified thickness must be plated none the less. One way to avoid having to plate would be to use copper foil which matches the desired thickness. However, copper can be measured using beta backscatter technology, similar to the method used to measure nickel and gold over contact fingers. A Strontium or other beta source is used with the measuring equipment. The equipment is calibrated using copper foils and bare laminate, or even an unplated printed circuit board, as standards. For applications where this measuring technique is required, the beta backscatter equipment representative should be consulted.

Micro-sectioning is an effective and accurate method of determining the thickness of plated metal. This is considered destructive testing, since a sample of the panel being plated is destroyed. To use this method either a test coupon, such as those of MIL-STD-275D, or simply a series of drilled and plated through holes can be placed on the panel. The section can be taken by removing a panel from the plating tank and punching out the coupon or test holes. The section is imbedded in a plastic epoxy or polyester and sanded to view a cross-section of the plated metal in the hole walls. When viewed under a microscope the copper thickness can be accurately measured with either a stage or eyepiece micrometer.

It is possible to measure the thickness of copper plated in a hole by measuring the resistivity across that plated hole. This technique is more suitable as a quality control check on completed boards, than for plating. The panel must have been etched already, and only holes which are not electrically parallel with other holes are usable.

TIN-LEAD

It is possible to determine thickness of plated tin-lead by measuring the hole diameter; however, it is not usually done. The thickness required is so thin as to be at the lower limit of detectability, especially when the extreme softness of plated tin-lead is considered. Since only 0.0003–0.0005 inch of the metal is required, by the time enough is plated to measure, too much has been deposited. There is another consideration with tin-lead; it is reflowed. Tin-lead is a soft, easily corroded metal when plated. The reflow, or fusing, operation melts the tin-lead and fuses it into the dense, corrosion resistant alloy known as solder. The molten solder has a surface tension which causes the metal to be thick-

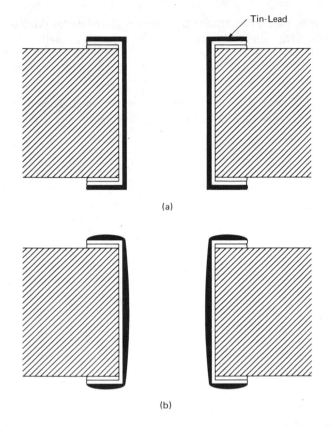

Fig. 13-2. (a) Plated tin-lead before reflow. (b) Plated tin-lead after reflow.

est in the center of circuitry and holes, and thinnest at edges; the net effect is to further reduce the plated hole size (Fig. 13-2). Most tin-lead thickness requirements are called out for the reflowed metal. Usually, 10–15 minutes of plating, at two-thirds the amperage used for plating copper, is sufficient to deposit and adequate amount of tin-lead. Micro-sectioning is an effective method for determining time and amperage effect on thickness.

NICKEL, TIN-NICKEL, TIN, AND SILVER

The thickness of all these metals can be easily determined by beta backscatter or magnetic techniques. A magnetic measurement (Nickelderm/Nickelscope) may be used to determine nickel. For unusual cases where these metals are being plated thicker, for use other than as an etch resist, the hole size measurement can be used to determine thickness of plated metal. Micro-sectioning is effective for determining thicknesses of these metals as well.

GOLD

The thickness of gold is almost universally determined by beta backscatter technique. It is possible to micro-section, but the deposits are so thin as to make visual readings unreliable.

14
Amperage Determination

In order that metal, especially copper, be plated to a desired thickness with an acceptable evenness and uniformity, the factors of current density must be thoroughly understood. As circuit density and complexity increase, and the tolerances of plated hole diameters decrease, the job of the plater becomes more difficult. Application of the understanding of current density and distribution will accomplish much more than mere attention to detail alone. A knowledge of the area to be plated and the recommended cathode current density of the plating bath is not sufficient to guarantee acceptability of the plating for printed circuits.

Current density is the amperage per square foot (a.s.f.) of plating area. Most manufacturers of plating baths for printed circuits formulate for and recommend about 30 a.s.f. for copper, and 15–20 a.s.f. for tin-lead. These values can only be used as a reference; the other factors involved with current density and current distribution must be given consideration for successful plating to be accomplished. The factors which affect current density and current distribution are:

1. Plating pattern geography.
2. Panel thickness and size of plated through holes.
3. Panel borders.
4. Plating racks.
5. Bath chemistry:

   - Concentration of metals and acids.
   - Concentration of organic leveling and brightening agents.
   - Concentration of contaminants.

6. Bath temperature.
7. Anode/cathode spacing.
8. Anode current density.
9. Anode depletion.
10. Plating bath agitation.
11. Cathode agitation.
12. Rectifier considerations.
13. Skill and experience of the plater.

GEOGRAPHY OF THE PLATING PATTERN

This is one of the most important considerations of plating. Involved are the physics of current distribution, and the physical layout (geography) of the circuit pattern. Figure 14-1 shows how electric current will tend to distribute itself on a bare copper panel racked for plating. The shaded portion shows the areas of highest current density. If a bare panel were to be plated at 30 a.s.f., which is the optimum current density recommended by most manufacturers of copper printed circuit plating baths, for 1 hour, the shaded area would measure thickest with plated copper. The center of the panel would measure the thinnest plating.

It is upon these facts of plating physics that printed circuit manufacturing is imposed. What is true for the bare panel carries over to each individual circuit pattern, and each location on the panel (Fig. 14-2). Circuits in the corner of a panel will plate at a faster rate than will circuits in the panel center.

The physical layout of a circuit pattern will present geographical configurations which will affect the current distribution and density. If a circuit is made of large land areas, and wide bus traces, but also contains narrow traces (0.010 inch or less) and isolated pads, not all areas of that circuit will plate at

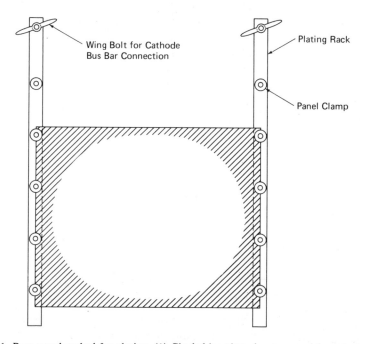

Fig. 14-1. Bare panel racked for plating. (1) Shaded locations denote areas of greatest current density; hence, fastest plating. (2) Panel center is area of lowest current density; slowest plating. (3) Plating rate is proportional to current density.

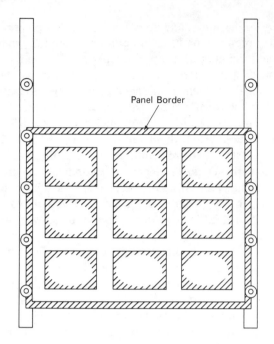

Fig. 14-2. Imaged panel racked for plating. Shaded locations denote areas of greatest current density. Panel borders and corner circuits plate the fastest.

the same rate. The isolated pads and narrow traces will tend to plate faster than the land areas and bus traces. If this circuit is plated at 30 a.s.f. the land areas and bus traces will plate somewhat less than 30 a.s.f., while the narrow traces and isolated pads will be plating at a rate above 30 a.s.f. This will probably cause no problems in the wider areas, but the narrow areas and isolated pads may plate so fast that they "burn" or become noduled. Burned plating occurs when copper is deposited faster than the atoms can be placed into a leveled matrix of metal; it deposits in bumps and nodules. The only way to prevent the nonuniform deposition from occurring is to lower the plating amperage to the point where current density in the narrow traces and isolated pads is at or below 30 a.s.f. This may mean that the overall current density for the panel may be as low as 15–20 a.s.f.

If the circuit pattern is uniform—no narrow traces (0.020 inch or wider), no exceptionally large land areas, no isolated pads, and overall land area evenly distributed throughout the circuit area—then plating at 30 a.s.f. will probably be accomplished with little problem. The plater will, however, always have to contend with higher plating at corners and edges of panels, plating being greatest at the corners and edges.

The tendency of a panel to plate fastest at the edges and corners, or a circuitry pattern to do the same, and to plate fastest along narrow traces and isolated pads, may present a problem in achieving plated through holes which have the required minimum amount of copper and still meet the tolerance range for the hole diameter after plating. This means that all the holes in the panel must be within 0.003 inch of the specified diameter, no matter where the holes are located on the panel and no matter how unbalanced the circuit pattern. The difficulty of plating within the diameter tolerance is increased when that tolerance is reduced. Hole diameter tolerances of ±0.002, or $+0.002$ -0.001, and even ±0.001 inch are common and becoming more so. The widespread use of automatic component insertion equipment and press fit connectors has created a demand for printed circuits with little tolerance in the plated through hole diameter. The plater or engineer charged with plating circuits to this tolerance will find it difficult or impossible to do so without plating well below the 30 a.s.f. which had formerly been considered optimum. The hole diameter tolerance, along with the evenness of the plating pattern, must be considered prior to setting amperage on the rectifier for plating.

Another problem created by uneven circuit pattern exists when one side of a panel is mostly land area, and the other side is traces and pads. The side with the traces and pads will tend to plate faster than will the opposite side. Also, the edges and corners will tend to plate much faster than will the center of the panel.

The plater and engineer must be aware of the plating physics and effect of circuit pattern and the demands of meeting hole diameter tolerances everywhere on the panel. Plating can be made more uniform throughout the panel by leaving "robber" borders along all four edges of each panel. Electric current is transmitted to the panel through the plating rack connections at the panel borders, and distributed around the panel along the borders. The wider the borders, the more uniform will be the distribution around the panel. This will help reduce a natural tendency to burn near the corners of panels. The use of plating racks which are bare metal will also aid current distribution at the panel borders, since electrical contact is made along the entire panel border, instead of just at the clip where the panel is fastened into the rack.

BATH CHEMISTRY

Bath chemistry affects the ability of a plating bath to deposit metal at a high or optimum current density. Every plating bath for printed circuit manufacturing has an optimum range for the plating bath constituents. For example, an acid copper sulfate plating bath may be formulated for optimum performance with the following concentration of constituents:

Copper (metal)	2.5 ounces per gallon
Sulfuric Acid	23.5 ounces per gallon
Chloride	80.0 parts per million
Brightener	0.5% by volume

Usually, a range is also quoted for the constituents. Should the metal fall below that range, the throwing power of the bath decreases; should metal content increase too far above the range, the conductivity decreases. A similar situation occurs for each of the constituents. All of these have an effect on the ability of the plating bath to operate at its optimum current density.

The presence or absence of organic contamination also affects the current density range at which a plating bath can operate. Organic contamination usually reduces the range. Brightener is a special class of organic contamination. When present, it causes the surface of the panel to polarize in the high current density areas (become electrically nonconductive). this causes the current to flow through the areas of the panel which had formerly been lower current density, until they too become polarized. Eventually, the polarized areas will again become conductive. The net effect is that the current is more evenly distributed across the panel, which produces more even plating with a leveled (bright) deposit. If the brightener concentration becomes too low, too little leveling is produced; too high, and the brightener becomes included in the deposit, making it brittle and perhaps dull or burned.

TEMPERATURE

Plating baths generally have an optimum temperature, which may be anywhere from 70 degrees (for acid copper or tin-lead) to 160 degrees Fahrenheit (for tin-nickel). Operating below the range may cause a reduced throwing power situation similar to insufficient brightener.

AGITATION

Good plating distribution requires good agitation of the plating bath, and even of the cathode (panel which is being plated). If there is no agitation, the plating bath becomes depleted of metal ions at the surface of the panel. This will quickly lead to burned plating if bath is being operated at the optimum current density. Agitation replenishes the metal ions so that plating may continue. Sometimes it may not be a good idea to use air agitation; in such cases, usually where tin or a foaming organic additive is present, the alternative is to supply agitation through either increased rate of filter pumping, or through agitation of the cathode bus bar.

ANODES

The distance between the anode and the cathode is important. If the printed circuit panel is too near the anode, that portion of the panel directly opposite the panel will plate the heaviest. A good rule of thumb is to place the anode and cathode bus bars at least 6–8 inches apart. This will attenuate most of the localized high current density just opposite the anodes, for a more uniform plating. It is possible to separate the anode and cathode even further. However, doing so increases the size of the tank, and the volume of plating bath required to fill the tank; this also creates a need for a large filter/pump to turn over the bath. The cathode agitation already mentioned will also alleviate the high plating rate opposite the anode.

The anodes themselves play an important part in the current density available to the plater, as well as in the distribution of that current. Most plating baths (acid copper, pyro copper, sulfamate nickel, and tin-lead) operate best with an anode to cathode ratio of 1:1 or 2:1. This means that for every square foot of plating area on the printed circuit panels, there should be 1 to 2 square feet of anode area. It also means that the anode current density range is between 50 and 100% of the cathode current density. Platers should know what the anode area is in their plating tanks, and what current they can plate at so as not to exceed that range. If the anode area is too much less than 1:1 with the cathode area, it will not corrode properly, the bath will fail to throw metal efficiently, and the dissolved metal content will drop with plating. Ultimately, lack of anode area will result in costly additions of metal salts, and perhaps a high brightener consumption rate. If it is known that anode area is low, then either anode area should be increased, or cathode current density should be kept lower than optimum.

The anode distribution will affect throwing power and have an effect on the ability of a plating bath to operate at peak current density; this is partially related to anode area. If bar anodes have been diminished drastically, or if the height of the slugs (chunks of anode metal) in the titanium baskets have dropped markedly below the surface of the plating tank, then the metal deposited in the areas of the cathode opposite the anodes will be greater than the metal deposited opposite the depleted areas of the anode hooks or baskets. The only way to even out the plating thickness will be to add anode area to replace that which has been diminished. This sounds sensible; but, when adequate metal is not being deposited for the elapsed time and amperage, many platers and engineers fail to look at the anodes, resorting to increased amperage instead.

OTHER CONSIDERATIONS

The rectifier must also be considered when determining the correct amperage for plating. If only a small quantity of panels, with little plating area, is to be plated it may be difficult to obtain correct amperage setting. The amperage may be so small that the rectifier cannot adequately resolve the setting. The net result may be plating which is either too high, and may therefore burn, or so low that extra plating time is required.

Another consideration which goes into making the final amperage determination is the skill and experience level of the plater. The plater is the one who best knows the peculiarities of the plating tanks, and is familiar with how best to plate difficult patterns without overplating or burning. The skill of the plater is difficult to quantify, but if the plater is experienced, an engineer or planner would do well to consult him/her when attempting to set amperage.

DETERMINING AMPERAGE

All of the above listed considerations must be considered when making a determination of what current density and amperage to plate a given job at. The actual calculation of plating amperage is accomplished using one of the three methods detailed below. A fourth method, voltage, will be briefly mentioned first.

Voltage

The voltage method allows the plater to load up a tank and increase plating amperage on the rectifier until the desired voltage has been obtained. That voltage will be one where a bright deposit is obtained most of the time. The drawback of this method is that the voltage for a given current density will fluctuate along with the acid and metal content of the tank; it will even fluctuate with variations in the brightener concentration. Furthermore, the voltage for a given current density will be different for each tank, since the plating chemistry will vary slightly from tank to tank. The degree of electrical contact attained at the plating rack/cathode interface and the anode hook/anode bus bar interface. An oxidized or dirty bus bar will cause a poorer connection to be made than to a clean bus bar. There is no way to obtain and keep a good control over any of these rapidly changing variables; hence, the use of rectifier voltage as a means of setting plating amperage is unreliable; but an aid which platers should always be aware of.

Of the three most common methods used for determining plating amperage, two are "guesstimations," and one is a good approximation.

METHOD 1

This method is called the area percentage, or % area, method. The formula is:

$$\text{Plating Amperage} = \frac{L \times W}{144} \times \text{\% Area} \times 2 \text{ sides} \times \text{a.s.f.} \times \text{No. of Panels.}$$

For 16 panels, 12×18 inches, with 35% area (see Fig. 14-3), the following calculation would be used for an acid copper bath. (*Note:* Area percentage is always expressed as a decimal equivalent: 70% = 0.70, 50% = 0.50, 35% = 0.35.)

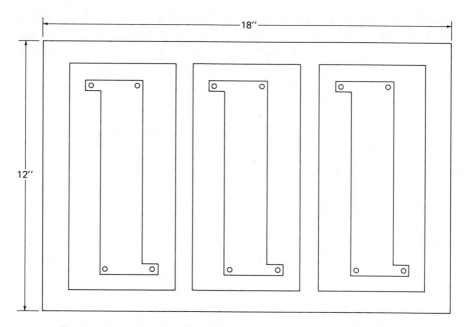

Fig. 14-3. Approximately 35% of this panel area is copper, and will be plated.

$$\text{Plating Amperage} = \frac{12 \times 18}{144} \times 0.35 \times 30 \text{ a.s.f.} \times 2 \times 16$$
$$= 504 \text{ amps.}$$

A guesstimation of the area of the panel which is available for plating must be made. If, for example, half of each side of a panel is resist, and the rest is copper, then the panel has 50% of its area available for plating. If three-quarters of one side and one-quarter of the other side is available for plating, that averages out to 50% of the total panel area.

Use the following optimum recommended current densities for calculating plating amperages, or consult the data sheets for a particular plating bath manufacturer: Acid

Acid Copper	30 a.s.f
Pyro Copper	25 a.s.f
Tin-Lead	20 a.s.f
Nickel	20 a.s.f

METHOD 2

This method is called the compressed area method. All the copper area on the panel must be pictured as being compressed into one end of the panel. The

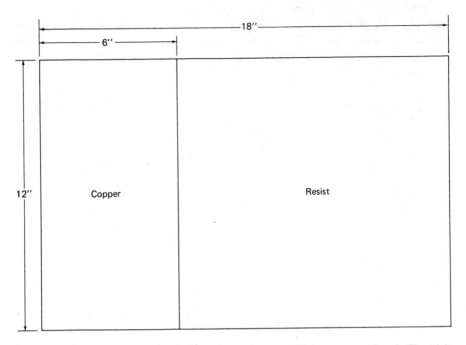

Fig. 14-4. Approximately one-third of panel area is copper, if the copper surface in Fig. 14-3 were to be compressed into one end of the panel.

length and width dimensions of this copper area are then measured and used to calculate amperage (Fig. 14-4):

$$\text{Plating Amperage} = \frac{L \times W}{144} \times 2 \text{ sides} \times \text{a.s.f.} \times \text{No. of Panels}$$

If there are 16 panels, each 18 × 12, and all the copper area is pictured as being compressed into 6 inches at the 12 inch end of the panel, then the amperage would be calculated as follows:

$$\text{Plating Amperage} = \frac{6 \times 12}{144} \times 2 \times 30 \times 16$$
$$= 480 \text{ amps.}$$

METHOD 3

This method is a better approximation of the plating area than are the two preceeding methods. It makes use of an area integrator, or area calculator. An accurate determination of the circuitry to be plated is obtained for a one-up circuit pattern using the area calculator. This area is then multiplied by the number of times the circuitry will be stepped and repeated on the panel. To this the border area of the panel and the border area between circuits are added. The sumation yields a good approximation of the area per panel to be plated, which is then multiplied by the number of panels, to obtain the plating amperage:

$$\text{Area} = \text{Area from Integrator} + \text{Panel Borders} + \text{Part Borders (inches)}$$

$$\text{Plating Amperage} = \frac{\text{Area}}{144} \times 2 \text{ sides} \times \text{a.s.f.}$$

NOTES

1. It should be remembered that the above methods of calculating plating amperage are good only for determining an approximate starting place. Everything else mentioned earlier must be considered before setting amperage on the rectifier.

2. Before plating, study the plating pattern: Isolated traces and pads in panel corners and along edges will plate faster than will those near the center. It may be necessary to reduce amperage below the calculated value in order to avoid burning or overplating in localized areas.

3. After plating for 10 or 15 minutes, pull several panels and look at the plating quality.

- Look for burning in the high current density areas. Reduce amperage if burnings is noted.
- Look for dullness in the center of the panel. Raise amperage a little if dullness is present, and there is no burning in corners.

4. Correct plating amperage is determined by evaluating several factors:

- Calculation base on guesstimation of area, or approximation.
- Considerations of the circuitry pattern to be plated.
- Observation of the quality of the plating.

5. Check the plating quality for amperage setting on the rectifier. Bright panel borders and racks, with the rest of the panel bright generally means amperage is correctly set, except for the following cases:

- Burning or noduling of isolated traces and pads means amperage setting should be reduced.

- Holes in corners and borders reducing in diameter faster than holes in the panel center means amperage should be reduced.

6. If amperage has been correctly determined for copper, and the subsequent metal to be plated is nickel or tin-lead, there is no need to go to great lengths to determine amperage; simply use ⅔ × Copper Amperage.

15

The Electroless Copper Process

The sole purpose of this process is to metallize the wall of drilled holes. This hole metallization provides an electrical connection between the sides of a panel, and to inner layers of multilayer boards. It is important that the plating supervisor train platers to understand this. Since metal in the hole is the requirement, the board must be processed at all times with this in mind.

DETECTING PROBLEMS

When inspecting panels as they are being processed along the electroless copper line, attention must be directed to the holes, not just the surface of the board. It often happens that the surface appears satisfactory, while little coverage is taking place inside the holes or along the edges. Panel edges are a more accurate reflection of copper deposition quality inside the holes, than is the surface. But, the edges are only a reflection, the holes must be looked at during processing. If there is a marginal condition in the electroless copper process (for example, low bath temperatures) it will show up inside the holes first, then on the board edges, and lastly on the surface. For this reason, Platers must be trained to look at the holes and not just the surfaces or edges.

DRYING IS REQUIRED

It is good to keep in mind that electroless copper deposits, from a room temperature bath, are only 15–20 millionths of an inch thick. When panels are removed from the electroless copper bath, they should be neutralized, rinsed, and dried as soon as possible to prevent oxidation. The drying process must be adequate to dry the holes as well as the surface. Typically, panels are dried during a mild scrubbing operation. The conveyor speed must be slow enough for water to be blown from the holes. Again, just as in processing panels down the line, the holes must be inspected; not just for voids, but for dryness. It is good practice to bake the panels gently, until dry.

THE NEED FOR SCRUBBING

Imaging may be applied to either the scrubbed or unscrubbed surface of a panel after electroless copper. It is not absolutely necessary that the surface be processed one way over the other. But there are some reasons to consider scrubbing:

1. Many screeners claim the scrubbed surface is one on which it is easier to spot image defects, and that in the long run it is easier on their eyes.
2. Sometimes tiny particles of electroless copper break off corners and edges. This can cause problems during imaging.
3. The scrubbed surface is esthetically more pleasing.
4. If the deposited copper is powdery and poorly adherent, it may be removed during the scrubbing operation. This is no substitute for reworking a panel which exhibits peeling during tape testing.

If the line is operated with good controls and procedures by properly trained operators, there is no good reason why the scrubbing operation could not be eliminated for most applications; this would reduce handling problems and increase daily throughput for electroless copper production.

DWELL TIME IN COPPER

It is common for platers to have misconceptions about how an electroless copper line can be run. Many feel it is perfectly all right to leave panels in the copper tank for an hour or more. This is not a good idea. The deposition rate for room temperature electroless copper decreases dramatically after 20 minutes. When panels are left for more than about 30 minutes, the deposit becomes spongy and powdery. If a (gloved) finger is run across the surface, a streak will be left where copper is removed. There is no advantage to leaving panels in the copper bath for more than 20 minutes. When this has occurred, panels must be scrubbed to remove the powdery deposit from the surface.

NEUTRALIZE CAUSTIC RESIDUE

Electroless copper is a caustic solution containing sodium hydroxide. Sodium hydroxide is hygroscopic, which means that it will pull moisture out of the air. If a pellet of this material is set on a counter top and exposed to the air, it will form a puddle of caustic water right on the counter top in just a few minutes. Caustic solutions like electroless copper baths are slippery to the touch and notoriously difficult to rinse off. The only fully reliable way to rinse caustic material is to neutralize first, then rinse. If panels are lifted from the copper bath, drained, and water rinsed only, the result will be a marginally clean surface. The degree of marginality is related to the quality of your rinse water, dwell time in the rinse, and the number of rinsing operations. Even if the boards are scrubbed, the surface will still be marginal.

What can this mean in the manufacturing operation? As soon as the panel is dried, it will start to pull water out of the air because of residual sodium hydroxide. The panel may never become wet or noticeably different to the eye or touch, but a problem may present itself down the line. Typically, excessive sodium hydroxide residues result in resist breakdown. On rare occasions breakdown may be due to isolated contamination on the panel. If a job has widespread resist breakdown it would be wise to examine the electroless copper neutralizing and rinsing procedure. If panels are (supposedly) neutralized, ask the platers if they have been omitting the step. If they have not, check the strength/quality of the acid. Usually, a 1% solution of sulfuric acid or phosphoric acid is all that is required for adequate rinse insurance.

Another indication of excess residue from sodium hydroxide is the presence of a dark, hazy, stained appearance on panel surfaces. The panel may look fine after scrubbing; but can take on a darker shade after sitting for a while.

ANALYZE REGULARLY

During the course of an operating shift, analyses should be conducted on a regular basis. Perhaps a full copper, sodium hydroxide, formaldehyde, pH check at the start of the day; and at least a copper or colorimetric test every few hours thereafter. A log should be established for panel square footage though the electroless copper line as well as the additions which were made during that processing. If the printed circuit shop does not have a laboratory to perform the analysis, a well trained plater can run the bath indefinitely using only simple colorimetric comparison kits supplied by most bath vendors. The plating supervisor should work with the plater to develop his/her ability to get a handle on throughput, colorimetry, and bath strength. Ideally, the bath is run down to about 70% near the end of the day; so the bath is weaker and more stable during periods of nonuse, for example, overnight, weekends, and holidays. It is also good to turn the heater off at the end of the shift, as the bath is more stable at lower temperatures.

AIR SUPPLY REQUIRED

Room temperature electroless copper baths require an air source which must be bubbled through the bath at all times. The volume of air need not be great; an aquarium air pump is all that is needed. The end of the air hose must be weighted, to lie on the tank bottom. This small amount of air is beneficial for two reasons: (1) air helps stabilize the copper and retard any tendency to plate out, and (2) air provides a small measure of solution agitation when the tank is not being used. During use, there is plenty of agitation by way of the plater physically moving the panels, or by a mechanical rack agitation system. However, when the bath is not in use, the small degree of agitation from the air pump helps provide more uniform heat distribution. This keeps the area near the heater from becoming the only "hot spot" in the tank. Temperature distribution becomes more important when the heater thermocouple sensor is located away from the heater, or on the bottom of the tank. Excessive air agitation should be avoided, as it will accelerate removal of formaldehyde (an important ingredient for deposition) and provide a false "sense of agitation." Platers may feel the panels are being adequately agitated when air is blowing through the tank in the same manner as is required for copper electroplating. At all times it should be remembered that only a trickle of air is required for good bath stability; too much air agitation removes the formaldehyde (which will extinguish the deposition reaction), and there is no substitute for adequate through-hole agitation.

THINK ABOUT THE BOARD BEING PROCESSED

Successful operation of an electroless copper line demands that consideration be given to the type of board to be run and the type(s) of operations that board will see after electroless copper.

If the board is a multilayer, or if it is a double sided board built on thick material (0.062 inch or thicker), there is a good chance of epoxy smear being inside the hole. During drilling, the drill bit reaches hundreds of degrees Fahrenheit. This melts the epoxy resin and smears it around the inside of the hole. For double sided work this can be a problem because electroless copper will not adhere well to smeared epoxy. It is not uncommon for electroless copper to blister and peel from a plated hole during the reflow operation. If thick panels are routinely run through smear removal, prior to electroless copper, the problem is avoided altogether. The hole metal blistering is always a potential for any board, regardless of thickness. However, as board thickness increases, the margin of confidence decreases that blistering will not occur. When boards are built on 0.093 inch or thicker material, blistering is almost assured if smear removal is not done.

The blistering problem is also a concern for multilayers, since many are 0.093 inch or thicker. But smear inside the hole is of concern for an additional reason on multilayers. This type of board has layers of circuitry laminated together. Electrical connections are made with those inner layers by the plated hole. If the plating is done on top of epoxy smear, no electrical connection can be made.

Teflon is another common material upon which circuits are printed. This material is very soft, and must be handled carefully; it is also extremely expensive. Platers should be cautioned to watch for this material on jobs circulating through their area of the shop. They must also be trained to process it with care. Certain types of electroless copper racking systems bite into the copper surface. No board should be loaded in such a rack without first ascertaining whether or not there is circuitry to be printed along that edge. Teflon is so soft that this is critical. There is another problem with Teflon, as well: electroless copper will not adhere to it any better than it will to smeared epoxy. Teflon is virtually inert to chemicals commonly used to process printed circuits. Yet it must be etched before electroless copper deposition, for good metal adhesion. There are compounds using sodium (Chemgrip, Tetra-Etch) which will do a good job of etching the teflon. When Teflon has been etched the electroless copper subsequently deposited in the holes should adhere very well. Double sided printed circuits can even be reflowed with confidence that hole metal will not blister.

Other than the substate used to construct the board, there are other points to consider for successful electroless copper plating. Some boards must pass thermal stress testing. This is floating a section of the board atop molten solder at 550 degrees Fahrenheit for 10 seconds. The section is then prepared for viewing under magnification. Poor drilling can cause problems. A dull drill bit will tend to punch its way through the laminate, rather than drill cleanly. This can create the problem of having to deposit over loose bundles of fibers, instead of over a smooth hole wall. It is difficult to get adequate coverage in cases like this. Voids and gaps in the plating, as well as plating which is easily cracked, are the result. The punching problem must be brought to the attention of the Drilling department for resolution, but once it has occurred, the plater is faced with successfully processing the boards. It should be mentioned that this is more common, and really only a problem, with multilayer drilling and processing, since hole quality is of paramount concern. For this reason, microsectioning of multilayer panels after smear removal and electroless copper is absolutely necessary.

The process which will successfully cover poor drilling also helps solve problems arising from thermal stress: hole wall pullaway. This, too, is more prevalent and of greater concern during multilayer processing. Good dimensional stability in printed circuits requires that prepreg and core epoxy be fully cured. This must be accomplished before electroless copper plating. The only way to ensure this is to bake all rigid and thin clad laminate before drilling or inner layer imaging. During multilayer processing etched layers are laminated together with prepreg. After multilayer lamination, panels should be baked to ensure complete cure of prepreg epoxy. A good baking cycle for freshly blanked laminate, or post lamination on multilayer, is the following: stack panels between steel pressplates, no more than one inch between plates; set oven to 300 degrees Fahrenheit and bake for 4 hours. If panels are layered to too great a height between pressplates, baking time must be increased several hours. When the baking cycle is complete, open doors to cool. This baking cycle will not only improve Z-axis dimensional stability, it will also make the epoxy less susceptible to chemical attack during smear removal. Z-axis stability is critical, since a panel which expands in this direction will crack the through hole plated copper. Fully cured prepreg will not be "eaten" away to the extent that cavities appear and glass fibers are bared.

The next step to overcoming hole wall pullaway during thermal stress has to do with the actual processing through electroless copper. After smear removal, the panels proceed directly into the first tank on the copper deposition line. The panels are then processed through each tank of the line until they come out of the copper bath. After copper, the panels are neutralized, rinsed and returned to the catalyst predip (sodium chloride or hydrochloric acid). They are reprocessed down the line, through copper, and neutralized. Only after the second run through copper deposition are they flashed with copper electroplate and microsectioned. The second pass through the line recatalyzes poorly covered areas of the hole (such as fiber bundles and cavities). An alternative procedure is to pull the panels from the accelerator bath, then return them to the catalyst predip. While the second method may catalyze difficult to cover areas, it does not make provisions for subsequent deposit over thinly covered areas.

It is rare that double sided boards will fail thermal stress testing if the freshly blanked panel is baked as recommended. When they do fail after baking, it is probably due to smear in the hole or factors related to the electroplated copper rendering it brittle.

Multilayer boards fail thermal stress testing for a variety of reasons: problems related to composites of copper laminates and inner layers of copper and fiberglass/epoxy, degree of epoxy resin cure, increased likelihood of drilling problems, smear and electroless coverage problems, and a variety of factors which accumulate during processing, including brittle copper due to electroplating difficulties. As long as the process and processor are sufficiently monitored, or trained, it is possible to operate with a minimum of scrap being produced during manufacturing because of problems in blanking, drilling, smear removal, and electroless copper deposition.

FLASH WITH ELECTROPLATED COPPER

Another point of good electroless copper processing which must not be overlooked is the need for flashing electroless copper plated panels with a thin layer of electroplated copper. There are two common situations where this is needed: (1) microsectioning the panel to determine the reliability of smear removal and hole coverage, since the 20 millionths of an inch of deposit from room temperature baths is not enough to see even with a microscope reliably; and (2) adding copper to the plated through hole for subsequent imaging with dry film photoresist. Generally, 0.0001 inch of copper electroplate is enough for either application. Dry film leaves a thin layer of adhesion promoter on the surface of the copper. If this adhesion promoter is not removed chemically, a copper-to-copper peeler will result during pattern plating. It is possible to remove the adhesion promoter by scrubbing with pumice after dry film imaging. This method is not advised because of the possibility of damaging the image, and the low reliability of completely removing adhesion promoter from all areas of the circuits.

USE PUMP AGITATION ON CATALYST

It is good to use a pump for turning over the catalyst bath. The catalyst and copper are the most critical of the baths on the line. Both are sensitive to temperature. In general, room temperature baths utilizing hydrochloric acid pre-dips, instead of sodium chloride, are less sensitive to low temperatures. No recommendation for running a hydrochloric acid system at a low temperature is intended. Rather, the acid system may have wider operating latitude than the salt system. If a shop has recently changed from the acid to the salt system, it should be kept in mind that temperature is *critical* with the salt system. The catalyst will require 10–40 degrees Fahrenheit more temperature for good operation than will the copper. Many acid catalyst systems will operate satisfactorily at 80 degrees Fahrenheit. Few, if any, salt catalyst systems will operate effectively below 85 degrees Fahrenheit. If the heater sits in the corner or along the side of a bath that has no agitation, a hot spot will develop. That hot spot will be right at the heater. If the thermocouple is located away from the heater, the heater will remain on virtually continuously. This will cause the bath to become very hot near the heater, and on the top of the bath. However, if the bath is stirred and the temperature measured, it may be too cool for successful chemical operation. The best heating will take place when the bath is turned over and temperature evenly distributed throughout the volume of the bath.

KEEPING FINGERPRINTS OFF PANEL SURFACE

Fingerprints are a problem in any plating operation. Few cleaner/conditioners will remove them adequately. If plating occurs over fingerprints, the result is apt to be a metal-to-metal peeler. The plating supervisor must train platers to handle panels to avoid fingerprints: wear gloves and handle panels by the edges only.

REWORK THROUGH ELECTROLESS COPPER

Reworking panels through electroless copper should be avoided; every pass down the deposition line will bring opportunities for voids, ring voids, and peelers. However, there are times when it cannot be avoided. When these situations arise, great care must be afforded for successful plating rework. Unless the platers are thoroughly experienced, the plating supervisor should oversee the task personally. Rework through electroless copper may be due to voided holes or peeling deposit. The procedure is slightly different, depending on the reason. If rework is necessary because of poor coverage the first time, it is not necessary to strip off the existing deposit, as long as no fingerprints have been placed on the panel surfaces. If the rework is for peeling deposit or if fingerprints have been placed on the panels, it will be necessary to chemically strip off the previously deposited copper. A procedure which works for reprocessing through electroless copper is the following:

1. Strip electroless copper in sulfuric peroxide or ammonium persulfate.
2. Rinse well.
3. Lightly scrub both sides of panels.
4. Reprocess the panels as if they were being run for the first time, with one exception: panels should only be immersed briefly in the micro-etch (ammonium persulfate or sulfuric/peroxide).

Any time panels are being reprocessed and the previously deposited electroless copper has not been etched off prior to deburring, do not immerse panels in a micro-etch. This will leave a smutty residue, from the catalyst, which will cause peeling.

Panels which are re-scrubbed and reprocessed through electroless copper have an increased chance for ring voids occurring at the lip of holes, especially the larger holes. If there is a great likelihood of this happening, the panels should not be scrubbed at the conclusion of the rework.

TROUBLESHOOTING ELECTROLESS COPPER

1. Dark (dull red, brown, or black) copper surface.

   a. Cold copper bath:

   - Check temperature of bath.
   - Check setting of controller.

   b. Inadequate operator processing:

   - Supervisor should oversee running of a load.
   - Talk with the plater, find out how load had been processed.

   c. Operator left panels in copper bath too long.
   d. Operator left panels in catalyst tank too long.
   e. Bath concentration too low: analyze and make needed additions.

2. Powdery deposit.

a. Operator left panels in copper bath too long.
b. Bath temperature exceeds recommendations of manufacturer.
c. Bath concentration is too high.
d. Catalyst has been dragged into copper bath causing excessive activity.

3. Bath is spontaneously plating out or excessively active.

a. Bath temperature is too high, cross-filter.
b. Bath concentration is too high, cross-filter: Bleed and feed with deionized water.
c. Panels or other metal have fallen into the bath, cross-filter.
d. Catalyst has been dragged into the bath, cross-filter or replace bath.

   - If bath does not settle down after cross filtering into a clean tank, dump and make up fresh.
   - Always leech sides and bottom of electroless copper tanks with concentrated hydrochloric acid.

e. Dust or particles of copper floating on bath surface, cross-filter.
f. No air agitation, cross-filter: Obtain air supply.

4. Poor hole or edge coverage.

a. Low catalyst or copper bath temperature.
b. pH too low in copper:

   - Analyze and make sodium hydroxide addition if required.
   - This condition may have resulted from the plater dragging accelerator into the copper bath, or from insufficient addition during normal replenishment.

c. Low catalyst or copper concentration: analyze.
d. Poor operator processing: Talk with plater, oversee running of another load.

5. Copper peels during tape testing.

a. Fingerprints, follow rework procedure.
b. Poor operator processing; talk to plater and oversee running of another load.

   c. Panels were left in copper bath too long: scrub.

   d. If there is only slight flaking during the tape testing, flash several panels with electroplated copper and retest.

6. Panel surface is mottled.

   a. Panels are being reworked without having been micro-etched.

   b. Insufficient time in micro-etch.

   c. Contaminated or non-deburred panel surface.

7. Panel surface is dark or mottled coming out of sulfuric/peroxide micro-etch.

   a. Copper build-up too high:

- Cross-filter, let cool and crystals will settle out.
- Cross-filter back to original tank.
- Analyze and make any needed addition.

   b. Temperature too low.

   c. Panels are being reworked, dark mottle is catalyst smut: Run finger over surface.

8. Mirror surface on top of catalyst: Catalyst has decomposed:

   a. Temperature too high.

   b. Chloride too low.

   c. Acid content too low.

   d. Water, instead of acid, added to raise level of acid based system.

   e. Bath was contaminated.

   f. *Note:* Notify the bath manufacturer of the problem and for suggestions on why it happened and how to resolve and prevent recurrence.

9. Most electroless copper problems are operator induced. Chemical analysis may confirm an imbalance somewhere. But unless proper training and supervision are given, the problem will recur.

Proper controls, good operator training by the supervisor, and adequate supervision on a continual basis are the secrets to trouble free operation of electroless copper. When problems do arise, review the basics of operation and talk to the platers first. Look for exotic causes only as a last resort.

ANALYTICAL INFORMATION

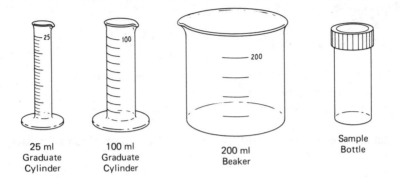

| 25 ml
Graduate
Cylinder | 100 ml
Graduate
Cylinder | 200 ml
Beaker | Sample
Bottle |

Catalyst Analysis

1. Rinse graduate cylinders, beaker and sample bottle with water.
2. Add 185 ml of dilute hydrochloric acid to the 200 ml beaker.
3. Use the 25 ml graduate cylinder to add 15 ml of Catalyst bath to the 185 ml of hydrochloric acid; stir.
4. Pour this into the sample bottle.
5. Compare sample bottle to the catalyst standards.
6. Make catalyst addition according to the chart below:

Catalyst Strength	Addition Needed for a 50 Gallon Bath
100%	No addition needed
90%	540 ml
80%	1081 ml
70%	1620 ml
60%	2160 ml
50%	2700 ml

Note: Dilute hydrochloric acid is made by adding 1 part of concentrated hydrochloric acid (37%) to 2 parts water.

| 10 ml
Graduate
Cylinder | 25 ml
Graduate
Cylinder | Sample
Bottle |

Color Standards: Analysis of Electroless Copper

1. Use the 10 ml graduate cylinder to add 5 ml of plating bath to the sample bottle.
2. Use the 25 ml graduate cylinder to add 25 ml of indicator to the same sample bottle.
3. Place lid on the sample bottle and shake.
4. Compare sample to the other color standards.
5. Make addition to the copper bath according to the chart below:

Bath Strength	Addition Needed for a 50 Gallon Bath
90%	½ gal. A, ½ gal. B
80%	1 gal. A, 1 gal. B
70%	1½ gal. A, 1½ gal. B
60%	2 gal. A, 2 gal. B
50%	2½ gal. A, 2½ gal. B

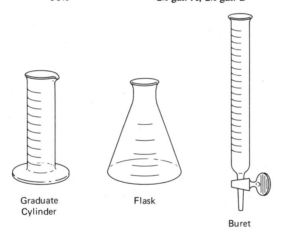

Graduate Cylinder   Flask   Buret

Electroless Copper Analysis

1. Rinse the 10 ml graduate cylinder and 250 ml flask with water.
2. Add 75 ml of water to the flask.
3. Add 10 ml of copper bath to the flask using the graduate cylinder.
4. Add 10 ml of ammonia buffer to the flask using the graduate cylinder.
5. Add 5 drops of P.A.N. to the flask, and stir.
6. Fill the buret with EDTA, and turn the buret handle to make sure the tip is unclogged. If no EDTA flows through the buret, unclog the tip.
7. Refill the buret and add EDTA to the flask through the buret, shaking the flask. Turn buret handle to OFF position when color of solution in the flask turns from purple to green.

8. Read the milliliters of EDTA used in the analysis, and make addition based on the chart below:

ml EDTA Used	Copper Strength	Addition Needed for a 50 Gallon Bath
9 ml	90%	½ gal. A, ½ gal. B
8 ml	80%	1 gal. A, 1 gal. B
7 ml	70%	1½ gal. A, 1½ gal. B
6 ml	60%	2 gal. A, 2 gal. B

Note: 0.05 Molar disodium EDTA, pH 9.5 ammonia buffer. Use self-zeroing 10 ml buret.

PROCEDURES

Flow Chart

1. Cleaner-conditioner.
2. Sulfuric acid/peroxide micro-etch.
3. Catalyst pre-dip.
4. Catalyst.
5. Accelerator.
6. Electroless copper.
7. 1% phosphoric acid.
8. Scrub and dry.

Operating Procedure

1. Look at the traveler of the job you wish to run.

   a. Are there any special instructions?
   b. Are the panels single sided or double sided?
   c. Does the traveler call for the panels to go through electroless copper?
   d. Are the panels multilayers?

Note: If panels are multilayers they are run through the smear removal line immediately prior to electroless copper. One of the panels is to be flashed in acid copper and microsectioned after electroless copper.

2. Examine the drawing and drill pattern on the panel. If you are using spring racks, you must determine how to load panels into springs without destroying copper surface where contact fingers will be located.
Note: If you are using basket racks, this step may be deleted.

3. Cleaner-conditioner:

a. Immerse panels 5–10 minutes, with agitation. Bath temperature should be at 160 degrees Fahrenheit.
b. Drain panels well.
c. Immersion rinse 30 seconds in each of two rinse tanks.
d. Drain well after rinse.

4. Sulfuric acid/peroxide micro-etch:

a. Immerse panels with agitation until they have a uniform matte pink surface.
b. Never walk away from this tank while there are panels in it.
c. If it takes more than one minute to etch panels, notify plating supervisor.
d. As you pull the panels out, look at the surfaces. *Note:* Shiny, patchy areas indicate unclean copper. The plating supervisor should be notified.
e. Drain well.
f. Immersion rinse 30 seconds in each of two rinse tanks.
g. Drain well.

5. Catalyst pre-dip:

a. Immerse panels at least 30 seconds. Bath temperature should be 100 degrees Fahrenheit.
b. Drain well. Immediately go into catalyst. Do not rinse after catalyst pre-dip.

6. Catalyst:

a. Immerse panels 5 minutes, with agitation. Bath temperature should be 100–110 degrees Fahrenheit.
b. Drain very well.
c. Immersion rinse 30 seconds in each of two rinse tanks.
d. Drain well after each rinse.

7. Accelerator:

a. Immerse panels 6 minutes with agitation.
b. Drain well.
c. Immersion rinse: 30 seconds in preceeding rinse tank then 30 seconds in rinse tank immediately following accelerator.
d. Drain well after rinsing.

8. Electroless copper bath:

a. Immerse panels 20 minutes with agitation. Bath temperature should be at 70–80 degrees Fahrenheit.

- Bath temperature is critical. Notify plating supervisor if it is not in range.
- Bath should turn lighter in color with bubbles after about 10 minutes. If it does not, notify plating supervisor.

b. Panel edges and holes should become covered with copper. If they are dark after 10 minutes, notify plating supervisor.
c. Drain well.
d. Immersion rinse 30 seconds in each of two rinse tanks.

9. 1% phosphoric acid:

a. Immerse panels for 5 seconds.
b. Drain.
c. Rinse for 5 seconds.

10. Scrubbing and drying panels:

a. Unrack the panels and place in the 1% phosphoric acid tray at the scrubbing machine.
b. Use 320 grit scrub brush or finer on the scrubbing machine. Follow scrubbing procedures.
c. As the panels come off the machine, place them in a holding rack. They must be dry and spot free.
d. This is a good time to look at the holes. Dark, voided holes are unacceptable. They indicate a process problem and should be immediately brought to the attention of the plating supervisor.
e. Oven dry several minutes.

Additional Information

1. Log in every job you run through electroless copper. This information is used to track bath loading and costs.

2. Always count panels when you have finished a job. Enter only the actual number counted. Sign off the traveler for electroless copper when job is completed.

3. When finished for the day, turn off heater and all electroless copper tanks, except cleaner-conditioner. Shut off rinse water.

4. Always wear safety glasses, rubber boots, gloves, and apron when operating the line for performing tank additions and changes.

5. Always analyze catalyst and electroless copper before using these tanks. Recheck them every 2 to 3 hours.

6. Get in the habit of checking temperatures on the baths. In some cases voids will result if temperature is even slightly low: catalyst and electroless copper are two examples.

7. Always look at the work in process. Get to know what the copper should look like, what the surfaces should look like, and what the holes and edges should look like as they come out of the different tanks. Each tank does something to the panels.

8. Never try to run panels that have not been deburred. All panels must be deburred prior to electroless copper.

9. Single sided panels never go through electroless copper. If the traveler calls for electroless copper on a single sided panel, show your supervisor.

10. Remember, the only purpose of the electroless copper line is to put copper in the holes. The panels must be agitated in each of the baths to get solution flowing through the holes!

11. Never put fingerprints on the surface of a panel—it may result in peeling at some stage in the operation. Always handle panels by the edges, just like your favorite record.

12. The sulfuric acid/hydrogen peroxide micro-etch generates heat as it is used. Be sure to shut heater off if the bath effervesces excessively or foams.

Reworking Panels Through Electroless Copper

1. Immerse panels in sulfuric acid/hydrogen peroxide micro-etch long enough to remove the deposit of electroless copper.

  a. Panel edges and holes will turn dark when deposit is removed.
  b. Drain well and rinse.

2. Run panels through scrubbing machine.

  a. Use maximum conveyor speed.
  b. Be sure that fingerprints are kept off panel surfaces.

3. Rack panels and process through electroless copper line.

  a. The only change is that the sulfuric acid/hydrogen peroxide micro-etch should be avoided altogether.
  b. Be sure to note on traveler that reworking is involved, and the quantity being reworked.

4. There is a great probability of ring voids occurring at the lips of holes, especially the larger holes, during electroless copper rework.

  a. Look at the lips of holes coming out of the electroless copper bath.
  b. If the copper on the surface appears to be worn, *do not scrub panels, dry them only.*

MAINTENANCE INFORMATION

Cleaner-Conditioner

Temperature: 160 degrees Fahrenheit.
Tank size: 113 Gallons.
Make-up: 3 Gallons solution A in deionized water.
Dump schedule: Mondays

Make-Up Procedure

1. Check availability of supply.
2. Turn off heater.
3. Wear safety glasses, boots, apron, and gloves.
4. Pump spent bath into cleaned plastic/plastic lined drum.
5. Rinse out tank.
6. Fill tank ¾ full with hot water.

   • Add 3 gallons solution A.
   • Fill tank to operating level with hot water.

7. Set heater at 160 degrees Fahrenheit.
8. Log in tank change.
9. Label spent drum: "Cleaner-Conditioner."

Note: It is a good idea to check warehouse for availability on Friday. If the cleaner-conditioner is in stock the heater can be turned off on Friday night. The cleaner-conditioner will then be cold for pumping into waste drum Monday morning.

Sulfuric Acid Hydrogen Peroxide Micro-Etch

Make-Up. 25 Gallons Concentrate, 10 Gallons Stabilized 35% Hydrogen Peroxide.
Temperature: 100–120 degrees Fahrenheit.
Tank size: 95 Gallons.

Dump schedule: None.
Cross-filter schedule: Fridays.

Make-Up Procedure:

1. Fill tank ⅓ full of deionized water.
2. Add 25 gallons of concentrate.
3. Add 10 gallons of stabilized 35% hydrogen peroxide.
4. Fill tank to operating level with deionized water.
5. Set heater at 100 degrees Fahrenheit.
6. Log in tank change.

Cross-Filter Procedure:

1. Turn heater OFF.
2. Wear safety glasses, boots, apron, gloves.
3. Pump into holding tank.
4. Wipe any residues from walls of tank.
5. Monday morning, pump cold micro-etch through 5 micron filter into micro-etch tank.
6. Set heater to 100 degrees Fahrenheit.
7. Any copper sulfate crystals in bottom of weekend holding tank should be placed in storage receptacle.

Note: The purpose of this step is to remove copper from the micro-etch bath.

Catalyst Pre-Dip

Temperature: 90 degrees Fahrenheit.
Tank size: 95 Gallons.
Dump schedule: Six weeks maximum.
Make-up: 225 pounds catalyst pre-dip in deionized water.

Make-Up Procedure:

1. Wear safety glasses, boots, apron and gloves.
2. Fill tank to within 5 inches of operating level with deionized water.
3. Set heater to 90 degrees Fahrenheit.
4. Add 225 pounds of catalyst pre-dip.
5. Stir until dissolved.
6. Add deionized water to operating level.
7. Log in tank change.

Replenishment:

1. Use hydrometer to measure specific gravity.
2. Make additions as follows:

Specific Gravity	Strength	Catalyst Pre-Dip
1.167	100%	No addition needed
1.152	90%	22 pounds
1.136	80%	44 pounds
1.120	70%	66 pounds

3. Stir addition until dissolved.
4. Log in addition made.

Catalyst

Temperature: 100–110 degrees Fahrenheit.
Tank size: 95 Gallons.
Dump schedule: None.
Make-up: 225 pounds catalyst pre-dip, 5.7 gallons catalyst in deionized water.

Make-Up Procedure:

1. Wear safety glasses, boots, apron, and gloves.
2. Fill tank ¾ full with deionized water.
3. Set heater at 100–110 degrees Fahrenheit.
4. Add 225 pounds of catalyst pre-dip; stir until dissolved.
5. Add 5.7 gallons catalyst; stir well.
6. Fill tank to operating level with deionized water.
7. Analyze with comparator.
8. Notify laboratory of new make-up; Laboratory will perform chemical analysis.
9. Log in tank make-up.
10. Change filter and turn pump ON.

Accelerator

Temperature: 70 degrees Fahrenheit.
Tank size: 95 Gallons.
Dump schedule: Mondays.
Make-up: 18 Gallons of accelerator in deionized water.

Make-Up Procedure:

1. Check warehouse for availability of supply.
2. Turn heater to OFF.
3. Wear safety glasses, boots, and apron.
4. Pump accelerator down the drain.
5. Rinse out tank.
6. Fill tank half full of deionized water.
7. Add 18 gallons of accelerator.
8. Fill tank to operating level with deionized water.
9. Set heater at 70 degrees Fahrenheit.
10. Log in tank change.

Electroless Copper

Temperature: 70–75 degrees Fahrenheit.
Tank size: 95 Gallons.
Dump schedule: None.
Cross-filter: Every Friday.
Make-up: 10 gallons solution A, 10 gallons solution B, remainder deionized water.

Make-Up Procedure:

1. Fill tank ⅔ full with deionized water.
2. Add 10 gallons A component, stir well.
3. Add 10 gallons B component, stir well.
4. Fill tank to operating level with deionized water.
5. Place air hose in tank.
6. Set heater at 70 degrees Fahrenheit.
7. Log in tank make-up.
8. Perform color standard analysis.
9. Notify laboratory of new make-up. They will perform complete chemical analysis.

Dump Procedure:

1. Check for availability of make-up supplies.
2. Wear safety glasses, boots, apron, and gloves.
3. Turn heater to OFF.
4. Pump bath into a cleaned plastic/plastic lined drum. Label drum "Spent Electroless Copper, Caustic."

5. Follow instruction under Cross-Filtering Procedure for tank cleaning procedure.

Cross-Filtering Procedure:

1. Wear safety glasses, boots, apron and gloves.
2. Turn heater to OFF.
3. Filter the bath through a 5 micron filter bag into a clean, empty holding tank.
4. Place the air hose into the plating bath.
5. Carefully lift quartz heater and scrape any deposited copper into the empty tank. Set the heater in adjacent rinse tank.
6. Unbolt heater guard and level sensor.

   a. Set heater guard in bottom of empty tank.
   b. Set level sensor in beaker of micro-etch to remove any deposited copper.

7. Fill the tank about half full with water.

   a. Pour in two gallons of concentrated nitric acid.
   b. Fill the tank to operating level with water.
   c. Any deposited copper should be etched off tank walls and bottom by Monday morning.

8. Monday morning, pump the spent nitric acid from the electroless copper tank into a holding tank.

   a. Thoroughly rinse out the tank and remove any trace of nitric acid.
   b. Wipe out inside of tank with a rag until it is dry.

9. Reassemble heater guard, heater, and level sensor in the cleaned tank.
10. Pump the electroless copper bath from the weekend holding tank into the cleaned tank.

   a. Place the air hose in the electroless copper bath, unless it has copper deposited on it.
   b. If the air hose has copper deposited on it, disconnect it from the air pump and immerse in nitric acid stripping tank.
   c. When cleaned, rinse well, reconnect to air pump, and place in electroless copper bath.

11. Set heater to 70–75 degrees Fahrenheit.

12. Use color standards to analyze and bring to 100% bath strength. Notify laboratory; lab will perform full chemical analysis.

13. Pump the spent nitric acid into the now empty weekend holding tank.

    a. When this tank has been cleaned of any deposited copper, the nitric acid should be pumped into a cleaned plastic/plastic lined drum for disposal.

    b. Label the drum "Nitric Acid, Corrosive, Oxidizer."

Alternative Cross-Filtering Procedure Using Ammonium Persulfate:

1. Wear safety glasses, boots, apron, and gloves.
2. Turn heater to OFF.
3. Filter the bath through a 5 micron filter bag into a clean holding tank.
4. Place the air hose into the plating bath.
5. Fill the tank about ¾ full of water.
6. Turn the heater ON and set at 80 degrees Fahrenheit.
7. Add about 50 pounds of ammonium persulfate. Stir until dissolved.
8. Add water to operating level. *Note:* Any deposited copper should be etched off by Monday.
9. Monday morning, pump the ammonium persulfate into another holding tank. *Turn heater to off before emptying tank.*
10. Rinse empty tank with water; then fill to operating level with water.
11. Add 10 gallons of sulfuric acid. Let stand for 30 minutes.
12. Pump sulfuric acid down the drain, rinse tank well.
13. Pump electroless copper bath into the clean tank through a 5 micron filter bag.
14. Place air hose in the electroless copper bath.

Note: If the air line has copper deposited on it, you must disconnect it from the air pump and immerse it in the nitric acid tank, then rinse it well, prior to placing it in the electroless copper bath.

15. Set heater to 70–75 degrees Fahrenheit.
16. Perform color standards analysis, bring bath to 100%.
17. Notify laboratory, for complete chemical analysis.
18. Pump the ammonium persulfate into the other empty holding tank.

HOW THE PROCESS WORKS

Cleaner-Conditioner

1. Removes light oils, such as finger prints, oxides, and stains.
2. Conditions hard to catalyze surfaces, such as plastics and glass.

3. Because it etches oxides from the copper surface, it becomes loaded with copper, and should not be dumped down the drain.

4. It is a caustic solution, which means that it is difficult to rinse. That is why there are two rinse tanks. It is very important to drain panels well at each stage of operation.

5. One can monitor the strength of the bath by tracking square footage of panels run through it, and by chemical analysis. Chemical analysis does not indicate how contaminated the bath has become, so tracking square footage is important.

Micro-Etch

1. Promotes good copper to copper bond by removing oxidation inhibitors on the copper foil and oxides.

2. In addition, if the etched surface has unetched splotches, those splotches indicate contamination.

3. Micro-etch is sulfuric acid, hydrogen peroxide, and stabilizers. The acid and peroxide are replenishable, so one can operate the bath indefinitely as long as one does two things:

    a. Periodically cool the bath (weekends) and remove the copper sulfate which builds up.

    b. Drain and rinse panels before micro-etch to prevent neutralization and contamination of the micro-etch.

Catalyst Pre-Dip

1. This is basically sodium chloride (salt) and an acid. The bath gets dragged into the catalyst to maintain chloride and acid strength.

2. This bath when heated also serves as a pre-heater of the panels prior to catalyst. Temperature of the catalyst is critical.

Catalyst

1. This bath makes a nonmetallic surface extremely receptive to an electroless copper deposit.

2. It is a dispersion of palladium particles surrounded by tin. These particles adhere to the surface of the panels. When the tin is removed (accelerator) the palladium is bared for the electroless deposit.

3. Temperature of this bath is critical.

4. If this bath gets dragged into the accelerator or the electroless copper

bath, it will contaminate each and cause the electroless copper to plate out on the wall of the tank.

5. Good draining and rinsing are absolutely necessary after this bath.

Accelerator

1. This is an acid conditioner that removes the tin from the particles of palladium of the catalyst.

2. If this bath gets dragged into the electroless copper tank, it will neutralize the electroless copper and contaminate it.

Electroless Copper Bath

1. The purpose of this bath is to deposit copper inside the holes of printed circuit boards.

2. The success of this bath is dependent upon three factors:

a. Chemical activity.
b. Mechanical agitation.
c. Temperature.

3. A deficiency of *any* of the three factors will result in poor coverage, thin deposits, voids and dark holes.

4. The chemical reaction is: Copper Ions + Formaldehyde + Caustic → Copper Metal + Formic Acid.

5. The bath strength must be up to par for all chemicals, or the reaction will be too slow or destructively fast.

6. Mechanical agitation is necessary to ensure rapid and uniform deposit, and that the deposit takes place inside the holes.

7. Temperature is important because the reaction will not occur if it is too low (much below 70 degrees Fahrenheit); and if the temperature is too high (much above 80 degrees Fahrenheit) the reaction will occur too fast resulting in a powdery deposit or spontaneous plate out on the wall of the tank and dust particles.

8. Bath contamination will result in the bath having a tendency to spontaneously plate-out.

9. Air agitation, provided by an aquarium pump, stabilizes the copper and helps prevent plate-out.

10. Cross filtering the bath into another tank is for several reasons:

a. To remove copper and dust particles.

  b. To etch deposited copper off the walls of the tank.
  c. To acid leach catalyst contamination from the walls of the tank.

11. Bath components are:

  a. A component—copper sulfate, formaldehyde, and stabilizers.
  b. B component—sodium hydroxide (caustic), and brighteners.

Parts of this material appeared in *Printed Circuit Fabrication* magazine.

16

Pattern Plating: Copper, Tin-Lead, and Other Metals

One of the major tasks of a plater is to understand the plating requirements of any given part number:

1. What metals are to be plated?
2. What thicknesses are required?
3. What are the critical plating tolerances?
4. What features of the pattern to be plated will need special attention?
5. What current density range is likely to be required to successfully plate the job?
6. Is the available plating set-up (tanks, bath, anodes, agitation, and rectifier) adequate to do the job?

If the plater understands these requirements and has determined the circumstances necessary to do the job, then the job is well on its way to being successfully completed. There are a number of other considerations involved with plating and running a plating area that cannot be neglected. Among these considerations are:

1. Daily operating routine.
2. Maintenance procedures.
3. Troubleshooting.
4. Training.
5. Understanding how the processes work.

DAILY OPERATING ROUTINE

There should be a set routine used to open the shop, close the shop, operate each facet of the plating functions, and make a smooth-transition from one shift to the next. If there is, then the routine can be taught to all plating personnel on each shift. Only when there is one routine being used to govern shop procedures can the routine be fine tuned, and consistency be brought to practices and product. An established routine can facilitate training, get personnel into good work habits, and prevent the type of difficulties which arise when people are not sure of what they are supposed to be doing.

Any established routine must cover these topics:

1. Opening routine for the day.
2. Getting a job ready for plating.
3. Pre-plate cleaning.
4. Plating tank operation.
5. Completing a job.
6. Shift change and/or closing routine.

Opening Routine for the Day

This comprises all the activities which are necessary to begin work, and should be performed in a prioritized sequence according to safety and urgency. When making up the list keep in mind such essentials as detailed below:

1. Turn on lights in all work areas.
2. Turn on all required circuit breakers.
3. Check the bath level of all tanks which are heated. It is good practice to bring all heated baths to operating level, first thing in the morning. This will avoid operating delays caused by the bath cooling down from water being added just prior to use, which is what will happen if the tanks are not topped off as early as possible in the working day. Also, topping off at this time will help avoid fires caused by low levels in heated tanks.
4. Check all timers for proper setting and verify that they are in ON position.
5. Check that all tanks which require heating are ON. As the rounds are being made, it is helpful to check the bath temperatures. Just because the controller light is on, there is no guarantee that one or more of the heating elements is not burned out. Much time can be wasted waiting for a tank to heat up, when the heating elements are not functioning properly. For this reason, the supervisor should carry a pocket thermometer in case the heated tank does not have an indicating controller for the heater.
6. Turn on power and water to solvent stills, if there are any.
7. Open valves and turn on pumps and agitation systems for all plating tanks.
8. Turn on plating rectifiers.
9. Check the quality of baths on the pre-plate cleaning line, and the electroless copper line; look for obvious contamination or signs that the time to change the bath has arrived. Check the tank change log to verify that all tank changes have been made on schedule.
10. A plater should be assigned to inventory all chemicals and supplies which will be necessary for the day's operation:

a. Tank change chemicals.
b. Anodes.

  c. Rubber gloves.
  d. Pumice.
  e. Filters.
  f. Etch.

11. Once supplies have been inventoried and obtained from the warehouse, a plater should be assigned to change any tanks which are due.

12. Clean anode and cathode bus bars.

  a. Monday morning is a good time to check all anodes for depletion and replacement.
  b. Tin-lead tanks should be checked for anodes which have fallen from the hooks and are lying on the bottom of the tank.

13. Remove plating racks from the nitric acid stripping bath, and add any other racks which are still in need of stripping.

14. Check pH meter of neutralizing sump for the pollution control system.

15. Note the general cleanliness and safety condition of the shop.

These are all basic for getting a plating area operating at the beginning of a work day. All of them should be done before any work is started. If work is begun without these activities being completed, then they will very likely be totally neglected until a problem, such as peeling plating, or burned plating from either a cold bath or lack of agitation, is noticed. If the basic necessities are taken care of in a routine manner, they are less likely to be forgotten, and less likely to cause loss of production when they are noticed. Once these have been taken care of, or at least assigned by the supervisor, then the next round of activities can be attended to.

Getting a Job Ready for Plating

The supervisor and/or plater should review the travelers of all the jobs waiting to be plated: due dates, metals to be plated, special notes written on the traveler, any jobs listed as being "hot." Jobs which may already be racked for plating should be checked for any required hole plugging. Jobs which are awaiting plating and require hold plugging should be assigned for plugging.

Once a job has been selected for plating the panels on that job should be looked at as they are being racked up. Important things to note are obvious contamination on the surface, scratches, or other obvious surface or imaging conditions which will result in scrap if the panels are plated. Platers must be trained to look at panel surfaces and images and understand what they are seeing; they should also be trained to stop, and not process, discrepant panels.

It is good practice to notice how much plating area there is on each side of

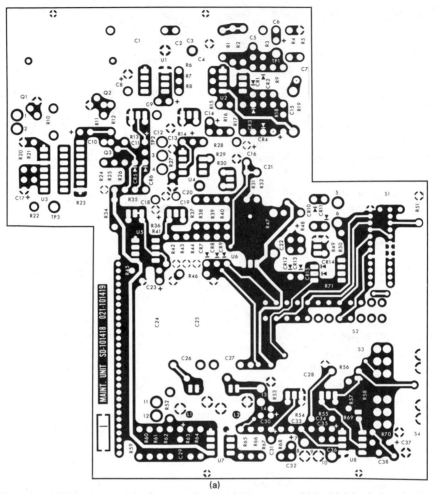

Fig. 16-1. (a) Mostly metal land area to be plated (component side). (b) Mostly low density circuitry, little metal area to be plated (circuit side).

the panel. If one side is mostly land area, and the other side circuitry, the panels should be flip-flopped for racking. Flip-flopping will provide a more even current distribution through the anode bus bars. You want to avoid situations where one anode bus is carrying 100 amps, for example, and the other anode bus carries 400 amps. Flip-flopping will assure that each bus will carry equal amperage (Fig. 16-1).

A plating log should always be kept. Every job can be logged into one book, and tracked according to the tank in which it is plated, or a separate log can be kept for each plating tank (Fig. 16-2). Once the job has been entered, the traveler and drawing of the part number should be consulted, and the geog-

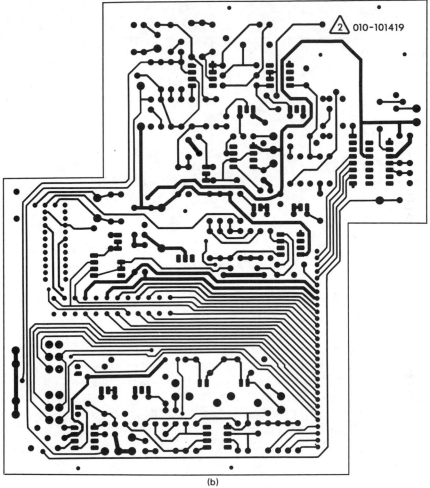

(b)

Fig. 16-1. (*Continued*)

raphy of the plating image studied. It is necessary to look for special notes written by the planning engineer, and for notes included in the drawing. No job should ever be plated until the following information has been determined:

1. What metals are to be plated?
2. What are the thicknesses of the metals to be plated?
3. Are there any restrictions as to where the metals are to be plated—such as solder only in the holes, with the circuitry to be bare copper?
4. Are there any unusually tightly toleranced plated through holes? The plater should use the holes which have the tightest tolerance as the holes

Job No.	Date	Plater	Copper Amperage	Time		Hole Sizes		Panel Quantity	Tin-Lead Amperage
				In	Out	Begin	End		

Fig. 16-2. A plating logbook should contain information useful to the plater. The logbook serves as a tool for organizing the work, and as a record which can be consulted in the future.

he/she uses to determine thickness of plated metal, as these will be the most critical.

5. Are there isolated holes which are likely to cause a plating problem?

Once this has been done the plater should measure the diameter of the hole or holes which will be used to determine thickness of the plated metal. The measured diameter is then written in the log book as the starting reference diameter. The last step remaining before beginning the plating operation is to determine the plating amperage. The procedure for doing this is discussed in the chapter on amperage determination.

Pre-Plate Cleaning

The plater must be aware of whether the circuit pattern has been applied with screen printed resist, or with dry film photoresist. There are great, and critical, differences in the cleaning procedures which must be used for photoresist imaging. Prior to processing photoresist imaged panels, the chapter on dry film photoresist should be reviewed, with special attention to the pre-plate cleaning instructions. There are several points to pay attention to. The panels must have been copper electroplated before dry film lamination. Voids in the plated through holes will occur if this has not been done. Dry filmed panels require etching of the copper surface prior to electroplating. If this is neglected all copper subsequently plated will peel and the panels will be scrap. Platers must be trained to etch the copper surface adequately, until it is matte pink, and to look for voids before the panels are actually placed in the electroplating tank. Etching the copper surface and looking for voids afterward are two key steps critical to successfully plating with dry film photoresist imaged panels.

At this time, since the first step of cleaning is about to begin, the rinse water valves should be turned on. Enough plating racks of panels should be loaded into the soak cleaner tank for one load in the copper plating tank.

1. Acid soak cleaner (for alkaline soluble screening resists):

a. Immerse panels for 2 to 10 minutes. Immersion time actually depends on the specifications of the cleaner as called out in the product data sheet supplied by the manufacturer of the cleaner, as well as the operating temperature and concentration. However, 2–10 minutes is a common time range.
b. Drain well, when removing. Failure to drain will result in the rapid depletion, through dragout, of the cleaning bath.
c. Spray rinse panels

   • Only one panel at a time should be placed into the spray rinse tank. Placing more in will prevent adequate spray from impinging on the surface of the panel. Most soak cleaners are viscous and difficult to rinse; inadequate rinsing may cause severe pitting of the plated metal, and eventual contamination of the plating bath.
   • Adequate spray rinse has been achieved when the panel stops sudsing. The panel must be agitated up and down through the spray, for the full length of the plating rack.

d. Immersion rinse:

   • Panels should be agitated up and down 3 or 4 times.
   • It is a favorite practice of many platers to slam the panels into the bottom of the rinse tank, or to let them drop: this should be avoided at all costs, as the tank bottom will wear and/or a seam split.

2. Sulfuric acid.

a. Immerse panels 1–3 minutes. Generally, immersion time here is not critical.
b. Drain briefly, to avoid excessive dragout.
c. Panels may now go either directly into the acid copper sulfate plating-tank, or be rinsed first, and then proceed directly into the plating tank.
d. Prior to removing panels from the acid, be sure the plating rectifier has been turned on.

3. Acid copper tank.

a. Load the racks into the copper plating tank.
b. Tighten wing bolts on the plating rack, so as to attain an excellent electrical connection for plating.
c. Care must be used to avoid striking the plating racks on either the anodes, or on other racks of panels.
d. Raise rectifier amperage to approximate, predetermined value.
e. Set timer.
f. Log time into plating log.
g. Check panels every 10–20 minutes.

- Notice condition of plating.
- Check hole sizes, in the panel center, and in the corners and edges.
- Adjust plating amperage upward or down, as needed.
- Log plating amperage into plating log.

h. When desired plating thickness has been attained:

- Reduce amperage to a low value.
- Loosen wing bolts of the plating racks on the cathode bus bar.
- Remove panel racks from the plating tank. Let panels drain well, over the tank, to prevent dragout.
- Rinse panels well.

4. Place the panels into the acid holding tank which is appropriate for the next metal to be plated:

a. Fluoboric acid—tin/lead.
b. Sulfuric acid—tin.
c. Sulfuric acid—nickel.
d. Hydrochloric acid—tin/nickel.

From this acid, the panels will either go directly into the next plating tank, or be rinsed first; and then proceed into the next plating tank. As a general rule of thumb the plating amperage for the next operation, which is most often tin-lead, can be determined by using ⅔ of the copper plating amperage. For metals other than tin-lead, ½–⅔ of the copper plating amperage may be used. There should be no need to recalculate square footage or current density. The amperage used for plating copper and the subsequent metal should be written into the plating log, along with the time in each plating tank and the final hole size after copper plating.

At the conclusion of the second plating operation the amperage should be reduced to a low value, the wing bolts of the plating racks loosened, and the panels removed. Panels should be drained over the plating tank from which they are being removed, and then rinsed prior to unracking.

When a job is completed, the plater should count the panel quantity and write this number in the job traveler, and then sign the traveler with the date of completion. If panels were damaged or scrapped by the plater for any reason, the panel quantity and reason should be listed on the traveler next to the plating operation in which the loss occurred. Platers should be encouraged to list scrap in this manner; everyone produces scrap from time to time. By tracking scrap on the traveler all work can be accounted for. Any panels scrapped should be deducted from the final quantity which is signed out by the plater.

Anytime panels are removed from a plating tank the plater should examine the quality of the plating:

Copper Plating

1. Are the racks bright, or dull?
2. Is the copper bright?
3. Are any areas of the panel burning; holes, traces, corners?
4. Are any pits or nodules present?
5. Is step-plating occurring anywhere on the panel, especially the contact fingers?
6. Is there any haziness along the resist, or at the holes?

Tin-Lead

1. Are the panel surfaces being adequately covered?
2. Are any areas shiny?
3. Is copper showing through anywhere?
4. Is treeing occurring anywhere?
5. Is there gassing at the panel surface or at the racks?
6. Is the tin-lead unusually dark or streaky?

These are not glib questions; any of these conditions may occur at any time. They may even occur on panels which appear to be plating well, when not closely examined. The plater must not only ask him/herself these questions, he/she must thoughtfully study the panel surface while asking them. Many adverse conditions go unnoticed until too late because the plating is not being studied closely.

It is good practice for the plater to get into the habit of noticing the voltage at which a job is being plated. The voltage is a good indication of how well a job is plating. High voltage can indicate many conditions:

1. Too many amps for the panel area being plated.
2. Insufficient anode area, due to:

a. Too much panel area.
b. Normal depletion.
c. Anodes which have fallen from the bus bar (common for tin-lead anodes which have not had the hooks soldered into the anode).
d. Anodes which are too short for the tank and panel size.

3. Anode polarization (inability to conduct electricity), due to:

a. Too high chloride content in copper bath.
b. Too low plating bath temperature.
c. Lack of anode area or too high plating amperage.
d. Chemical contamination.

4. Lack of brightener.
5. Lack of bath agitation or cathode agitation.
6. Increase in plating amperage due to carelessness, or from defective equipment.
7. Loss of conductivity at one or more electrical connections from the rectifier to the plating system, due to corrosion or other reason.

The plater should keep an eye on the level of plating bath in the plating tank. Normal operation results in a reduction of the plating bath level from dragout and evaporation. Since panels may not always be placed in exactly the same location in the plating rack, the height of the panel relative to the tank surface changes. The last thing any plater wants to discover is that the top of a panel did not get plated. Also, tanks develop leaks, hose connections loosen, and hoses slip off pipes and valves to develop leaks; any of these situations may cause a slow or sudden drop in the level of a plating bath.

As loads of panels are added to the tank and logged into the plating log, the plater must keep track of the amp-hours of plating in each tank. Brightener must be added to copper plating tanks several times a day. It is usually added on the basis of amp-hours; for instance, 300 millileters of brightener may be added for every 1000 amp-hours. If the plater neglects to make the addition, the brightener concentration of the bath may drop to the point where extremely poor plating occurs.

Shift Change in the Plating Shop

It is good practice to have the relief shift begin work about 15 minutes before the end of the previous shift. This allows time for the platers to exchange information on any of the jobs being plated. The relief plater and the plater about to leave should inspect the panels which are in each of the plating tanks. There is less chance of poor plating going unnoticed, or of overplating occurring when both platers inspect each tank together. Information about how each tank and piece of equipment is functioning can be transmitted, logs signed, and the need for amp-hour brightener additions noted.

Shutdown

Just as in beginning work for the day, there are a number of important considerations which must not be neglected or left to chance at the end of the day. Establishing a written routine will ensure that these considerations will not be overlooked; the routine also can be taught with greater ease and comprehension if it is written down. Among any list of considerations should be:

1. Do not load panels into a plating tank which cannot be completed before the final shift ends.
2. Complete all log entries, and sign travelers.
3. Turn off air agitation to plating tanks.
4. Turn off rectifiers to all plating tanks.
5. Turn off all filter/pumps, and close their valves.
6. Turn off all rinse water valves.
7. Set all timers for heated tanks.
8. Turn off power to all heated tanks, either at the timer or at the controller and circuit breaker (if no timer is being used).
9. Top off all plating baths and heated tanks with water.
10. Clean all bus bars.
11. Unrack plated panels.
12. Remove plating racks from the nitric acid that have been stripped clean.
13. Load all plated-on racks into the nitric acid.
14. Turn off ventilation system.
15. Turn off power and water to any solvent stills, or other powered equipment which should not be left on.
16. Rinse off and wipe down all equipment and tanks.
17. Mop any floors which need to be kept clean.
18. Verify that all barrels are labeled, and stored with bungs in place, and in the proper location.
19. Leave notes for the opening shift pertinent to safety, and productivity.
20. Lock all doors, close windows.
21. Turn off appropriate circuit breakers.
22. Turn off all lights, except those which are required to be on.

MAINTENANCE INFORMATION

There are many aspects of maintenance which must not be taken for granted. These aspects include:

1. The safe operation, and good repair, of the plating area.
2. The necessity of prompt and routine changing of the cleaning lines and filters.
3. Calibration of rectifiers and heater controllers.
4. Cleaning electrical connections to the bus bars.
5. Anode inspection and replacement.
6. Chemical analysis of the plating baths, and prompt addition of required chemicals.
7. Routine use of the Hull cell to monitor brighteners, peptone, impurities, and plating performance; together with purification procedures such as dummy plating and carbon treatment.
8. Proper disposal and storage of waste chemicals.
9. Maintaining adequate pin gauges for measuring hole diameter.
10. Thorough and ongoing training program for all plating personnel.

Safe Operation and Good Repair of the Plating Area

This is a manyfold consideration. A plating area has got to be one of the most potentially hazardous work areas in any company. There is the potential for fires, physical damage due to hazardous chemicals, electrical hazards from the presence of wet floors and electrical equipment, and the presence of trainees, along with the experienced workers, all going about their jobs.

Most plating floors are made of wooden duct boards built over a sump. This arrangement is called a wet floor. Because of continuous exposure to acids, caustic chemicals, and water, the duct boards tend to break. Broken duct boards must be replaced as soon as possible. Until they have been replaced, a marker must be placed over the broken board(s).

The safe handling of chemicals (changing tanks, pumping chemicals, making additions, moving heavy barrels, and the wearing of safety clothes and equipment) is too often neglected in the smaller printed circuit shops. A written procedure must be established for every task involving chemicals. This procedure must then be taught to all personnel, and its use enforced as the only acceptable procedure. Among the considerations must be:

1. Safety glasses must be worn at all times, by all personnel in the plating area. Even people entering the area on a tour must be provided with safety glasses. Too often, there is reluctance on the part of platers to wear safety glasses, and too little resolve on the part of the supervisor and the company to

enforce their use. By choosing comfortable glasses which are propery fitted to the employee, and firm reminders by all plant personnel, the wearing of safety glasses will become a routine part of life in the plating shop.

2. Rubber gloves, boots, and aprons must be worn to protect hands and clothes. The body can be damaged through prolonged exposure to hazardous chemicals. Just because no immediate effects are noticed by platers not wearing these clothes items, does not mean that their use should be neglected.

3. Many people do not know how to safely pick up a tray-type rack of printed circuits. The tray should be gripped at the front, tilted back, and then lifted from the underside with the other hand. A rack should always be carried with the end which is furthest from the person tilted downward. People should not try to carry racks flat or parallel to the ground. To carry racks parallel to the ground is difficult and results in panels flopping back and forth and falling from the rack.

4. The proper use of barrel dollies should be demonstrated to all platers. It should be emphasized that any person should feel free to ask for assistance if they feel they need it when moving heavy barrels.

5. Procedures for pumping, pouring and transporting concentrated acids such as sulfuric acid, hydrochloric acid, nitric acid, and fluoboric acid should be established, written, and demonstrated to all plating personnel. Other platers should be trained to keep watch over anyone else who is handling concentrated acids. This is to protect themselves from someone else's carelessness; and to protect and provide aid to the other person. People handling these acids may need immediate assistance should something go wrong; also, they might spill acid on themselves without being aware of it.

6. *General rule for mixing acids.* ALWAYS ADD THE ACID TO WATER: NEVER ADD WATER TO AN ACID. Acids may react violently when water is added to them.

7. Showers and eyewashes must be present and easily accessible to all personnel in the plating shop. They should be tested daily.

8. Keep hoses and chemical containers out of walkways.

9. When cleaning bus bars on plating tanks, the plater must not let water splash on other platers or equipment. Also, when topping off or adding water to any tank, the plater should never walk away from that tank without shutting off the water and removing the hose. To allow platers to walk away from tanks which are having water added to them is to ensure that, sooner or later, a plating tank is going to overfill and spill.

10. A written procedure should be established for changing chemical holding tanks, such as cleaners, stripper, and acids. This procedure must be demonstrated and enforced for safety. A simple item, such as turning off the heater in a heated tank which is about to be pumped out or drained will prevent a fire.

Also, all too often, platers will energize the motor to a barrel pump without securing the end of the hose attached to that pump. A hose which is flapping around spraying the chemical being pumped is far too dangerous. Platers must be taught to hold the end of a hose which is attached to a pump about to be energized.

The proper guidelines for draining or pumping out a holding tank can prevent a major pollution violation or hazard from taking place. Some tanks can be dumped to the sump, others must be pumped into drums for hauling away; while still others may be dumped only after neutralization of acid or caustic content.

11. Fire extinguishers must be present, easily identifiable, and fully charged. Also, the chemical content of the extinguisher must be rated for handling the types of fires likely to occur in a plating area: chemical fires, burning liquids, electrical fires.

12. Provisions must be present for neutralizing acid or caustic spills. An absorbent substance, such as saw dust, should be handy for soaking up spills on floors. This is an aid for containing and cleaning spills, and may be necessary to prevent a major pollution or safety violation.

13. All personnel must be educated about the chemicals and equipment of the plating area. They must be informed as to the nature of the chemical (acid, caustic, solvent, etc.), how to safely process the chemicals, and what to do to protect themselves in case of contact with the chemical, or accident with the equipment.

All chemical tanks and containers must be clearly marked as to what chemical is present, and what the nature of that chemical is. This type of chemical safety is sorely lacking in many plating shops.

All personnel must understand what to do in case of an accident or emergency.

Prompt and Routine Changing of Cleaning Chemicals and Filters

Considering the price of chemicals in a cleaning line, such as cleaners which may cost $20 a gallon, it is understandable that many printed circuit manufacturers are reluctant to change chemicals any more often than is required. But how are the chemical changing criteria established?

The potency of many cleaning chemicals can be established by simple titration (volumetric chemical analysis); many chemical manufacturers even provide these kits with their chemicals. The potency of a chemical does not tell the whole story of how well it will function as a cleaner; some criteria must be established for discerning when the concentration of contamination has reached the point that the chemical bath must be dumped, and a new bath made up. The chemical data sheet for each bath, the manufacturer's representative, and the experience of the plater are generally sufficient to establish all necessary criteria. Once the criteria are established they should be written down for all to follow. Where practical, a schedule of tank change and a tank change log should also be set up. The tank change log is the key to assuring that all cleaning bath changes are made, and not neglected. The end of a work day is a good time to change cleaning chemicals, as little productive time will be lost.

The complete set of instructions for each tank should be kept in the change log, so that no plater ever has to go searching for the information, and all platers perform the change in exactly the same way. The following is an example of the type of information which should be kept with the change log:

1. Acid Soak Cleaner:

Make-up: 15 gallon Brand XYZ cleaner with tap water.
Temperature: 70–125 degrees Fahrenheit.
Tank size: 100 gallons operating capacity.
Dump schedule: Weekly

Make-up procedures:

1. Wear safety glasses, boots, apron, and gloves.
2. Check availability of make-up chemicals before dumping.
3. Turn heater to OFF.
4. Drain tank to sump.
5. Fill tank ¾ full with tap water.
6. Add 15 gallons of Brand XYZ cleaner.
7. Fill tank to operating capacity with tap water.
8. Set heater to 100 degrees Fahrenheit.
9. Log in tank change.

2. Sulfuric Acid:

Make-up: 15 gallons concentrated sulfuric acid, reagent grade.
Temperature: Room temperature.
Tank size: 100 gallons operating capacity.
Dump schedule: Weekly.

Make-up procedure:

1. Check availability of make-up chemicals before dumping.
2. Wear safety glasses, boots, apron, and gloves.
3. Neutralize with soda ash, and dump to sump.
4. Fill tank ¾ full with tap water.
5. Add 15 gallons of sulfuric acid, slowly.
6. Stir after adding acid.

7. Fill tank to operating level with tap water.
8. Log in tank change.

3. Fluoboric Acid:

Make-up: 20 gallons concentrated fluoboric acid.
Temperature: room temperature.
Tank size: 100 gallons operating capacity.
Dump schedule: Weekly.

Make-up procedure:

1. Check availability of make-up chemicals before dumping.
2. Wear safety glasses, boots, apron, and gloves.
3. Pump into plastic/plastic lined barrel, or neutralize and dump to sump.
4. Fill tank ⅝ full with tap water.
5. Add 20 gallons concentrated fluoboric acid.
6. Fill tank to operating level with tap water.
7. Log in tank change.

It should be mentioned that failure to change cleaning chemicals can lead to a number of problems. Soak cleaner which has become diluted and polluted will not adequately clean the copper surface of the panels which are being cleaned for copper plating. This cleaner is called upon to perform a very big job. It must remove fingerprints, hand lotions and oxides which have built up on the surface: the proper functioning of this bath is critical. When it has become spent, copper-to-copper peelers will show up. The loss of even one panel may be more than enough to offset the savings from trying to squeeze another day out of the tank before changing. If plating for the day has already begun and this cleaner was due for change, the change should be made, even if it will cause a delay in production.

The change of sulfuric acid and fluoboric acid is similarly critical. These acids are dragged into the plating tank to maintain acid strength in the electroplating tank. They also remove oxides, and may serve as a temporary holding tank for the panels. If they have become diluted, then the acid which is dragged out of the plating tank will not properly be replenished. These tanks also serve to prevent contamination of the plating tank. The sulfuric acid is the final tank before the copper electroplating tank. Any cleaner which was not rinsed properly from the panel surface may still be removed by soaking in this acid. If this acid becomes contaminated with cleaner, the only way to prevent that cleaner from getting dragged into the electroplating tank, is to change it regularly.

The fluoboric acid serves a similar function in preventing residuals of the copper plating bath from getting into the tin-lead plating bath. In addition, it removes oxides from the plated copper surface. This means that the fluoboric acid actually becomes contaminated with copper ions from the copper oxide which it removes. Eventually, this acid will turn slightly to distinctly greenish blue from the build-up of copper. If the fluoboric acid bath is not changed regularly, and often, copper will get dragged into the tin-lead plating bath and cause other problems, such as dewetting during reflow.

Filters belong on any list of items requiring regular change. Filters remove most of the particulate matter which accumulates in a plating tank. The types of particles which find their way into plating tanks include:

1. Plastic particles which chip off the plating tanks from panels banging into it, and which are dragged in from the rinse tanks.
2. Dust and dirt from the plating environment.
3. Anode film and chunks from torn anode bags, which get into the plating bath by platers adding so much water that the bath level is higher than the anode bag.
4. Rubber plugs which fall out of non-plated through holes.
5. Particles of plating resist which breakdown and come off the panel.

The filter is the number one defense against any of these causing nodules and plugged holes. It is possible for the filter to become so clogged with debris that the pressure is great enough for the solution to flow around the filter seals, effectively eliminating all filter action and redepositing the debris from the filter back into the bath. This is why filters must be changed on a periodic basis. It is possible to place a pressure gauge on the filter chamber to track pressure loss over the life of the filter. If this is done, a given pressure loss can be set as the value at which filters will be changed. Another reason for changing filters regularly is that some of the filtered debris may break up, or totally dissolve, thus polluting the plating bath.

There is another type of filter which removes dissolved organic matter, a carbon pack filter. This type of filter will remove brightener or other dissolved organic matter. It is very handy to run with on small tanks, such as contact finger plating tanks. When dissolved organics need removing from larger plating tanks, either a large carbolation cannister can be placed in series with the filter/pump or the entire bath can be batch carbon treated.

The periodic changing of filters also provides the opportunity to remove any air which may have been sucked into the cannister. This is primarily a problem on copper plating tanks, since air is used to provide agitation. The pump intake must be located on the bottom of the tank, below the air sparger (Fig. 16-3). It is common to direct the air sparger nozzles downward, to the tank bottom. This stirs up any debris which may be lying on the bottom of the tank; the stirred up debris can then be removed by filter action. However, it increases

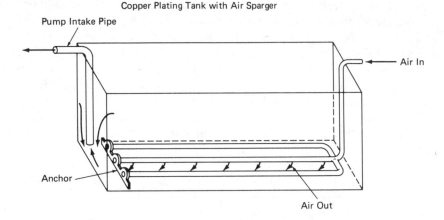

Fig. 16-3. Copper plating tank with air sparger: (1) Air from each leg of sparger is directed downward, to stir debris for removal by filter/pump. (2) There should be a sparging leg beneath each cathode bus bar. (3) The anchor firmly positions sparger 1 to 2 inches off tank bottom. (4) Pump intake must be situated away from, and beneath, air sparger: otherwise filter chamber will fill with air.

the likelihood of sucking air into the pumping system. If the pump intake is located anywhere but at the bottom of the tank, the pump will most probably suck up enough air to completely fill the filter chamber cannister. When this becomes filled with air, virtually no filtering action is taking place. If filter cartridges are routinely clean and/or dry when they are changed, this indicates the presence of excess air and little filtering action.

The problem of air intake can be eliminated by installing an air bleed valve at the top of the filter chamber. The valve can be periodically cracked open to release the air. A plastic hose can be attached to the valve, with the other end protruding into one of the anode filter bags. This will allow air to be continuously bled off (Fig. 16-4).

Some filtering systems can use filter cartridges which are slightly (about ½ inch) shorter than optimum. It is common for filter suppliers to ship printed circuit shops 19.5 inch filters, instead of 20 inch filters; or 9.5 inch filters, instead of 10 inch filters. The plater should always know what he has received, and know whether or not he/she can use the shorter filter. If the filter chamber is equipped with an open ended nut for holding the filter in place, there is little problem with the shorter filters. If, however, the chamber is equipped with a cap type of nut which will bottom out when using a shorter filter, then the shorter filter can not be used under any circumstances.

Acid plating baths must use only Dynel or polypropylene filters, either the spun or the woven fiber. If the spun fiber is used, the filter can be loaded directly into the filter cannister. If the woven type of fiber is used, the filter

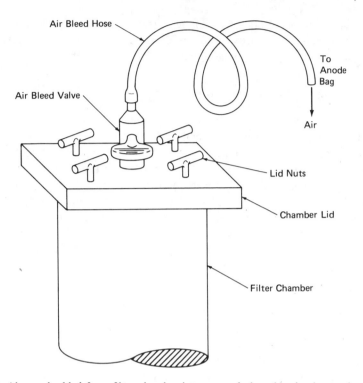

Fig. 16-4. Air can be bled from filter chamber into an anode bag; bleed valve need only be cracked open.

must be rinsed prior to loading into the chamber. Woven fibers are coated with a sizing chemical (soaplike) to let the fibers weave tighter together. This soap must be rinsed out with hot water. Most of the sizing can be removed by soaking the filters for 15 minutes in hot water; notice that this water will suds up. All that is necessary now is a good final rinse. A hose can be inserted into the core, with one end of the cartridge flat against a duct board or a sink. Steaming hot water is then forced through the filter for 30–60 seconds; the cartridge is then turned over and the process repeated. Failure to rinse the sizing will result in a heavy foam on the surface of the plating tank, and eventually pitting will occur if filters are repeatedly changed without removing the sizing.

Filters are available with a variety of weave porosities. Commonly used are filters with porosities of 5–25 microns or more. The finer the porosity used, the better the filter action. However, the power of the filter pump must be considered. All plating baths require adequate filtering and circulation. A plating bath such as Watts or sulfamic nickel is very viscous. Nickel baths often require pore sizes of 20 microns or larger; finer pore sizes reduce bath circulation to an unacceptably low rate. Acid copper plating baths generally operate well with 5 micron filters. Pyro copper baths, because of viscocity considerations, may require upwards of 10 micron pore sizes. The plater must always consider bath turnover rate when choosing filters and pumps.

Calibration of Rectifiers and Heater Controllers

Rectifiers and heater controllers for the plating line provide two of the best "handles" the plater can have for controlling the plating operations. It is critical that these instruments indicate correct values. Yet these pieces of equipment are often exposed to unfriendly environments: the corrosive atmosphere of the plating shop, and acid from the gloves of platers as buttons are pushed and knobs turned. It is vital that these be placed under a calibration program for at least twice yearly calibration. It is common for the calibration service to discover major inaccuracies, which can then be dealt with by the maintenance department. The quality of the product, and the safety of the plating shop are dependent upon accurate rectifier and heater readings.

Clean Electrical Connections to the Bus Bars

The bus bars and the rectifier connections to them are exposed to an extremely corrosive environment. They have acids repeatedly poured on them. When the plating bath evaporates, it leaves crystals of plating salts. These salts will prevent a good electrical connection from forming when the plating rack is suspended from the bus bars. In some cases the connection is so poor that virtually no plating occurs at all. Wire brushes and water must be used to clean these bus bars at least once a day—in the morning. However, any time the connection is yielding poor throwing action at the cathode, or the bus bars are visibly crusty with crystals or oxidation, they must be cleaned thoroughly.

The rectifier leads are bolted to the bus bars; they, too, become corroded like the terminals of a battery. At least once a week they must be disconnected and cleaned with a wire brush. Failure to do so may lead to virtually no electricity passing through a given bus bar. This means that the bulk of the plating is occurring through the other buses. A situation like this means there is absolutely no control over cathode area and plating rate. The plater can periodically feel the cables which are connected to the buses. If the temperatures of the cables are not approximately the same, then uneven plating is taking place because of inadequate electrical connections. This can be verified by testing the electrical current passing through the buses with tong-type ammeters.

Anode Inspection and Replacement

In order that plating proceed continuously and with consistancy, it is necessary to inspect and replace the anodes on a regular basis; once a week works fine for most applications. Anodes generally come in the form of bars, which hang from the anode bus bar by hooks, or they come as chunks, which are dropped into titanium baskets which hang from the anode bus bars. The plater need only lift the hook to check the anode bar, or measure the depth of the anode chunks beneath the surface of the plating bath.

The advantages of anode baskets with chunk type copper anodes are:

1. Theoretically it is easier to adjust and control anode area, since the dimensions of the baskets remain constant.
2. The surface area of copper available for corrosion is greater than that of bar type anodes.
3. Copper plating is most critically affected by variations in anode area, and by introduction of large areas of unfilmed anode. The use of anode baskets means that a pail of copper chunks can be kept in the plating area, which platers can draw from to keep the anode baskets topped off, and the anode area constant.
4. The continuous and inconvenient cleaning of anode bags to place on new anodes is eliminated, as is the time consuming and inconvenient removal of used bags from old anodes which are being removed from the plating tank.
5. The expense of anode bags is virtually eliminated, since the anode bags on the titanium baskets rarely need replacing.

The chief advantage of bar anodes is that it is easier to reposition anode area for special plating situations, such as plating panels which are mostly land area on one side, and low density circuitry on the other. Also, if the copper concentration has built up to the top of the range, it can be lowered merely by letting the anodes be consumed, then spreading out the remainder. With basket anodes, letting the copper be consumed would result in uneven plating occurring from the top to the bottom of the panel. Of course, it is possible to remove an anode basket to reduce anode area.

There are some restrictions on the hooks which can be used to hang anodes. Copper plating tanks require titanium hooks; any other metal will result in metallic ion contamination of the plating bath. Tin-lead anodes require Monel hooks; titanium or other metal would quickly dissolve into the plating bath. The Monel hooks for tin-lead should be soldered into the anode before it is placed into the plating bath. If the hook is not soldered, plating solution and the weight and softness of the tin-lead will cause the anode to fall off the hook. Falling anodes are extremely inconvenient, and are a major cause of cracks in polypropylene plating tanks and subsequent loss of costly plating bath and productivity.

Anodes must be cleaned before placing them into either an anode bag or into the plating tank. Any number of chemical cleaning methods work well. A good pumice scrub with a brush, followed by thorough rinse works well also. If an anode has oil or grease on the surface, as sometimes results from drilling and tapping holes for the hooks or cutting the anode, the oil or grease must be completely removed prior to normal anode cleaning. The anode bags must be cleaned in a manner similar to filter cartridges, and for the same reason. They can be soaked in hot water repeatedly. As they are being transferred from one soak to the next, they must be wrung out.

All properly corroding sacrificial anodes for acid copper plating develop an anode film. The purpose of the anode bag is to prevent this film from getting into the plating bath. Should the anode bag become torn, as often happens when it is struck by panels being loaded into the plating bath, the anode film will contaminate the plating bath and may result in nodules or other problems, such as de-wetting of solder during reflow. A torn anode bag should be replaced as soon as it is discovered. When removing bags from old anodes, the bags must be inspected for tears and holes before they are reused.

New bar anodes should be placed on the center anode bus bar, for a three anode bus plating tank; the older anodes on the center bus bar can be moved over to the outer bus bars. This will assure optimum anode consumption; the center anode bus works about twice as hard as either of the outer buses.

Tin-lead anodes should be checked for depletion just like any other bar anode. There is another requirement which is equally important: tin-lead anodes tend to fall off the anode hooks. They must be lifted from the bus bar and the hook turned to verify that a secure connection still exists. The hook need only be turned enough to apply firm pressure. This operation can only be neglected if the hooks have been soldered into the anode (a propane torch works well) and shop history proves that the soldering was successful. There is too much at stake—a crack in the plating tank, loss of the plating bath, and a major pollution problem and fire potential—not to check the security of the anodes on their hooks.

The anode may lose the reliability of its electrical connection to the anode bus bar because of oxidation of the hook or of the bar and build-up of plating salts. The electrical connection may be reestablished by sliding the anode hook back and forth one-quarter inch or so. This need only be done if a visual inspection raises a question about the connection, or if there is a metal throwing power problem.

When replacing anodes, pulling out the old ones, or just cleaning existing anodes, it is good practice to examine the corrosion characteristics and anode film of the anode.

1. Anodes which are corroding in a dumbell shape, with a largely uncorroded bottom end furthest from the hook, indicate that the anode is longer than necessary for the plating conditions and panel lengths. The ideal anode length is

2–6 inches shorter than the panel when both are suspended in the plating bath. An anode which is too long adds nothing to plating quality or uniformity, and wastes metal.

2. The surface of the corroding anode should be fairly smooth. Pitted surfaces, or surfaces with crevices, indicate inadequate anode area and the possibility of impurities dissolved in the anode metal, or uneven distribution of alloys or additives (such as phosphorous in copper).

3. A black anode film on tin-lead anodes, especially when that film is thick, indicates the presence of various contaminating elements in the anode. Tin-lead anodes should only be purchased from reputable suppliers of plating anodes and metals. Anodes made from old printing lead or reclaimed solder should be avoided at all costs.

4. The surfaces of copper anodes for acid copper sulfate baths should be black. With anodes which have just been removed from the plating bath the blackness should remain on the copper even after the anode has been rubbed with a finger. Running a finger over the anode surface may remove the thicker portion of the film; but if shiny copper is exposed, the film has not formed properly. A properly corroding copper anode will form a tenacious black anode film. Beneath the anode film is a layer of tiny copper particles, called copper fines. The solid bar of copper of the anode, or copper chunk, breaks down into these copper fines before dissolving into the plating bath through the anode film. The copper fines lie on the smooth surface of the corroding anode. If the anode film is not black, is not tenacious, and there are no copper fines, then there is a throwing power problem: chemical imbalance, too high temperature, insufficient chloride, low brightener, or possibly organic contamination.

5. A silver film on the anode indicates that chloride content is too high, and the anode is on the verge of polarizing, or becoming electrically nonconductive.

6. A reddish film on a copper anode indicates nonformation of a proper anode film, which is required for good throwing and anode corrosion. The film is cuprous oxide, and indicates that the anode has polarized. This film will dissolve by itself as the anode sits in the plating bath; and unless the conditions which caused it to form (low brightener with high anode current density, or low temperature) are corrected, it will reform.

7. A yellow or greenish-yellow anode film on a copper anode indicates that either phosphorous content is too low, or the wrong (or contaminated) brightener has been added to the plating bath.

Chemical Analysis and Additions

The reliability of any plating system is directly related to the control which is maintained over the chemistry of the system. A reliable plating system demands:

1. Full chemical analysis at least once a week.
2. Prompt addition of all chemicals required to bring the system to optimum parameters.
3. Operation of the system in such a manner that chemical additions and adjustments be required as little as possible.

The weekly chemical analysis report serves as kind of a score card on the skill level and understanding of the plating personnel about the processes they operate. Large additions of metal salts may indicate lack of anode area. Imbalance of tin/lead ratio in a tin-lead tank, despite continuous additions of salts, may indicate lack of anode area and incorrect plating current density, as alloys are sensitive to current density for maintenance of metallic ratios. Large additions of extra brightening agents indicate that better control over amp-hour tracking and routine brightener adds are needed. Even acid additions, such as sulfuric acid to copper baths and fluoboric acid to tin-lead baths, indicate that dragout is too high and better draining upon completion of electroplating is needed; it may even indicate a need to raise the concentration of acid in the pre-plate cleaning line holding tank. If a company uses the weekly chemical analysis report merely to obtain information on how much of what chemicals to add, a great opportunity to gain a handle on their plating system is being lost.

Printed circuit manufacturers who possess an in-house laboratory have a real advantage over manufacturers lacking this facility. The in-house laboratory offers immediate results on analysis, immediate Hull cell testing of the analyzed plating bath—and the ability to correlate all of this with the current performance of any and all of the plating tanks. When analyses are performed by an in-house chemist, the chemist and other plating personnel have the opportunity to see first hand what the plating looks like for any given and known chemical composition.

It is always nice when a printed circuit shop can afford to have a full-time chemist to handle analyses and chemical maintenance duties. It is not, however, necessary to hire a full-time chemist. The printed circuit industry is blessed with a wealth of technical talent available in the form of technical sales/service representatives. Many of the representatives who service the plating/printed circuit industry are chemists. The corporate offices of these companies are staffed with laboratories and chemists to aid the printed circuit manufacturer with virtually any technical problem that may be encountered. Almost all of these people are ready, willing, and able to help the printed circuit manufacturer set up a laboratory and train the plating/engineering personnel in standard analytical procedures. All that may be necessary for the printed circuit manufacturer to do is make the necessary commitment.

It is not uncommon to find platers who have learned to perform simple volumetric and gravimetric analyses. Such people are usually committed to excellence in plating, and to gaining a fuller understanding of plating principles. One of the best tools a printed circuit manufacturer has to encourage learning and adherence to procedures by the plating personnel is to offer them the opportunity to get their feet wet in a laboratory. If this task seems overwhelming to those charged with running a plating area, they can always hire a college chemistry student part-time to set up and run the program.

Once the analysis has been made, and a chemical addition determined and calculated, the addition should be made without excessive delay. The plater should keep in mind the idea of conformance to optimum chemical composition. The ranges published in most data sheets for chemical content may in fact yield plated metal; but the plating time, center of the hole, and land area to hole throwing power ratio will vary as the bath concentration varies. There are some principles which platers should follow to prevent sudden shocks to the plating bath:

1. Do not add chemicals while panels are being plated.

2. Do not add large volumes of chemicals all at one time; spread a large chemical addition over several loads of plating. This will afford the opportunity to evaluate the effect of the addition on plating.

3. If the pattern of chemical additions suddenly changes, question the results of the analysis.

4. Consult the weekly analysis report, and make a list of the types and quantities of chemicals to be added. Sign off the list after each addition.

5. Add only the purest plating chemicals available. Use reagent or C.P. (Chemically Pure) grade chemicals for the plating baths. Technical grades are good for stripping and cleaning, but not for plating.

6. If a chemical which is about to be added has an unusual color or smell, do not add that chemical to the plating bath:

   a. Have the suspicious chemical returned to vendor, or analyzed.
   b. Do not be afraid to open another container of the chemical: the entire plating bath may be at stake.

7. Do not add chemicals to a plating tank where the level of the plating bath is at the shoulder of the anode. If the level of the bath rises above the anode shoulder, there is the possibility of chemical attack on the anode/hook connection, and of contaminating the plating bath with matter from inside the anode bag.

Plating Bath Control with the Hull Cell

Without a doubt, the greatest tool available to help understand plating baths is the Hull cell. It is common knowledge among people who work with plating systems that electroplating baths are very much like people: they have person-alities which change periodically; some are better than others; they respond in very unusual manners at times; and at other times they may be ill and in need of diagnosis and therapy to maintain or regain health. The Hull cell is the tool of choice to make sense of these situations.

Some chemicals must be added continuously throughout the day to copper plating and other baths; others need be added weekly or even less often. The Hull cell is the tool of choice to gauge the need for addition agents. An excess of addition agents is also easily determined by use of the Hull cell. There are some chemicals which tend to accumulate in plating baths: spent addition agents, leached-out resist, unwanted metallic ions. All plating baths have a certain degree of tolerance for organic and metallic contamination. The Hull cell can be used to track, over a period of time, the build-up of these contam-inants. By tracking contamination level it is possible to decide upon the opti-mum level of contamination which can be tolerated without adverse effects on the plated product, and therefore to decide upon the correct time for preventive purification measures, such as carbolation, carbon treatment, and dummy electroplating.

The Hull cell is a miniature plating tank designed to plate on a brass panel. The panel is situated at an angle to the anode (Fig. 16-5); the result is a record of the plating bath performance across a wide range of current densities. Each plating bath to be evaluated has an anode for that metal, or alloy, which goes into the Hull cell along with the plating bath. The healthier a plating bath is, the wider the current density range at which acceptable plating will be obtained (Fig. 16-6). Conditions which tend to reduce the acceptable plating range (bright range for copper, nickel, and some other metals) are undesirable conditions, and they can be tested for with the Hull cell. Some examples of undesirable plating conditions are:

1. Too high or too low a temperature.
2. Too much or too little brightener.

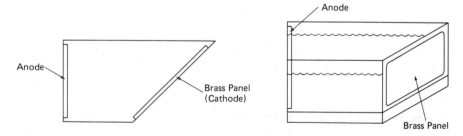

Fig. 16-5. Brass cathode sets at an angle to anode; which sets up a current density gradient at the cathode.

Hull Cell Control of Copper
Operating Conditions: Temp. 75°F, Current 2 Amps, Time 10 Mins., Air Agitation.

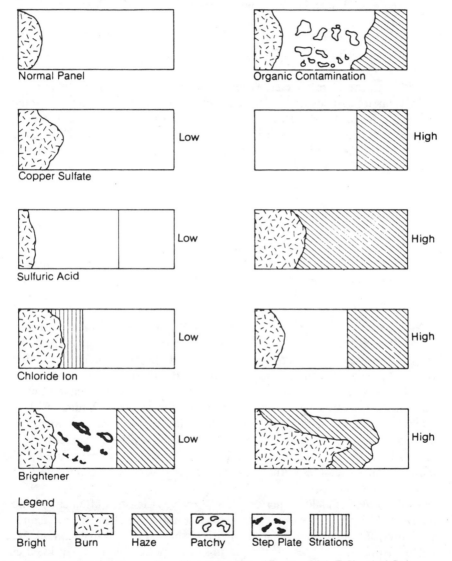

Fig. 16-6. Hull cell diagram for acid copper sulfate. (Courtesy M & T Chemical Co.)

3. Too much or too little metal, acid, chloride, or other bath constituent.
4. Too much or too little peptone in a tin-lead bath.
5. Too much organic contamination (addition agents and resist breakdown products).
6. Too great a metallic ion contamination (copper in tin-lead is a common problem).
7. Too high or too low a pH.
8. Metallic composition of alloy not in balance.

Although there are some conditions which do not show up in the Hull cell, such as photoresist contamination of a nickel bath, these limitations are minor. There are instances where the Hull cell can be used to prove the health of a plating bath; in fact, this is exactly what the desired outcome of a Hull cell test should be for the well run plating department. However, there are times when a plating problem is evident, and the problem is not due to the plating bath itself:

1. Lack of agitation.
2. Rectifier malfunction.
3. Incorrect amperage determination.
4. Poor pre-plate cleaning, or other contamination of the panel surface.
5. Insufficient anode area, or contaminated anode.
6. Poor electrical connection.

When situations like this arise, it is helpful to eliminate as many variables as possible, and that is where the Hull cell comes in. Each aspect in the lists above can be tested for by performing Hull cell plating tests. The more experience the plater has in running Hull cell evaluations, the more information he/she will gain from each test; and the fewer tests that person will need to perform to arrive at valid conclusions. Once the plater is familiar with the effects any given condition has on the Hull cell test panel, the need to perform confirming tests is reduced. There are times when a plating problem may be caused by more than one problem, all or none of which may be related to the plating bath. When these situations arise, there is usually a need to run a battery of Hull cell tests.

Hull cell tests should be performed routinely, at least as often as chemical analyses. The person performing the Hull cell tests, the chemist, the engineer, the plater, should always look at the plating which comes from each tank evaluated by the Hull cell test. If the plating being performed on the printed circuits is optimum, the plater should be able to correlate this to the Hull cell. If the printed circuit is coming out step plated, the plater should be able to correlate this to the Hull cell test panel. The Hull cell loses much of its preventive and diagnostic power when performed by someone in a laboratory who never sets foot in the plating shop, and who never looks at the production plating.

The basic steps for performing a Hull Cell are given below:

1. Place the correct anode in the Hull Cell, for the bath which is to be evaluated.

2. Clean a brass Hull cell panel:

   a. A back-and-forth pumice scrub using a bristle brush works well.
   b. Rinse the panel well, and set it in a beaker of acid (20–50% hydrochloric or sulfuric acid).

3. When the clean panel is setting in acid, draw a sample of the plating bath and pour into the Hull cell. The test should be run at the temperature of the plating bath in the shop area.

4. Rinse the Hull cell panel well, and set in the Hull cell.

5. Immediately turn on agitation, if it is going to be used, and apply electric current from a D.C. rectifier, negative lead to the brass panel, positive lead to the anode.

6. Set current at desired amperage, and timer at desired time:

   a. 5–10 minutes is ample for most applications. 10 minutes is required to test for anode polarization.
   b. The following are useful amperage settings:

Copper	1 or 2 amps
Tin-lead	1 amp
Nickel	1 amp
Tin-nickel	1 amp
Gold	1 amp

7. At the conclusion of the plating time, remove brass panel. A water rinse, followed by acetone rinse with air/oven dry, will prevent undesirable streaking or tarnishing of the plated metal.

8. Write the date, amperage, and plating time on the back of the panel for later reference.

The test has now been completed, all that remains is to interpret the results. Most manufacturers of plating baths publish diagrams of what the Hull cell panel should look like for a number of conditions. This should be referred to. If none is available, the plater may perform a number of tests by changing the bath chemistry and running Hull cell panels. This type of testing is highly recommended.

It is not necessary to exactly duplicate the agitation conditions of the plating tank in the Hull cell. Generally, a simple back-and-forth motion across the brass panel (using a glass or plastic rod) is adequate. Air agitation in the Hull cell is available, but really contributes nothing to the results.

In addition to comparison to other Hull cells, the completed Hull cell test panel can be read for current density information using the Hull cell current density ruler. This ruler is laid across the panel. The scale corresponding to the plating amperage used for the test is calibrated to denote what current density each area of the test panel was plated at, the highest current density at the left edge, the lowest at the right edge. By using this scale, it is possible to get a handle on the range of current densities at which optimum plating can be expected.

There is a limit to just how literally the test panel can be interpreted. Because the test panel displays a bright plating range in acid copper, for example, of 1–80 amps per square foot does not mean that the production panels may successfully be plated at 80 amps per square foot. However, if the test panel is routinely bright up to 80 a.s.f., and then creeps down to being bright only up to 50 a.s.f., the test panel indicates the need to add extra brightener to the plating bath. This can be verified by adding a small amount of brightener to the Hull cell plating bath (equivalent to perhaps 500 ml of brightener being added to the production plating bath from which the Hull cell sample was drawn), and then running another Hull cell test panel under identical conditions.

Virtually any procedure which can be performed to the production plating bath can be performed to a small sample of that bath, and the results evaluated with the Hull cell. It is far better, and more cost effective, to perform a drastic operation to a small sample of plating bath and then perform a Hull cell test to that sample, than to run the risk of wasting time and endangering the plating bath.

When Hull cell testing is performed regularly, the plater develops a record of the performance of the plating baths across a wide range of current densities. A gradual reduction of brightness range in copper (or other bath), or the gradual movement of streaky or dull plating into the plating range used in production (10–30 a.s.f. for most baths), indicates the gradual depletion of addition agent, or gradual build-up of contamination. By observing the change weekly with the Hull cell, the plater can keep the addition agent in balance by making corrective additions, and keep the contamination in check by carbolating or dummy plating. Thus, the use of the Hull cell can prevent normal conditions (depletion and contamination) from causing downtime and adverse plating on the product. Catastrophic changes to the plating bath, such as gross contamination, can also be detected, and the near exact causes determined.

Adequate Pin Gauges

The chief tool used by the plater to determine when adequate metal has been deposited in the holes is the pin gauge. There are two systems available: the discrete pin system and the tapered pin system. The discrete pin system is far more accurate than the tapered pin, but it is more difficult to maintain. The tapered pin system is more easy to maintain, but is less accurate. Some of the problems associated with these systems are discussed in the chapter on determining the thickness of plated metal.

When it comes down to resolving the diameter of a plated through hole there is nothing available which is more accurate or easier to use than the individual pin. A set of pins has one pin for each 0.001 inch of hole diameter. The pin is inserted into the hole, and the pin which easily fits in the hole with just a slightly snug fit is the correct pin for that diameter hole.

The disadvantages of these pins are that platers frequently do not return them to the properly labeled slot in the pin box, and the pins tend to rust. Most platers understand the necessity of keeping the pins in the correct hole in the pin gauge box, but the pins do get lost, and they do get mixed up. The problem of losing pins, and of putting them back incorrectly, can be virtually eliminated by training the platers (1) never to let the pin out of their hand until they return it to the hole in the box from which they removed it, and (2) never to remove more than one pin at a time from the pin gauge box. It is so easy to pull out one pin, then another, and so on, until the correct one has been found. This is fine, but, a pin should never be laid down or placed anywhere except in the correct slot from which it was removed. Obviously, if there is only one empty hole in the pin gauge box, there can only be one correct slot into which the pin can be returned. It is easy to forget the diameter of a pin once it has been removed. If several pins are removed and laid down on a table top while the plater searches for the pin which fits the hole, the probability of correctly returning all the pins to the pin gauge box is very low.

The second major problem with pin gauges is that they rust. Pins are handled by platers with acid on their hands, they are pushed through holes which are being plated in an acid plating bath, and they are stored in pin gauge boxes with acid still on them. The best the plater can hope for is to reduce the amount of acid on the panel by rinsing it well before measuring the plated through holes. The likelihood of removing all the acid is low, and the likelihood of storing the pins free from moisture is even lower. Thus, the pins will rust.

Since there is always going to be a potential for getting the pins mixed up, and the pins are always going to rust, the plater must deal with the situation. If stainless steel pins can be found, they should be bought; this will diminish the rust problem. Platers must rinse panels before measuring the holes; and no more than one pin should be removed from the gauge box at a time. Mixed up pins, or pins not returned to the correct hole in the gauge box, are noticeable. It is really not that difficult to notice an aberration in the diameters of a row of consecutively increasing pins. When the plater notices an aberration, he/she must use a micrometer to sort the pins correctly. It is good practice to place a label on the end of each pin, like a flag, that has the pin diameter written on it; and to routinely check the pins with a micrometer. Also, since pins get lost, and since they rust, extra pins should be kept on hand by the plating supervisor.

Whenever the Plating or Inspection department has replaced an old set of pins with a new set, the old pins should still be kept around, just to replace lost pins.

Tapered pin gauges eliminate the problem of getting the pins mixed up, and, since they are usually stainless steel, they are not as susceptible to rust as are discrete pins. However, these tapered gauges come out of calibration easily. Even though a calibration block (this is a flat, or otherwise, piece of metal with holes of known diameters drilled in it) may be immediately available to the plater, calibration is a real problem. Also, the diameter measured may vary from plater to plater for the same hole. Measured diameter is highly dependent on the force with which the tapered shaft is forced into a hole. Measured diameter is also dependent upon the angle of the shaft to the surface of the panel as it is inserted into the hole. The shaft must be perpendicular to the panel surface for accurate readings. The force of insertion and the degree of perpendicularity are subjective, and will vary widely from plater to plater, and even from one day to the next for the same plater. For these reasons, the tapered pin shaft should be banned from the plating shop.

Waste Storage and Disposal

All plating shops generate huge volumes of hazardous chemical wastes in the form of acid, caustics, corrosives, heavy metal solutions, solvent, and oxidizing chemicals. Almost all of these represent a health hazard to the worker, and an environmental hazard to the rest of the locality in which these chemicals are used, stored, and disposed of. The wisest step which can be taken is to train all workers in the safe handling of these chemicals, and in proper techniques of storage and disposal. There are volumes written on this topic; all that will be mentioned here are some basic but often overlooked common sense considerations.

Safe handling must be built into the procedures for all processes where these materials are used; protective clothing, care in handling, concern for others, and awareness of the dangers and of emergency procedures for accidents are essential. Platers must be trained from the day they start work in the procedures to be used. These must be built in to the written procedures for all processes. Some to remember are given below:

1. Always wear rubber gloves, safety glasses, and aprons (or other protective clothes such as a smock or lab coat), and even rubber boots. This is the first line of defense against accidental bodily contact. It takes only one splash to cause blindness.

Many platers do not like the idea of having to wear cumbersome safety gear. It takes a real commitment on the part of management to bring about the routine use of such items as safety glasses, and even gloves. However, once their use is a firmly established fact of life at the company, most platers will quickly adapt. The company should take care when choosing the safety equipment to be worn; comfort must be one of the considerations. Also, some safety glasses, for instance, hinder side vision because of the type of side shield they use. Some glasses even have a tendency to slide off the nose bridge too easily. When side vision is good, and the glasses stay on, there is less objection to their use. There are similar considerations in the fitting and comfort of gloves.

2. The safest method of handling acids and caustics should be decided upon, and that method established as the one way these chemicals will be handled. The method should be demonstrated to all personnel, and its use enforced. This will bring about consistency in handling. When acids and caustics are handled any other way, all personnel will know to be on the lookout for unsafe conditions.

3. The procedure to be followed for disposing of each chemical should be written into the process procedure. There should never be doubt in anyone's mind as to how to dispose of a tank full of chemicals. If the bath can be neutralized, then dumped to the sewer, everyone should know. Some chemicals need several stages of treatment before they are fit to be dumped. Each of these steps should be written out for all to see. All the equipment and processing chemicals must be at hand.

4. Each plating tank, holding tank, and barrel must be labeled with the name of the chemical within. If a barrel is empty and needed to store a waste chemical, that barrel must be thoroughly rinsed out. When the waste chemical is placed inside, a new label must be affixed to the barrel. That label should name the contents as "Waste" or "Spent," acid, caustic or oxidizer, and the chemical name. All barrels must have the bungs kept in place at all times.

Thought must be given to the type of material a drum is made of before a chemical is placed inside for storage or hauling. Corrosive chemicals should only be stored in plastic drums or plastic lined steel drums.

Also, concentrated sulfuric acid, for instance, should never be added to a drum which has water, or an aqueous chemical, inside. Sulfuric acid reacts violently with water. There are other dangerous chemical combinations. Ammonium persulfate will react violently if accidently mixed with a caustic material, as when electroless copper and ammonium persulfate are accidentally mixed in a drum because of poor labeling.

5. Barrels should be stored in an area slightly isolated from the work area. They should be stored in rows of identical material: acids, caustics, solvents, for easy identification, and to prevent a mix-up.

6. A master list of all chemicals used and their waste disposal considerations should be kept on hand. The barrel storage area is an excellent location to keep a copy.

Training Program

The skill level of the plater is important to accomplishing the tasks required of the plating department. Once a plating line is set up it is fairly easy to teach an individual how to shuffle panels down the line; such individuals are called tank jockeys, and should not be confused with platers. A plater is an individual who understands how to set up the line, maintain it, and make it work well for accomplishing the full spectrum of printed circuits he/she will be called upon to plate.

Unfortunately, in many printed circuit shops, there is either little time to train platers with a full understanding of the science and art of their trade, or there is no one capable of presenting this material effectively. Too often in plating and other aspects of printed circuit manufacture individuals must teach themselves and repeat the same mistakes that everyone before them made. In short, they end up reinventing the wheel. All companies should have a training program to prevent wheel reinvention as the primary training mode, and to teach the basics of the trade. Some of the requirements for a training program are:

1. Written process procedures. These can be posted in the work area for all to read and refer to.

2. Manufacturer's data sheets on each of the plating baths should be available for platers to refer to; and they should be encouraged to refer to them.

3. New personnel should be shown around the printed circuit shop on their first day. The entire process of printed circuit manufacturing should be explained to them, from Production Control and Planning to Shipping.

4. The basic considerations of reading blueprints and job travelers should be demonstrated and explained to all new personnel. It should never be taken for granted that new personnel understand these things.

5. No plater should be set loose to work on their own until they have had explained to them, and been trained in, every facet of the job:

a. Reading job travelers and blueprints.
b. Plating area determination.
c. Considerations for determining amperage and thickness of plated metal.
d. Safe work habits and safety gear.
e. Look at panels before and after processing them, and understand what their eyes see.
f. Processing panels from racking and cleaning through correct operating procedures of the electroplating tanks.

6. Platers should be encouraged to learn other aspects of their job, such as:

a. Micro-sectioning of plated through holes.
b. Analysis of the plating baths.
c. Process control using the Hull cell, dummy plating, carbon treatment, and other techniques.

No one should forget that the plater who understands all aspects of his/her job will be more productive in output and quality than will the tank jockey. Printed circuit shops which suffer from the tank jockey syndrome (high downtime of the plating baths, low productivity, spotty quality, frequent burning, overplating and underplating, safety problems, and other glitches in operation) should seek training for their platers. Most platers are ready, willing, and hungry to learn; they must be encouraged by management to do so.

What sources of expertise and knowledge are available for training?

The sales and technical service people from chemical suppliers are usually willing to put themselves out to train and work with plating personnel. It often happens that representatives of competing vendors for your business will even come in house to set up written procedures and a training program. This is an excellent source of expertise.

Data sheets and other publications of the chemical manufacturer are a storehouse of information. These typically cover every aspect of set up, maintenance, operation, and troubleshooting.

The American Electroplater's Society has some information on printed circuit and other plating technology which will be of value to the printed circuit plater. The American Electroplater's Society appears to be devoting increasing attention to the demands of printed circuit plating.

The I.P.C. (formerly called the Institute of Printed Circuits) in Evanston, Illinois, publishes many papers and other documents on printed circuit plating. Their annual convention offers platers the opportunity to keep abreast of their technology.

Trade shows, like NEPCON and Circuit Expo, offer papers on aspects of printed circuit plating and other topics.

Magazines like *Printed Circuit Fabrication, Circuits Manufacturing, Insulation/Circuits,* and *Electronics Packaging and Production* regularly publish articles of interest to the printed circuit plater.

Management and supervisors should always remember that they must train personnel to carry on with understanding when supervision is absent. What value is a supervisor if the quality of his/her department goes on the skids every time he/she is not present for a day or more? This is the importance of training.

OTHER CONSIDERATIONS

How Electroplating Works

Acid copper is typical of plating baths which use sacrificial anodes. Fig. 16-7 shows a cross-sectional view of a copper plating tank and plating bath. Bars of copper are suspended from the anode bus bar. Printed circuit panels to be

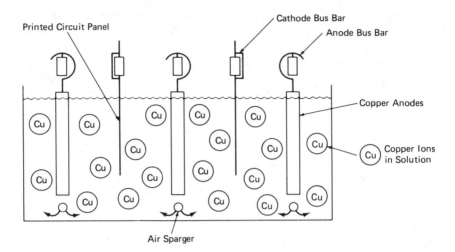

Fig. 16-7. View of a copper plating tank: (1) A negative charge applied to the cathode bus bar causes copper ions (Cu) to leave solution and plate on to the printed circuits. (2) A positive charge applied to the anode bus bar causes copper metal to corrode off the anodes and dissolve into plating solution as copper ions (Cu). (3) The net effect is to "throw" copper from the anode to the cathode.

plated are suspended from the cathode bus bar. Both the panel and the copper anode are immersed in a solution of sulfuric acid and copper sulfate. The sulfuric acid provides electrical conductivity to the plating bath. The copper sulfate provides the metal for plating.

A rectifier is connected across the anode and cathode bus bars, with the negative terminal at the cathode bus, and the positive terminal at the anode bus. When voltage is applied copper ions deposit onto the surface of the cathode, which is the printed circuit. At the same time copper dissolves off the copper anode bars into the plating bath, to become copper ions. This process is referred to as *throwing metal* from the anode to the cathode.

If good control is maintained over the anode area, the concentration of metal will remain virtually constant, even after weeks of plating. If anode area is not sufficient, the metal content will deplete and additions of the metal salt will be required. Metal salt additions are the most expensive method of adding metal for plating. It is far cheaper to provide adequate anode area. If the metal content increases, without the addition of metal salts, as the weeks go by the presence of excess anode area is indicated.

Copper, in the case of copper plating, is always dissolving off the anode. The copper content of a plating bath will increase just by letting the tank lie fallow. If a copper plating tank is going to be shut down for an extended period of time (many weeks or months) it is good practice to remove the anodes from the bath.

What Are Carbolation and Batch Carbon Treatment?

Carbon filtering and treatments refer to the ability of activated carbon to remove organic materials from a plating bath. There are some organic materials which are necessary and desirable to have in the bath, such as brighteners and grain refiners (peptone in the case of tin-lead plating). However, excess brighteners and grain refiners, or the breakdown products of these compounds, and other undesirable organic compounds (plating resist, anode film, sizing agents) are detrimental to quality plating. These compounds build up in the plating bath through normal operation and need to be removed periodically.

Small organic molecules adsorb onto the surface of activated (pure, cleaned, and baked) carbon. The larger, more complex the organic molecule, the less is its tendency to adhere to the carbon. One technique used to make organic contamination easier to adsorb onto carbon is to add hydrogen peroxide or potassium permanganate to the plating bath. These chemicals attack and breakdown organic molecules. Heat is added, not only to help the peroxide or permanganate, but to aid adsorption of the organic molecules onto the carbon.

There are two commonly used methods employing activated carbon: batch treatment and carbolation. When only a slight excess of normally desirable organic compounds needs to be removed, or when only the broken-down organic molecules are to be removed, carbolation is generally used. When all organics are to be removed, batch treatment is generally used.

Carbolation consists of adding a large cannister of activated carbon, usually granulated carbon, in series with the filtering system. The plating bath is then pumped through both the filter and the carbolation cannister. Heat is usually not applied, unless the bath being carbolated is usually heated anyway.

Batch carbon treatment is usually a more involved procedure:

1. The plating bath is pumped into a separate holding tank.
2. Hydrogen peroxide is added (about 500 milliliters per 100 gallons of plating bath). Note: never add hydrogen peroxide to tin contain plating baths.
3. The bath is heated to 120–140 degrees Fahrenheit for several hours, until the hydrogen peroxide has been burned off and effervescence is no longer evident on the bath surface.
4. Activated carbon (granular or powder, 5 pounds per 100 gallons of plating bath) is added and stirred.
5. After several hours, the bath is pumped back into the plating tank.

When carbon has been added directly to a plating bath, special procedures must be used to remove it. the filter used to pump the treated and carbon-loaded bath back into the plating tank must be coated with a layer of diatomaceous earth, or some other nondissolving filter aid. The filter pump should recirculate through a small bucket of water. While this is being done, one or two cups of filtering aid should be added to the water. This will cause cloudiness; when the cloudiness has disappeared, the filter is coated with filter aid. The filter aid effectively increases the ability of the filter to remove small particles. Without turning off the pump, set the bucket in the holding tank with the plating bath, and submerge it. The pump intake must be in the bucket all the time, so that pumping action is constant and unbroken. If the pump is shut off or pumping action lost, even for a second, the filter aid will come off the filter cartridge. The entire process will have to be repeated after the filter is cleaned of the old filter aid.

An alternative to adding carbon to the plating bath is to place the carbolation cannister in series with a filter pump, and recirculate the plating bath through the cannister. The cannister must be capable of holding 5 pounds of granulated activated carbon per 100 gallons of plating bath in the holding tank. The recirculation can be stopped after several hours or when the plating bath is the same color as a new bath, minus addition agents. The bath is now ready to be pumped back into the cleaned plating tank.

Whether performing a batch treatment or periodic carbolation, always use fresh carbon in the cannister. If insufficient carbon is present, or if used carbon has not been replaced with fresh, the contamination already adsorbed onto the carbon will be dislodged from the carbon and recontaminate the plating bath. If recirculation does not clear the plating bath, so that it looks like a fresh make-up within just a few hours, the cannister should be recharged with fresh carbon and recirculation continued.

The advantage of powdered carbon is that it has greater surface area available for adsorbing organic molecules than does granulated carbon. There is no questioning this fact. However, powdered carbon is very undesirable to work with. The slightest draft picks it up and carries it all over the work area. Should it land in a plating bath, nodules will result. The carbon tends to cover every square inch of the environment where it is being used, including covering the persons working with it. Powdered activated carbon is also difficult to filter, even with the use of filter aid. For these reasons, it should be banned from use in the plating shop. Only dustless, granulated activated carbon should be brought into a plating shop. The activated carbon should be purchasd only from a plating supply house.

Whenever a carbolation cannister is being used, the carbon should be rinsed with water while in the cannister. After the cannister is charged, water should be recirculated through it, then drained. This will remove particles of carbon dust from the charge and prevent them from getting into the plating bath.

After the plating bath has been returned to its plating tank, it should be chemically analyzed, and tested with a Hull cell. There is always a small volume of plating bath lost, which has to be made up. The treated bath can be evaluated with the Hull cell to determine the effectiveness of the carbon treatment, before addition agents are added to the plating bath.

Chemicals All Platers Should Be Familiar With

1. Copper; chemical symbol: Cu.
2. Tin; chemical symbol: Sn.
3. Lead; chemical symbol: Pb.
4. Tin-lead, solder; chemical formula: SnPb.
5. Nickel; chemical symbol: Ni.
6. Gold; chemical symbol: Au.
7. Silver; chemical symbol: Ag.
8. Palladium; chemical symbol: Pd.
9. Chloride, the ion of chlorine; chemical symbol: Cl^-.
10. Hydrochloric acid; chemical formula: HCl.
11. Sulfuric acid; chemical formula: H_2SO_4.
12. Fluoboric acid; chemical formula: HBF_4.
13. Boric acid; chemical formula: HBO_3.
14. Sodium hydroxide, lye, or caustic soda; chemical formula: NaOH.
15. Potassium hydroxide, lye, or caustic potash; chemical formula: KOH.
16. Soda ash, anhydrous sodium carbonate; chemical formula: Na_2CO_3 or sodium carbonate monohydrate, chemical formula: $Na_2CO_3 \cdot H_2O$.
17. Copper sulfate, copper salt; as a blue crystal it is copper sulfate penta-hydrate, chemical formula: $CuSO_4 \cdot 5H_2O$.

Troubleshooting

Troubleshooting is an art and a science which is good for all platers, chemists, and engineers to learn. There is really no such thing as having a problem and looking up the cure in a book. Any problem may be caused by any number of causes. From time to time, a problem may be the result of several adverse situations, or marginal conditions, manifesting at one time.

When a problem does occur, observation and reasoning are the tools used to discover the etiology. Most problems can be avoided altogether just by maintaining good control over the processes, and by training. Obvious causes of a problem should be sought first. Platers should always be consulted, as they are the ones who work most closely with the process. Cleanliness is next to holiness in plating. Contamination of the plated surface can result in many detrimental plating conditions, from metal peeling to pitting. Always start with a clean surface, always keep the cleaning chemicals clean, and rinse well. Peeling of plated metal is never caused by the plating bath. Peeling is always caused by plating on a contaminated, or oxidized, surface—no matter what metal is peeling, or from what metal the plating is peeling. There are no conditions in a plating bath which will result in peeling.

Observing the platers may also turn up clues to a problem. Sometimes people will forget important considerations, or fall into a bad habit. A problem may just be due to something like this. If a plater, for instance, places two panels is a spray rinse tank, only one side of each panel is likely to be rinsed well. It may be that both sides are being rinsed well enough 99% of the time to prevent pitting, but the situation is only marginal and a pitting problem is just around the corner.

It may be necessary to perform tests, to run test panels and Hull cells, to evaluate hypotheses. It is possible to completely carbon/hydrogen peroxide treat a few hundred milliliters of plating bath, and run Hull cell tests before and after. It is also possible to make laboratory additions to a sample of the plating bath, and evaluate the results with the Hull cell.

Keep a notebook whenever a problem arises, and make notes on what has been done to eliminate the problem and the results of those steps. Refer to this notebook whenever a problem arises. Never forget, there is always at least one reason for every problem. Do not be afraid to discover what the reasons are.

Tin-Lead Plating

The basic purpose of tin-lead plating is to provide an etch resist to protect the copper plated circuitry. The tin-lead also helps to preserve the solderability of the printed circuit during subsequent use. Tin-lead is a dull, porous, easily oxidized metal when plated. The process of fusing, also called reflow, melts the tin-lead for a few seconds, and fuses it into the shiny, dense, corrosion resistant alloy known as solder. For this reason, tin-lead plating is also called solder plating—although the tin-lead is technically not solder until after fusing.

Tin-lead plating tanks do not usually have a lot of agitation. This is to protect the tin (stannous tin, Sn^{2+}) from oxidizing (to stannic tin, Sn^{4+}). It is helpful to have two pumping systems for tin-lead plating tanks. Since air cannot be used to agitate, the plating tank is dependent on pumps to provide agitation. Plating tanks which have two pumps will throw better than tanks with only one pump. Although two pumps may be used, only one need have a filter cannister attached. Sometimes it is useful to have a cathode rod agitator set up to move the printed circuits while they are being plated.

Tin-lead baths do not have brighteners; instead they have peptone (or a peptone substitute) to act as a grain refiner. Peptone is hydrolyzed (chemically decomposed) beef protein. Additions of peptone should be kept to a minimum. It is not added on the basis of amp-hours, like copper brightener is. Peptone should only be added on the basis of Hull cell results. When a one amp Hull cell panel (run with no agitation) shows gassing or treeing near the plating range of 15–25 amps per square foot, it is time to make a small (no more than 250 milliliters per 100 gallons) addition of peptone. If gassing at the cathode and/or treeing (burning) occurs in the plating tank, check to be sure that the pumps are on, the valves are open, and the amperage is set to the desired value. If all this checks out, then a peptone addition to the plating tank is indicated.

Tin-lead plating baths are sensitive to contamination. Peptone and copper are the most common contaminants. Too much peptone will cause out-gassing (solder bumps or pimples) during fusing. Too much copper will cause de-wetting. The peptone can be kept in check by adding only small amounts, only when needed. Periodic carbolation, four hours at least once a month, will keep peptone and its breakdown products in check. After carbolation, it will be necessary to perform a Hull cell and make an addition of fresh peptone. Copper can be kept under control by dummy plating the tanks about four hours a week at 3–5 amps per square foot. It is critical that all panels be thoroughly rinsed free of sulfuric acid prior to loading into the fluoboric acid soak or the tin-lead plating tank. Sulfuric acid will cause a white precipitate in the tin-lead tank; this precipitate is lead sulfate. It is possible to remove all the lead and ruin the plating bath by adding enough contaminating sulfuric acid.

An anode bag or two of boric acid is usually suspended on the corners of tin-lead tanks. The boric acid will retard the decomposition of fluoboric acid to hydrofluoric acid. This is important for good plating and for safety reasons, as hydrofluoric acid is extremely dangerous. Platers must keep these bags full, and check them at least once a week.

The optimum deposit of tin-lead is a uniformly light, dull, matte grey. If the deposit comes out dark, it may need a little peptone: run a Hull cell before adding any, however. If the deposit comes out streaky, it should be dummy plated. The plater should always check the solder after reflow, as the reflowed deposit tells something of the quality of the plated deposit.

17
Contact Finger Plating

Electrical connection to printed circuit boards can be made through various connectors, directly soldered connections, and contact fingers. One of the advantages of contact fingers is the ease with which a printed circuit may be installed and removed from a piece of electronic equipment; the circuit board can simply be pushed in and pulled out. The requirements for simplicity, ease of use, and reliability have placed several restrictions on the fabrication of the contact finger. The contact finger must be free of oxidation and corrosion, and remain so for the expected life of the equipment. It must also be wear resistant (hard). These needs are generally satisfied by using gold plated on top of nickel.

Gold (24 karat or 99.8%) is used because of its superior electrical conductivity, and because it will not oxidize or become electrically passive. However, if the gold is plated directly on top of copper, the copper will migrate into the gold, causing it to tarnish (due to copper oxidation) and lose its electrical integrity. In order that this migration be prevented, nickel is plated as a barrier coat between the gold and the copper. No migration of copper into gold will occur, and the electrical resistivity of the gold is preserved for all time. The plated nickel is harder than the copper or the gold, and adds greatly to the wear resistance of the contact fingers.

Contact fingers are plated after the printed circuit has been pattern plated, the resist stripped, and the copper foil etched off the epoxy laminate. The general procedure is as follows:

1. Tape circuit to expose the contact fingers.
2. Strip tin-lead.
3. Scrub the bared copper.
4. Plate nickel.
5. Plate gold.
6. Remove tape from printed circuit.

The above sequence may be carried out using manual (Fig. 17-1) procedures or it may be carried out with the use of an automatic plating line (Fig. 17-2). Since the late 1970s, automatic plating lines for contact finger plating have come into their own. This piece of equipment not only improves throughput and productivity, it improves the distribution of gold metal plated across the row of contact fingers, to conserve precious gold.

AUTOMATIC CONTACT FINGER PLATING

Automatic equipment requires that the printed circuit panel be taped off prior to plating. The taped off panel is loaded onto the machine and run down the processing line in a vertical manner. The panel is immersed in the plating solution, which is directed at the contact fingers through spray nozzles at high velocity. There are no sacrificial anodes; the anodes for both nickel and gold are inert. As the metal is plated from solution, it must be replenished through addition of the metal salts. The panels are processed through tin-lead stripping, scrubbing, rinsing, and nickel and gold plating in just a few minutes. The nickel and gold are plated at about 100 amps per square foot rather than at 5–15 amps per square foot. The automatic plating equipment offers several advantages over manual plating which cannot be ignored:

1. Gold savings.
2. Increased productivity.
3. Higher quality.
4. Enclosed module.
5. Labor savings.

Gold Savings

Typical examples of gold distributions obtained from a manual contact plating operation are shown in Fig. 17-3. For a minimum requirement of 30 microinches, it is common to find measurements of 40–50 micro inches at the center of a panel, and 10–20 micro inches more at the edges of the panel. Even if the process is fine tuned (for a long panel quantity run) so that 30–35 micro inches is obtained at the panel center, gold is still wasted from excess plating at the panel edges. The best which can be hoped for is to optimize control of plating chemistry (temperature, pH, specific gravity, gold content, brightener, and impurity level), and design of plating racks, plating tanks, filter pump and sparging system, rectifier, and electrical connections. Automatic contact finger plating equipment achieves virtually exact and uniform gold deposition on every panel in a job, and everywhere gold is measured on the row of contact fingers. The cost of gold is a major monthly expense at any printed circuit shop. Even a small shop may spend $30,000 a month on gold. If uniform gold deposition would allow savings of 10–30% (and often more) on gold consumption for the same number of panels, that printed circuit shop will save $3,000–$9,000 per month ($36,000–$108,000 per year).

Increased Productivity

The panels are loaded, either automatically or manually, onto the conveyorized, vertical, in-line plating machine. The conveyor speed sets the number of

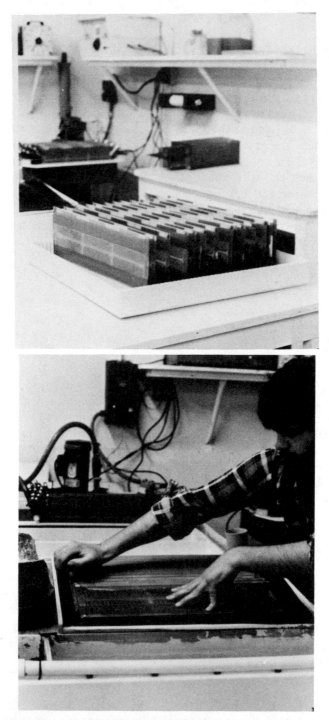

(a)

(b)

Fig. 17-1. (a) Tin-lead being stripped from contact fingers. (b) Using a brush and pumice to scrub copper contact fingers. (c) Nickel plating, as a barrier coat between the copper and gold. (d) Gold plating.

(c)

(d)

Fig. 17-1. (*Continued*)

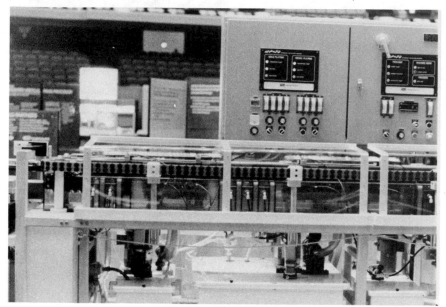

(a)

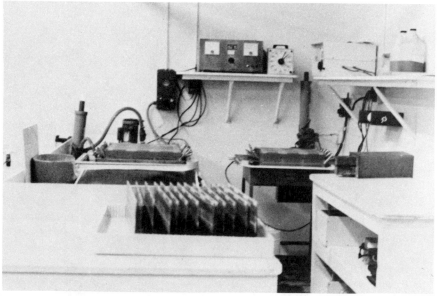

(b)

Fig. 17-2. (a) Typical automatic contact finger plating line. (b) Manual contact finger plating line.

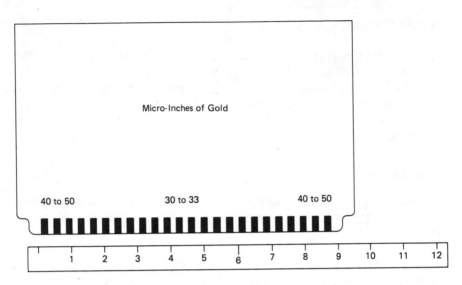

Fig. 17-3. Thickness distribution of plated gold. Measuring linear inches of contact fingers for area and amperage calculations.

panels which can be plated in an hour. The limiting speed is typically determined by the thickness of nickel required. Even at the lowest conveyor speed settings, it is common to achieve throughputs of 120–150 panels (18 inches across the contact fingers) per hour. This value can jump dramatically higher for nickel thickness requirements lower than 300 microinches. All of this can be done with only one plater (for machines equipped with automatic loading/unloading devices) to monitor and operate the machine. The rapid plating is partly responsible for increased throughput, but the even and uniform distribution across each panel also contributes, since there is no need for drastic overplating to reach minimum requirements.

Higher Quality

Automatic plating for contact fingers results in higher quality than can be achieved with manual plating, for several reasons. The panels spend less time sitting in tin-lead strippers. There is reduced tape seepage, and less attack upon the epoxy/glass substrate. All of the steps, except for taping and detaping, are performed automatically. There are absolutely no handling operations. The multitude of operations involved with manual plating (scrubbing, rinsing, racking, unracking, and moving from one tank to the next) is a major cause of damaged circuitry. These operations are performed by a machine, for exactness and consistency.

Enclosed Module

Each of the processing tanks is an enclosed module, almost totally isolated from the environment. This means that access to the gold bath is difficult. This results in improved security and reduced likelihood of contamination. The other baths are also isolated, which means that neither the bath nor the environment is likely to be contaminated. Automatic contact finger plating machines, and their environments can be maintained immaculately clean.

Labor Savings

With an automatic contact finger plater, what was done on two shifts, can be accomplished on one. What formerly required two to four platers, can be accomplished with one.

An automatic contact finger plating machine is best run in a shop which has a chemical laboratory. Since throughput is extremely high, and plating is rapid, there is a need to maintain process control at optimum levels. Although the need for chemical additions to nickel and gold can be made on the basis of plating rate (time/amperage required to achieve a given thickness), greater overall consistency is achieved by maintaining pH, specific gravity, acid and metal contents at a uniform level: and this is best monitored with an in-house chemical laboratory.

MANUAL CONTACT FINGER PLATING

When manual contact finger plating is being used, big dividends will be paid to the shop which chooses to optimize chemistry, equipment, and plater training and supervision. One fact which must be kept in mind is that two nickel tanks are generally required to feed one gold tank. If the gold tank lies fallow while nickel is being plated, much productivity is lost. Since plating time in nickel is a major factor affecting throughput, and since gold must be conserved through optimum distribution, the design and capacities of both tanks must be given careful consideration.

CHEMICAL CONSIDERATIONS

The performance and efficiency of a gold plating bath is drastically affected by several factors (Fig. 17-4).

pH. Generally, the efficiency (weight of gold plated per amp-minute) increases with higher pH. For most printed circuit hard cyanide gold, there is a slight plateau between about pH 3.8 and 4.8. Because hydrogen gas is

Cathode Efficiency Data

Optimum Conditions

Gold Content	- 1 oz./gal.
Conducting Salts	- 400 g/gal.
pH	- 4.2
Temperature	- 35°C
Current Density	- 10 ASF
Cobalt Content	- 125 ppm

I. *Efficiency Vs. Temperature*

Temperature	Cathode Efficiency %
25	52.2
35	55.8
40	55.2
45	54.9
50	53.0

II. *Efficiency Vs. pH*

pH	Cathode Efficiency %
3.9	46.2
4.2	55.8
4.5	58.1
5.0	62.8

III. *Efficiency Vs. Gold Content*

Gold Content Troy Oz./Gal.	Cathode Efficiency %
0.5	41.3
1.0	55.8
1.5	59.5
2.0	69.8

IV. *Efficiency Vs. Cobalt Content*

Cobalt Content (ppm)	Cathode Efficiency %
50	68.5
80	60.2
125	55.8
180	54.3

V. *Efficiency Vs. Current Density*

	Cathode Efficiency %
4	44.8
7	52.8
10	55.8

Fig. 17-4. Factors affecting efficiency of gold plating. (Courtesy Engelhard Corporation.)

evolved, the pH of gold tends to increase with plating time. The advantage of operating within the pH plateau is that efficiency and plating rate are more consistent; and consistency in gold plating is critical to cost effective operation. pH also affects the rate at which the cobalt brightener is plated out in the deposit. The optimum pH is specified by the manufacturer of the gold plating bath, but the pH values nearest the center of the plateau will generally provide the most uniform cobalt content in the plated gold. pH must be checked at

Effect of Operating Parameters on Co-Deposition of Cobalt

Optimum Conditions

Gold Content	1 oz./gal.
Conducting Salts	400 g/gal.
pH	4.2
Temperature	35°C
Current Density	10 ASF
Cobalt Content	125 ppm

I. *Percent Cobalt in Deposit vs. Current Density*

Current Density (ASF)	% Cobalt
4	0.2
7	0.29
10	0.29
15	0.25

II. *Percent Cobalt in Deposit vs. pH*

pH	% Cobalt
3.9	0.40
4.2	0.29
4.5	0.26
5.0	0.19

III. *Percent Cobalt in Deposit vs. Gold Content*

Gold Content (oz./gal.)	% Cobalt
0.5	0.27
1.0	0.29
1.5	0.33
2.0	0.29

IV. *Percent Cobalt in Deposit vs. Temperature*

Temperature °C	% Cobalt
25	0.29
35	0.29
40	0.24
50	0.22

V. *Percent Cobalt in Deposit vs. ppm Cobalt in Electrolyte*

ppm Cobalt in Electrolyte	% Cobalt
50	0.11
80	0.18
125	0.29
180	0.36

Chart courtesy of Engelhard Corporation.

least once per shift, and adjusted daily with chemicals recommended by the bath manufacturer. It is better to try to keep pH near the optimum recommended value than to let it slide around within the operating window through lax pH additions.

Temperature. Most baths operate between 60–70 and 100 degrees Fahrenheit. As with all other parameters, that temperature which is recommended as

Throwing Power

I. *pH Variation*
   A. *Solution Composition & Operation*

Gold Metal	-	1.0 oz. tr./gal.
Cobalt Metal	-	125 ppm
Sp. Gravity	-	15° Be
Temperature	-	35° to 37° C
Current	-	1 amp (10 ASF)
Time	-	5 minutes

   B. *Results*

pH	Throwing Power %
5.0	52.0
4.8	52.0
4.5	45.9
4.2	56.0
4.0	52.7

II. *Co Content Variation*
   A. *Solution Composition & Operation*

Gold Metal	-	1.0 oz. tr./gal.
Sp. Gravity	-	15° Be
pH	-	4.2
Temperature	-	35°–37°C
Current	-	1 amp (10 ASF)
Time	-	5 minutes

   B. *Results*

Co Content ppm	Throwing Power %
75	56.0
100	50.4
125	56.0
150	53.2
175	55.5

Chart courtesy of Engelhard Corporation.

optimum by the bath manufacturer should be the only temperature used, for efficiency of the plating bath and consistency of deposit quality.

Gold Content.   The efficiency of a gold bath increases with higher metal content. Gold is costly enough that its content is a major plating consideration. Good uniform plating can be achieved with gold metal in the range of 1.0–1.2 troy ounces per gallon.

Brightener and Impurities.   Printed circuit plating baths for contact fingers typically use cobalt and/or nickel as hardening agents. Their content in the deposit, which determines hardness or brittleness, is affected by pH, specific gravity, temperature, and plating current density. The build-up or presence of excess contaminants in a gold plating bath tends to reduce the current density range at which bright deposits may be obtained. The brightener, or hardener,

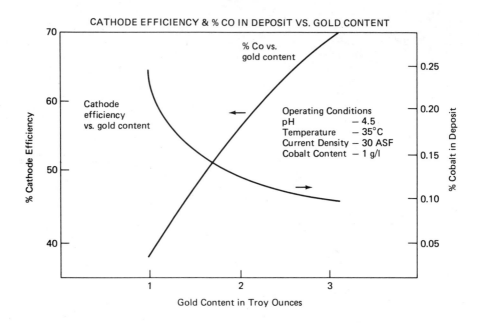

CATHODE EFFICIENCY & % CO IN DEPOSIT VS. GOLD CONTENT

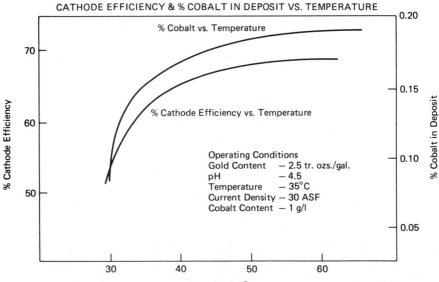

CATHODE EFFICIENCY & % COBALT IN DEPOSIT VS. TEMPERATURE

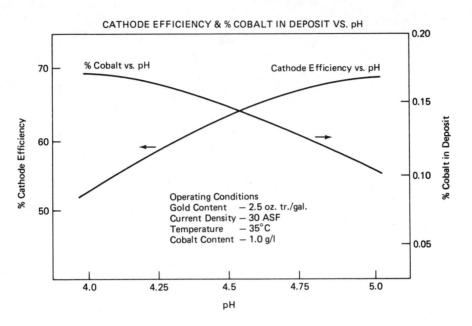

CATHODE EFFICIENCY & % COBALT IN DEPOSIT VS. pH

Operating Conditions
Gold Content — 2.5 oz. tr./gal.
Current Density — 30 ASF
Temperature — 35°C
Cobalt Content — 1.0 g/l

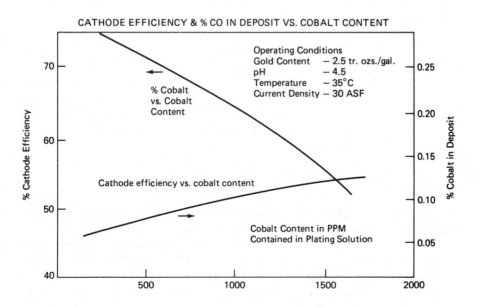

CATHODE EFFICIENCY & % CO IN DEPOSIT VS. COBALT CONTENT

Operating Conditions
Gold Content — 2.5 tr. ozs./gal.
pH — 4.5
Temperature — 35°C
Current Density — 30 ASF

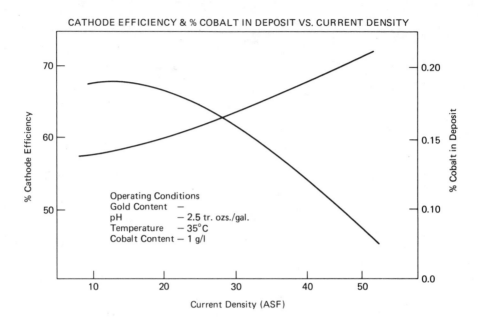

CATHODE EFFICIENCY & % COBALT IN DEPOSIT VS. CURRENT DENSITY

must not be added in excess quantity, as it cannot be easily removed. Organic contamination can be controlled by running a carbon-pack filter cartridge periodically, or at all times.

Specific Gravity. The throwing, or covering, power of a bath is affected by the specific gravity. This parameter is monitored with the use of a hydrometer. Specific gravity tends to decrease during plating, from dragout and other losses. It is controlled by additions of monopotassium phosphate, or other chemical as recommended by the bath manufacturer. The specific gravity variation will affect the content of cobalt and other contaminants (desirable and undesirable) in the deposit.

Nickel Plating

The factors which affect gold plating also affect nickel plating. Fortunately, the cost of nickel metal is not the consideration that the cost of gold is. However, the gold is deposited on top of nickel, so the nickel plating bath operation is virtually as important as the operation of gold plating. The two common plating baths used for plating nickel in printed circuit shops are the Watts nickel, and the nickel sulfamate. The nickel sulfamate bath has come into wide application since the mid-1970s, and is most common in printed circuit shops. A common make-up for this bath is:

Nickel (metal)	10 oz/gal
Nickel Chloride	1 oz/gal
Boric acid	5 oz/gal
pH	3.8
Anti-pit agent	0.2%
Stress reducer (leveling agent)	0.05%

The pH can be maintained with additions of nickel carbonate (to increase pH) and sulfamic acid (to decrease pH). pH will tend to increase during the course of plating because of hydrogen gas formation. The most common cause of decreasing pH is the dragging of acid from the cleaning line into the bath.

Boric acid is controlled by keeping an anode bag full of boric acid immersed in the tank. This chemical acts as a buffer at the cathode surface.

Anti-pitting agent is a surfactant which dissolves hydrogen bubbles at the cathode surface, and helps to wet the surface being plated.

The stress reducer, which also acts as a leveling agent (brightener), reduces the tensile stress which is otherwise found in plated nickel, and promotes hardness. This type of additive is good for printed circuit nickel contact fingers.

Nickel sulfamate is the primary source of dissolved nickel metal for plating. As the plating progresses, nickel in solution is plated out and replenished by anode corrosion.

Nickel chloride is a minor source of dissolved nickel metal. Its primary function is to provide chloride for proper anode corrosion. The chloride reduces anode polarization, and increases bath conductivity. The chloride also reduces plating potential at the cathode. When chloride is below recommended concentration, throwing power is low, and the bath may deplete itself of nickel.

Generally, throwing power is greater with increased temperature, pH, and nickel and chloride concentrations. Throwing power decreases with decreases in the parameters just mentioned, and with increases in the levels of impurities, such as brightener and resist breakdown products and metallic contamination. The level of contamination can be determined with the aid of a Hull cell. Organic contamination will show up as a patchy dullness. Metallic contamination shows up as a light haziness in the plating range. Dummy plating at 5 amps per square foot will reduce most forms of metallic contamination. The effectiveness of electrolytic purification can be determined by performing another Hull cell at the conclusion of several hours of dummy plating. Organic contamination can be removed by placing carbon pack filters in the filter cannister for several hours. A subsequent Hull cell will determine the need for further carbon treatment.

PLATING TANK CONSIDERATIONS FOR CONTACT FINGER PLATING

The plating tanks should be made of rubberized steel or polypropylene. Polypropylene is preferable, since even a slight hole in the rubber lining of a steel tank would present a contamination threat. Ideally, the tanks are deep enough for an anode to cathode separation of 4–6 inches. This electrode distance will allow a more uniform deposition of plated metal. All contact finger plating tanks should spill over into a deeper solution sump. The use of a spillway will assure that the level of the bath is constant. This is necessary to assure that all panels, when racked for plating, will sit at the same level in the tanks. If the level of a contact finger plating tank changes there is a possibility that parts of the contact finger will not get plated, and sometimes the solution level will be over the top of the plating tape.

Agitation must be through solution pumps together with a sparging system. Air should not be used for contact finger plating, although it may be best for other nickel plating applications. Air agitation may cause nickel to be splashed onto the printed circuit. Nickel plating solutions are viscous, and sometimes require the use of two pumps to supply adequate filtration and agitation. Because of the high solution viscosity filters with 20–50 micron pores may have to be used. Gold, too, is a viscous solution. However, 10 micron filters (polypropylene or Dynel) may be used. It is common practice to run only carbon pack filters in gold baths. *Note:* if the decision has been made to use air agitation, consult the manufacturer's data sheet for the nickel plating bath being used, as it may be necessary to use a nonfoaming anti-pitting agent.

The pumps should be connected to a sparging system, so that all solution returned to the plating tank is directed upward at the contact fingers being plated. This will assure excellent solution movement for uniform plating. The solution must not be directed downward in nickel plating baths, as it will disturb the anode film and cause particles to be distributed throughout the plating bath.

Anodes can be made from either platinized titanium screen mesh (preferred), or from graphite carbon. The anode lies along the tank bottom and is electrically connected to positive polarity on two ends for maximum electrical uniformity. The anode for gold plating is totally inert. Nickel rounds can be spread over the anode screen in nickel plating tanks; these rounds will provide a sacrificial anode and should prevent having to make metal salt additions, other than required to replenish dragout. Usually, the anode electrical connections are at the left and right sides of the plating tanks. The cathode bus bars run along the front and back edges of the tank, and are connected to the negative electrical polarity of the rectifier. The plating rack sits across the plating tank, resting on the two cathode bus bars. Sometimes the cathode electrical connection is not made at these bus bars. The electrical connection may be applied directly to the plating rack via a clamp. Thus the clamp may be attached to the plating rack so that the rack is electrically hot as the panels are loaded into the plating bath. The idea is to prevent galvanic deposition of metal, which is not very adherent. However, either method for making electrical contact to the cathode works very well; galvanic deposition will not prove to be a problem because of the potentials involved.

Rectifiers used should be capable of 10 volts, and have an adequate amperage rating so that heat will not be a problem. It is convenient if the rectifiers are equipped with timers. All rectifiers for plating precious metals should be equipped with amp-minute meters. This is necessary for several reasons. By performing chemical analyses, and correlating the analysis with amp-minute readings, it is possible to determine the amp-minute/troy ounce constant. This is the number of amp-minutes of plating for each troy ounce of gold consumed. Tracking this number on a regular basis will allow the chemist to determine his/her degree of success in controlling the critical process parameters for the precious metal bath; a wide variation in this number means that pH, temperature, metal content, current density, or specific gravity are not under optimum control. Wild fluctuations may indicate significant losses of the plating bath; whether through leaks or theft. Once the constant is determined, gold can be added on the basis of amp-minute reading. Thus, for a constant of 500 amp-minutes per troy ounce of gold, three ounces of gold plating salts are needed after 1500 amp-minutes of plating.

APPEARANCE OF THE GOLD DEPOSIT

The gold deposit's appearance is affected mostly by the appearance of the nickel which is plated beneath it. If the nickel is dull or hazy, so will the gold be. If the nickel has pitting, so will the gold. The gold deposit is so thin that many defects do not have ample latitude to develop. About the only detrimental visual characteristics likely to be observed in gold contact fingers is a dark, burned deposit; this will have a very dark brown or deep yellow color. This condition may be caused by:

1. Too high a current density.
2. Too low a current density (for aged baths).
3. Too cold an operating temperature.
4. Insufficient solution agitation, such as occurs when the pump has not been turned on.

Another condition frequently associated with gold plating is that of peeling. There are several reasons which cause gold to peel off contact fingers. When peeling has occurred, it is important to determine if the peeling occurred because (1) the gold peeled off the nickel, (2) the nickel peeled off the copper, (3) the plated copper peeled off the base copper, or (4) the gold peeled off gold.

Gold peels off nickel because of insufficient activation of the nickel prior to plating. Nickel metal may passivate as soon as 10–20 seconds after it has been removed from a nickel plating bath. If there is any delay in moving racks of panels from the nickel tank to the gold tank, it is wise to set the rack of panels in 50% by volume water and concentrated hydrochloric acid for one or two minutes. This will provide adequate activation for subsequent gold plating. Gold peeling from a nickel surface is always due to the nickel surface being contaminated or passive; gold peeling is never due to a condition of the plating solution in either the nickel or the gold tank.

If gold has peeled from the contact fingers because of a nickel/copper peeler, the copper was insufficiently cleaned prior to plating. Two common causes for this condition are: (1) the copper oxidized prior to being loaded into the nickel plating tank, and (2) the tin-lead was not properly stripped from the contact fingers prior to nickel plating. After tin-lead stripping, the contact fingers are scrubbed with pumice, or an imbedded grit pad (either manually, or mounted on a wheel). This scrubbing operation will not make up for poor stripping. Should some tin-lead be left on the copper surface of the contact finger, a nickel/copper peeler will result. This condition sometimes is visible as a "bubble" beneath the plated nickel/gold, usually at the top of the contact finger next to the plater's tape. Oxidation prior to nickel plating occurs when there is a delay in getting the racked panels into the nickel tank. If the panels are set in a water rinse for more than a few seconds, oxidation will occur. Nickel/copper peelers are never due to a condition of the nickel plating solution.

If the gold peeled from the contact finger because of a copper/copper peeler, then the base copper was not sufficiently clean during the pattern plating operation. Before further contact finger plating is performed, a panel which has been through the plating operation should be reflowed either in hot oil or under infrared radiation. After reflow, the plating adhesion should be tested either with tape or by digging at the circuitry with a razor blade knife. The reflow operation may be all that is needed to promote adhesion of the copper/copper bond. Should peeling persist after reflow, it would be pointless to continue with contact finger plating; as the panels will ultimately result in scrap.

There are times when the gold will actually peel or break off gold. This is not a common situation, and when it does occur the plated gold is probably more than 100 microinches thick. There are limits to the thickness at which hard gold can be plated without cracking. It is not necessarily true that a printed circuit which has been overplated with gold is just as good as one plated to the required thickness. Gold which cracks off of itself can cause a short circuit. The gold bath manufacturer's recommendations for gold content and level of brightener should be followed with the same exactness as is given to brightener level in a copper plating bath.

APPEARANCE OF THE NICKEL DEPOSIT

Nickel is plated to a greater thickness than is gold. Because of this the quality of the deposit is of great importance, since it will have a drastic effect upon the visual appearance of the contact finger. Some of the common conditions of the nickel plate are:

1. Burned plating.
2. Hazy plating.
3. Dull plating, or patchy dullness.
4. Step plated appearance.
5. Pitting.
6. Nodules.
7. Peeling or "bubbles" beneath the plated nickel.

Burned plating can result from too high a current density, and shows up as noduled deposit in the high current density areas at the ends of the panel. Burning may also be due to depletion of the stress reducer/grain refining agent. Sometimes burning may be heralded by the appearance of dullness. A Hull cell performed before and after stress reducer has been added to a liter of plating bath will determine if the dullness or burning was due to depletion of this addition agent. Plating from a bath which is too far below the recommended temperature range will also cause dullness and burning. Lack of agitation, as occurs when a filter is clogged, the pump valves closed, or the pump motor is turned off, will contribute to burning.

Patchy dull plating generally indicates the presence of excess organic impurities. This can be confirmed by carbon treating a liter or two of plating bath in the laboratory, adding a proportionate amount of addition agents, then performing another Hull cell. Excess organics can also be determined by running carbon pack filters in the filter canister for several hours and then performing a Hull cell. There are certain types of organic impurities which can build up in a nickel plating bath (such as photoresist leach-out) which will not show up as haziness during contact finger plating or Hull cell plating, and which will cause pitting. This kind of contamination may show up as low throwing along right edge of Hull cell panel.

Haziness indicates the presence of metallic impurities. Whenever this is suspected a Hull cell should be run; the Hull cell should also exhibit haziness in the plating area. Metallic contamination can be confirmed by dummy plating blank panels at 5 a.s.f. for an hour or two, then performing another Hull cell. Both metallic and organic impurities are a fact of life in any plating operation. Routine carbon filtration and low current density dummy plating should be planned into the preventative maintenance of all plating shops.

Step plating is very rare in nickel plating baths. When step plating is noticed on contact fingers, it is generally due to step plating in the electroplated copper, and not in the nickel. Step plated contact fingers must not be ignored. When noticed they must immediately be brought to the attention of the plating supervisor or chemist. The proper course of action would be to inspect the plating of all copper currently in progress, and the copper of contact fingers immediately after tin-lead stripping.

Pitting in nickel contact finger plating may be due to two things: (1) a high level of organic impurities, or (2) depletion of the surfactant anti-pitting agent. The anti-pitting agent will have absolutely no effect on reducing pitting which is caused by excessive organic contamination. Pits which result from depletion of surfactant are small and round, and may be confined to the highest current density areas of the panel. These can be alleviated simply by adding a dose (per bath manufacturer's data sheet) of anti-pitting agent. Pits which result from organic contamination are large, and often have an irregular shape. Pits of this type may be accompanied by a smoky appearance of the nickel, with "smoke" emanating from the pits. Pits of this type are randomly distributed over the contact finger area.

Nodules in a plating bath indicate particulate matter floating in solution. The most common source of particulate matter in most plating baths is the anode film. The anode film of a nickel bath can be disturbed by directing a stream from the pump at the bottom of the tank, or by stirring the bath. When nodules are due to this source, letting the bath sit for an hour or so will eliminate the nodules, as the particles resettle to the tank bottom.

Bubbles beneath the nickel plate and nickel/copper peeling are discussed under "Appearance of the Gold Deposit."

AMPERAGE DETERMINATION FOR CONTACT FINGER PLATING

Amperage is determined by approximating the area to be plated, then multiplying this square-footage value by the desired current density. One method for approximating area is to assume that each linear inch of contact fingers (Fig. 17-3) is one square inch of plating area, for double sided boards. For single sided boards, assume that two linear inches of contact fingers is one square inch of plating area.

Thus, if 10 panels are to be plated, and each has seven linear inches of contact fingers, the area to be plated is approximately $7 \times 10 = 70$ square inches, or 0.486 square feet. Gold and nickel can usually be plated successfully at 10 amps per square feet. If problems occur in the deposit, such as burning or dull plating, the amperage may be decreased or increased as needed. As plating baths age, the range of current density for bright plating narrows, and it is possible to obtain a dull deposit at low current densities. (This is a more common limitation in gold plating than in nickel plating.)

A general formula for determining the plating amperage for double sided contact finger panels is:

$$\text{Amperage} = (\text{Linear Inches} \times \text{Panel Quantity} \times \text{ASF})/144$$

where ASF = amps per square foot. The actual plating amperage for any plating operation must be determined by a combination of calculation and observation. The plated product must always be examined for the need to make adjustments in amperage. No formula can take into account complicated plating geometries, or nonideal conditions in the plating tank.

CONTACT FINGER PLATING PROCEDURES

Prior to Plating

Before beginning any plating operation there is a routine which should be firmly established as a way of life in the printed circuit shop. For contact finger plating that routine should include the following preparation guidelines.

1. The plater should read the job work order traveler and blueprint for the part number. The plating requirements for the following must be understood:

- Types of metals to be plated.
- Thicknesses of metals to be plated.
- Areas to be plated.

The platers must never assume that the contact fingers are even supposed to be nickel/gold plated. Some printed circuits have only solder, or tin-nickel, or nickel plated contact fingers.

Note: Never plate gold on top of copper, even if gold is the only metal specified. Nickel should always be plated as a barrier coat.

2. Sometimes there are plated through holes either in the contact fingers, or at the top of the contact fingers. The contact fingers should not be taped off until the blueprint has been consulted for information on how high up on the contacts metal is to be plated.

3. The contact finger plater must look at each panel which is to be plated to verify that a solid plating bar exists from the panel borders to the contact fingers. If the plating bar is missing, it must be painted-in using silver conductive ink. If this is not done, the contact fingers may not plate, or they may only plate on one side of the panel. For multiple up panels, the plating bar must extend to all the circuits.

4. The plating amperage for each bath must be determined before the panels are loaded into the tank.

5. The plater must verify that all plating baths are up to temperature, all filter pumps are on, all plating baths are filled to their proper operating solution level, and all rectifiers have been turned on.

6. When panels have been taped off, they must be pressed with a roller tape pressing machine. It may be necessary, if leakage beneath the tape has been a problem, to pre-heat the taped panels prior to roller pressing. The roller pressing should be used even if the panels have been taped and pressed by an automatic taping machine. If taped panels have sat for more than a day after being pressed, it is good practice to re-press the panels before plating. If the tape is not tightly pressed to the circuitry, tin-lead stripper and plating solution will wick under the tape and leave a dark line of corroded circuitry, and perhaps even exposed copper. This is an unacceptable condition.

7. Always check the bus bars of the nickel and gold plating tanks before beginning to plate. They should be clean. The level of solution must also be at the proper operating level. Add water if needed.

Plating Procedure

1. Select enough panels to fill one plating rack, and set them in the slots of the tin-lead stripping tank.

    a. Be sure that no stripper splashes into the slots, as it will damage the panel at that location.

    b. After one or two minutes, or when the tin-lead has been completely stripped from the contact fingers, remove the panels two at a time from the stripper, being careful not to drip stripper on the other panels or in the slots. Set the panels in a rinsing tank.

2. Scrub the contact fingers. Either of two methods may be used:

    a. Method 1. Manual pumice scrubbing:

- Lay five panels at a time on a flat surface so that the contact fingers are layered like the steps of a staircase.
- Wet the bristles of a brush and press them into an acid activated pumice scrubbing compound.
- Scrub the contact fingers with a back and forth motion on each step of the layers.
- Rinse the panels to remove the bulk of the pumice.
- Carefully turn the panels over, and scrub the contact fingers on the other side.

- Rinse the bulk of the pumice from the contacts and from the surface of the panels.
- Immersion rinse the panels.

b. Method 2. Machine scrubbing:

- Firmly press the panels, one at a time, into the spinning wheels of an imbedded grit scrubbing machine.
- Inspect contact fingers of each panel to verify that they are being cleanly scrubbed.
- Set scrubbed panels in an immersion rinse tank.

3. Loading the plating racks.

a. Set the racks over the acid loading tank. The acid loading tank should contain 1–1½ inches of 50% sulfuric acid, or 50 : 50 by volume concentrated hydrochloric acid. Hydrochloric acid is a better activator of nickel, if it becomes necessary to use this tank for nickel activation prior to gold plating.
b. Firmly clip one panel into each of the loading positions of the rack. Tighten clips as needed.
c. Loading the panels while the contact fingers are immersed in acid is critical for ensuring that the copper to be plated on is not oxidized.

4. When rack is loaded, lift it from the acid loading tank and dip the contact fingers into a water rinsing tank for a second or two.

5. Set the rack of panels directly on the bus bars of the nickel plating tank.

a. In some cases, the rectifier lead may be clipped directly to the rack before it is lifted from the acid loading tank.
b. As soon as the rack has been set on the tank, look at the level of the plating tape in the bath. The plating bath should not be above the top of the tape. If it is, immediately raise the panels in the rack.

6. Set the predetermined plating amperage and timer immediately. In cases where more than one rack of panels is being loaded at a time, as soon as the first rack is set on the nickel tank some slight amperage should be reading on the rectifier. No panels should ever sit in a plating tank without amperage.

7. After 10 minutes, remove a panel from the center of the rack and measure the thickness of nickel.

a. Also look at the quality of nickel plated.

b. Adjust amperage or plating time as needed to achieve quality plating and desired nickel thickness.

c. When returning measured panel to the plating rack, reactivate the nickel surface with a pumice/machine scrub.

8. The time and amperage for each job plated should be logged into the contact finger logbook, along with job number and panel quantity. Also log in the reading of the amp-minute meter on the gold rectifier.

9. When sufficient nickel thickness has been achieved, reduce amperage.

a. Lift the rack of panels from the nickel tank.

b. Quickly rinse the contact fingers in immersion rinse tank.

c. Set rack on gold plating tank.

10. As soon as rack has been set on Gold tank:

a. Look at the level of the tape in the plating bath; adjust panel height in bath if necessary.

b. Set the gold plating rectifier to predetermined amperage.

c. Set the timer.

11. After 5 minutes, remove one panel from center of rack and measure the gold thickness.

a. Also look at the plated gold, and adjust time and amperage to achieve quality plating of the required thickness.

b. Lightly scrub the gold contact fingers before placing the panel back into the plating rack.

c. Log in the time and amperage used to achieve gold plating.

12. At the conclusion of gold plating:

a. Reduce amperage.

b. Lift the rack from the gold tank, letting it drain over the tank until bulk of the gold plating bath has stopped running off the contact fingers.

13. Set the rack on the gold dragout/rinsing tank for unloading. The gold dragout/rinsing tank should be equipped with an ion exchange resin and pump to reclaim gold from the dragout.

14. At the conclusion of the job:

a. Count all panels plated, and log job into the contact finger plating logbook.

b. Log in the amp-minute reading.

c. Complete the required entries on the job work order traveler.

Additional Notes

1. Always keep the bus bars wiped clean after every load of panels.
2. Always check the level of plating bath in the tank, and make water additions as needed.
3. Never walk away from the tin-lead stripping tank while there are panels in it. Excessive immersion of taped-off contact fingers results in seepage of the tin-lead stripper beneath the tape.
4. If a rack of panels cannot be placed directly into the gold plating tank after they have been nickel plated, set the panels in the acid loading tank for one or two minutes to activate the nickel, just before the panels do go into the gold tank.

Maintenance Information

1. Acid Loading Tank.   This bath can be made up once a day, or left for longer periods. If the acid is not changed once a day, an addition of a gallon of acid will assure that it does not lose its potency. It need only be changed when it takes on a green color, or when obviously contaminated. This bath can be neutralized with soda ash, then dumped to the sewer (check applicable government regulations).

Tin-Lead Stripper.   It is prudent to change this bath once a shift, or more often if needed. Failure to change it often results in longer immersion times and incomplete tin-lead stripping. Longer immersion time increases the probability of stripper wicking up beneath the tape to damage the underlying tin-lead. Incomplete stripping of the tin-lead, due to depleted stripper, increases the probability of copper/nickel peelers or bubbles beneath the nickel. All these conditions must be avoided. The loss of even one panel more than balances the cost of changing the bath. This bath must not be dumped to the sewer, because of metal content. It must be pumped into a plastic or plastic lined drum for waste hauling (check applicable government regulations).

18
Resist Stripping

Resist stripping is an area with enormous potential for causing scrap. It is common for many printed circuit manufacturers to ignore the stripping operation as an area which should be brought under the same controls as the other processes in the wet processing area. Whether the manufacturer is using a screen printed resist or a photopolymer, the methods, chemicals, and parameters for the stripping operation must be understood, optimized, and performed uniformly by all personnel on all shifts.

SCREEN PRINTED RESISTS

The three general classifications of screen printing resists are vinyl resists, alkaline soluble resists, and ultraviolet curable alkaline soluble resists. Alkaline soluble resists may also be stripped in the strippers used for vinyl inks; however, vinyl inks will not strip in alkaline strippers.

Vinyl inks require chlorinated hydrocarbon solvents (such as trichloroethylene and methylene chloride) or aromatic hydrocarbon solvents (such as toluene) for stripping. Simple immersion of the imaged panel in the stripping solution will cause the ink to dissolve completely; spray stripping in stationary or conveyorized equipment also works well. Since these solvents are volatile, and evaporation occurs rapidly, it is good practice to have a series of stripping tanks, each with progressively cleaner solvent. The final operation in the stripping process must be to rinse solvent from the panel surface with a water spray, before the solvent has had a chance to dry. Drying of solvent will leave a thin scum on the copper surface. Spray stripping equipment, whether conveyorized or stationary, should be the same as is used for developing solvent developing photoresist (see Chapter 10, Dry Film Photoresist, Figs. 10-20, and 10-21).

In order that the panels may see progressively cleaner solvent as they progress through the stripping operation, the solvent baths must be rotated. As soon as the first solvent tank (or chamber for a spray operation) becomes loaded with resist, it is discarded and replaced with the solvent from the second tank; and so on, for as many tanks as are being used. The last solvent tank which the panels see prior to water rinsing, is the only tank which is made up of fresh solvent. All solvent tanks must be kept topped off with solvent, as solvent evaporates from them. Allowing the solvent to evaporate without replenishing has the effect of loading the solvent with resist; what resist is already in the solvent will increase in concentration. After stripping and the final water spray rinse, it is good practice to lightly scrub the panels and inspect for residual resist or scum before etching the copper.

Solvents used for stripping vinyl resist are expensive. This fact alone is probably the major contributing factor in the demise of vinyl resists, and there are still other costs associated with processing these solvents. Chlorinated and aromatic hydrocarbons cost $4 and $5 per gallon. All of them give off noxious fumes which create a health hazard. It is common for personnel working with them to complain of headaches. Aromatic hydrocarbons are extremely flammable and pose a continuous threat to any company using them. There are pollution considerations involved with using these solvents. No longer can they be simply washed and dumped down the drain. There are storage and hauling regulations which must be complied with.

The alkaline soluble plating resists are much less demanding to work with for health, fire, and pollution control considerations. In fact, for sheer ease of processing, alkaline soluble screening resists are unsurpassed by all other imaging methods. These resists strip well by tank immersion and spray methods. Since the stripper is aqueous (sodium hydroxide with or without butyl cellosolve or other aqueous diluted solvent) scum formation is not a critical consideration, as long as the panels are kept wet at all times.

Alkaline strippers are made of either sodium hydroxide (buffered or not) or sodium hydroxide with a small amount (10% or less) of butyl cellosolve or butyl carbitol. The water miscible solvent is a surfactant which aids the stripping ability of the sodium hydroxide; the solvent also tends to increase the loading capacity of the stripper. Because of the sodium hydroxide, a strong caustic chemical, protective clothes (safety glasses, rubber boots, apron and gloves) should be worn at all times.

Like solvent strippers, the alkaline strippers do strip well at room temperature. However, the stripping rate, completeness, and stripper loading capacity increase dramatically at elevated temperatures. 120–140 degrees Fahrenheit is a good temperature range for optimum stripping. If a spray stripping unit is being used which is made from PVC plastic, the temperature may have to be limited to 125 degrees Fahrenheit. As with solvent strippers, the final procedure should be to spray rinse water onto the panels. Sometimes this is followed by immersion in dilute (1–2%) acid, followed by another water rinse. The acid will neutralize the stripper for free rinsing; it will also remove oxidation from the copper surface.

Loading Stripper with Resist

There is a definite limit to how much resist can be stripped using a given batch of chemicals. Sooner or later the stripping bath will lose its potency, and require a fresh make-up. Strippers for the alkaline soluble resists tend to have a higher loading factor than do strippers for vinyl resist. The alkaline strippers are affected by temperature, pH, and the addition of surfactants such as butyl cellosolve. However, whenever any stripper reaches its loading capacity, it does a poor job. When a stripper becomes overloaded, it only partially strips; the rest of the resist is a soft pasty mush. When this happens in conveyorized stripping, the resist which is not removed is picked up by the rollers of the conveyor and redeposited on panels passing through the equipment. Even if the stripping bath has been changed, if the softened, mushy resist has been picked up on the rollers, those rollers will have to be cleaned. For this reason, when the stripper shows signs of high loading, it should be changed.

Tell-tale signs to watch for when stripping are:

1. The presence of softened, partially stripped resist.
2. Longer immersion time, or the need for slower conveyor speed setting.
3. The presence of a light film on the panel, which is the same color as the resist.
4. The presence of resist along the edges of traces, when viewed under magnification.
5. The presence of unetched copper on the panel surface, or along trace edges, on panels which have been stripped and etched; this may be visible only with the aid of magnification.

Ultraviolet curable alkaline soluble resists strip best in a heated sodium hydroxide/butyl cellosolve stripping solution. However, these resists strip in a manner similar to photoresists; they come off as particles. The stripping of these resists should be conducted in the same manner as photoresist.

PHOTORESIST

Photoresist stripping is slightly more complicated than is the stripping of screen printed resist. The resist does not truly dissolve into the stripper; it blisters and falls off in pieces and strands. Even the solvent developing and stripping photoresists do not truly dissolve into the stripper. It is common for particles of photoresist to remain along the edges of circuitry after stripping and rinsing. The problems of solvent developing photoresist discussed in the chapter on photoresist imaging apply to stripping photoresist with solvents.

1. Rapid solvent evaporation from the panel surface, which tends to cause scum.
2. Water condensing in the solvent (which is cold because of excessive evaporation).
3. Water causing scum during final rinsing, if not performed fast enough.
4. The need for progressively cleaner solvent baths, leading up to the final water rinse.

All photoresists may be stripped in solvents, even the alkaline soluble. However, only alkaline stripping will be discussed. Because photoresist strips as particles and strands, it is more difficult to strip using spray and conveyorized equipment. Such equipment should be fitted with a coarse filter; even this will tend to clog up. The best approach to follow when stripping photoresist is to strip the bulk of the resist in a dip tank. Only after the bulk has been removed should the panel be spray stripped to complete the job. Photoresist must always be stripped in a heated solution: 125–140 degrees is adequate for most applications. Photoresist will not usually strip (in less than 15 minutes) if the temperature is under 120 degrees Fahrenheit.

The basic steps for stripping photoresist are the same as those for stripping screen printed resist. The difference is in the need to inspect the stripped circuitry, and the critical need for scrubbing the stripped panel as an integral part of the stripping routine. After the bulk of the photoresist has been stripped in an immersion tank, the panel may be sprayed with heated stripper. This spray operation will remove most of the residual resist. However, where fine line circuitry is involved, or where the plating has mushroomed over the photoresist, the circuitry must be scrubbed lightly for mechanical cleaning of resist residues. Perhaps the best scrubbing action available is from the Tru-Scrub, because of its ability to duplicate, with greater consistency, tedious and meticulous fine motion brush scrubbing (Fig. 18-1). Immediately after stripping, panels should be run through a Tru-Scrub machine using a stainless steel brush. Exceptionally fine line or high density circuitry should be run at 50% of maximum conveyor speed. In difficult cases, they may have to be run twice; the second time the panel is run through the scrubbing machine, it should be rotated 90 degrees. Immediately after the panels are scrubbed, a 7–10 $\times$ magnifying eyepiece should be used to inspect the completeness of the stripping operation. If no resist is noticed, the panels are ready for either etching or touch-up prior to etching.

It is common practice to immerse newly etched panels in a bath of electroless tin. Sometimes this step is not performed until just prior to the reflow operation. Panels should never be immersed in an electroless tin or tin-lead bath until they have been inspected for completeness of resist removal. If there is resist residue on the copper surface prior to copper etching, the copper will remain on the panel after etching. This is most common along the edges of traces, and gives the circuitry a ragged appearance which sometimes resembles resist breakdown. Photoresist imaged panels must be examined after stripping and after etching for the presence of this condition. If it is noticed, the panels must be re-stripped, re-scrubbed, and then re-etched. This condition can be virtually eliminated by following good stripping practices; most of the time when this problem has occurred, it has done so because someone has short circuited the stripping operation. If there is unetched copper and the panel is immersed in electroless tin, the tin will cover the copper, and further stripping and etching will only result in overetching the circuitry.

Fig. 18-1. Tru-Scrub: an oscillating brush scrubber.

The heavier the loading of resist in the stripper, the greater will be the likelihood of stripping/etching problems with photoresist. If a plater is having difficulty removing the bulk of the photoresist, he/she can be certain that there is resist residue along the edges of the circuitry. When this situation is noticed, the stripper must be changed immediately. It can be made up with hot water for immediate use.

DISPOSING OF SPENT STRIPPER

The chlorinated hydrocarbon type strippers can almost always be reclaimed by distillation. Printed circuit manufacturers who use these solvents in high volume should also be equipped to distill them. When no provisions have been made for distilling the solvent, the printed circuit manufacturer should drive the best deal possible from solvent reclaimers. Sludge from the bottom of a still has no value, and must be hauled away for disposal.

When flammable aromatic hydrocarbons have been used for resist stripping (and this should be discouraged, because of fire hazard), no attempt should be made to salvage the solvent by distillation. It is even doubtful that the solvent can be safely retained for hauling.

Alkaline strippers are considerably easier to dispose of. In many instances, they can simply be neutralized with an acid and dumped to the sewer. If the stripper contains photoresist, or a UV curable resist, the stripper should be acidified to remove the resist, then neutralized prior to dumping. In all cases, the disposal instructions of the stripper manufacturer, and the local, state, and Federal water and waste regulations should be followed. If in doubt as to the exact details, consult the stripper's manufacturer, and the representative for the local water pollution control authority.

19
Etching

Etching is the process which actually transforms an image into a circuit. Even though a panel may be plated, or a layer imaged, no electrical circuit has been formed until the base copper foil has been etched from the laminate. There are many demands placed upon modern etching chemistry and equipment.

THE CHALLENGES OF ETCHING TODAY

Reduce Undercut and Overhang: Improve Etch Factor

Undercutting produces overhang. Undercutting is the removal of copper from the sidewall of the conductor as the copper foil is etched. Generally, the longer the printed circuit panel is in the etchant, the greater will be the undercutting; high speed, lower pH, alkaline ammonia etchants produce less undercut (Fig. 19-1). As undercut and overhand are reduced, the etch factor increases. Etch factor is the ratio of total copper thickness to the distance of undercut (Fig. 19-2):

$$\text{Etch Factor} = \text{Total Copper Thickness}/\text{Undercut}.$$

A high etch factor, which means low undercut and low overhang, generally indicates the ability to hold fine lines with tight conductor spacing. Etchants and etching conditions which result in high etch factor also result in good conductor edges, approaching the artwork for acuity of reproduction. In contrast, low etch factors result in circuitry which has more ragged traces. Excessive overhang of plated etch resist, whether tin-lead, tin, tin-nickel, or nickel, causes problems with short circuiting. As the overhanging conductor breaks off, it forms an electrical bridge between two points in the circuitry. These metal slivers are the leading cause of electrical short circuits. Reducing overhang can only help increase overall board yields.

Reduce Attack on Etch Resist

Ideally, the etchant would not attack the etch resist at all. However, the etch resist must be able to last for several passes through the etching operation.

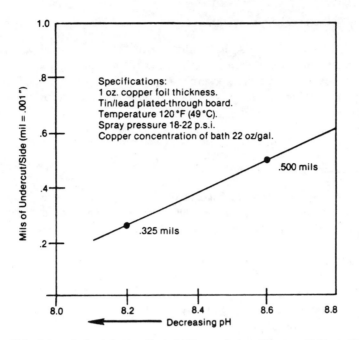

Fig. 19-1. Ammoniacle etchants: effect of pH on undercut. (Courtesy Philip A. Hunt Chemical Co.)

Etches which vigorously attack the etch resist will result in ragged traces. Also, whenever the resist is a little thin, perhaps due to a scratch, a vigorous etch will cause an etch-out at that location. There are times when it is necessary to re-etch a panel or layer. The etchant must not be so strong that it severely attacks the resist on the second pass through.

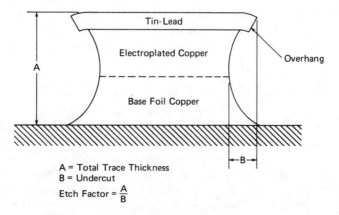

Fig. 19-2. Etch factor.

Reduce Attack on Laminate

The etchant must not be so harsh that it attacks the laminate surface upon only a brief exposure. Etchants which attack the laminate will cause measling and weave exposure. Weave exposure is a major defect which destroys electrical insulation. With acidic etchants such as chromic acid and sulfuric acid/hydrogen peroxide it is important that exposure time and bath temperature be under control.

Reduce Costs

There are many factors which go in to the total cost picture of an etching system. It is good to evaluate them when considering changes in etching chemistry and equipment. The costs factors include:

1. Cost of initial make-up.
2. Cost of replenishing chemicals.
3. Copper loading capacity.
4. Cost of any required brighteners.
5. Ease of operation, control, and chemical addition.
6. Restrictions on the types of work which can be processed.

The initial make-up cost will be amortized over the life of the bath, and is of little concern unless the cost is simply outrageous, or short bath life is expected. The cost of replenishment chemicals and the copper loading capacity are of greater importance, since they drastically affect the direct operating cost. A chemical etching system which has a copper loading factor of 7.5 ounces of copper per gallon will require replenishment about three times as often as a system which operates at 24 ounces per gallon. Generally, the more often additions are required, the more difficult the system is to operate unless replenishment is automatically controlled, using an inexpensive control system. If any special brightening operations for the tin-lead are required, they add to the control and operation problems and expenses.

Increase Speed

As long as the etching solution is evenly distributed over the surfaces being etched and the etch resist is not unduly attacked, the quality of the etching (reduced undercut and sharpness of traces) increases with faster etching speeds. Faster etching speeds also mean thicker foil weights can be etched. Faster etching conditions also mean higher throughput. Etching is not usually a bottleneck operation. However, when there are many layers to be etched, as well as double sided and completed multilayer panels, a greater demand is placed on etching time.

Increase Consistency of Etching Rate from Panel to Panel

The more consistent the etch rate on consecutive panels, the easier it is to obtain uniformly etched panels. All of the conditions which affect the etch rate must be brought under control so that no drastic differences exist from one panel to the next. This can be accomplished by monitoring copper content, pH, bath chemical strength, temperature, evenness of solution flow (sparging system or spray nozzles and agitation for spray nozzles). Greater consistency will result when all of these are near the optimum recommended values of the bath manufacturer, and not just kept within a rather broad operating range. If chemical additions are required every 25 panels, then the additions must be made.

Increase Evenness of Etching Rate on Entire Surface of Panel

The evenness of etching from one side of a panel to the next, and from one location on the panel to another, is brought about by subjecting the surfaces to an even flow of etchant. If a batch tank is being used for etching, the solution flow must be uniform everywhere in the tank. There should be a sparging system for solution movement. A vertical agitation system should also be used for slight panel movement to even out solution flow. It is good practice to study the solution movement in a tank which has no panels in it. This will allow verification of uniform flow. If clogged nozzles are suspected, they must be checked before running work through the system. Spray etching systems must be checked to verify that the nozzles have not become clogged, and that the manifold agitation system is working. This agitation system moves entire banks of spray nozzles and is very effective in achieving a uniform spray pattern across the width of the conveyor.

Increase Ability to Safely Handle and Etch Thin Foils and Thin Laminates

The equipment for etching layers must be capable of gently and reliably handling thin laminates. This is accomplished by letting rollers hold the layers against the conveyor wheels for conveyorized etching system, and by the method of racking for batch type etching systems. There is a tendency for thin laminates to wrap around rollers and conveyor wheels, due to their thinness. When thin laminates are etched in a batch operation, they tend to warp and fall out of racks. Racks for thin laminates must support the laminate across its width and length, and at each end to keep it securely in the rack.

Thin foils, such as ½ or ¼ ounce, present another problem. The etcher must hold the laminate securely enough that no scrapes or scratches occur. Thin foil will not stand up to mechanical abuse the way that 1 ounce foil does. If laminate with thin foil flutters around in a loose rack, the fluttering action may be enough to scratch through the foil. One problem of running thin laminates through conveyorized etchers is that they sometimes become stuck inside, with subsequent layers being shoved beneath the stuck layer by action of the conveyor. If this happens on ½ or ¼ ounce foil clad laminate, the foil may tear.

When running thin laminates and laminates with thin foil, the operator must be aware of the potential for damage, and the limitations of the equipment. The operator must not hesitate to devise alternative processing schemes to allow him/her to process these types of laminates.

Prevent Sludging (or Crystallization) and Stabilize:

- Copper content.
- pH or acid/base content.
- Thermal conditions.
- Loss of etchant efficiency due to fumes or decomposition.

In many instances, the operating parameters cannot be stabilized without controlling the loss of fumes. Alkaline ammonia baths tend to increase in pH, or decrease, depending on the adequacy of ventilation. If pH rises with little abatement over the course of a week or so, it would be wise to check vents, fans, and dampers to increase ventilation. Also, a decrease in ventilation may be necessary if pH continues to drop. The sulfuric acid/hydrogen peroxide etching systems have a similar problem, but this is related to the stability of the hydrogen peroxide. If the etching bath begins to bubble at an accelerated rate, it will be necessary to stop etching and resolve the condition which is causing the excessive bubbling: high copper content with low acid, or high temperature.*

Sludging is a problem which will cause a major loss of productivity, and which indicates poor control over the stability of the etching chemistry. Sludging occurs in ammoniacal etchant because there is too much copper for the amount of solubilizing chemicals. It frequently happens that the process is operating fine one day, only to sludge out the next. What may have occurred in these situations is excessive venting of ammonia. When the specific gravity reaches the upper limit of the etchant, any drop in ammonia (or pH) can result in the copper-ammonia complex falling out of solution. Some shops will keep pH up by adding aqua ammonia. If continuous additions of aqua ammonia have to be made, sludge can occur simply because too much water is being added to the system. If the pH drops regularly, and no adjustment can be made

---

*CONCISE CHEMICAL DICTIONARY, 9th Edition, p 452 Gessner G. Hanley

to the ventilating system, then ammonia should be added in gaseous form. A cylinder of anhydrous ammonia can be plumbed into the primary etchant tank. This will allow ammonia addition without upsetting the solubility of the copper-ammonia complex, which results when water is added.

A similar situation can exist in sulfuric acid/hydrogen peroxide etching systems. As the acid depletes, the ability of the solution to hold copper increases. In fact, when there is little sulfuric acid present, the copper sulfate concentration can exceed 15 ounces per gallon in the etching solution; even chilling the bath will not drop the copper sulfate from solution. However, when sulfuric acid is added to bring the acid content back to 100%, all of a sudden crystals of copper sulfate pentahydrate will drop out. The crystals can drop out so heavily that the system cannot handle the load. When this happens, the etching operation must be shut down and the etchant pumped out for separation of the crystals, and perhaps unclogging of etchant feed lines.

Reduce Pollution Control Problems

Water pollution due to copper is a common problem with printed circuit manufacturing. The use of alkaline ammonia etchants adds slightly to the problem, since the copper is complexed with the ammonia and cannot easily be removed by ion exchange or alkaline precipitation. As the panels leave the primary spray etching chamber they are heavily loaded with copper bearing solution. The solution contains up to 30 ounces/gallon of complexed copper. One of the advantages of alkaline ammonia etching is the high copper loading factor; but this is a liability when it comes time to prevent the copper from getting into rinse water effluent. A large portion of that copper is removed during the secondary spray operation. Here (Fig. 19-3) the etchant which is sprayed onto the panels is non-copper bearing replenisher. When the panels pass into water

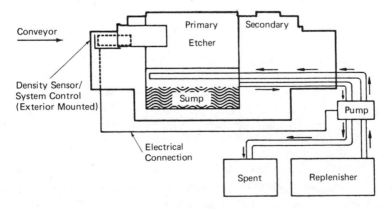

Fig. 19-3. Ammoniacle etching system. (Courtesy Philip A. Hunt Chemical Co.)

rinse, a greatly reduced amount of copper goes to the drain. However, some ammonia does enter the waste water system, complexing any copper which is present. An air knife can be used to strip excess ammonia from the panels before water rinse, but this practice reduces the ability of the water to freely rinse copper and etching salts.

The water pollution problems of a sulfuric acid/hydrogen peroxide system are slightly less. These systems contain no complexed copper. Also, the copper content is only a fraction of that for ammoniated etchant, which reduces the burden on the water rinsing and pollution control systems.

Increase Ease of Operation

The easier an operation is to perform, and the easier a piece of equipment is to operate, the greater is the likelihood of obtaining better quality and through-put. Ideally, all that need be done is to determine a conveyor setting, or an immersion time, and that will remain very much constant for a given copper weight. Of course, this can only be done when there is excellent control over all other chemical, thermal, and mechanical conditions. If copper content, bath strength, and temperature vary over wide ranges, it is more difficult to obtain ease of operation, since the operator must study the panels and the condition of the etching solution, and make a judgment for each load.

Obviously it is important to use equipment and processes which lend themselves well to consistent operating conditions, and automatic control of various bath parameters. Systems which can be operated continuously, with automatic control over chemistry and temperature, have advantages over systems where the operator must analyze the solution, stop the process and make an addition, or wait for the bath to cool down.

Increase Ease of Chemical Additions

Part of making a chemical system easier to use is to make it easy to add required chemicals, especially if chemical additions are required virtually continuously, as they are for etching systems. When additions can be made with ease several benefits result: (1) the additions can be made more often, without disrupting the operation; (2) the additions require less effort and will tend to be more accurate; and (3) the operating chemistry will stay nearer to optimum, instead of just being within the operating range.

Decrease Dependency on Operator Skill

The less the etching procedure is dependent on operator judgment, attention, and operation, the greater will be process control, consistency of quality, output, and ease of operation. If the bath strength can be monitored with a hydrometer, and that hydrometer is perpetually floating in the etchant in a position for all to see, the operator will be more aware of the bath strength than if he/she has to stop the operation and perform a color analysis or titration. If the bath can be bled and fed with replenisher by simply turning on a pump, chemical additions are going to be made more frequently than if the operator has to stop the operation, put on chemical safety clothing, and siphon hazardous chemicals out of a drum in order to make the chemical addition. If the determination of bath strength and the bleed-and-feed chemical addition can be fully automated, then the bath strength can be kept close to optimum with no effort on the part of the operator. The determination of bath strength and the addition of replenisher chemicals are two of the most important steps to be followed in attaining good process control. In processes which are heavily operator dependent, most difficulties result directly from failure of the operator to perform analysis and addition tasks efficiently.

Reduce Reflow Problems

One of the reasons tin-lead is such an interesting alloy to work with is that almost all of the problems associated with it, such as thinness of deposit, de-wetting, and outgassing or solder bumps, can occur as a result of (1) the copper upon which the tin-lead is applied, (2) a condition inside the tin-lead plating bath, or (3) any number of operations the tin-lead will see before it is reflowed. One of the detrimental conditions which the tin-lead will see is poor cleanliness, or failure to remove salts of the etchant after copper etching. When sulfuric acid/hydrogen peroxide etchants are used, these problems are not so great. A simple dragout immersion after etching, followed immediately by a water rinse, usually proves sufficient. The etchant should not be allowed to dry on the surface of the etched panels.

The use of alkaline ammonia etchants does complicate the cleanliness problem a little. Two of the leading causes of de-wetting is high copper concentration in the tin-lead and the presence of salts of any kind on the tin-lead surface. The residues of alkaline etching are not completely soluble in water. The copper-ammonia complex is soluble in the presence of ammonia or hydrochloric acid. The secondary etching chamber functions to rinse copper-ammonia complex from the panel surfaces. However, the residues left are only slightly soluble in water. It is unlikely that adequate reflow quality can be maintained without a post-etching or pre-reflow cleaning operation to remove etchant residues. Simply soaking panels in 3–5% ammonia water or hydrochloric acid as they come out of the etcher will greatly enhance the reflow quality. The fact that this soak turns blue is evidence that it is needed. This soak must be changed every 25 panels, or so.

Improve Quality of Etched Traces and Circuitry

When all etching parameters are at optimum, and the equipment is functioning as it was designed to, the quality of the circuitry will be the best attainable.

Why is it that etched circuitry sometimes appears ragged, pitted, or cut? When the conditions of the first paragraph are met, the etching time required to remove the copper foil is at a minimum, and the lowest etch factor will be attained. As conditions deviate from optimum, so does etching time, undercut, overhang, and attack on the etch resist. This can lead to a number of problems.

As the overhang increases, so does the likelihood of slivers from the etch resist breaking off. When the slivers break off, the copper conductor which was formerly protected by that overhang becomes subjected to further etching. The result is ragged traces, and even broken traces. Too vigorous an application of etchant, whether through immersion agitation or spray pressure, can cause overhang to break off, with ragged over-etched traces the result. Circuit defects caused by poor handling, such as scratches in the etch resist, become magnified. A scratch which does not quite cut all the way through the etch resist may not cause a "cut trace" or etch-out. However, if the etchant is allowed to attack the resist excessively, either through prolonged exposure or through chemical imbalance, there is increased chance of that shallow scratch causing an etchout.

Poor resist stripping often goes unnoticed as a cause of poor circuit quality. Sometimes resist is left clinging to the sides of traces and circuitry after stripping. If this is not detected by the etching operator the result will be copper remaining on the surface everywhere resist is present. If the etched circuitry is not examined with magnification the resulting raggedy looking traces may appear to be the result of resist breakdown, where plating beneath the resist has occurred. There is a tendency on the part of many operators to simply re-etch the panels; this will do no good. When this condition is noticed, the stripper must be changed and the panels restripped. Only after restripping in fresh solution should the panel be re-etched. During the second etching operation only a fraction of the etching time should be used; any more will result in over-etching and ragged traces.

Ability to Handle Laminate with Uneven Foil Weights, Such as 2/1

One of the requirements of etching for multilayer manufacturing is the need to etch laminate which has copper foils of differing weights on the same piece of laminate. It is common to have 1 ounce copper on an outer layer, with 2 ounce copper on the inner layers. Spray etching is the only reliable and economical method of meeting this requirement, since it is possible to shut completely one spray manifold.

Ability to Handle Temperature Sensitive Laminates

There are some laminates, such as polyethylene, which have stresses built into them during their manufacturing operation. Such laminates tend to be sensitive to temperature. Should they be etched above 90 degrees Fahrenheit, the stresses will be annealed and the material will lose mechanical rigidity, and warp. For such conditions, it may be necessary to maintain a bath of ferric chloride, which will operate well at lower temperatures.

GENERAL ETCHING PROCEDURES

Conveyorized Alkaline Ammonia

There are a number of points of consideration which should be attended to every day, before running work through the etcher:

1. Temperature of etchant.
2. Condition of filter screen. There are usually one or more screens used to filter large objects, such as sludge, from the spray manifolds. These should be pulled out every morning. If they are caked with sludge a water hose should be enough to clean them off.
3. Baume (specific gravity) shoud be within range.
4. pH should also be within range.
Note: If the Baume is near the top of the range, and the pH near the bottom, the solution must be replenished before etching.
5. Notice the pH and color of finishing solution. If it is blue, or has white particles precipitated, it will require changing prior to etching work. Finishing solution with a pH above 1 is less effective.
6. Turn on ventilation fan motors. Notice the setting of the ventilation dampers. They should be closed at the end of the day, and opened during the working day. When they are set to OPEN, it may be that fully opened is not the desired setting. If pH has a tendency to decrease from day to day, regardless of replenishment, then dampers should be set to only a partially opened position.
7. Turn on water spray rinse and notice the flow through the nozzles. If water appears to be restricted, with all valves wide open, the filter probably needs to be changed. If spray pressure appears to be restricted after changing the water filter, the individual nozzles will have to be checked for obstructions.
8. Notice the general condition of the etcher. It should be cleaned and wiped free of etchant every night during shutdown. Look at the gears which turn the conveyor wheels. They must be free of etchant build-up. The drip pan where panels are fed into the machine, likewise, must be free of etchant build-up.
9. Check the level of etchant in both the spent drum and the replenisher drum. If it is low in either, then another drum must be brought in. Should the spent drum fill up without being noticed, etchant will spill over. Should the replenisher drum empty and not be noticed, the sump level will be decreased as etchant is pumped out, and the Baume will rise without abatement.
10. When the spray pumps are turned on, pressure must rise to the desired

value immediately. 16–24 psig is the commonly used range of spray pressures, with the upper manifold set a pound or two higher than the lower. If pressure does not rise immediately:

- Shut pump motors off.
- Check setting of valves to spray manifolds. If they are closed, open them.
- Check level of etchant in the etcher sump. If it is low, add to it from the barrel which contains spent etchant, or from a barrel of starter solution. Never bring up the level of etchant in the etcher with replenisher! Adding replenisher will cause Baume to drop and pH to climb, possibly pushing both parameters out of safe operating range.
- Recheck filter screens for sludge.
- Only when all of the above have been checked and it has been verified that they are within correct operating guidelines should be pumps be re-energized.

Note: All etching operators must be trained to check spray pressure immediately upon energizing pumps, and to shut them off if pressure is absent. This is necessary to avoid damaging the pumps.

Only when all of the above has been completed should the etching operator attempt to run work through the etcher.

11. If a trough containing diluted ammonia water or hydrochloric acid is being used as a soak to remove etchant residues, the operator should prepare the trough at this time. 10 gallons is sufficient to soak a load of 25 panels.

12. A job should be selected for etching, and the job work order traveler reviewed prior to running any of the panels.

- Are there any special notes written on the traveler for the etching operator?
- Do the panels require second touch-up before etching? Has this been performed yet?
- What is the weight of copper to be etched? Is it the same on both sides of the panel?
- Will rack marks cause etch out on panel borders needed for tip plating?

13. The operator should log the job into the etching logbook. The etching logbook should tabulate the following information:

- Customer or job number.
- Weight of copper for each side.
- Number of panels or layers.
- Conveyor setting used to etch the job.
- Operator's name or initials.
- Date.

The logbook serves as a record of process operating history. This will help the operators gain an understanding of what conveyor setting to use for each weight of foil, and for layers versus panels. If there is a change in conveyor speed, after a long operating history with little change, the operator can be alerted for potential problems with either the equipment or the etching chemistry. The logbook provides a guide to what conveyor setting should be used when starting a new job.

14. After beginning the job, the operator should turn on all spray pumps for the etchant, water rinse, and finishing solution.

15. Then one panel is run at the conveyor setting used for the last job of the same foil weight. Before placing the panel on the conveyor the operator must examine it to verify that it has been cleanly stripped. A magnifying eyepiece must be used for viewing. The first article must be evaluated by the operator after etching. This should be done with adequate lighting and magnification.

- Is there copper remaining in broad areas of the panel? This may show up as a brown film on the laminate.
- Are the traces overetched? They must not be ragged, and if an ungloved finger is run across the circuitry, the edges of the circuitry must not be too sharp. It will take experience to develop a feeling of what is "too sharp" and what is not.
- Do the traces exhibit apparent resist breakdown? If they do, the panel must be closely examined to verify that the breakdown is not in fact unetched copper due to poor stripping. If the unetched copper is due to poor stripping, the panels must be restripped in fresh stripping solution.
- If the traces exhibit breakdown or if they are ragged after etching, this first article panel should be compared to the unetched panels. Any defect noted on the unetched panel must be brought to the attention of the plating supervisor and Quality Assurance. The etching operator, like everyone else in manufacturing, must be trained to spot undesirable conditions and to bring them immediately to the attention of those who need to know.
- If the panel appears to be slightly over-, or underetched, adjust conveyor speed.

16. When a panel has been etched correctly, the rest of the job is ready to be run. It is good practice to run panels in lots of 25. The operator should continue to examine panels as they come from the etcher.

17. The operator should also keep an eye on the hydrometer. If the hydrometer is not a permanently affixed item, he/she must stop the operation and check Baume. If replenishment is not automatic, then the replenishing pump must be manually turned on to bleed and feed to system. This pump should be attached to a timer, so that over-replenishment does not occur.

18. Every 25 panels the soak water must be changed.

19. When the job has been completed, the panels must be counted, and both the work order traveler and the logbook entries properly completed.

Some shops routinely immerse etched panels in electroless tin. This covers the sidewalls of the conductors to eliminate copper showing after reflow. Panels should never be immersed in electroless tin until it has been verified that no unetched copper remains on the panel.

The chemistry of high speed formulations of alkaline ammonia conveyorized spray systems is controlled automatically. The control units, generally supplied free of charge by the etchant manufacturer, sense specific gravity and use this signal to activate or deactivate dual pumps for bleeding and feeding etchant. Since the specific gravity (Baume) is controlled within narrow operating limits, 24.5 ± 0.5 degrees Baume,* the etch rate is very stable. Temperature fluctuations due to exothermic reactions rarely present a problem. The etch rate varies linearly with temperature, unlike that of sulfuric acid/hydrogen peroxide.

Sulfuric Acid/Hydrogen Peroxide Batch Etching

When operating with this type of etching system, both chemistry and equipment, more attention to details of operation, analysis, chemical addition, and control are required on the part of the operator. Fig. 19-4 shows the layout of typical batch operation equipment. The components are:

1. Control module.
2. Work tank. This is the immersion tank which holds the rack of panels.
3. Drop-out drum (crystallization drum). It is here that the copper sulfate pentahydrate crystals are collected, to remove copper from the system.
4. Etchant grow drum. For containing excess etchant which develops as a result of chemical additions.
5. Chill tank. It is here that the etching solution is cooled for copper crystal removal.
6. Refrigeration unit. This cools the coils which are immersed in the chill tank.
7. Chill tower. This is the metal rack which supports the chill tank. It may also be used to support the refrigeration unit; however, this is commonly located on the roof.
8. Crystal extraction drum.

The etchant recirculates from the work tank to the chill tank to the drop-out drum and back into the work tank. The copper is removed from the etchant as it is cooled. Ideally, the copper crystals drop out in the drum provided for that reason. The system can operate indefinitely by adding make-up chemicals

---

*Philip A. Hunt Chemical Corp. TECHNICAL BULLETIN No. 50

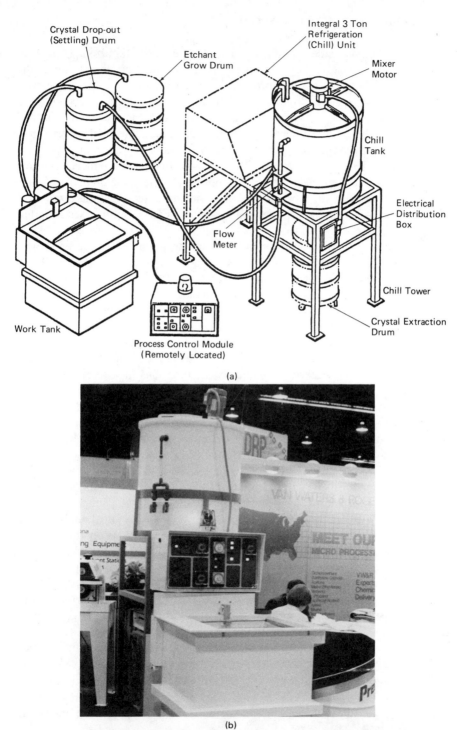

Fig. 19-4. (a) Sulfuric acid/hydrogen peroxide etching system. (b) Batch etcher for sulfuric acid/hydrogen peroxide etching. Courtesy, Equipment Specialties, Inc.

of (1) hydrogen peroxide, (2) sulfuric acid, (3) peroxide stabilizers, (4) inhibitors to reduce attack on the etch resist (tin-lead for most printed circuit applications), and (5) a host of other proprietary ingredients, including* rate exhaltants, undercut control additives, and foam or anti-foam agents (depending on whether immersion or spray is being used). Although the etchant is recirculated through the chilling apparatus, it is also recirculated inside the work tank. Sufficient and uniform etchant flow is critical to achieving acceptable etching results. The rack, when immersed, rides on a cam which helps to meet agitation requirements. At conclusion of the etching cycle the rack of panels being etched is lifted from the work tank, with brief draining, and immersed in a dragout tank. After dragout immersion the rack of panels is immersed in water and/or water spray rinsed with a hose.

When the system is under excellent control and maintenance, and a steady work load is being processed which does not overload the thermal capabilities of the system, the factor which controls the amount of etching on the panels is the immersion time. The immersion time is analogous to conveyor speed setting for controlling etch when a conveyorized spray system is being used. When a company elects to buy a sulfuric acid/hydrogen peroxide etching system, after using alkaline ammonia chemistry in conveyorized spray equipment, there is a training and learning curve which must be recognized and dealt with adequately from the very start. This equipment functions in a totally different manner than operators may have been used to before. The chemistry is different, and so are the methods of control. There is no bleed-and-feed operation for controlling etch rate. The etching reaction is far more exothermic (heat evolving) than are the reactions for alkaline ammonia etching. Heat and copper are byproducts which must be removed at a steady rate.

Since there is no bleed of spent etchant and feed with fresh in the manner of alkaline ammonia, separate additions of chemicals must be made periodically. The need for additions cannot be made on the basis of a simple indicator such as a hydrometer; it must be made on the basis of chemical analysis. An experienced operator will be able to make periodic additions on the basis of etching time required, and on the basis of square footage as tracked in the etching logbook. For most applications, the additions will be two part, one for hydrogen peroxide and stabilizers, and one for sulfuric acid and inhibitors. These will usually be depleted according to a ratio, such as 3:1, or 4:1. This must be experimentally determined for each bath manufacturers' chemicals. Determining this ratio will simplify the analysis/addition problems.

Depletion of chemicals poses another problem—the etch rate and copper content will vary with the concentrations of the primary ingredients (Fig. 19-5). The solubility of copper will also vary with the concentration of the ingre-

---

*Problem Solving With Peroxide Sulfuric Final Etchant, Buz Steger, Jan 1983 PRINTED CIRCUIT FABRICATION

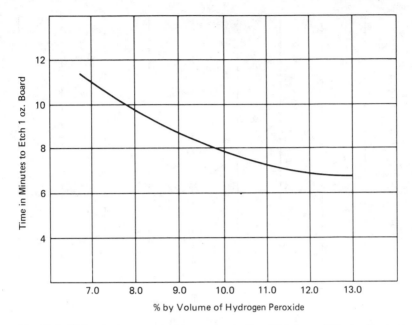

Fig. 19-5. Etch rate versus replenisher concentration. (Courtesy of Electrochemical Co.)

dients, especially sulfuric acid. At 100% bath strength, for example, one bath will maintain copper solubility of 7.5 ounces/gallon at an operating temperature of 115 degrees Fahrenheit. When the bath reaches 60 degrees Fahrenheit in the chill tank, the copper content drops down to 5.3 oz/gal as a result of crystallization. The system will continue to work fine as long as hydrogen peroxide, sulfuric acid, and temperature (operating and chilling) are maintained. In practice, as the sulfuric acid depletes, the copper content rises. With severe depletion of acid, the copper content will climb. When an addition of sulfuric acid is finally made, copper sulfate pentahydrate crystals can spontaneously precipitate throughout the system. For this reason, it is important to keep chemical additions small but frequent. Also, copper is catalytic in the decomposition of hydrogen peroxide.* Running the bath chemistry too far from optimum may result in the presence of more copper than the system can handle, hence spontaneous crystallization. Copper solubility too far above what is recommended by the bath manufacturer can result in spontaneous and uncontrolled decomposition of the hydrogen peroxide. This is what is happening when the bath bubbles uncontrollably, even when it is not being used for etching.

Thermal operating parameters must be understood and not exceeded. Each manufacturer has a recommended loading factor which will allow continuous

---

*Op Cit Gessner G. Hanley

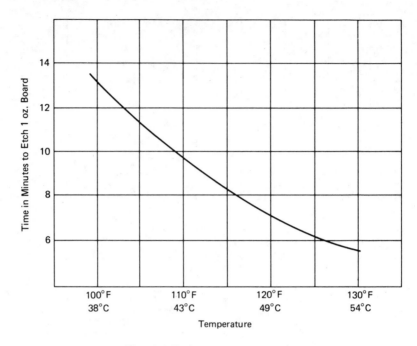

Fig. 19-6. Etch rate versus temperature.

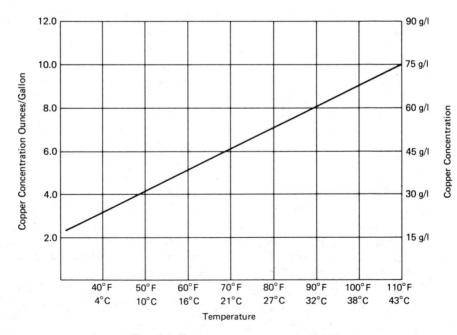

Fig. 19-7. Temperature versus copper solubility.

operation without exceeding the ability of the system to remove heat. Excessive heat interferes with a controlled etch rate and destabilizes the hydrogen peroxide. The loading capability of an etching system will have to be determined experimentally. The maximum loading capacity is the total square footage of copper which can be etched per load on a continuous basis, without exceeding the ability of the system to remove heat and copper. One possible result of exceeding the loading capacity is that copper content (and temperature) may build to the point that copper will spontaneously crystallize when the system is allowed to rest, such as overnight. Loading will be a function of the size of the work tank, and the size of the refrigeration unit. If loading becomes a problem, the capacity can be increased by increasing the refrigeration capacity.

The total volume of etching solution will tend to grow as the system is used day after day. Chemical potency is maintained by addition, not by bleed-and-feed. Copper is removed by crystallization, not by solution removal. As the drop-out drum is filled with crystals, it is replaced with another drum. The filled drums can be sold for the copper value, and the copper reclaimed. There are some systems which do not crystallize the copper to remove it; they plate the copper from the etchant. Such an operation can sell the copper thus plated. Most of these systems have solution control problems which may be overcome with further research. The buildup of certain metallic contaminates (Table 19-1) should not be allowed to exceed the limits specified by the bath manufacturer. Obviously, it will be difficult to maintain nickel below 1 ppm when nickel is being used as an etch resist. One advantage of having to remove etchant to counter volume growth is that it bleeds contaminants from the system.

Maintaining Control

This bath is not like other batch processes, such as electroless copper, which can be run to 30% or more depletion without problems. Although the bath may actually perform at 60–70% of bath strength, the copper content increases and there is increased probability of spontaneous and uncontrolled decomposition of the hydrogen peroxide. When a bath is partially depleted, and the copper

Table 19-1. Maximum Recommended Concentration of Metalic Contaminants.

Hydrochloric Acid (as HCl)	5 ppm
Nitrate (as NO_3)	5 ppm
Iron (as Fe)	42 ppm
Antimony (as Sb)	1 ppm
Selenium (as Se)	20 ppm
Manganese (as Mn)	0.2 ppm
Nickel (as Ni)	1 ppm

content has grown dramatically above 10–12 ounces/gallon, restoring chemical potency will cause precipitation of more crystals than the system can handle. The need for chemical control and frequent additions must be trained into every etching operator. As an aid in maintaining bath strength the following steps can be taken:

1. Determine the ratio of component A (sulfuric acid plus inhibitors) and component B (hydrogen peroxide plus stabilizers) which must be added to keep the bath in balance.

2. Analyze the bath frequently for both components, as part of gathering the information for No. 1 above, and track the etching time as a function of bath strength. The most important values are those which are obtained for 90–100% bath strength, since the bath is going to be most trouble free when bath strength is maintained at or above 90%.

3. Make a chart for the etching operator to follow which lists the replenishment volume of each component, as a function of etching time.

There are simple titrations which are performed to test for copper content, sulfuric acid, and hydrogen peroxide. These procedures are listed in the data sheets supplied by the bath manufacturer. The manufacturer may even supply the glassware and chemicals for the tests, as well as training for the operators.

Sulfuric acid/hydrogen peroxide etchant is used in conveyorized etchers. However, etchers constructed to handle this etchant tend to cost considerably more than either batch etchers or conveyorized etchers for alkaline ammonia. Among the reasons for the added cost are that conveyorized spray equipment costs more than batch processing equipment, and an etcher which requires a chilling tank and refrigeration unit will cost more than etchers which require neither. The following prices are typical and for comparison only of etchers with about 100 gallon work tanks:

Conveyorized spray (alkaline ammonia)	$35,000
Conveyorized spray (sulfuric/peroxide)	$50,000
Batch immersion (sulfuric/peroxide)	$24,000

OTHER ETCHING CONSIDERATIONS

Etching Thin Laminates, Inner Layers, and Small Panels

Thin laminates can present a problem when conveyorized equipment is used, as the laminate may tend to get hung up inside the etcher. An easy way to prevent this is to make a train of laminate. The thin laminate is simply taped end-to-end using plater's tape (used for taping off contact fingers for plating). The first piece of thin laminate is taped onto a panel of heavier laminate which pulls the train through the etcher (Fig. 19-8).

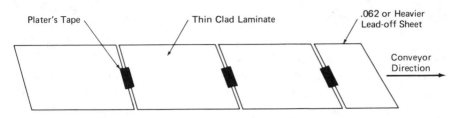

Fig. 19-8. Tape train for etching thin laminates in a conveyorized spray etcher.

Thin clads for multilayers present another point of interest. It is common for these materials to have 1 ounce (or ½ ounce) copper on one side, and 2 ounce copper on the other. There are two ways this can be etched. The layers can be taped back to back so that copper of identical weights is exposed to the etchant. After etching one side, the layers are parted and retaped to etch the other side. Layers such as this can also be etched simply by shutting feed valves to the lower spray manifold. If the area around the tooling holes is not already protected by photoresist, it is good to use plater's tape to protect this copper. When the tooling holes have a copper land area surrounding them, about 1 × 1 inch, there is less chance of the laminate tearing during the booking operation and lamination.

Small panels can easily get lost in conveyorized etchers. They can be processed safely by making a tape-train, as is done for thin laminates, or by taping the small panel onto the surface of a larger panel. Of course, the latter method would only etch one side at a time.

The Advantages of Thin Foils

Thin foils of ½ ounce or less can be very helpful in reducing undercut, attaining fine lines with high density, and in using differential etching for bare copper circuitry. If ½ ounce foil is used, the panel will spend half as much time exposed to the etchant as would be required for using 1 ounce foil. This means that the sidewall of conductors are exposed to the etchant half the time, and undercutting is greatly reduced. One of the problems of achieving fine line circuitry (0.005 inch or less) is being able to etch the foil without etching off the conductors. ½ ounce copper foil also increases throughput by 100%, and cuts the cost of etchant in half. However, it must be handled carefully, as it is easily punctured.

Differential etching occurs when the plated copper is used as the etch resist. To do this reliably, the plated copper must be considerably thicker than the foil copper. Generally, plated copper must be equal to or greater than foil copper, to hold circuit acuity. If ½ ounce foil is being etched, the attack on the plated copper will be only about half as much as if 1 ounce foil had been used.

Non-Tin-Lead Etch Resists for Alkaline Ammonia Etching

Plated metals such as nickel, tin-nickel alloy, and tin hold up very well to alkaline ammonia etchants. Nickel is subject to attack by the fresh replenishing etchant which enters the system through the secondary spray operation used to reduce copper on the panel surface before water rinsing. Prior to etching, when nickel is the etch resist, the secondary spray must be turned off and left off long enough for even droplets to cease falling from the spray nozzles. Even brief contact with replenisher will leave an etched, spotted, and non-shiny surface on the nickel plate. This is not a problem with either tin-nickel or tin.

There are occasions when it is desirable to use immersion tin or black oxide as an etch resist.* Both of these surfaces hold up well to alkaline ammonia etchants. When processing, it is critical that the surfaces not be damaged, as this will cause an etch-out at those locations. It is good practice to perform second touchup after stripping of the imaging resist.

Gold plated over nickel is an excellent etch resist, as is silver plated over nickel. When a shop uses both of these precious metals, it must take the precaution of not co-etching panels. The bath in which the gold is etched can be used to etch all resists except silver. Silver will deposit on the gold, and it may contaminate tin-lead. When it is necessary to etch silver plated panels, the etchant must be pumped out and replaced with a drum of etchant in which only silver panels are etched.

Organic resists are formulated to withstand most etching solutions. All will withstand common acid etchants. However, alkaline soluble screen printing inks are not formulated to work well in etchants with pH above 7.5.† Vinyl screen printing resist and solvent and semi-aqueous developing dry film photoresist work very well as etch resists. The fully aqueous developing photoresists should be given more careful consideration. Although they will probably work for most operations, the manufacturer's technical representatives should be consulted and evaluation samples run. Not all printed circuit shops maintain the same degree of process control. Since these resists do develop and strip in an alkaline medium, they should be evaluated first before running production. An excessively high pH creep in the etching system could destroy resist integrity.

---

*Solder Mask Over Bare Copper: Advantages And Manufacturing Techniques R. H. Clark, Aug. 1982, Circuits Manufacturing.
†Dexter Corp./Hysol Div. Bulletin PR3010/3011.

20
Tin-Lead Fusing

Fusing, or reflow, is the process of melting codeposited tin and lead just long enough to form the alloy called solder. (See Fig. 20-1.) Prior to fusing the surface of the tin-lead has few of the properties of solder, and is esthetically unappealing. Why is it that tin-lead is almost universally reflowed?

1. The appearance of the printed circuit is greatly enhanced, and the metal is denser.
2. The solderability of the printed circuit is improved over that of the code-posited tin-lead.
3. The solderability of the printed circuit is protected for future soldering.
4. Solder slivers are reduced, since overhanging tin-lead flows onto the surface of the conductor.

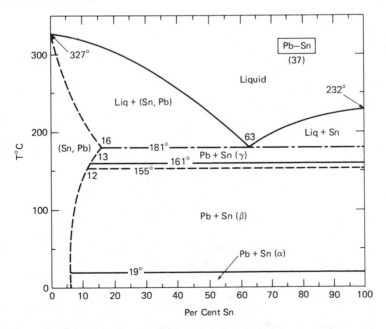

Fig. 20-1. Tin-lead phase diagram. This diagram shows what happens to solder's melting point as tin:lead ratio varies. There are temperatures and compositions where solid tin or lead are present with the liquid.

5. When reflowing is used in conjunction with tin immersion, the bare copper sides of the conductors also become covered with solder, so that no bare copper shows.

6. Reflow is a quality control step since the following become readily apparent:

- Defects in solder purity.
- Contamination of copper surface beneath the solder.
- Blow holes and other plated-through holes.

There are two widely used methods of fusing: hot oil immersion and the conveyorized infrared oven. Most printed circuit shops employ both of these methods. Another method, vapor phase reflow, is occasionally used. Vapor phase reflow consists of passing the panel through a chamber where heated vapors from a high boiling point fluid melt the solder. The hot oil method is widely used because it is simple to operate, inexpensive in terms of capital expenditures, very gentle on printed circuits, and generally does a good job. It is, however, slow. Conveyorized infrared ovens are very fast and perform well, but the cost of getting set up is higher than for hot oil.

HOT OIL REFLOW

Hot oil is used for a variety of applications:

1. When there is only a small panel quantity.
2. When the panels are exceptionally thick.
3. When there is an unusually large area of ground plane on one or both sides.
4. When multilayers, especially thick multilayers, are to be reflowed.
5. When the panels are made from teflon laminate.
6. When the panels are being re-reflowed for cosmetic reasons.

The basic steps of hot oil reflow are given below:

1. Prepare tin-lead surface for fusing.

   a. Immersion in a solder brightening or immersion tin bath.
   b. Water rinse, with optional scrub.
   c. Oven dry 15 minutes to several hours. Some shops like to bake panels, especially multilayers, for several hours to assure moisture removal to prevent delamination. This step is often not necessary, and should be performed only if shop experience finds it to be necessary.

2. Place panel in hot oil immersion tongs.

3. Immerse panel in fluxing solution. This step can be eliminated if panels are fused within a few hours of the tin-lead preparation step; and the shop experience verifies that fluxing adds nothing to the quality of the reflowed tin-lead.

4. Immerse in pre-heating oil (Fig. 20-2). This is an optional step. Some shops like to immerse panels in hot oil which is only 250–300 degrees Fahrenheit, before immersing in the fusing tank which is around 425 degrees Fahrenheit.

5. Immerse in fusing oil.

a. Hold the panel parallel with the surface about two inches above it. Dip one edge beneath the oil and rock the panel until it is completely immersed in the oil.

b. The panel is immersed about one or two inches below the oil's surface and gently rocked back and forth, to force solution through the holes and across the surface.

c. As the operator rocks the panel, he/she will notice the tin-lead begin to melt. The panel should be immersed two to three seconds passed the point where are of the tin-lead appears to have melted.

6. Removing panels from hot oil (Fig. 20-3.)

Fig. 20-2. Hot oil reflow station.

(a)

(b)

Fig. 20-3. (a) Correct angle for draining oil after fusing. (b) Incorrect angle for draining oil: molten solder may flow to edges of circuitry.

a. The reflowed panel is lifted from the oil one edge at a time. The panel should be held several inches above the oil at 30 degrees, to let the oil drain for a few seconds.
b. Do not hold the reflowed panel at 90 degrees to the oil for draining, as this will cause molten solder to flow to the edges of the circuitry.

7. Set the reflowed panel in a sink or tank of water.
8. Repeat this operation until about 25 panels have been reflowed.

a. It is good practice to look at both sides of the first one or two panels after reflow, to verify thoroughness and quality of the operation. The holes should also be checked.
b. Care must be used to keep from scratching panels as they are laid in the water sink.

9. Panels must be rinsed well. The oil residue is a viscous gel, and very difficult to rinse.

a. Sponge both surfaces of each panel as they are removed from the cooling sink.
b. The panels must go through a series of water rinses, where both sides are sponged. The final rinse should be hot water. It helps to use a dishwashing detergent to remove this oil.
c. Care must be used to assure cleanliness of the holes; oil must be removed from them.

Hot oil fusing is slower than conveyorized reflow, but it is more gentle to the panels and may result in better looking solder and laminate surfaces. One common error in processing is inadequate oil removal in the post-fusing cleaning steps. Failure to remove this oil will result in solderability problems when the board is wave soldered after stuffing with components. The adequacy of the oil cleaning can be judged by pouring acetone over the surface of the panel. Only one surface need have acetone poured over it, since it is actually the cleanliness of the holes which is important. The acetone should be collected in a glass beaker. The beaker is then heated to drive off the acetone. If there is a heavy residue of oil in the bottom, cleaning was inadequate.

Conveyorized Infrared Oven Fusing

This is the fastest form of tin-lead reflow available. Ovens are available in a variety of sophistications (Fig. 20-4). One of the chief advantages of hot oil has been the ease with which thick panels, and panels with large ground areas can be reflowed. Modern infrared ovens are equipped with lengthy preheating chambers, and have stable heating profiles. With them, the more difficult

Fig. 20-4. Conveyorized infrared (I.R.) fusing oven, with aqueous cleaning module.

reflowing jobs can be more easily processed. The job of cleaning the panel after fusing is somewhat easier with infrared ovens, since only flux residues must be removed, not an oily gel. Modern fluxes are thinner and more soluble in water for freer rinsing.

Fusing ovens are available with cooling, cleaning, and drying chambers added on as accessories. Thus, all an operator need do is set the panel on the conveyor and monitor the product and process. The basic oven performs three functions: fluxing, preheating, and fusing. A fully automatic fusing set-up will flux, preheat, fuse, cool, rinse the flux residue, and dry the panels—all in one straight line. Any printed circuit shop which is running more than prototypes should avail themselves of this opportunity to obtain so much work so fast and reliably, with no manual operations other than loading and unloading the system.

The basic steps involve setting infrared heating lamps at the manufacturer's recommended value, monitoring flux to recommended specific gravity, and monitoring reflow quality through visual examination of the solder on the surface and in the holes; attention must also be paid to the edges of the panels. Ideally the panel will be reflowed, and the solder on the panel edges will not have been completely fused. Fused solder on panel boarders indicates the need for higher conveyor speed. Conveyor speed is the chief control over the process. With thick panels, there may be a tendency for the edges of the panels to singe. If it is not possible to fuse a panel without singeing the laminate, then the panels must be fused using hot oil. If panel edges do darken, then conveyor speed and even lamp amperage may have to be adjusted.

If only the conveyorized oven is being used, and there is no rinsing unit the fused panels will have to be placed in a trough of water to cool. The trough is located at the end of the conveyor. As panels exit the oven, they are lifted off the conveyor and set into the water trough. Panels must be lifted with two hands, one on either side of the panel. If only one hand is used, there is a tendency to pick up the panels at the corner, which then becomes bent since the laminate is quite warm and pliable. Panels must be set in the water carefully, to avoid striking other panels. Careless unloading practices after reflow are a major cause of damaged panels. Reflow operators must be taught to handle panels as if they cost hundreds of dollars each; which in fact they probably do. The trough water should probably be changed every 25 panels, or less. As long as the water remains uncloudy, the flux residue can be easily rinsed off with a final spray rinse. If the water in the cooling trough becomes cloudy, only 70% isopropyl alcohol will cut the flux residue. Any time a tried panel exhibits an oily appearance when held up to light for examination, that panel will have to be cleaned by immersion and scrubbing with isopropyl alcohol. When panels are inspected during fusing, the presence of measles, tiny white dots in the laminate, indicates the need for faster conveyor speed, as the panels have been exposed to too much heat.

Reflow conveyor speed is set according to the appearance of the fused panel. The need for a slower conveyor speed is indicated by the presence of unfused tin-lead (cold spots) in the circuitry. The need for faster conveyor speed is indicated by:

1. Darkened laminate.
2. Completely fused solder on the panel borders. It is good to have cold spots in the border areas.
3. Measles in the laminate.

PROBLEMS

1. Sometimes the reflowed tin-lead appears dull. The dullness, as opposed to a cold spot, indicates thin tin-lead. The solder in the holes may or may not also be dull.

2. De-wetting, extreme unevenness in the solder thickness, with or without bare copper showing at the edges of the circuitry, can be caused by a number of conditions.

   a. De-wetting which is slight and noticeable mostly in wide bus traces and ground planes indicates copper contamination of the tin-lead plating bath. This can be remedied by dummy plating the bath for several hours at 3–5 amps per square foot. If the de-wetting has appeared gradually, and is light, this is probably the cause. Copper tends to build up during normal operation of the tin-lead bath, and must be plated out routinely.
   b. De-wetting which is gross on one side of a panel, and totally absent on

the other side, indicates careless pre-fusing preparation of the panel. Perhaps the operator forgot to scrub the side which de-wetted.

c. De-wetting which appears suddenly indicates poor preparation of the panels prior to fusing. If one job, or set of panels, has no evidence of de-wetting, and the next is horribly de-wetted, rerun the panels through preparation.

d. De-wetting can occur on panels which have simply set around for too long a period of time after etching. Panels should be fused as soon as possible after etching. Copper can be oxidized beneath unfused tin-lead, and cause de-wetting.

3. Outgassing. Outgassing appears as tiny bumps or ridges of bumps in the fused solder. It is caused by organic contamination of the tin-lead bath. The organic contaminant has become included in the deposited metal. The only cure for outgassing is carbon purification of the plating bath. The tin-lead plating bath must be recirculated through a cannister containing about 5 pounds of activated carbon per 100 gallons of bath. When pure, the plating bath should be water white, or slightly yellow. A Hull cell can be performed to verify removal of contamination and to indicate sufficient replenishment of peptone.

4. Blow holes which usually appear inside the plated through hole indicate the presence of entrapped moisture or other contaminant. Bake remaining panels before continuing with the fusing operation. Four hours at 200 degrees Fahrenheit is sufficient.

5. Delamination or blistering of the laminate. When this occurs on a multilayer, it may indicate improper curing of the prepreg and the presence of moisture or other volitiles. All remaining panels should be baked at 300 degrees Fahrenheit for about two hours. At the conclusion of this curing step the panels should be reflowed. If the delamination or blistering occurs in double sided laminate, the panels should be reflowed in hot oil, and the laminate manufacturer notified.

Other Notes

1. Panels should never be reflowed prior to plating of the contact fingers. Fused solder is difficult to strip during tin-lead removal for plating.

2. If the panels are immersed in an immersion tin bath prior to fusing, the tin deposited on the gold of the contact fingers will burn off during fusing.

3. Hot oil need not be changed when it turns black. The oil darkens with exposure to air, and should not be left uncovered when not in use. As long as the reflowed solder has a good appearance, the oil is acceptable no matter how dark it is.

SECTION SIX
MULTILAYER PRINTED CIRCUITS

21
Multilayer Processing

Multilayer printed circuits contain layers of circuitry which are carefully registered and laminated together. Interconnections to the various layers of circuitry are made through the plated through hole. Multilayer circuitry is a topic of great interest to both the designer and the manufacturer of printed circuits. The advent of multilayer circuits increased the number of electrical interconnections which could be made in a given board area. The ability to make a high number of interconnections is of paramount importance when integrated circuits are to be inserted into the printed circuit board. With the need for greater circuit density comes the need for a greater understanding of just what can be achieved with multilayer circuit design, multilayer materials, and fabrication equipment, processes, and techniques.

In the past, printed circuits (no matter how complex) have been pretty much devices for making electrical connections from point A to point B. The multilayer printed circuit added the ability to make connections to ground and voltage planes. These internal copper planes also permitted signal carrying circuits to be electrically shielded from the noise of other circuits and electrical fields. Another benefit added by the multilayer is the ability to bury signal carrying traces internally, so that only copper land area is visible on the outer two layers. There are several advantages of this type of construction. When the traces are located on external layers, they are subject to damage through abrasion. A mere scrape, which would not greatly affect the function of a ground plane or shield, could devastate thin copper circuitry and render the entire circuit scrap. Also, by placing the signal traces on an internal layer, the dimensions of the circuitry are determined by the thickness of the base copper foil, and the resist used to image the layer for etching. However, when the circuitry is on the external layers the dimensions are determined by the plated copper with tin-lead etch resist, the shape of the reflowed tin-lead, the thickness of the base copper foil, and the undercutting action of the etchant.

It is the challenge of the printed circuit industry (designers, manufacturers, and suppliers of chemicals, materials, and equipment) to manufacturer multilayer printed circuits which accomplish the following:

1. Provide more layers of interconnected circuitry.
2. Establish interconnections through smaller, more densely packed holes.

3. Route finer traces, which are closer together, and route more of them between the land areas around holes (pads).
4. Manufacture thicker multilayers.
5. Manufacture thinner multilayers.
6. Manufacture multilayers with circuitry that conforms to the artwork with ever increasing tight tolerances (± 0.001 inch or less).
7. Manufacture multilayers with more uniformity in the thickness of the laminate, and uniformity of the overall thickness of the finished product.
8. Manufacture multilayers with less warpage.
9. Manufacture multilayers with greater dimensional stability in the X-, Y-, and Z-axes.
10. Manufacture multilayers with greater integrity of the plated through hole.
11. Manufacturer multilayers which will withstand higher operating temperatures, and greater variations in the range of operating temperatures.
12. Manufacture multilayers with greater resistance to environmental effects of chemicals, humidity, and temperature (as above).
13 Manufacture multilayers with lower dielectric constant and greater uniformity of dielectric constant.
14. Manufacture multilayers from a greater variety of materials: epoxy, modified epoxy, polyimide, polyolefin, Teflon (DuPont), Kevlar (DuPont), quartz/epoxy, metal core materials, and other substates.
15. Manufacture multilayers with improved layer-to-layer registration.
16. Manufacture multilayers with heat sinks bonded on.
17. Manuacture multilayers which are more reliable.
18. Manufacture multilayers with greater productivity, and fewer steps.
19. Manufacture multilayers which meet the requirements of MIL-P-55110C.
20. Manufacture flexible printed circuits.
21. Manufacture rigid-flex circuits.
22. Manufacture stripline microwave circuitry.
23. And, of course, manufacture more layers for a given thickness.

GENERAL PROCESSING STEPS

Multilayer printed circuits are fabricated by etching circuitry onto one or more sides of thin copper clad laminate, generally using dry film photoresist as the imaging medium. In this chapter, and throughout the book, MIL-P-55110C is often referenced over other specifications because of its importance in building printed circuits for military applications.

1. Tooling holes are punched or drilled into both the artwork and the thin copper clad laminate, to assure good registration during imaging and layer lamination.

2. The layers are then imaged with photoresist and touched up.

3. All exposed copper is chemically etched off, and the photoresist stripped from the remaining copper.

4. The layers are now inspected for completeness of etching, accuracy of imaging, and defects in the circuitry. If any rework is needed, it is performed.

5. The layers are almost ready for lamination into complete panels. However, it is generally good practice to grow a layer of black copper oxide onto the copper circuitry. The black oxide improves the adhesion of the epoxy prepreg to the copper of the circuitry by providing a rough surface for the epoxy to grip. This is done by immersing the cleaned panels into a hot tank of oxidizing chemicals.

6. After black oxide coating, the layers are rinsed and baked. This baking dries the layers and removes absorbed moisture from the laminate prior to lamination.

7. The layers are now ready for lamination. Prepreg should already have been cut to panel size and tooling holes punched or drilled.

8. During lay-up for lamination the layers are arranged into properly sequenced and layer oriented stacks. Prepreg is interleaved between the layers. As the layers and prepreg are stacked, they are pinned onto a caul plate which is used to hold them during the lamination operation. When a stack has been completed another caul plate is pinned on top. This operation is called booking.

9. After a book has been made it is set in the center of the plattens of the multilayer lamination press. It is common for the press to have been preheated. Pressure and temperature are applied for a given cure time.

10. At the completion of the cure time, the press is opened and the book transferred to a cold press for cooling. The hot press is now ready to accept another book for lamination.

11. After the cold press cycle has been completed the book is taken apart by knocking out the pins holding the caul plates together.

12. Epoxy flash which has flowed from the panels edges is trimmed.

13. It is good practice to further bake the newly laminated multilayer panels to ensure a complete curing of the prepreg. Insufficiently cured prepreg will cause a lot of other problems during subsequent processing. After post-lamination baking, the multilayer is almost ready for processing as a double sided panel.

14. The panel must be drilled, but before drilling it is necessary to verify the registration of the layers. This can be done by two different methods. One method is to simply etch the copper from both sides of a laminated panel. This will bare inner layer circuitry; or at least the clearance holes and land areas for holes which can be inspected for registration. Another method which can be used to verify inner layer registration of multilayer panels is to x-ray the panel.

15. Drilling is the next operation. It is good practice to use the etched panel,

or an inner layer, as a first article to check the drilling registration before proceeding with drilling panels which may be worth several hundred dollars. The etched panel is superior to an inner layer, since it will have seen the same thermal conditions as the other panels.

16. After drilling, the multilayers are not quite ready for handling as a double sided panel. Drilling frequently smears epoxy around the inside of the hole. This is rarely a problem for double sided boards. However, smear inside the hole of a multilayer printed circuit must be removed, as it will prevent an electrical connection from forming to inner layer circuitry by the plated through hole. Smear removal is performed just before electroless copper plating of the holes. This operation is sometimes referred to as etchback, since the epoxy fiberglass is etched back from the edge of the hole at the inner layers.

17. After smear removal (or etchback), copper is chemically deposited inside the holes. It is good practice to run drilled test coupons on all multilayer panels. These coupons can be micro-sectioned after electroless copper deposition to verify the completeness of the smear removal process. If the smear has not been removed, a good electrical connection cannot be formed to the inner layers, and the panels will be scrap if they are not reprocessed.

18. After this micro-sectioning operation, the panels may be processed as double sided panels, with a few exceptions which will be discussed in the following pages.

PLANNING THE CONSTRUCTION OF A MULTILAYER PRINTED CIRCUIT

Construction Requirements

In order that the finished multilayer meet the design requirements, it must be put together correctly. When planning the construction of a multilayer job for production, some of the points to consider are:

1. Overall thickness and tolerance.
2. Hole sizes and accuracy of drilling.
3. Minimum annular ring requirements.
4. Layer-to-layer registration requirement.
5. Minimum trace width, and minimum air gap between conductors.
6. Customer specified dielectric thicknesses and layer spacing.
7. Number of layers.
8. Copper thickness, inner and outer layer.
9. Conformance requirements of circuitry dimensions to either the artwork or the blueprint.
10. Layer sequencing.
11. Requirements for etchback.
12. Type of laminate.

13. Testing and certification requirements.
14. Processing parameters, such as temperature, pressure, and cure time-during lamination.
15. Power and ground clearances at holes.

Meeting Thickness Requirements

The nominal thickness of a panel is the desired thickness. Since it is unlikely that any material or panel is going to be made exactly to nominal thickness, an acceptable tolerance of deviation must be spelled out on the blueprint. A typical tolerance for the thickness of a 0.062 inch thick multilayer is 0.062 $\pm$ 0.007 inch. The thicker the panel, the greater the tolerance should be.

There are numerous factors which affect the thickness of the panel. Some of these are discussed below:

1. Where the thickness is to be measured. Thickness may be measured at the contact fingers, the surface of the laminate, or metal-to-metal across the conductors. Each of these locations will yield a different value. Should the location for taking the thickness measurement not be spelled out in the blueprint, the following guidelines may be used: (1) Boards with contact fingers should be measured at those contact fingers; (2) boards without contact fingers should be measured metal-to-metal across the conductors.

2. Thickness of copper foil. The copper foil thickness used on inner layers may or may not be called out by the customer on the blueprint. If it is, then this is the foil weight which must be used. Inner layer foil is usually 1 ounce or 2 ounce copper. Foil weight, such as 1 ounce, means that the foil all by itself weighs 1 ounce per square foot. 1 ounce copper foil is 0.00144 inches thick (1.44 mils). There is a linear relationship between foil weight and thickness; thus, 2 ounce foil is 0.00288 inch thick, and ½ ounce foil is 0.00072 inches thick. There is more discretion which can be used by the printed circuit manufacturer when it comes to outer foil requirements, since conductors on the outer layers are subject to plating.

Also, copper foil on inner layer ground planes will contribute to over all thickness of the panel. If the inner layer circuitry is composed of signal traces, the copper will become encapsulated by the prepreg and will not contribute to panel thickness. Care must be used in planning, since encapsulated circuitry may violate dielectric spacing requirements.

3. Thickness of metals to be plated. Thickness of plated metal is generally specified as the amount of metal inside the through plated hole. Occasionally the actual conductor thickness is called out, e.g. total thickness of conductor copper to be 1 ounce or 2 ounce. For every 0.001 inch of copper plated inside the hole, there is about 0.001–0.0015 inch of copper plated on the surface of the panel. Thus, if a customer requires 0.001 inch of hole wall copper (Fig. 21-1), about 0.001–0.0015 inch of copper will be plated on the conductor surface.

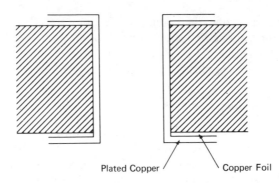

Plated Copper   Copper Foil

Fig. 21-1. Copper plating rate is 1.0–1.5 times faster on the surface than in the hole.

It is common practice to plate 0.0012–0.0015 inch of copper in the hole, for a hole wall requirement of 0.001 inch. The planning engineer can safely assume that 1 ounce of copper (0.00144 inch) will be plated on the conductor surface for every 0.001 inch of copper required in the hole. Tin-lead requirements are generally 0.3–0.5 mils (0.0003–0.0005 inch).

4. Thin copper clad laminate. (Thin copper clad laminate, or thin clad, refers to laminate with a core thickness of 0.030 inch or thinner.) The stated thickness of thin clad laminate refers to the thickness of the core material—the epoxy/fiberglass—and does not include the copper thickness. Thus, a thin clad laminate with a stated thickness of 0.022, with 1 ounce copper on each side, is actually about 0.022 + 0.00288 = 0.025 inch rounded off to the nearest mil, when measured with a micrometer. The allowable tolerance variation for this type of material depends upon the nominal thickness, and the tolerance class; see Table 21-1, taken from MIL-P-13949F. If four sheets of 0.022 inch thin clad are used in a multilayer, the cumulative tolerances of ±0.003 inch must be taken into consideration. If the overall thickness of the multilayer panel must be built to tight tolerances, it may be necessary for the printed circuit manufacturer to presort the thin clad laminate stock according to measured thickness.

5. Prepreg resins and glass fabric style. Prepregs manufactured by the var-

Table 21-1 Thickness and tolerance.

THICKNESS OF BASE LAMINATE (INCH)	TOLERANCE OF BASE LAMINATE (± INCH)	
	CLASS 1	CLASS 2
0.001 to 0.0045	0.0010	0.00075
0.0056 to 0.006	0.0015	0.0010
0.0061 to 0.012	0.0020	0.0015
0.013 to 0.020	0.0025	0.0020
0.021 to 0.030	0.0030	0.0025

ious companies differ significantly in performance, processing parameters, and thicknesses before and after pressing. The manufacturer of each prepreg used by a printed circuit shop should be consulted for optimum processing parameters of their materials. Prepregs from more than one company should never be used on the same lot of multilayers. When a multilayer job is being run, the planning engineer and the lamination engineer or technician should know exactly which prepreg is being used.

Epoxy resins differ in gel time and percentage flow. The temperature, pressure, and rate of temperature and pressure increase during the lamination cycle also affect overall thickness, and thickness distribution from one side of the panel to the other. Glass cloth differs in the weaves, yarn styles, number of threads per inch, thickness, and ability to hold resin, and allow resin to flow during the gel state. In general, prepregs with higher resin contents are found on the thinner glass cloths (104, 108, 112). These prepregs should not be used to build up dielectric thicknesses greater than .014 inch. Another reason to avoid using thinner glass styles for building up dielectric thickness is cost. Thick prepregs cost just about the same as thin prepregs. If three layers of a thin prepreg are used to build dielectric thickness which could be achieved with one layer of thicker material, then the prepreg cost of that printed circuit is triple what it could be.

The greater the resin content, the greater will be the Z-axis expansion (thickness instability), the resin flow, layer slippage, and nonuniformity of panel thickness. Resin rich prepregs are best used for flowing around and bonding to the copper circuitry. The planning engineer should use a mixture of thicker glass styles and core materials for building up dielectric thickness, and thinner styles for their resin richness and better flow and adhesion against the copper circuitry.

The style of glass can also have a marked effect on the ability to hold thickness tolerance. A glass style such as 1528, used for building up dielectric thickness, is superior to the more common 7628.* 1528 has two bundles of yarn in each strand of weave, which are wound in opposite directions. 7628 has the identical number of fibers in each bundle, but they are wound into one large bundle in the same direction. When the 1528 fabric is coated with B stage epoxy resin during prepreg manufacturing the resin flows with greater ease around the yarn bundles; more importantly, the resin flows into the bundles and encapsulates each fiber. The resin has greater difficulty penetrating the bundles of 7628, and there is little encapsulation of the fibers of those bundles. The 7628 fabric is more likely to compress and yield poor Z-axis dimensional stability than is the 1528. Consult Table 21-2 for a comparison of the before

---

*Trade Offs In Choosing The Right Epoxy/Glass Pre-preg For Printed Circuits, Bob Carter Lamination Technology Corp.

**Table 21.2. Comparison of Prepreg Thickness
Before and After Pressing and Curing.**

	104	113	7628
OAK	0.003/0.0015	0.0055/0.0035	0.008/0.006
NORPLEX	0.0025/0.0018	0.0055/0.0038	0.009/0.007
FORTIN	0.003/0.002	0.005/0.004	0.008/0.007

and after pressing thicknesses of several fabrics from different prepreg manufacturers.

6. Physical characteristics of the circuitry. The copper circuitry affects the thickness of the finished board. If an inner layer is mostly land area, such as a ground or voltage plane, obviously the copper thickness has a direct effect on the finished thickness of the panel. However, if the inner layer is only sparsely populated with low density signal traces, the effect of these conductors on the overall thickness is not so obvious. Isolated conductors will bury themselves in the prepreg, especially if the build-up of prepreg is very thick. When this is the case, the conductors have little effect on the finished board thickness. The effects of ground planes and isolated conductors are best dealt with by the planning engineer when he/she has the artwork for the layers present for examination. Since 7628 prepreg has neither the ability to conform well to signal traces, nor the resin content and flowability of the thinner fabric styles, it is not a good fabric style to have next to signal layers; especially high density signal layers. Several layers of isolated signal traces (low density circuitry) may mean an extra sheet of prepreg is required for panel thickness.

The planning engineer must take all of these considerations into account when planning the construction of the multilayer. It is helpful to have the artwork for the inner layer circuitry, or at least a blueprint of them, to refer to. If the printed circuit shop uses laminate and prepreg from more than one source, copies of all data sheets must be on hand, including data on the recommended processing parameters. The planning engineer would do well to talk with other personnel in manufacturing to gauge the correlation between planned construction and the finished product.

Annular Ring, Layer-to-Layer Registration, and Layer Spacing

The ability to hold annular ring requirements (size of the copper land area circumscribed about the drilled hole) is a function of the circuitry registration from one side of a thin clad layer to the other, and from one layer to another across the prepreg bonding sheets, and the ability to accurately drill a hole. To understand the requirements for annular ring and registration it is helpful to know how these are measured. Annular ring and layer-to-layer registration are measured by performing two micro-sections, one along the longest and one along the shortest sides of the panel. Inner layers require a minimum of 0.002

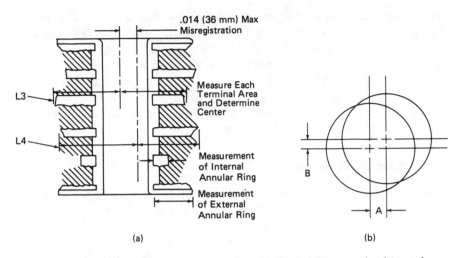

Fig. 21-2. (a) Method for measuring annular ring. (b) Method for measuring layer-to-layer misregistration.

inch annular ring. Outer layers are permitted 0.005 inch; however, it is permissible to have 0.004 inch annular ring on one side if it is due to drill splay (the tendency of a drill bit to drill at an angle), and if the other side does have 0.005 inch annular ring.

The micro-sections are viewed at 50–100× magnification. Registration is measured from the center of the hole to the outer edge of the annular ring, for the two length and width micro-sections (Fig. 21-2). Annular ring is measured differently, depending on whether it is inner or outer layer. Inner layer annular ring is measured from the edge of the drilled hole (where laminate copper and plated copper meet) to the edge of the annular ring. Outer layer annular ring is measured from the edge of the hole wall plating to the edge of the annular ring. The layer registration which has the greatest deviation is the critical measurement; since it, too, will have an impact on acceptability if it is greater than 0.014 inch. The layer registration is measured as the shift in the centers of the annular rings from one layer to another.

The planning engineer must be aware of the annular ring requirement, and the likelihood of it being met given the customer's artwork, thickness requirements, and dielectric spacing requirements. The artwork must be checked during incoming artwork inspection to verify layer-to-layer registration, and pad size for plated through and clearance holes. If the artwork pad is 0.050 inch for a hole which will be drilled at 0.045 inch, the printed circuit manufacturer is left with $(0.050 - 0.045)/2 = 0.0025$ inch annular ring as the maximum possible; there is no room for manufacturing tolerances due to imaging, etching, and lamination. Clearly, incoming artwork such as this must be rejected, with no attempt made to build the multilayer circuit.

Artwork is generally pinned to the laminate for dry film imaging of the layers. If the tooling holes for registration of the artwork and layers for lamination are drilled, a ±0.001, 0.002 inch location error is introduced. It is possible to punch these tooling holes faster, and with greater accuracy than can be achieved with drilling. Punched hole locations for registration can be routinely placed within better than 0.5 mil accuracy, assuming the same environmental controls as are used for drilling. All multilayer manufacturer's should avail themselves of the speed and repeatable accuracy of punching tooling holes for artwork and thin clad inner layer laminate.

Layer Spacing and Sequencing

Generally, there will be a chart such as is shown in Fig. 21-3, which shows how the layers are to be sequenced. When this is the only set of instructions provided for sequencing and spacing, the printed circuit manufacturer is pretty much free to construct the board as the planning engineer sees fit, and using the materials available already in stock. However, there may be notes listed which specify what the laminate requirements are.

It frequently happens, especially on designs of six layers or more, that for each layer the thickness of laminate, copper, and prepreg are specified (Fig. 21-4.) This is often the situation when the design requirements call for controlled impedance in the multilayer circuitry and spacing. When the printed circuit planning engineer encounters such tight design requirements, he/she should discuss the construction and fabrication with the multilayer lamination department. If the board has not been overspecified (a call to the customer's design or quality department will establish if this is the case), then great care must be used in selecting the laminate and designing the lay-up.

The number of layers being laminated, in itself, does not necessarily have an adverse affect upon registration. Indeed, copper is an easier material to drill through, than is fiberglass/epoxy. Copper not only acts as a heat sink, to keep the drill bit cool, it will also not contribute to drill splay in the manner that bundles of glass fibers do. In addition, the copper circuitry acts as a drag to layer slippage.

It is poor practice to use prepreg to build up dielectric thickness, as already stated. The only real requirements on the minimum thickness are that it be sufficient to withstand the voltage testing for arcing resistance, and that there be at least two sheets of prepreg, for a minimum of 0.0035 inch of pressed thickness between conductors on adjacent layers.

The ability to achieve good layer and drilling registration is also dependent upon being able to maintain good registration during lamination. During lamination pressing the layers are floating on a soft gel of epoxy resin under high pressure (200–400 psi or more). All attempts to reduce the tendency of layers to shift under these circumstances must be made. Some of the options available are given below.

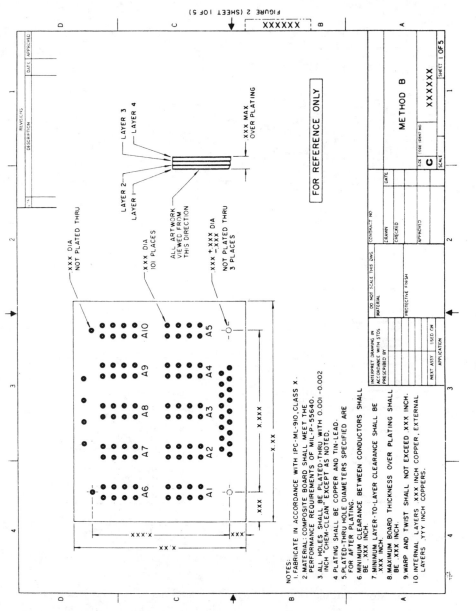

Fig. 21-3. (Courtesy of I.P.C.)

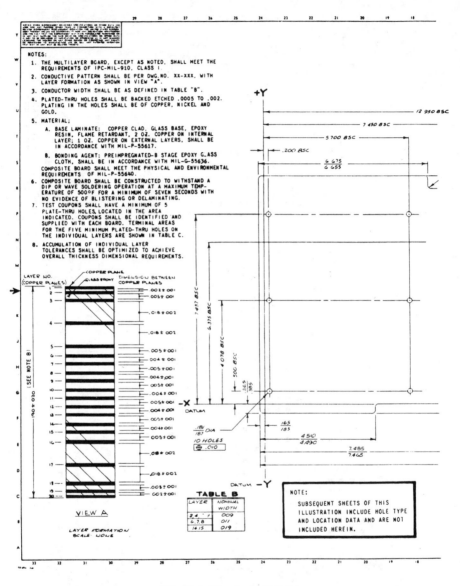

Fig. 21-4. (Courtesy of I.P.C.)

1. Use of steel caul plates for tooling. Steel (such as Picard 420) caul plates exhibit a thermal coefficient of expansion which is closest to that of copper clad epoxy fiberglass. This will reduce sheer forces exerted on the layers during lamination.

2. Use of copper lands on inner layers where ever possible. Designers should

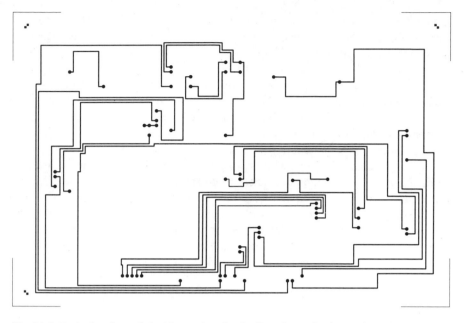

Fig. 21-5. Resin dam is needed with very low density inner layer circuitry such as this; too little copper is present to reduce layer shifting during lamination.

add as much copper land as possible to fill areas of low circuit density. However, this is something the circuit manufacturer has no control over; he/she must live with the circuit as designed. There are areas where copper can be added outside of the circuit in the form of resin dams. The resin dam should be used to reduce layer slippage during lamination, to provide conditions which will yield even overall panel thickness, and to channel air rich resin out of the circuit area. A layer pattern such as that shown in Fig. 21-5 should have a wide resin dam all around the four sides. There are different schools of thought about what form a resin dam should take; two examples are shown in Fig. 21-6. The advantages of the solid dam are that there is more copper, less resin will

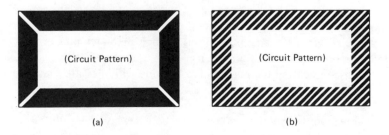

Fig. 21-6. (a) Solid borders with vented corners. (b) Fully vented border.

flow since it is restricted, and the solid borders add helpful rigidness to the layer for ease of processing. A vented border provides for slightly increased resin flow, while still adding needed copper to reduce layer slippage. It is possible to maintain a solid dam near low density circuitry, and vent the dam near large land areas; to promote evenness of lamination thickness and resin flow for reduced layer slippage.

3. Use of large borders. Large borders mean using a panel size which is larger than the size required to encompass the circuitry. A panel size which is greatly larger than the circuitry (at least 3 inches on all four sides) allows the use of larger copper borders, with or without venting. Even if the dams are not vented, a large border allows sufficient room for air rich resin to flow and carry the air from the circuit area, lessening the chances of blisters and delamination.

4. Use thicker glass fabrics, such as 1528, with lower resin content and reduced flow, for dielectric build-up, and only use thinner, more resin rich glass fabrics for flowing around and filling in areas of circuitry. Dielectric build-ups with 113 or thinner fabrics should be avoided when possible.

5. Use of thicker thin-clad laminate to build up dielectric thickness. Obviously, cured laminate is not going to flow, and layer slippage will have been reduced.

6. Lower temperature during cure, and less pressure will also reduce slippage. However, the temperature and pressure recommended by the laminate manufacturer are the parameters which should be used.

7. Where possible use thicker laminates with copper around the tooling holes. When etching inner layers, tape or resist should cover the tooling holes. This will result in less damage and tearing of the laminate at the tooling holes during booking and lamination. When the strentgh of the laminate at the tooling holes is assured, one source of layer slippage will have been reduced greatly.

8. Temperature profile across the platens of the lamination press should be very uniform. If there is a wide variation in the temperature, the rate of cure will not be even, nor will the rate of resin flow.

9. Caul plates and platens should be very flat and of uniform thickness.

10. There are differing schools of thought on the best number of tooling pins to use. Some experts feel that more than four pins will be difficult to insert accurately, resulting in booking problems and damage to the laminate. Others feel there is little problem and the added pins help prevent slippage, especially when thin clad laminates are being used. ¼ inch diameter pins are generally considered more stable than ⅛ inch pins.

11. The use of floating tooling pins for booking will alleviate some tendency for layer slippage.

12. Use an even build up of prepreg from one side of an inner layer to the other. In other words, the dielectric spacing between layers should be symmetrical (Fig. 21-7). Nonsymmetrical dielectric may be a design requirement of

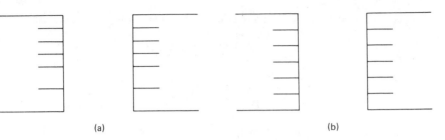

Fig. 21-7. (a) Unbalanced construction. (b) Balanced construction.

(a)

(b)

the printed circuit; when this is the case, the planning engineer should contact the customer to explain the difficulty, and that such dielectric spacing is a major contributor to warpage.

Drilling also has an effect on registration and annular ring. As already mentioned, it does little good to have excellent alignment and layer registration, only to have an annular ring problem caused by poor artwork, the hole pad being too small to do the job. But, there are two other drilling factors which affect the annular ring and hole registration: (1) drill splay, and (2) drill placement.

Drill splay is the tendency of a drill bit to go off at an angle, once it strikes the panel. It can be caused by drill wander; the drill bit "walking" around when it strikes the surface to be drilled, and it can be caused by other factors such as panel warpage and excessive drill bit vibration or slippage. One technique which will help to alleviate splay is to use an entry material, especially a thick (0.015 inch) aluminum material. The entry is softer than copper and allows the bit to enter immediately, instead of allowing it to wander. Warpage can be limited somewhat by using double stick tape when loading panels on the drill table.

The accurate placement of a hole, in an X-Y coordinate grid is enhanced by accurately calibrated and properly maintained N/C drilling and programming equipment. It is also important to remove sources of vibration from the drilling table and drill spindles. This is one of the reasons a granite slab is used for the drilling platform. In addition, dampers on the spindles, reduced feed rates, and reduced spindle travel in the Z-axis (up and down movement) should be considered. For many applications, there is no need for full spindle travel; it should be set to lift from the drilling table for only a short distance, like half an inch versus a full inch of travel. When drilling small holes, such as 0.032 inch or smaller, it is good to reduce the feed rate. The rpm can also be reduced according to calculation to achieve a given surface square footage. The greater the time between hits, the better the bit can rid itself of vibration prior to entering

Surface Feet = The diameter of a circle (in feet) with the diameter of the drill bit

$$\text{Surface Feet/minute} = \frac{\text{rpm} \times \pi \times 0}{12}$$

$$\frac{\text{Surface Feet/min} \times 144}{\pi \times (\text{Drill Radius})^2} = \text{rpm}$$

Fig. 21-8. Important drilling parameters.

the laminate again. The smaller the hole to be drilled, and the more accurate that hole location must be, the better it will be to slow feed rate (Fig. 21-8).

Aspect Ratio of the Plated Through Hole

The aspect ratio is the ratio of the board thickness to the diameter of the plated through hole. The specification IPC-ML-910A recommends a 3:1 ratio as the minimum for Class I printed circuits. When the diameter of the plated through hole gets down to about one third of the board thickness, some undesireable effects are noticed:

1. The ability of fluids, such as electroless copper and concentrated sulfuric acid (for etchback or smear removal), to adequately flow through the hole is seriously hampered. If the smear removal and etchback reagents cannot freely pass through the hole their ability to perform the intended function is undermined; the result may be increased probability of poor inner layer connection and poor copper coverage.

2. It is difficult to throw copper or other metal by electrodeposition into such a restricted opening. The result is apt to be unacceptably thin copper plated on the hole wall of high aspect holes. Although the copper may be adequate at the lip of the holes, it may be only 50%, or less, inside the hole (Fig. 21-9).

3. The difficulties encountered in processing metal on the hole walls may lead to testing failures, such as barrel cracking and apparent resin recession during thermal stress testing.

4. Chip removal during drilling also becomes more difficult and excessive heat generation occurs. The additional heat may be more than enough to cause excessive smear.

5. As the aspect ratio of the plated through hole approaches 3:1, there is a tendency of the solder in these holes to plug the hole during solder reflow. This can be a serious situation for the board manufacturer which is actually not due to the plating process. If at all possible, the diameter of the smallest plated through hole should be kept at least 0.005 inch (for 0.062 inch thick laminate) to 0.010 inch or better (for 0.093 inch or more thick) above the 3:1 aspect ratio.

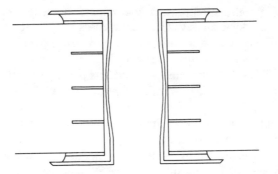

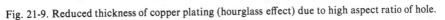

Fig. 21-9. Reduced thickness of copper plating (hourglass effect) due to high aspect ratio of hole.

Minimum Trace Width and Spacing Between Conductors

Assuming optimum photoresist imaging, as opposed to screen printing the image, there is no good reason why multilayers of 0.005 inch traces and spaces cannot be manufactured routinely. Fortunately, the manufacturer does not always have to manufacture to such tight requirements. Traces and spaces of 0.008 inch should be well within the routine processing requirements of all multilayer manufacturers. The 0.008 inch space allows the manufacturer to even meet the rather loose layer-to-layer registration shift of 0.014 inch allowed by MIL-P-55110C.

Manufacturing printed circuitry with less than 0.008 inch traces and spaces involves overcoming more complicated problems, since there is not much tolerace to play with. Some of these problems are:

1. Artwork must be virtually perfect as it arrives from the customer. The acceptability guidelines used during incoming artwork inspection must be much stiffer than those used to build less demanding circuits. Touch up, such as pin holes, nicks, protrusions, etc., become a major task when the circuitry is composed of 0.005 inch traces and spaces. The tight spacing often may be less than needed to fit a pen point or knife blade. If the artwork requires more than simply adding company logo and a date code, it should be returned to the customer. The printed circuit manufacturer who decides to tackle artwork with tight circuitry in need of a great amount of touch-up should be very wary. If the customer needs a circuit built to exacting tolerance, then the customer should be ready to provide artwork made to those same standards.

2. The normal touch-up after imaging and before etching or plating can be just as difficult as the artwork touch-up. The phototools used in production must be kept scrupulously clean. The dry film imaging area and all the equipment must be kept clean, and checked for dirt routinely during processing of tightly toleranced printed circuits.

3. The drilling registration must be the best attainable. Every mil of spacing tolerance lost to poor drill registration increases the chances of producing a scrap circuit. These circuits must be drilled only in one panel high stacks; until successful experience indicates otherwise. Vibration must be kept to a minimum on the drilling machine, and reduced hits per minute must be used.

4. Tooling holes for inner layer alignment and artwork registration must be punched, not drilled. The front-to-back and layer-to-layer registration demands of tightly toleranced circuitry require the extra 0.0005–0.001 inch of accuracy attained by punching.

5. Care must be exercised by all people in the manufacturing and inspection processes. Any scratch in the tin-lead before etching could result in the copper being etched through at this point; all circuits must be second touched up prior to etching.

6. The outer layer copper should be of a reduced weight, such as ½ ounce. This will limit undercutting during the etching process. However, the same thin copper which can make etching fine lines easier can also present rather severe handling problems. Thin foil cannot withstand scratches well, the foil is easily pierced. All personnel must be warned of the difficulties of handling thin foil, and special attention drawn to its presence whenever it is being used.

7. It is generally good practice to run first articles for critical operations such as lamination, drilling (drill an etched panel after lamination), etching, and routing, on all printed circuits where so much extra effort is being expended. The same multilayer first particle panel which is being used to check lamination registration and drilling registration, can also be used to run the routing first article.

8. It is common, in this type of circuitry, to find tight dimensional agreement required between the finished circuitry and the artwork. When this is the case, the printed circuit manufacturer must measure not only the artwork, but also the dry film photoresist imaged panel, before etching even the first layer. If the imaged layer meets circuit tolerance, it should be etched and measured again. The completed circuit trace width requirements must be met before the entire job is processed.

9. High intensity U.V. light sources must be used to expose dry film, for optimum acuteness of resist to artwork.

In general, tightly toleranced circuits can be manufactured without special equipment (other than what has been listed so far), as long as that manufacturer:

- Understands the requirements of the circuit.
- Understands the capabilities of the processing equipment.
- Understands the capabilities of the processing personnel.
- Makes no assumptions that all is going well—this is something which must be verified at each step in the manufacturing process by inprocess and first article inspection.

PROCESSING OPERATIONS

Laminate Selection and Prepreg

The actual laminate to be used is specified on the job work order traveler, as is the prepreg type and construction sequence. Fig. 21-10 shows an example of how the information from the planning engineer is transmitted. Although the planning engineer is the only one who should be authorized to write the requirements on the traveler, everyone involved with the manufacturing operation should read the requirements, study the blueprints, and verify in their own mind that the instructions are correct. This should be done for any type of printed circuit being built; however, it is critical for multilayer fabrication, since so much time and money are involved. The manufacturing instructions should list:

1. Laminate. Each laminate to be used, and where it is to be used.
2. Copper thickness and orientation, for each laminate (1/0,1/1,2/1,2/2, etc.)
3. Prepreg. Each style of fabric, and the number of sheets, and sheet orientation for each dielectric requirement in the printed circuit panel must be spelled out. The brand of prepreg being used should also be listed, since this will have an effect on the completed thickness, and the temperature and pressure of lamination. This must not be left to the discretion of the technician performing the stack-up for lamination.
4. Overall thickness and thickness of each dielectric build up, after lamination, should be dimensioned. If this is not done on the blueprint, it must be written on the construction diagram of the traveler.

A very important aspect of the glass fabric, especially laminate is the warp and weft (Fig. 21-11). These are the directions of fiber bundles in the weave. The warp directions on all laminate in a printed circuit board must match. Crossing warp and weft directions leads to severe warpage in the laminated panel. No amount of baking will remove this type of warpage. Warpage can also be due to uneven dielectric constants, and differences in direction of circuitry, and even uneven temperature profile of the press platens, but these can all be alleviated to a great extent by a baking operation between caul plates or other heavy objects. Caul plates of metals whose thermal coefficient of expansion is not very close to that of laminate will cause severe warp.

The printed circuit manufacturer must never take for granted the direction of warp and weft in prepreg or laminate. Laminate must be purchased with arrows printed on the foil which depict the warp direction. All laminate must be sheared into panel size, and punched with tooling holes, with the arrows aligned in the same direction. Although the layers are to be punched with the

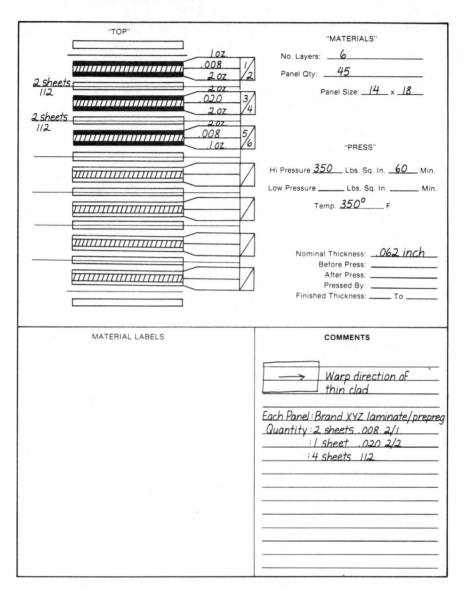

Fig. 21-10. Important multilayer information.

arrows aligned in the same direction, the punching must not be performed until later, during the imaging process. It is permissible to align laminate with arrows in opposite (180 degrees out) directions, but never 90 degrees out. Warp direction of prepreg does not play as important a roll as with laminates.

It is not a good idea to use laminate, even one piece, when the warp direction is not known. Unfortunately, it is all too common when a piece of laminate has

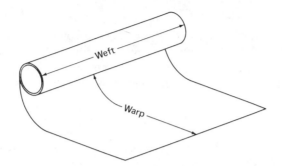

Fig. 21-11. Prepreg warp.

been damaged or lost, just to cut another piece, or use a piece which is already cut, with little regard for warp direction. The importance of proper warp alignment must be trained into all processing personnel. Whenever laminate is sheared, a drawing of the panel with the direction of warp shown by an arrow must be made on the job work order traveler. This will identify the panel dimensions in which the warp runs.

Some type of identification scheme must be used to identify the laminate according to copper thickness, and orientation of copper thickness. If laminate has 1 ounce copper on one side, and 2 ounce copper on the other, the layer must be marked, by number scribing or notching, to indicate copper weight. The identification scheme must also indicate core thickness, to avoid uncertainty and the need to "mic" each layer for thickness prior to photoresist lamination.

Laminate Preparation for Imaging

The layers must be free of oxides, fingerprints, dirt, and other contaminants before imaging. Numerous techniques are available. Generally, a combination of chemical and mechanical cleaning methods are used. Chemical cleaning is used to assure removal of oils and other substances which could also contaminate the mechanical cleaning process. Hot alkaline cleaners, or vapor degreasing, work very well.

Mechanical cleaning can be accomplished by oscillating brushes, rotating brushes with imbedded grit, or pumice scrubbing. All these methods are also used to clean thick laminate for double sided printed circuits, too. However, when scrubbing thin clad laminate, there are restrictions due to the tendency of thin clad materials to be damaged by mechanical equipment. Any operation which will result in severe warping or bending must be avoided. Even nicking or bending of the corners must be avoided, as it will lead to other problems further along in the manufacturing operation on conveyorized equipment.

Pumice scrubbing, by arm motion with a hand held brush, or by a hand held rotary brush, works very well and is gentle on the laminate. However, when great quantities of layers are to be processed, it becomes impractical. The Tru-Scrub machine, which duplicates oscillating, rotatory arm motions can also be used with pumice. There are numerous other pumice scrubbing machines available as well. When pumice is used, adequate water spray rinse must assure its removal.

When chemical cleaning is being used exclusively, with no mechanical scrubbing operation, then a micro-etchant such as sulfuric acid/hydrogen peroxide must be used after the hot alkaline cleaner. The surface should be etched to a matte pink, and be very uniform. Areas of shiny copper indicate the presence of contamination on the copper foil.

Lamination with Dry Film

The cleaned layers should be baked to dryness at 200 degrees Fahrenheit or so for 10–20 minutes in a circulating air oven. If the dry film photoresist laminator is a hot roller variety, the panels can be removed from the oven, stacked, and laminated at room temperature. If the laminator is a hot shoe variety, the panels should be laminated with photoresist while they are hot. This means that no more than 6–10 at a time can be removed from the oven for lamination.

Dry Film Photoresist Lamination

After the cleaned and dried layers have been removed from the oven, they need to be sorted and stacked. The sorting should be done so that only layers of a given core thickness and copper weights are in the stack. The laminates with mixed copper weights, such as 2/1, must be stacked so that only the lighter foil weight faces up. The sorted and stacked layers are now ready to be dry film laminated.

The photoresist laminator should be preheated and loaded with photoresist of proper width, to correspond to layer dimensions. Resist thicknesses of 0.001–0.0015 inch are commonly used. Ideally, lamination is performed by two people, one to feed layers in, and another to cut and trim excess photoresist from the layers as they come off the machine. All four edges should be trimmed of excess resist, or particles will break off and interfere with the exposing operation.

After all layers have been laminated and restacked as they were, they are almost ready for tooling hole punching. Before punching tooling holes, the technician must decide how the layers are to be oriented. This information can be obtained by looking at the layer diagram on the traveler. The tooling holes will be used to align the layers. Care must be taken to ensure that the layers which are supposed to be 1 ounce on side A, and 2 ounce on side B, are punched to reflect the correct copper orientation. It may be, for instance, that 10 layers will be punched with the 2 ounce side up, and 10 punched with the 2 ounce side facing down. When the layers have been further sorted to reflect these needs, they are ready for punching.

The pins which do the punching should be set for the correct configuration for the panel size being run. The layers are inserted one at a time and punched. It is important that only layers which have already been laminated with photoresist be punched. If the layers are punched prior to laminating with photoresist, the photoresist subsequently applied will restrict the diameter of the tooling hole, hence, the ability to register artwork using those holes.

Phototool Requirements

The photo department should already have produced the phototools (working film) needed to image the layers. It is good practice for the multilayer technician to compare this set of artwork with both the layer diagram and the blueprints for accuracy of layer sequence, and for accuracy of tooling holes which have been punched into the phototools. The artwork should be stacked up, with the emulsions oriented as if they were emulsion-to-emulsion against the photoresist covered copper surface. When they have been thus oriented, all of the tooling holes must align, and all of the hole pads of the artwork must also align. When artwork registration pins are inserted into the tooling holes of the stacked up artwork, all the circuitry holes should align within ± 0.002–0.003 inch. If the registration is off more than this, or if the tooling holes do not align properly, the artwork must not be used.

The phototools should be oriented so that the emulsion will lie next to the cover sheet of the photoresist for exposure. The images must be negative for inner layers, and positive for outer layers. If the emulsion is away from the panel (facing up) when pinned to the layer, it is incorrect. Although it might very well produce an acceptable circuit, there will be some loss of definition, resolution, and ability of the imaged layer to match the dimensional widths of the circuitry. The technician should also inspect for the need to perform touch up on the phototools before beginning to use the artwork.

Artwork should be stored in a temperature and humidity controlled environment; typically 70 degrees Fahrenheit at 50% relative humidity. If the artwork has been exposed to a different temperature and humidity, allow it to come to equilibrium for several hours before using it to expose high density, tightly toleranced circuit images. The temperature of the phototool and the vacuum frame of the exposure source will affect the dimensional stability of the circuit image (Fig. 21-12). If the circuitry being imaged has wide tolerances, slight instability of the image on the phototool may not be too critical. However, if the circuitry has small annular rings, narrow clearances at power and ground, and the panel size is large, dimensional stability is a critical consideration. If a large number of exposures are being made, it is good practice to allow an

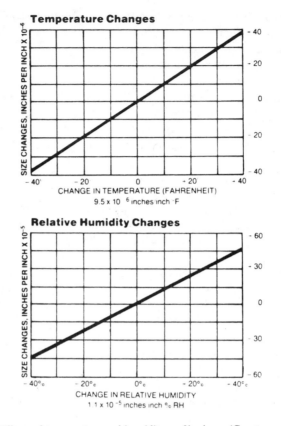

Fig. 21-12. Effects of temperature and humidity on film base. (Courtesy of DuPont)

experimentally determined amount of time to elapse between exposures, sufficient to control temperature of the contact frame.

Exposing and Developing the Photoresist

The procedures discussed in the chapter on Photoresist Imaging should be understood and followed. When exposing both sides of a layer, the artwork for one of those sides can be taped against the bottom of the contact frame, with the emulsion facing upwards. The layer is laid against this phototool, and the other phototool laid against the other side of the layer. A pin registration system, as described in the chapter on Artwork Registration, must be used. If a large number of panels are being fabricated, and a double drawer contact printer is being used, it is advisable to have the Photo department prepare two sets of phototools. When this is being done, two technicians should be working together, one per drawer and set of layers. Layers which have inner layer circuitry on one side and an outer layer on the other side should be coated with photoresist on both sides. However, during exposure, no phototool will be used to expose the outer layer. The outer layer will be processed with exposed photoresist covering the entire side to keep it from being etched. Thick adhesive fume tape can also be used to coat the outer layers.

If the core material of the layers is very thin, it may be wise to tape the layers end-to-end for developing in a conveyorized developer. Taping into a "train" helps prevent layers from sliding over one another and getting caught in the conveyor rollers. After they have been developed, they should be placed in racks which have added support in the form of strips of laminate or drilling back-up material. This support is needed to keep the layers from bending at the edges while in the racks. Developed layers are ready for image touch-up prior to etching.

During touch-up, all circuitry must be inspected closely. It is common practice to inspect one article of each layer, and identify defects. Each subsequent layer will be examined at the areas noted on the first article, and touched up if needed. Defects such as these, if they occur in the same location on more than one layer, are called "repeaters," and are attributed to defects in the phototool. It must not be assumed that repeaters are the only defects which will be found. Both sides of every layer must be inspected for isolated defects. There are special pens available which contain a resist ink that is suitable for touching up. Fine brushes and a non-alkaline soluble ink resist also work well for touching up, and are superior for getting into tight spaces.

Etching the Inner Layers

Inner layer etching is different from etching double sided panels, for several reasons. The laminate is much thinner. This presents problems of just getting it through conveyorized etching equipment. The end-to-end taping method is a must for conveyorized equipment. For batch equipment, if the layers are too thin, they bend easily. Racks must be designed for supporting the layer edges, as well as the centers. Another problem encountered when etching inner layers, is differing copper foil weights, from one side of the layer to the other. This difficulty can be overcome by taping layers back to back, so that only copper foil of identical weights is exposed to the spray etchant. Also, it is possible to turn off the etchant spray nozzles on the lower manifold, so that only one side of the layer is etched at a time. A first article should always be run, to set conveyor speed. The situation is not quite so easy when using batch etching systems, as etchant will still penetrate between the sides which are taped together.

When etching, it is common practice to place tape over the tooling holes. When the layers are etched, and the tape removed, a copper land area will remain around the tooling hole, for added strength during booking. After etching, the layers must be rinsed, and the resist stripped. It is good practice to inspect the layers after etching and resist stripping, as the layers have been through a significant amount of processing and handling. After layer inspection, they are ready for black oxide treatment. Layer inspection may be performed after black oxide. In fact, this may even be preferable since black is easier to see than is bare copper.

Black Oxide Treatment

The black oxide must cover all areas of the copper circuitry with an even coating. Areas of noncoating, or reduced coating, indicate contamination of the copper. The layers must, therefore, be cleaned well before attempting black oxide treatment. Like the preparation of layers prior to photoresist lamination, the cleaning is typically done using a combination of chemical and mechanical cleaning. A hot alkaline cleaner, followed by an acid immersion and pumice scrubbing or other mechanical scrub.

A number of rack and clipping schemes may be used to hold the layers for black oxide immersion. A load of cleaned layers is immersed in the hot tank of chemicals which will grow a layer of black oxide onto the copper. A typical immersion time is 2–3 minutes at 200 to 212 degrees Fahrenheit. There are many proprietary black oxide baths on the market. A typical formulation is given in Table 21-3. Once the layers have been immersed, they should not be lifted from the bath until immersion time has elapsed; exposure to air will inhibit development of an acceptable oxide layer.

The speed of the deposit will vary with the age of the bath. The deposition rate will slow as the bath ages and becomes depleted. As the bath ages, it becomes contaminated with resist, epoxy, and copper. All of these contaminants have an effect on the rate at which the black oxide is deposited. Although there are chemical analytical, and other measurements available for determining bath potency, the only reliable device for appraising bath strength is the amount of immersion time needed to obtain a dense, black layer of oxide. As the immersion time creeps upward from 2 to 3 minutes, the bath will need replenishing. Replenishment can be performed by scooping a mixture of the required chemicals directly into the bath. One method for operating the bath after a fresh make-up is to add more chemicals as needed to maintain immersion time. When 100% of the make-up volume has been added, let the bath deplete itself. When cool, it may be pumped out, and made afresh. Safety glasses, boots, apron, and gloves should be worn at all times when running the line, and when making chemical adds. The bath is extremely toxic and corrosive, including the fumes. All safety and handling instructions of the bath manufacturer should be followed.

At the conclusion of immersion time, the layers are removed from the black oxide bath and immersed (with no draining time when leaving the black oxide

Table 21-3. Typical Formulation for a Black Oxide Bath Operated at 210°F.

1. Sodium hydroxide	16 ounces/gal
2. Sodium chlorite	8 ounces/gal

bath) in 150 degree Fahrenheit water for a minute. The layers are then rinsed well in an immersion water rinse tank, and baked to dryness. When the layers are going to be processed through lamination immediately, they should be allowed to bake for an additional hour or two. This extra baking is recommended by most laminate manufacturers and assures removal of moisture which has been absorbed into the laminate. The black oxide bath attacks the butter coat epoxy, baring the weave and inner resin for increased moisture absorption. When the layers are inspected for coverage, any layers with bare copper should be stripped of black oxide (50% water and concentrated hydrochloric acid), and run again through the black oxide tank. Do not let people place fingerprints on black oxide coated layers.

Cutting Prepreg

Prepreg is a messy material to work with because it sheds fibers of glass and resin crystals. The prepreg cutting and storage area should be isolated from the rest of the multilayer preparation area. If large quantities of a given panel size are being run, it is handy to have a prepreg cutting machine. This same machine, if cleaned thoroughly, can also be used for cutting copper foil, for printed circuit shops which apply the outer layers as foil. It is good practice for all personnel handling prepreg, sheets of foil, and inner layers to wear clean white gloves.

Prepreg can be cut easily with a razor blade knife. A roll of the desired material is suspended over one end of a table, the fabric is unrolled to the panel dimension, and cut with the knife. Enough prepreg must be pulled out for it to be cut in the warp direction matching that of the laminate. If cutting in the direction across the warp (weft direction) is needed, a paper cutter can be used to slice off pieces cut in the warp direction.

Once the sheets have been cut, they must be taken to the dry film imaging room for punching tooling holes. The punching pins must be set to the configuration for the panel size. Prepreg can be punched in stack thicknesses of 0.030–0.050 inches with little problem. A template which denotes hole and location can be used to drill the tooling holes, since their size is not critical.

The prepreg quantity required per panel is generally written on the job work order traveler. This required number of sheets should be verified by the multilayer technician.

The total pressed thickness of prepreg required to build the panel is used to calculate the number of sheets of prepreg per panel by dividing the total thickness of pressed prepreg required by the thickness of the pressed prepreg per sheet as noted in the data sheet for that particular material, or as determined experimentally:

Pressed Prepreg Required/Thickness of Pressed Sheet
= Number of Sheets per Panel.

It must never be forgotten that prepregs of the same fabric style, like 112, obtained from different manufacturers, will not necessarily have the same pressed thickness. The data sheets for each manufacturer must be consulted, and the manufacturer of the prepreg must always be known. It is poor practice to use prepregs of the same style from more than one manufacturer, in the same panel. The multilayer technician must be aware of: (1) where on the panel the overall thickness measurement is to be taken, and (2) the fact that metal plating will add 0.003–0.005 inches of thickness to the overall dimensions of the panel. Also, copper thickness and encapsulation of signal traces must be kept in mind.

At the same time prepreg is being cut, release sheet must also be cut. Tedlar (DuPont) is commonly used. Enough must be cut to place one sheet between every panel in the book, and between the caul plate and the panel. Release sheet is cut 2 inches longer in length and width than the panel dimensions. It is not necessary to punch the release sheets with tooling holes.

Preparation for Booking

Before actually performing the booking operation, the technician must be thoroughly organized. All materials and equipment must be on hand and ready to be stacked into books for lamination. This entails:

1. Sorting layers according to numbers.
2. Orienting tooling holes in each stack of identical layers.
3. Verifying correctness of tooling holes for orienting layers.
4. Gathering release sheets and punched prepreg for booking.
5. Gathering caul plates for booking.

Stacks for booking can be prepared as follows:

1. Take an outer layer, any outer layer, and lay it face down on the table.
2. The required number of sheets of prepreg is now laid on top of this, so that tooling holes align.

3. The next layer in the sequence is laid on top of this prepreg.

4. The required number of sheets of prepreg are now laid on top of this layer, with tooling holes aligned.

5. And so on, until all the layers for a complete panel have been stacked. The layers must be stacked atop the prepreg according to numerical sequence, and not according to the tooling holes. The technician must refer to the layer diagram drawn on the work order traveler. When this is done, all the tooling holes should be aligned. If not, then either the layers were incorrectly punched, or they have not been oriented correctly. When it has been determined that the layers do align correctly when sequenced according to layer chart, all the layers of the job should be so stacked and oriented with prepreg. The books for lamination will be prepared from these stacks. Tedlar (DuPont) must be placed between all panels, and each item should be brushed free of debris prior to pinning to caul plates.

Caul Plates

The caul plates for the given panel size should be set on the table and inspected for cleanliness. If the plates have been abused, and burrs are sticking up on the edges, these burrs should be filed off before the plates are used. There should be no chunks of resin on the surfaces or in the tooling holes. Each pair of caul plates will have a number stamped on the edges, to identify plates of a given pair. There will also be an arrow, or other marker, to indicate which are the inside faces of the plates (Fig. 21-13).

Only one caul plate of the pair will have bushings for holding registration pins. The registration pins align the layers and hold the book together. Typically, there are two lengths of pins, ½ short and ¾ long. The ¾ long pins are used to get the maximum number of panels into each book. The shorter pins are used when there are not enough panels to fill a book. If the longer pins are used when there are an insufficient number of panels to warrant using them; they will stick out the other side of the top caul plate. If a book like this is placed into the press, the pins will damage the platens.

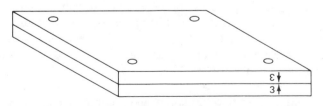

Fig. 21-13. Correctly oriented set of caul plates.

Loading a Book

1. Place the caul plate which has the pin bushings on the table; the direction arrow should be pointing upwards.

2. Place one pin (of the length which reflects how thick the final book is to be) in each of the bushings. The pins can be gently tapped down, firmly in place, with a hammer.

3. Situate the properly stacked layers with prepreg near the caul plate, together with the release sheets. White cotton or polyester gloves should be worn.

4. Use a soft brush to brush the caul plate surface free of dust or other particles.

5. Brush off one side of a piece of release sheet, and lay it on the caul plate. The release sheet can be pushed over the pins, to lie flat and wrinkle free on the plate; also brush off the other side of this sheet.

6. Brush the top outer layer of the stack, and lay this side down against the caul plate. The layer will have to be fitted over the registration pins, and pressed firmly against caul plate.

7. Brush the side of that layer which is now showing.

8. Brush the piece of prepreg from the same stack, lay this piece brushed side down over the pins of the caul plate; and brush the other side.

9. If there is more prepreg on the stack, load this onto the caul plate in a like manner.

10. Also load other layers in a like manner, until one complete multilayer panel has been loaded onto the caul plate.

11. Next, brush a sheet of release sheet, and lay this side on top of the outer layer of the panel on the caul plate; brush the other side of the release sheet.

12. Continue in this manner, until the book has been filled. Typically, a book will accommodate six panels, with an over all thickness of 0.062 inch each.

13. When the book has been filled, brush one side of a release sheet, and lay that side down on top of the outer layer of the last completed panel; brush the other side.

14. Brush the inside of the top caul plate; this is the side with the arrow pointing to it. Lay this plate over the pins on the stacked book. The arrows will be facing each other, and the tooling holes and pins will align.

15. Use the wooden butt of a hammer to gently tap the top caul plate and secure it to the pins. *Note:* If the pins stick up slightly above the surface of the top caul plate, use a punch and hammer to gently pound the pins into the bushings of the lower caul plate. Pins which stick above the surface of the caul plate must not be loaded into the press.

16. It is good practice to count layers and prepreg sheets as you book; quantities and sequences must always be identical.

17. When one book has been completed, another should be filled. The book-

ing continues until all the books are filled, until enough books have been made to load the press, one book per opening in the press, or until all of the layers have been used up. Depending upon the requirements for the multilayer circuitry and tolerances, it might be wise to run a first article of one panel, or one book, to verify post-lamination registration.

OTHER CONSIDERATIONS

The Press

The press (see Fig. 21-14) provides the pressure and temperature for lamination; it also controls the rise in temperature and pressure. There are many manufacturers of presses, and differences between presses: hydraulic, mechanical, electric heating, steam heating, single opening, multi-opening, etc. But, the purposes remain the same. Specific information on operation and cycles must be obtained from the press manufacturer.

The temperature, pressure, and rate of increase in both of these which are most suited to the prepreg and laminate being used should be obtained from

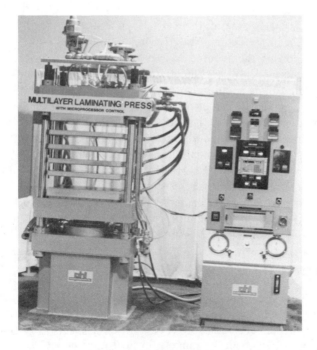

Fig. 21-14. Multilayer press.

the manufacturer of the materials. The examples quoted here are only typical values for a given set of materials.

It is common for multilayer facilities to be equipped with a cold press. The cold press is used to cool newly laminated panels under pressure. Typically, one cold press will service two hot presses. When two hot presses are being serviced by one cold press, the cold press may be mounted on a track for movement. The cold press is located directly in back of the hot press, and books are pulled, or pushed, from the hot to the cold very fast. It is possible to build multilayer panels without using a cold press; however, this would mean that newly laminated panels would have to sit in the books until cool, which could take several hours. Sometime cold water is circulated in the press platens. This provides the same result as a cold press; however, lamination time is lost while in the cooling cycle.

Pressing Material

It is common practice to use some sort of pressing material between the platens and the caul plates; just as a pressing material (the release sheet) is used between the caul plate and the layers. The pressing material fulfills a number of needs. It helps to distribute the pressure of lamination more uniformly over the caul plate and panel surfaces. Although the caul plates and platens are, ideally, perfectly flat, in practice there is no such thing. Pressing material helps to compensate for imperfection. Also, the pressing material delays the heating of the layers, allowing the uncured resin more time to flow, and the temperature to spread more evenly over the plates. This helps assure air removal and adequate flow into tight circuit areas and crevices. The pressing material also helps to keep the platens clean of resin which has flowed out of the book, and which would otherwise bond to the platen.

One of the most common and cheapest forms of pressing material is Kraft paper. There is an abundant supply in every shipment of laminate. For this reason, it is most commonly used. However, silicone rubber pads are finding wider use today. The silicone is heat resistant and does an excellent job of meeting all the tasks which pressing material is called upon to perform. It is also expensive. As the tolerances of multilayers become narrower, it is perhaps a good idea for the manufacturer of printed circuits to evaluate this material.

Heating the Press

Most presses have water and electric power running to them. All water valves and electrical breaker switches should be turned on. The heater control and pressure set points should be double checked for correct settings. The platens should be open. Allow about 30 minutes for heat up time.

Loading the Press

Before anything is loaded into the press, the operator should look at:

- The heating dials, to verify that they are set at desired the value (350 degrees for this example).
- The heating cycle timers: Preheat—15 minutes.
  Material Cure—60 minutes.
  Cooling—12 minutes.

1. Be sure adequate pressing material is on hand (Kraft paper) in about 5 foot pieces. Use six pieces per book, three on top and three below.

2. Place three pieces of Kraft paper on each of the platens; let excess paper lie out the back end into the cooling press.

3. Set one book on the Kraft paper of each platen. The book should be centered on the platen. It will be necessary to hold the Kraft paper with one hand, to keep it from sliding, while the book is pushed into the center of the platen. The book must be viewed from the side of the press, and centered along this axis also.

4. Place three more pieces of Kraft paper on top of the book, with excess lying out the back.

5. Loading of other platens should proceed likewise. It is important that the press be loaded and closed expeditiously but carefully. Once the first book has been loaded, loading of the other books must proceed quickly.

6. Press the CLOSE button(s) and hold until each platen has closed. Generally, the hydraulic pump motor will kick off automatically.

7. As soon as the motor kicks off, the heating cycle should start automatically, with preheating. The press should cycle automatically through the entire temperature cycle, except that the cooling cycle will not begin until the books are loaded into the cooling press, and it is closed.

8. When the press closes, pressure will begin to rise. Monitor it, and adjust it for the size of caul plates used in the book. The pressure control knob is used to make the adjustment. When the desired pressure is achieved, the control knob is locked into position. It is good practice to check the pressure setting periodically, as it may wander slightly. For an operating pressure of 350 psi, a chart can be made up for each size of caul plate. The pressures shown in Table 21-4 are calculated for 350 psi, using the formula:

$$\text{Length} \times \text{Width} \times \text{Operating Pressure} = \text{Total Pressure}$$

9. When a loud buzzer goes off, this signals the end of the material cure cycle.

**Table 21-4. Total Pressure on
Caul Plates (L × W × 350 psi).**

CAUL PLATE SIZE	TOTAL PRESSURE
12 × 12	50,000 lbs.
12 × 16	67,000
14 × 16	78,000
16 × 18	100,000
18 × 24	150,000

10. Press the OPEN button, and hold until pressure has dropped, and the platens are fully opened.

11. Unload the press:

a. Use the Kraft paper to pull the books into the cooling press.
b. Books must be centered squarely, just as they were for the cure cycle.
c. Turn the cold press switch to the CLOSE position. Hold this switch until the hydraulic motor pump kicks off, and the cooling cycle timer kicks on.
d. At the conclusion of the cooling cycle, the cooling press will open automatically.

12. Unload the cooling press.

13. Remove top layer of Kraft paper. It can probably be reused, if it has not been damaged.

14. Place the books on a table for unpinning, with the bushing side facing upwards.

15. Use a hammer and punch to knock the pins out, letting them fall into a receptable. Use care, so that pins do not become jammed in the bushings.

16. Remove panels from the caul plates, which can now be reused for other books.

The panels will probably be stuck together when coming out of the caul plates. The resin flashing can be loosened, and the panels pulled apart; this should be done over a garbage can. A hand shear can be used to cut the flashing from each of the laminated panels. The release sheet is not usually reused, as it is contaminated with resin flashing.

It is good practice to punch numbers into the corners of each panel, to identify the job to which it belongs. This should be done before any further processing is carried out. An unidentified panel is virtually the same thing as scrap. After identifying the panels, the thickness should be measured with a micrometer, and logged into the lamination logbook. If the panels are being run under the provisions of MIL-P-55110C, a serial number should also be punched onto the panel.

The epoxy resin is not necessarily fully cured; in fact, it probably is not. It is good practice to bake all freshly laminated panels prior to drilling. A bake of 4–8 hours at 300 degrees Fahrenheit is generally sufficient. The time can be adjusted as the quality records from micro-sectioning indicate. If the prepreg shows a higher rate of attack during smear removal (after drilling), than does the core material, it would be wise to increase the baking time, not decrease. When baking, the panels should be placed in stacks of no more than one inch in height; unless they are separated by metal plates and spacers, for uniform and adequate heat distribution.

There are other conditions which indicate more baking is needed. If delamination is noted at any step in the manufacturing process, further processing of the panels must be stopped. All the panels of that lot must be baked immediately, at 300 degrees Fahrenheit for at least 4 hours. After this baking operation, one panel can be tested by running it through the process which resulted in the delamination. If no further delamination is noticed, chances are good that the rest of the panels can be processed with confidence. It may be that the panels have slipped through without the post-lamination bake to complete the cure; for this reason, a post-lamination logbook must be maintained at the oven, and a slot provided on the job work order traveler for the technician to sign off.

During the smear removal and/or etchback operation after drilling, microsections should be performed to panels after electroless copper. This is necessary to verify adequacy of the operation, as well as to check the copper coverage and drilling quality. If the prepreg shows a higher rate of attack by the chemicals, the need for increased post-lamination cure is indicated. It is good practice to stop further processing and rebake the materials. The rate of chemical attack on fully cured prepreg will not exceed that of the core material. Continued processing will result in panels which will almost certainly fail thermal stress testing, as required in Group A inspection per MIL-P-55110C (10 second solder float at 550 degrees Fahrenheit).

Miltilayer panels should be micro-sectioned at several steps in the manufacturing process. Military circuits are run with the test coupons of MIL-STD-275D. Although these are required coupons, it is good practice to run additional coupons. The other coupons should be run on all commercial boards, and can be run on military boards as well. These coupons need only have plated through holes with inner layer connections. They are run strictly for the quality evaluations of the printed circuit shop. Critical steps for micro-section are given below:

1. After smear removal/etchback. This is usually done in conjunction with electroless copper, since it is critical to micro-section after electroless copper. There is little point in trying to micro-section after smear removal/etchback

without electroless copper to outline the hole wall and define the inner layer/ hole wall interface.

2. During pattern plating. At this stage it is important to know the thickness of metal being plated, tin-lead as well as copper. Miltilayer printed circuit panels have a lot more money tied up in them; good practice is to micro-section during the plating process to protect that investment.

3. After etching and tin-lead reflow. Military specifications require micro-sectioning. Thickness of reflowed solder, solderability, degree of etchback and verification of smear removal, thermal stress evaluation, annular ring, Z-axis expansion, integrity of barrel copper, layer-to-layer registration, and other items can only be checked through micro-sectioning. If the micro-section samples are failing thermal stress testing, it would be wise to rebake all the panels. If panels routinely fail thermal stress testing, it would be good practice to begin baking multilayer panels after drilling, before smear removal. Other steps in the chemical processing should also be evaluated. It should be verified that the thin clad laminate is being baked sufficiently after being sheared into panel form.

Other Information

Black oxide is not the only treatment available for improving the bond between the prepreg and the copper during lamination. There are similar treatments which result in a brown coating, and there is a red oxide treatment available. There is also a zinc treatment available, which is similar to the treatment performed to copper foil prior to manufacturing laminate. The red oxide and zinc treatments are reportedly superior for bonding of polyimide multilayers.

It is possible to purchase thin clad laminate which has been treated on the outer surfaces of the copper foil with the same treatment used on the surface already bonded. The advantage of using this already treated foil is that black oxide can be avoided altogether. The laminate can even be purchased cut to panel size. The only preparation steps needed before dry film imaging are the incoming bake, and vapor degreasing or alkaline cleaning. After imaging, the resist is stripped and the layers rinsed and baked. After inspection, the layers are then ready for lamination.

The use of foil for outer layers in the manufacturing operations will save money, and reduce the number of layers being processed. Operating in this mode will significantly reduce the throughput time, and reduce defects and losses from handling.

FR-4 epoxy/fiberglass is not the only material laminated into multilayer circuits. Polyimide has applications in chip carrying devices, and in circuits operating at elevated temperatures. Also, it is reported to be superior to other materials for little Z-axis expansion. Polyimide processing is straightforward, but requires great attention to detail. The laminating temperature, pressure, and cure time are greater than required for FR-4. Polyimide is a harder material and requires that fewer hits be made per drill bit. Should smear occur in the hole, plasma etching is required to remove it. Teflon is also used in multilayer construction. Processing of Teflon multilayers also requires greater attention to detail. Laminate manufacturers are generally quite willing to send technical representatives out to work with printed circuit manufacturers who are interested in laminating these materials.

22
Etchback and Smear Removal

Etchback is the etching back of laminate resin from inside the drilled hole. Generally, this is done to bare the copper land of inner layer terminal areas on multilayer printed circuit boards. When drilling printed circuit boards, the drill bit can reach several hundred degrees Fahrenheit. Whenever drill bit temperature exceeds the glass transition temperature of the laminate resin, there is every possibility of resin smear occurring. The drill bit melts the resin and smears it around the inside of the hole. If the smear covers the inner layer terminal area it will prevent an electrical connection from forming during the electroless copper plated-through hole process. Also, the resin smear can be quite smooth; thus preventing good mechanical adhesion from forming between the hole wall and the electroless copper. It is the function of the smear removal process to remove smear from the hole wall to allow inner layer electrical connections to be formed, and to allow formation of adequate copper/hole wall adhesion. It is possible to have smear removal without having the resin etched back; however, the terms etchback and smear removal are often used interchangeably. "Etchback" is desirable on multilayer printed circuits, and has no meaning when dealing with double sided printed circuits. "Smear removal," however, is desirable not only for multilayers, but also for double sided circuits. Smear can be a major cause of voiding on thick double sided panels, especially panels which are 0.093–0.125 inch thick.

There is another aspect of etchback which is not a concern for smear removal. After resin has been etched back (MIL-P-55110C requires 0.00002–0.003 inch of etchback) the glass fibers remain exposed. It is good etchback practice to dissolve the bare fibers. This is performed using hydrofluoric acid. Hydrofluoric acid has fallen into some disuse because of its extreme potential hazard to operators. As a replacement, ammonium bifluoride is used. Ammonium bifluoride will dissolve the glass fibers to an appreciable extent. It will also etch the glass slightly so that good adhesion is achieved with the electroless copper.

TECHNIQUES FOR ETCHBACK AND SMEAR REMOVAL

There are three methods for performing etchback which are currently used: (1) chromic acid, (2) concentrated sulfuric acid, and (3) plasma. The chief resin which must be removed by etchback is epoxy. Concentrated sulfuric acid and chromic acid work well for epoxy removal. Each of these methods has advantages or disadvantages which make them desirable or undesirable based on the capabilities of the individual shop. Chromic acid is a replenishable, steady state operation with a fairly uniform rate of resin removal. Its chief disadvantage is the water pollution control problem presented by ionic chrome. Concentrated sulfuric acid is fairly cheap to set up, but the etch rate varies with the purity of the acid. Since concentrated sulfuric acid is extremely hygroscopic (it absorbs water from the air) the acid has a limited life. In recent years, plasma (charged gas) has become a viable method of performing smear removal and etchback. A plasma operation is fairly clean and does not contribute to water pollution, nor does it present the potential operator health hazards of chromic or sulfuric acid. In addition, acids will not remove smear from polyimide, which is being used increasingly in multilayer construction. Plasma must be used if smear occurs in polyimide printed circuits. The disadvantage of the plasma system is the initial cost of a plasma etchback set-up. Liquid chemical etchback systems are being developed at this time.

CHROMIC ACID

Chromic acid etchback chemistry lends itself well to continuous operation. The bath can be analyzed by simple titration (instructions are supplied by the bath manufacturer) and additions made to maintain activity. Unlike concentrated sulfuric acid, which is a solvent for epoxy, the bath does not become heavily loaded with resin. The epoxy is oxidized to carbon dioxide and water by the Cr^{6+} chromium ion. The only limiting factor on bath life is the build up of Cr^{3+}, which takes weeks to months before seriously affecting bath potency. Chromic acid is operated in conjunction with either hydrofluoric acid or ammonium bifluoride, and with a chrome reducer. The reducer functions to neutralize the very active Cr^{6+} chromium ion to the relatively inactive Cr^{3+}. Should the Cr^{6+} not be neutralized it will contaminate the other baths down stream and reduce their ability to condition the plated through hole for electroless copper deposition.

Operation

Before panels are processed through chromic acid some preliminary steps must be taken. The panels should be run through a deburring/sanding machine, just as if they were double sided panels coming from drilling. Next, the black oxide should be stripped from the outer layers. This will keep the etchback baths from becoming heavily loaded with copper. The black oxide can be removed by immersion in a 50:50 mixture of concentrated hydrochloric acid and water. The panels should then be baked to dryness, as some manufacturer's of chromic acid etchback chemistry recommend that there be no moisture in the drilled holes prior to immersion in chromic acid. When the panels have been deburred, stripped of black oxide and baked, they can be racked for etchback, followed immediately by electroless copper.

Before commencing etchback, the operators must be wearing safety glasses or goggles, rubber apron, boots, and gloves. Splashing of any of these baths must be avoided: all are extremely hazardous. Should any of these solutions come in contact with the skin, the skin must be rinsed immediately and neutralized with a mild basic solution such as sodium bicarbonate or sodium carbonate monohydrate (soda ash).

The basic processing steps are given below:

1. Chromic acid. The panels must be agitated through the holes while immersed. Typical immersion times are 1–4 minutes, at 140–150 degrees Fahrenheit. Panels should be drained well over this tank at conclusion of immersion. Good draining will help reduce the burden on the water pollution control system.

2. Dragout. This tank is important in keeping chromium from entering the waste water effluent system. The panels must be agitated for 30–60 seconds to force water through the holes. Solution from the dragout tank can be used to maintain volume of the chromic acid bath, since heated baths need to have the water replenished. Drain the panels well over the dragout tank.

3. Water rinse. After dragout, the panels are immersed and agitated in rinse water. Either a single or double cascading configuration may be used. Drain well when leaving the rinse tank(s).

4. Chrome reducer. This bath reduces residual chrome from Cr^{6+} to Cr^{3+}, as stated previously. The panels should be allowed several minutes of immersion time, with agitation. This tank must be analyzed by titration at least daily, and should not be allowed to become depleted. The effectiveness of this operation will greatly affect electroless copper deposition. Drain well.

5. Immersion rinse. Either single or double cascading tanks. Agitate and drain well.

6. Hydrofluoric acid/ammonium bifluoride. This should be made up according to the instructions of the chromic acid bath manufacturer. Typical values are 1–1½ pounds/gallon of ammonium bifluoride, or 10–20% concentrated hydrofluoric acid in water. These baths are extremely hazardous. Immersion times of 2–5 minutes are adequate, and the panels must be agitated through the holes. Drain well.

7. Immersion rinse, with agitation, for several minutes. Drain well.

8. Electroless copper line. The panels are now ready to proceed through the electroless copper process. They may go immediately from the rinse of Step No. 7 above, into the cleaner/conditioner of the electroless copper line. Gen-

erally, the best copper deposit is obtained if the panels are not dried or reracked prior to electroless copper deposition.

SULFURIC ACID

Panels should be deburred before etchback, since they will continue on to electroless copper. The removal of black oxide prior to etchback is optional. The panels must be baked to dryness before immersion in the concentrated sulfuric acid. This system operates on a totally different chemistry than that of chromic acid, and the requirements must be understood and never be forgotten by operating personnel. The epoxy is dissolved by the un-ionized concentrated sulfuric acid, with the generation of epoxy residues. The rate of dissolution is dependent upon the purity of the acid. The acid must be analyzed daily before use, and the process engineer must devise a chart for immersion time versus acid purity. The immersion times must also reflect the degree of etchback, as measured by micro-sectional analysis. Immersion times will vary from between 5 and 6 seconds up to 45–60 seconds. The epoxy residues hold the potential of contaminating the holes, if not rinsed quickly and thoroughly. Because of epoxy loading and moisture absorption, this bath has a short life when used in continuous operations.

While immersion time for chromic acid may be several minutes, immersion time for concentrate sulfuric acid is 5–60 seconds. The attack by sulfuric acid is very severe. There is little room for error in immersion time. Should the prepreg be insufficiently cured, it will be etched back to a far greater extent than will the core material. When immersion time is up, the panels must be swiftly removed and immersed in water for rinsing: the panels must not be allowed to drain over the acid tank. The residue laden acid must be rinsed free of the holes before the acid can absorb much moisture from the air. Any delay in rinsing can result in deposition of the epoxy inside the holes as a contaminant.

The tanks which contain the concentrated acid and the hydrofluoric acid should be kept covered whenever the line is not in use. When the line is to be run, these lids should be removed—and care taken to avoid accidental splashing of water into the concentrated sulfuric acid, as an explosion of acid will result. All protective safety clothes must be worn.

The basic processing steps are given below:

1. Rack panels for etchback and electroless copper. The panels must be agitated through the holes. However, since concentrated sulfuric acid is so viscous, good agitation is difficult to achieve. The operator must be aware of the bath purity, and the required immersion time for that degree of purity. It is helpful to have a chart posted in the work area for ready reference. At the concluson

of immersion time, lift panels swiftly and smoothly (to avoid splashing acid), and immerse in the water rinse tank.

2. Rinsing. The panels must be vigorously agitated through the holes as soon as they enter the water. The intent is to blow the epoxy laden acid from the holes before the residue sets up. Drain well.

3. Neutralizer/conditioner. This bath contains caustic chemicals, solvents, and surfactants. The purpose is to remove epoxy char and debris from the etch-back holes, and to neutralize the acid. The net result should be improved bonding of electroless copper to the hole wall. The potency of this bath should be monitored daily by simple acid titration. Additions can be made to the extent recommended by the bath manufacturer. Vigorous agitation through the holes is required for several minutes. The bath is generally operated around 125 degrees Fahrenheit. Drain well.

4. Rinsing. Agitate through the holes, drain well.

5. Hydrofluoric acid/ammonium bifluoride. Use the same make-up and immersion instructions recommended under chromic acid. Drain well.

6. Rinse. Immersion in a cascade rinse.

7. Electroless copper cleaner/conditioner. Panels should be immersed in this bath to neutralize residual acid in the recessed areas of the hole wall. This is typically a hot alkaline bath. There is nothing to prevent processing the panels immediately down the rest of the electroless copper line. After the cleaner/conditioner step, the hole wall is as receptive as it will ever be for copper deposition. However, the concentrated sulfuric acid does become heavily laden with epoxy, and there is a potential for epoxy residue to be redeposited inside the hole. Because of this potential, it is good practice to stop chemical processing with Step No. 7. The panels should be scrubbed and baked for an hour or so at 250–300 degrees. This will dry the panels and remove entrapped solvents and moisture. The holes can now be inspected by random sampling for adequate smear removal without redeposition of epoxy. Should residue or charred debris be found, the panels must be re-processed with a brief immersion in fresh concentrated sulfuric acid.

PLASMA

This is by far the cleanest, least producing of hazardous waste, and safest to the operator of all available etchback/smear removal techniques.

What is plasma? Plasma is a highly energized mixture of electrons and charged molecules (ions). A slight oversimplification is to call plasma "ionized gas." Under very low pressure one or more gases, such as oxygen or oxygen and carbon tetrafluoride, are passed through a powerful and concentrated electromagnetic field. This mixture of electrons and ionized molecules is extremely reactive, especially toward organic materials, such as resin and resin smear. As this reactive mixture passes through the drilled holes, the resin is etched back.

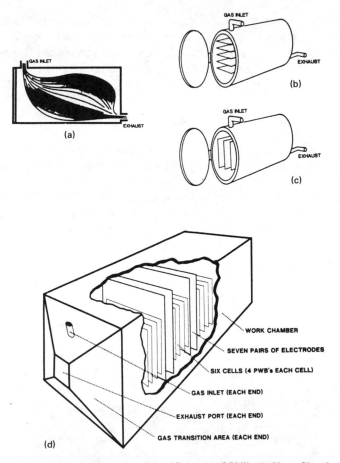

Fig. 22-1. Layout of plasma chambers. (Courtesy of Philip A. Hunt Chemical Co.)

The resin breakdown products and spent gas, which is laden with the breakdown products, are exhausted to the atmosphere. The gases are in the plasma state only while they are between the electrode plates. Fig. 22-1 shows the layout of panels in several types of plasma chambers.

Panels can be processed for etchback almost immediately after being removed from the drilling equipment. It is always advisable to bake panels prior to etchback in plasma equipment (1–2 hours at 250–300 degrees Fahrenheit). This assures adequate cure of the prepreg and removal of all moisture. The presence of moisture or contaminating solvents and oils can prevent resin removal in a controlled manner. The rate of etchback is determined by (1) etching time; (2) power level; (3) number of panels in the chamber, and their thickness; and (4) gas mixture.

A further advantage of plasma etchback is that it is currently the only reliable method of performing etchback and smear removal on polyimide resin. This feature will (as polyimide multilayers become more widely used) undoubtedly foster advances in plasma etchback technology, reduce the future cost of plasma equipment, and result in more widespread use of this technique. The chief disadvantage of plasma is the cost; some units cost more than $140,000.

OTHER ETCHBACK CONSIDERATIONS

1. After panels have been through the etchback/smear removal process, it may be desirable to inspect the adequacy of the smear removal prior to processing through electroless copper. This can be done by inspecting for shiny copper rings inside the holes. If the rings (annular rings of the inner layer terminal areas) look dull, this indicates inadequate smear removal. The inner rings should be bright and shiny all around the circumference of the hole.

2. After etchback and electroless copper it is good practice to have one or more of the panels plated in electrolytic copper. About 0.0001–0.0002 inch of copper should be plated. These panel plated boards should be microsectioned at the test coupons. The microsection is the only reliable method for checking drilling quality, cure of prepreg, adequacy of smear removal and etchback, and coverage of electroless copper. Figs. 22-2A through 22-2E represent cases of:

A. Poor drilling quality, as represented by loose bundles of glass fibers.
B. Insufficient cure of prepreg, as represented by adequate etchback on core

Fig. 22-2(a). Loose fiber bundles, caused by poor drilling.

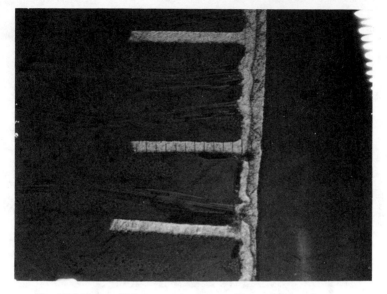

Fig. 22-2(b). Insufficient prepreg cure causing cracked barrel copper.

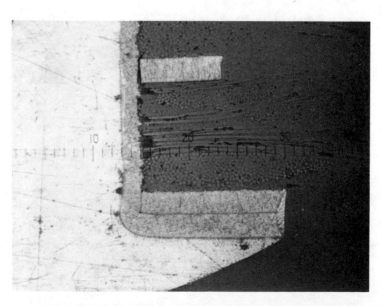

Fig. 22-2(c). Epoxy smear preventing inner layer connection.

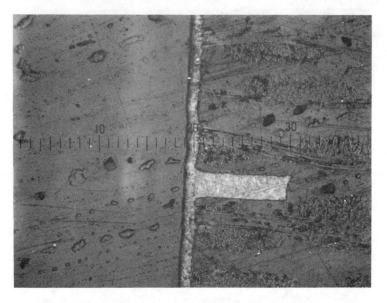

Fig. 22-2(d). No inner layer smear, good connection.

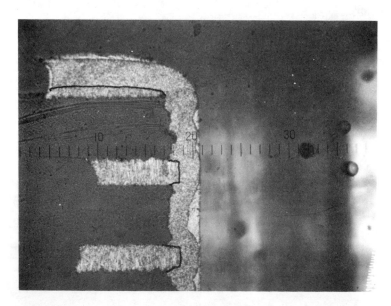

Fig. 22-2(e).

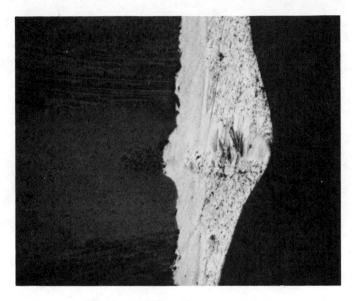

Fig. 22-2(f). Protruding glass fiber.

material, and excessive etchback on prepreg. Insufficient cure of prepreg can be distinguished from poor drill quality (rough holes) and over etchback, by uneven resin attack on opposite sides of the copper: one side is prepreg, the other core material.

C. Epoxy resin smear at an inner layer, preventing an electrical connection from forming.
D. Virtually no etchback, even though there is no evidence of resin smear.
E. Good etchback, no resin smear.
F. Protruding glass fibers.

SECTION SEVEN
PROCESS CONTROL

23
The Laboratory

The in-house laboratory is an indispensable tool for any modern printed circuit facility. The support which can be offered to virtually all of the manufacturing operations is too great for a well run shop to do without. The process of manufacturing printed circuits is one of processing chemicals and materials: chemistry of electro and electroless metal deposition, metal etching, photochemistry and photolithography, drilling and machining, as well as other aspects of plastics fabrication. The laboratory can be used for monitoring and maintaining many of the most critical processes, including plating and all of the other wet processes associated with that area, multilayer fabrication, imaging, drilling, quality control, pollution control, and even fabrication. The laboratory is just as indispensable in providing support for customer relations in evaluating failures which may or may not be due to the quality of the printed circuit fabrication.

The basic functional areas of the laboratory are those of chemical analysis and material analysis. The chemicals and materials which require evaluation are the incoming chemicals and materials, the chemicals and materials being used to perform work on the printed circuits, the chemicals and materials out of which the printed circuit is being manufactured, as well as the final product of all the chemical and material processing: the printed circuit.

USES OF THE LABORATORY

Plating and Wet Processing

Plating baths (and the other wet processing systems) are sensitive to changes in chemical concentration, temperature, content of addition agents, and contamination from organic and metallic sources. Changes in all of these occur continuously, and all have an impact on the visual and metallurgical quality of the plating. The thickness of the plating, degrees of coverage, and wetting of solder, which are indicative of process performance, cannot be fully evaluated without a laboratory for performing micro-sectional analysis. The degree of understanding and control of these considerations relates directly to the ability of the manufacturer to produce a quality printed circuit.

Electroplating. Electroplating baths require periodic analysis of their metal content, acid content, level of addition agents, pH, as well as determination of the balance of anions (like sulfate and chloride for nickel plating) and cations for alloy plating (like tin-lead and tin-nickel). Some of the most common printed circuit plating baths, and the important species and other considerations which must be kept under control, are given below:

1. Acid copper sulfate:

   - Copper metal.
   - Sulfuric acid.
   - Chloride.
   - Brightener.
   - Temperature.
   - Organic contamination.

2. Pyro copper:

   - Copper metal.
   - Tetrapotassium pyrophosphate.
   - Ammonia.
   - Brightener.
   - pH.
   - Temperature.

3. Tin-lead:

   - Tin metal.
   - Lead metal.
   - Tin:lead ratio.
   - Fluoboric acid.
   - Peptone.
   - Boric acid must also be present, but is not usually analyzed for.
   - Copper, as a contaminant.
   - Organic contamination.

4. Nickel (Watts or sulfamate baths):

   - Nickel metal.
   - Chloride.
   - Sulfate (Watts nickel).
   - Boric acid.
   - Brightener.
   - Anti-pitting agent.
   - pH.

- Temperature.
- Metallic contamination.
- Organic contamination.

5. Tin-nickel:

- Tin metal (stannous tin).
- Nickel metal.
- Tin:nickel ratio.
- Fluorine.
- pH.
- Temperature.
- Metallic contamination.
- Organic contamination.

6. Bright acid tin:

- Tin metal.
- Fluoboric acid.
- Peptone.
- Boric acid, must be present, but is usually not analyzed for.
- Brightener.
- Brightener carrier agent.
- Temperature.
- Metallic contamination.
- Organic contamination.

7. Gold (cyanide or chloride baths):

- Gold metal.
- pH.
- Temperature.
- Organic contamination.

The chemical content does not tell the whole story of each bath. There are considerations such as covering power and land-to-hole throwing power, current density range of brightness, brittleness, voids, noduling, and other defects of the plated metal which must be evaluated.

Electroless Plating.   Electroless, or immersion, plating baths typically require analysis and addition at least once a day; several times a day is considered standard for electroless copper. These baths also may require the presence of an entire cleaning, conditioning, and catalyzing line, with tanks requiring continuous analysis and addition for acceptable performance. Some of the more common baths are listed below, with their associated considerations:

1. Electroless copper:

   - Copper metal.
   - Sodium hydroxide.
   - Formaldehyde.
   - pH.
   - Temperature.

The appearance of this bath is important, but not for esthetic reasons alone. The appearance gives some indication of how well the bath is covering inside the holes; this is important, since the sole reason for this bath is to deposit copper inside the drilled holes of the printed circuit board (and occasionally on the edges of edge plated printed circuits). Chemical analysis alone is not sufficient to determine the quality of the plated metal.

In addition to the copper bath, there are cleaners, catalysts, conditioners, pre-catalysts, and etchants which require just as much attention to analysis as the copper:

Cleaner/conditioner:

- Caustic content.
- Copper content.
- Temperature.
- pH.

Micro-etchant:

- Sulfuric Acid
- Hydrogen Peroxide
- Temperature
- Copper Content

These are the important parameters for sulfuric acid/hydrogen peroxide etchant. Ammonium persulfate or sodium persulfate are much less important as a micro-etchant for electroless copper.

Pre-catalyst:

- Chloride normality.
- Acid normality.
- Specific gravity.

This is an important bath from the standpoint of helping to maintain the chloride and acid content of the palladium catalyst, which is an important and extremely expensive bath.

Palladium catalyst:

- Palladium content.
- Tin content.
- Chloride content.
- Temperature.
- Specific Gravity
- Acid Normality

This bath must be kept within proper operating range; failure to do so will result in voided holes on the finished printed circuit. If the acid or chloride normality fall too low, the palladium will drop out of solution and the bath will be ruined.

Accelerator:

- Acid normality.
- Temperature.
- Copper Content

2. Immersion tin: This bath may be used as part of the preparation for tin-lead fusing, as a touch up procedure to cover bare copper, or as an etch resist after copper plating. Tin concentration, pH, and temperature are critical for situations where the bath is being used as an etch resist; however, this is becoming one of the more important uses of the bath:

- Tin metal.
- pH.
- Temperature.
- Copper is detrimental, but is usually not analyzed for.

3. Electroless nickel. This bath requires a cleaning and catalyzing line. However, it is not that often used to manufacture printed circuits. When this bath is used, the cleaning line for electroless copper may be used:

- Nickel metal.
- Acid.
- pH.
- Temperature.

Smear Removal and Etchback.   These are processes which are carried out in the plating area, usually as part of running multilayers or, for smear removal, a thick double sided board, through electroless copper. During drilling, the hot drill bit may melt the epoxy of the laminate and smear it around the inside of drilled holes. The epoxy smear inside a hole will cause an adhesion problem for electroless copper on thick (0.093–0.250 inch) double sided material. For multilayer panels, the smear will prevent an electrical connection from forming to the inner layer circuitry. Concentrated sulfuric acid (or chromic acid) removes this smear. The same acid will etch back epoxy from around the annular ring of the inner layer copper. Etchback improves the reliability of the inner layer electrical connection and helps prevent barrel cracking of the plated through hole copper. However, the process can create problems if potency and purity of the chemicals on this line are not maintained:

1. Concentrated sulfuric acid: Percentage sulfuric acid.
2. Neutralizer/Conditioner:

  ● Percentage caustic.
  ● Temperature.
  ● pH.

3. Ammonium bifluoride/hydrofluoric acid.

Copper Etchant.   Alkaline ammoniacal etchants are probably the most widely used today. They operate fairly trouble free, and addition and bath control can be maintained by specific gravity activated sensors. The most important parameters which need to be tracked are:

1. Temperature.
2. pH.
3. Baume (specific gravity).

Sulfuric Acid/Hydrogen Peroxide Etchants.   These are becoming more widely used, primarily because of water pollution control regulations. These etchants demand a little more care than do ammoniacal etchants. The temperature tends to increase, and, as with electroless copper, periodic analysis and chemical additions must be made to maintain the bath strength. There is no easily monitored parameter which can be automatically monitored and used for bath control. Since the copper is removed continuously, there is no real need to track it as part of the daily routine. Important parameters which need to be tracked are:

1. Hydrogen peroxide.
2. Sulfuric acid.
3. Temperature.

Another factor which should be monitored in the etching process—indeed, it is required for printed circuits produced per MIL-P-55110C—is the undercut of the circuitry. The best method for monitoring undercut is to microsection the circuitry.

Black Oxide.   This is little more than a tank of very hot aqueous chemicals. However, the rate at which the black oxide is applied to the copper is dependent upon the temperature, and the concentration of active chemicals. It is, perhaps, going overboard to analyze for the components in this bath. As long as the temperature is checked to verify that it is up to the operating temperature, the only other control needed is the immersion time required to obtain the desired level of oxide. If the immersion time extends passed the desired value, the technician need only scoop in replenisher until immersion time has reduced to an acceptable value.

Multilayer Processes

The smear removal, etchback, and electroless copper operations are critical for successful multilayer manufacturing. Chemical analysis and monitoring can be used to keep these in top performing condition, but micro-sectioning is absolutely required to inspect the quality and success of those operations, and to help set immersion time parameters.

There are several other aspects of multilayer quality which can only be monitored by performing and evaluating micro-sections of the holes:

1. Layer registration.
2. Copper thickness of inner and outer layers.
3. Dielectric thickness.
4. Laminate voids.
5. Z-axis expansion.
6. Resin recession.
7. Fiber protrusion.
8. Barrel cracking of the copper, and other defects in the through plated hole.
9. Delamination and failure analysis.

There are times when it is necessary to evaluate the prepreg characteristics. This may be desired any time a lamination problem arises, or when prepreg has been stored for a long period of time or stored under less than desirable circumstances. Incoming inspection may also require the characterization of the prepreg. There are times when it is desirable to inspect the weave of the glass in the prepreg as well as the thin clad laminate; and the glass ply construction used by the laminate manufacturer. Some of the more important characteristics commonly evaluated in prepreg are:

1. Percentage volatiles.
2. Percentage resin.
3. Gel time.
4. Percentage flow.

Imaging Quality

The quality of the imaging process, especially dry film photoresist, can be more fully understood with the aid of a micro-section:

1. Photoresist foot.
2. Mushrooming (resist overplating).
3. Resist scumming of sidewalls.
4. Ragged traces, which may or may not be due to over etching.

Drilling Quality

Poor drilling quality can cause several problems; yet it may be difficult to nail down the exact cause of a problem without the aid of micro-sectioning:

1. Noduling.
2. Fiber bundles.
3. Burrs or debris in holes.
4. Voids.
5. Cracked copper in the through plated hole.

When exotic materials are under consideration, such as polyimide, modified epoxies, FR-5, or teflon, it is desirable to fully evaluate the drill action inside the hole. The situation is the same for drilling multilayers, and such factors as stack height, drill feed and speed, and the number of hits per drill bit are dependent on the thickness of the boards, the number of layers, and the ounces of copper being drilled through. A lot of problems down the line from drilling and lamination (apparent Z-axis expansion, barrel cracking, cracking at the inner layer connection, smear, copper tear-outs, loose fibers, etc.) can be traced to the tendency of a drill bit to punch, rather than cut, its way through the board; the best way to determine the degree of punching is by evaluating micro-sections.

Quality Control

The quality of modern printed circuits requires product monitoring with micro-sections, and chemical monitoring of the process chemicals, baths, and equipment. No manufacturer can meet military requirements for MIL-P-55110C without a micro-sectioning facility; and no manufacturer can meet the requirements of MIL-I-45208 without making extensive use of a laboratory for chemical analysis of the plating baths.

Process Histories

One of the benefits of a laboratory is the ability to develop process histories. Thorough record keeping of test results allows the chemist or engineer to develop a feel for the level of performance of any of the processes that is not possible with only visual observation of process and product. For instance, a slow increase in the copper content of a given plating tank will have no sudden effect upon plating quality or parameters. It does, however, indicate the need to reduce anode area, just as the gradual drop of copper content—or continuous need for addition of copper sulfate—indicates the need to increase anode area. Uneven thickness of plated copper from one side of a board to the other (for an even plating pattern) can indicate an uneven anode distribution, or the need to replenish anodes, or it can mean an uneven distribution of amperage through the bus bars, possibly due to poor electrical connections. This type of observation would be made with micro-sectioning. The exact cause of a distribution problem may be due to any number of factors. However, the problem would likely go unnoticed without routine evaluation of microsections.

Using a Laboratory

There is a routine which will develop at printed circuit shops which use an in-house laboratory. That routine will vary depending on the staffing of the laboratory, and the load placed upon it by manufacturing. The type of process equipment, construction of plating tanks, size of plating tanks, hours of operation, and manual versus automated operation will have a major effect on the need to analyze the plating bath; both chemically and with a Hull cell. The land-to-hole throwing power, and the time required to plate a mil of copper will vary greatly from one copper plating bath to the next; it will even vary for the same bath depending upon the ambient temperature and the care with which the brightener is controlled. Generally, however, if the plating tank is of adequate size, the anode/cathode spacing is adequate (6–8 inches for plating with a D.C. rectifier), and the anode area adequate and maintained, there should be no need to analyze the chemistry of a plating bath more than once a week. Automated plating lines, and high speed plating lines may be an exception to this.

Certainly, if a shop is having difficulty maintaining the chemistry of any of its plating baths, it should analyze as often as necessary to establish good control and a degree of confidence. Electroless copper, electroless nickel, electroless tin-lead, and hydrogen peroxide/sulfuric acid pattern etching lines will also demand frequent chemical analysis.

The Hull cell is needed to monitor the level of organic additives in all plating baths, as well as the build-up of metallic and organic contamination which will affect the plating. Again, for most manual plating applications, there should be no need for more than a weekly Hull cell determination of plating bath performance. As with chemical analysis, the need for more than a weekly Hull cell indicates the need for greater process control and/or operator training.

Routine use of the Hull cell will allow the chemist to determine whether or not a certain condition is normal for a given plating bath, or indicates the need for investigation. The Hull cell also permits the chemist to track the decline in brightener, or the creeping of dullness into the plating range, or the gradual build-up of ionic contamination in a nickel bath. By tracking these conditions, corrective action can be initiated before a plating problem impacts manufacturing.

Routine

Generally, the first thing the chemist does is to verify that all of the temperatures, pH's, and specific gravities of the plating lines are at optimum values, or at least within operating range per specification of the bath manufacturer. A good technique for tracking this information is to write up a daily control sheet (Fig. 23-1). Properly filling out a control sheet will ensure that no important parameter is overlooked. Also, tracking the information for the control log affords the opportunity to verify that the plating baths are at their operating levels. If it is necessary to add water, doing so now will afford adquate mixing time before sampling the baths for chemical and Hull cell analysis. It should be noted that even though analyses may be carried out only once a week, the control sheet should be logged daily.

The third act of the day, if it is the day analyses are performed, will be to sample the plating baths. The chemist should use a long sampling tube, so that sample bottles can be filled with plating bath taken from a number of locations and depths within the plating tank. Each sample bottle must be identified with the tank whose sample it will hold: Copper #1, Copper #2, etc. The sample obtained can be used for chemical anaysis as well as for the Hull cell analysis.

It is perhaps best to perform the Hull cell analysis first, since this is representative of how the plating bath will be performing. The information from the Hull cell analysis can be used to recommend, for instance, an additional 500 ml of brightener be added to a copper tank, or that a tin-lead tank be dummy plated for copper removal prior to running production; this is the type of information a chemist should get to manufacturing as soon as possible. By observing the need for brightener in a copper plating bath on a Monday morning, the chemist might prevent burning or step plating on the first load of production run; and this is the point of having a laboratory in-house, better product quality with less down time.

After performing the Hull cell testing, the samples can be used to analyze

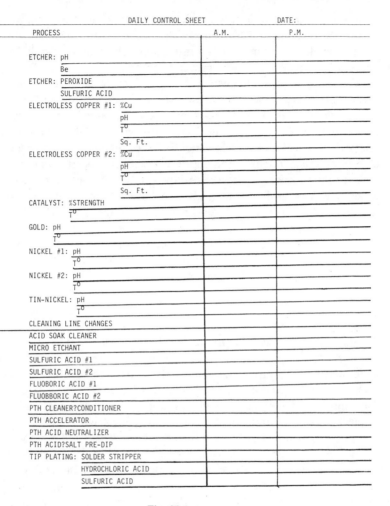

DAILY CONTROL SHEET		DATE:
PROCESS	A.M.	P.M.
ETCHER: pH		
Be		
ETCHER: PEROXIDE		
SULFURIC ACID		
ELECTROLESS COPPER #1: %Cu		
pH		
T°		
Sq. Ft.		
ELECTROLESS COPPER #2: %Cu		
pH		
T°		
Sq. Ft.		
CATALYST: %STRENGTH		
T°		
GOLD: pH		
T°		
NICKEL #1: pH		
T°		
NICKEL #2: pH		
T°		
TIN-NICKEL: pH		
T°		
CLEANING LINE CHANGES		
ACID SOAK CLEANER		
MICRO ETCHANT		
SULFURIC ACID #1		
SULFURIC ACID #2		
FLUOBORIC ACID #1		
FLUOBBORIC ACID #2		
PTH CLEANER?CONDITIONER		
PTH ACCELERATOR		
PTH ACID NEUTRALIZER		
PTH ACID?SALT PRE-DIP		
TIP PLATING: SOLDER STRIPPER		
HYDROCHLORIC ACID		
SULFURIC ACID		

Fig. 23-1.

for all other chemical concentrations. After analysis, a list of any required additions can be made and submitted to the plating supervisor to make at his/her earliest convenience. The chemist will normally track the amp-minute reading on the gold plating rectifier, and calculate the amp-minute/troy ounce constant, the rate at which gold is being consumed in plating. If the gold content is maintained at near optimum (through frequent additions of gold salts based on amp-minutes), pH and temperature maintained near optimum, then the amp-minute/troy ounce constant should be consistent from one week to the next. The constant will be in the neighborhood of 500–550 amp-minutes per troy ounce of gold. Any sudden drop in this constant should be investigated, as it indicates either a leak in the tank or a pump connection, or a possible theft. When there has been a decrease in the amp-minute constant (which exceeds the history developed by past determinations) the analysis should be performed again, and the reading on the rectifier rechecked.

With Hull cell and chemical analyses out of the way, the chemist can turn his/her attention to the performance of the smear removal/etchback, and electroless copper lines (The plating department should, of course, always be using copper and catalyst color standards). This will typically be a micro-sectional analysis of a panel or panels which have been run through these lines. The quality of smear removal, the depth of etchback, and the quality of the electroless copper coverage, and the drilling quality can be determined by electrolytically copper flashing a panel which has been processed, and then micro-sectioning a set of holes for viewing under a microscope. If inadequacies are noted for any of these processes, the chemist should spread the word as soon as possible. If all appears to be well, the chemist can move on to other tasks; such as waste water analysis, examination of incomming materials, failure analysis of unacceptable product, etc.

Troubleshooting

The laboratory can provide immediate answers to questions which arise when there is a plating problem, or other chemical process problem. In some cases, as for plating problems, a chemical analysis may indicate that all required constituents are at optimum values. When this happens, it is necessary to analyze for impurities. However, the quickest and most reliable method for checking impurities is to perform a Hull cell. All plating baths contain organic and metallic impurities; however, only the Hull cell can tell whether or not the impurities are present to the extent that they are causing a problem.

If impurities are suspected, and indeed present per the Hull cell, it is good to confirm that a recommended course of purification will cure the problem. The laboratory will be of use here also. If a metallic impurity is suspected, the plating bath can be dummy plated for several hours. A subsequent Hull cell panel will provide evidence of whether or not the metal has been removed sufficiently that it will not impact production quality.

If an organic impurity is indicated, it is good to perform a carbon treatment on a small sample of the plating bath, before all the time and expense of the carbon treatment is performed on the entire plating volume. A quart of plating bath can be fully treated with heat, hydrogen peroxide, and carbon. This purified sample can then be Hull cell tested, and retested after addition of additives.

The laboratory testing is a good technique to prevent wasting of time and money in pursuit of a dead end solution to a problem. If the course of action tested yields a positive result, the chemist can have confidence that the action he/she recommends will not be in vain. Many printed circuit and other plating shops needlessly dump plating baths or follow some other potentially fruitless course of action without having all the facts at their disposal, facts which only a laboratory can supply, and which would be expensive (in time and money) to obtain from an outside laboratory.

ANALYTICAL PROCEDURES

The following are analysis procedures together with equipment and chemical lists for those procedures. These have been chosen for their simplicity and accuracy; and are typical of analyses found in the data sheets for many plating baths.

Acid Copper Sulfate

Equipment Required:

1. 5 ml pipet.
2. 25 ml pipet.
3. 250 Erlenmeyer flask.
4. 30 ml dropper bottles (2) with droppers.
5. Magnetic stirrer (optional).
6. 50 ml self zeroing buret.
7. Buret stand.

Chemicals Required:

1. 28% ammonium hydroxide, reagent grade.
2. 20% nitric acid, reagent grade.
3. 1N Sodium Hydroxide.
4. 0.1N EDTA, disodium salt.
5. 0.1N silver nitrate.
6. P.A.N. indicator (0.1 gram P.A.N. in 100 ml methanol).
7. Methyl orange Indicator (0.1 gram methyl orange in 100 ml water).
8. 0.01N mercuric nitrate. Carefully measure 1.083 grams of mercuric oxide (HgO). Mix 10 ml of concentrated nitric acid with 10 ml of water. Add all of 1.083 grams of mercuric oxide to the diluted nitric acid, and stir until all the red mercuric oxide has dissolved. Dilute this to 1000 ml with water.

Procedures:

Copper

1. Place 100 ml of deionized water into a 250 ml Erlenmeyer flask.
2. Pipet a 5 ml aliquot of the plating bath to the 100 ml of water.

3. Add enough 28% ammonium hydroxide to just turn solution deep blue, this will be about 2–5 ml.
4. Add 10 drops P.A.N. indicator; stir well.
5. Add 0.1N EDTA, with stirring, until solution just turns light green.

Calculation:

$$(\text{ml EDTA}) \times 0.1696 = \text{Copper metal ounces/gal}$$

Sulfuric Acid

1. Place 100 ml of deionized water into a 250 ml Erlenmeyer flask.
2. Add 5 drops of methyl orange indicator.
3. Pipet 5 ml aliquot of plating bath into the flask.
4. Add 1N sodium hydroxide, with stirring, until solution just turns light yellow.

Calculation:

$$(\text{ml sodium hydroxide}) \times 1.31 = \text{sulfuric acid ounces/gal}$$

$$(\text{ml sodium hydroxide}) \times 0.574 = \% \text{ sulfuric acid.}$$

Chloride:

1. Place 50 ml of 20% nitric acid in 250 ml. Erlenmeyer flask.
2. Pipet 25 ml aliquot of plating bath into flask.
3. Add 3 drops of 0.1N silver nitrate, stir well.
4. Immediately add 0.01N mercuric nitrate until solution turns completely clear.

Calculation:

$$(\text{ml. mercuric nitrate}) \times 14.2 = \text{chloride ppm.}$$

The silver chloride will turn the solution cloudy; during titration with mercuric nitrate, there are two stages of cloudiness. The solution will turn almost clear, light cloudiness; once this has been achieved, an additional amount of mercuric nitrate will have to be added until the solution is thoroughly clear.

Pyro Copper

Chemicals Required:

1. 28% ammonium hydroxide.
2. Sulfamic acid.
3. 50% sulfuric acid.
4. Potassium hydroxide beads.
5. 15% potassium iodate.
6. 12.5% zinc sulfate.
7. 0.1N sodium hydroxide.
8. 0.1N sulfuric acid.
9. 0.1N sodium thiosulfate.
10. 0.1N disodium EDTA.
11. 1% starch indicator.
12. Methyl red indicator (1 g methyl red in 50 ml water and 50 ml isopropanol).
13. pH 4 buffer.
14. Variac D.C. power source.

Equipment Required:

1. 250 ml Erlenmeyer flask
2. 250 ml, 400 ml beakers.
3. 50 ml buret.
4. 100 ml volumetric flask.
5. 75 mm glass funnel.
6. Magnetic stirrer.
7. 5 ml, 10 ml, 25 ml pipets.
8. Micro Kjeldal condenser, with heat source.
9. pH meter.

Procedures:

Copper:

1. Place 5 ml aliquot of plating bath into 250 ml Erlenmeyer flask.
2. Add 3 grams of sulfamic acid.
3. Add 20 ml of 50% sulfuric acid, under fume hood.
4. Heat 10 minutes.
5. Dilute to 100 ml with deionized water.
6. Heat for 5 minutes; cool.
7. Add 10 ml of 15% potassium iodide.

8. Add 3 ml of 1% starch.
9. Titrate with 0.1N sodium thiosulfate, until solution turns white.

Calculation:

$$\text{(ml thiosulfate)} \times 0.169 = \text{Cu oz/gal.}$$

Pyrophosphate:

1. Place 10 ml aliquot of plating bath in 100 ml volumetric flask; dilute to 100 ml.
2. Pipet 10 ml of solution from volumetric flask into 250 ml beaker.
3. Add 50 ml of water and a stir bar.
4. Adjust pH to 3.8 dropwise with diluted sulfuric acid.
5. Add 25 ml of 12.5% zinc sulfate.
6. Titrate with 0.1N sodium hydroxide to pH 3.8.

Calculation:

$$\text{(ml sodium hydroxide)} \times 1.17 + \text{(oz/gal Cu)} \times 1.37$$
$$= \text{pyrophosphate. oz/gal}$$

Ammonia

1. Pipet 25 ml of 0.1N sulfuric acid into 400 ml beaker.
2. Add about 125 ml deionized water.
3. Add 5 drops methyl red indicator (1g in 50 ml water + 50 ml isopropanol).
4. Place under micro-Kjeldal condenser, so that tip is about ½ inch below surface.
5. Pipet 10 ml of plating bath into Kjeldal, rinsing funnel with 5 ml deionized water.
6. Add 1 gram potassium hydroxide beads with deionized water flushing; quickly seal Kjeldal stopcock.
7. Apply moderate heat with D.C. variac; sufficient to distill about 20 ml in 7 minutes.
8. Lower 400 ml beaker until condenser tip is above solution level, rinse tip into beaker.
9. Turn off variac power source.
10. Titrate solution with 0.1N sodium hydroxide, from red to yellow.

Calculation:

$$[\text{(Normality of sulfuric acid)} \times 25 - \text{(ml sodium hydroxide)}$$
$$\times \text{(normality of sodium hydroxide)]} \times 0.23 = \text{ammonia oz/gal}$$

or

$$[2.5 - 0.1 \times (\text{ml sodium hydroxide})] \times 0.23 = \text{ammonia oz/gal.}$$

Note: Always record pH of plating bath at time of analysis.

Tin-Lead

Chemicals Required:

1. 20% sulfuric acid, reagent grade.
2. Concentrated hydrochloric acid, reagent grade.
3. 0.1N iodine.
4. 1N sodium hydroxide.
5. 0.1% methyl orange indicator, aqueous.
6. 1% starch indicator.
7. Sodium bicarbonate.

Equipment Required:

1. 5 ml pipet.
2. 250 ml beaker.
3. 500 ml vacuum flask, with rubber vacuum hose.
4. 50 ml Gouch crucible, low form, with vacuum flask adaptor.
5. 50 ml buret, self zeroing.
6. Buret stand.
7. Aspirator.
8. Magnetic stirrer/hot plate.
9. Dessicator.
10. Drying oven.
11. Analytical balance.

Procedures:

Tin:

1. Add 50 ml of concentrated hydrochloric acid to 100 ml of water in a 250 ml beaker.
2. Add 5 ml aliquot of plating bath.
3. Add 1 gram sodium bicarbonate.
4. Add 10 ml of 1% starch indicator.
5. Titrate immediately with 0.1N Iodine, to a dark endpoint (purple or black).

Calculation:

$$(ml\ Iodine) \times 0.16 = Tin\ metal,\ ounces\ per\ gallon.$$

Lead:

1. Place 100 ml of 20% sulfuric acid in a 250 ml beaker.
2. Add 5 ml of plating bath.
3. Heat for 10 minutes, cool for 60 minutes.
4. Filter through a dried and weighed Gooch crucible, medium porosity.
5. Rinse well with deionized water.
6. Oven dry 1 hour at 200 degrees Fahrenheit.
7. Let cool 30 minutes in a dessicator.
8. Weigh on an analytical balance.

Calculation:

$$(Wt.\ crucible + lead\ sulfate) - (Wt.\ crucible) = mg\ lead\ sulfate$$

$$(mg\ lead\ sulfate) \times 18.2 = Lead\ metal\ ounces\ per\ gallon$$

Fluoboric acid:

1. Place 100 ml deionized water in 250 ml Erlenmeyer flask.
2. Add 5 ml plating bath.
3. Add 3 drops methyl orange indicator.
4. Titrate with 1N sodium hydroxide to cloudy endpoint.

Calculation:

$$(ml\ sodium\ hydroxide) \times 2.2 = fluoboric\ acid\ ounces\ per\ gallon.$$

Nickel

Chemicals Required:

1. Ammonia buffer (1 molar ammonium chloride adjusted to pH 9.5 with 28% ammonium hydroxide).
2. Murexide indicator (1:1 Murexide and potassium chloride).
3. Sodium chromate indicator (1% aqueous).
4. Mixed indicator (0.2% bromothymol blue and 1% bromocresol purple in ethanol).

5. 0.1N disodium EDTA.
6. 0.1N silver nitrate.
7. 0.1N sodium hydroxide.
8. Calcium carbonate.
9. Mannitol.
10. pH 4 buffer.

Equipment Required:

1. Pipets: 1 ml, 2 ml, 5 ml.
2. 10 ml graduate cylinder.
3. 250 ml beaker.
4. 250 ml Erlenmeyer flask.
5. 50 ml buret.
6. Buret stand.
7. Magnetic stirrer.
8. pH meter.
9. Spatula.

Procedure:

Nickel:

1. Place 100 ml deionized water in 250 ml flask.
2. Add 10 ml ammonia buffer.
3. Add 1 ml plating bath.
4. Add about 0.4 gram Murexide indicator (solution should be greenish-blue).
5. Titrate with 0.1N disodium EDTA to a purple endpoint.

Calculation:

$$(ml\ EDTA) \times 0.78 = nickel\ metal\ ounce/gal.$$

Boric acid:

Note: It may be necessary to heat sample of plating bath to redissolve boric acid prior to pipetting aliquot for this test, it will tend to clog pipet.

1. Place 2 ml of plating bath in 250 ml beaker.
2. Add 5 drops of mixed indicator.
3. Add spatulas of mannitol powder as needed to make a paste.
4. Titrate with 0.1N sodium hydroxide to a faint blue endpoint.

Calculation

$$\text{(ml sodium hydroxide)} \times 0.41 = \text{boric acid oz/gal.}$$

Nickel chloride:

1. Place 100 ml deionized water in 250 flask.
2. Add 5 ml of plating bath.
3. Add 1 ml sodium chromate indicator; if bath pH is below 4.0, add 1 gram of calcium carbonate.
4. Titrate with 0.1N silver nitrate to first faint permanent endpoint.

Calculation:

$$\text{(ml silver nitrate)} \times 0.319 = \text{nickel chloride oz/gal.}$$

Gold

Chemicals Required:

1. Concentrated sulfuric acid.
2. Concentrated nitric acid.
3. 50% hydrogen peroxide.

Equipment Required:

1. 400 ml beaker.
2. 50 ml graduate cylinder.
3. Hot plate.
4. Rubber policeman.
5. 500 ml vacuum flask, with hose.
6. Aspirator.
7. Drying oven.
8. Analytical oven.
9. 50 ml Gooch crucible, medium with crucible adaptor.
10. Dessicator.

Procedures:

1. Pipet 10 ml aliquot of plating bath into 400 ml beaker, under a fume hood.
2. Add 20 ml concentrated sulfuric acid.
3. Add 10 ml concentrated nitric acid.

4. Heat until brown fumes appear; the brown fumes will give way to copious white fumes of sulfurous oxide, as the yellow precipitate turns to gold metal.

5. When all of the yellow precipitate has disappeared, and solution is very dark, cool solution.

6. Rinse walls of beaker with 5 ml portions of 50% hydrogen peroxide, heat till copious fumes appear. If gold foil appears on walls of beaker, cool and rinse again with hydrogen peroxide, and reheat.

7. When all gold has precipitated to bottom of beaker, cool and add 75 ml of deionized water.

8. Filter through a dried and weighed 50 ml Gooch crucible.

Calculation:

$$(grams\ gold) \times 12.17 = gold\ troy\ oz/gal.$$

Note: Always record amp-minute reading on rectifier at time of sampling.

Electroless Copper

Chemical Required:

1. pH 9.5 ammonia buffer (70 grams ammonium chloride in 750 water, buffered to pH 9.5 with ammonium hydroxide).
2. pH 10 buffer.
3. P.A.N. indicator.
4. 0.1N disodium EDTA.
5. 0.1N iodine.
6. 0.1N hydrochloric acid.
7. 1M sodium sulfite.
8. 1% starch indicator.
9. 28% ammonium hydroxide.
10. Concentrated hydrochloric acid.

Equipment Required:

1. 5 ml, 20 ml pipet.
2. 250 ml beaker.
3. 10 ml graduate cylinder.
4. 50 ml buret.
5. pH meter.
6. Magnetic stirrer.

Procedure:

Copper:

0. Pipet 20 ml sample into 250 ml beaker.
1. Place 100 ml water in 250 beaker.
2. Add 25 ml of pH 9.5 ammonia buffer (70 grams ammonium chloride in 750 ml deionized water, adjusted to pH 9.5 with 28% ammonium hydroxide).
3. Add 10 drops of P.A.N. indicator.
4. Titrate with 0.1N disodium EDTA to a pale green endpoint.

Calculation:

$$(\text{ml. EDTA}) \times 10.4 = \text{Cu \%}$$

Sodium hydroxide and formaldehyde:

1. Place 50 ml deionized water in 250 ml beaker.
2. Add 5 ml aliquot of plating bath.
3. Use a pH meter calibrated to pH10 with buffer titrate to pH 10.2 with 0.1N hydrochloric acid.
4. Record milliliters of 0.1N hydrochloric acid; this value is used to calculate sodium hydroxide.
5. Add more 0.1N hydrochloric acid until pH 9.5 is reached.
6. Add 100 ml of 1M sodium sulfite (126 grams sodium sulfite in 800 ml deionized water, buffered to pH 9.5, diluted to 1000 ml with deionized water.
Note: pH should increase when sodium sulfite is added.
7. Titrate back to pH 9.5 with 0.1N hydrochloric acid.
8. Record milliliters used, for formaldehyde calculation.

Calculation:

$$(\text{ml hydrochloric acid}) \times 0.8 = \text{sodium hydroxide grams/liter}$$

$$(\text{ml hydrochloric acid}) \times 0.6 = \text{formaldehyde grams/liter}$$

MICRO-SECTIONING

There are several reasons for performing micro-sections; these include the need to measure the thickness of plated metal, the need to examine the quality of a plated through hole, the need to measure the registration of a multilayer, and the need to perform failure analysis. The micro-section is performed by cutting a small section from a printed circuit, encapsulating this section in epoxy or polyester plastic, polishing one surface of the section, and viewing it under magnification. It may also be necessary to lightly etch the surface to be viewed, with either ferric chloride or an ammoniacal etchant, to resolve and heighten grain boundaries of the various metals.

If the section is being prepared to measure dielectric thickness or thickness of plated metal, the section may be punched from the printed circuit. If, however, the section is being prepared to investigate the integrity of a plated through hole, the section must be cut with great care so that nothing in the preparation causes defects to appear in the section. If Z-axis expansion and cracking of barrel copper are to be studied, obviously punching the section from the board would tend to crack the copper and render the results obtained totally unreliable.

There is a great variety of micro-section encapsulation equipment available, as well as a variety of microscopes available for viewing the section once it has been prepared. Some equipment will encapsulate a section under heat and pressure, the result being a cylindrical section with the flat surfaces almost perfectly parallel to each other. This type of encapsulation equipment is the most expensive. A cheaper method is to use a liquid plastic (polyester or epoxy) which is poured into a cup which contains the section; the section is supported by a mounting spring (Fig. 23-2). The bottom surface of the cup pulls apart for easy removal of the cast section. This type of encapsulation is very quick and inexpensive, and lends itself well to casting multiple sections.

Virtually any type of microscope can be used for viewing the section. It is common to find inverted stage microscopes in use, however, since only one side

Fig. 23-2.

of the section need be flat, and the surfaces need not be perfectly parallel for viewing. The microscope should be equipped with a Polaroid camera, eyepiece micrometer, stage with vernier movements, light source, and rotating objectives for changing magnification.

Once a section has been identified for viewing, it can be cut from the panel or board using fairly course means, as long as no force is exerted directly upon the area to be viewed, and no cutting or shearing occurs within about one inch of that area. The coarsely cut section must next be secured in a slow cutting diamond saw for cutting closer (within about ⅛ inch) to the viewing area. Once this has been accomplished the section is mounted in a support spring and encapsulated.

If holes are to be viewed, the section must be mounted for grinding and polishing so that the holes may be viewed with the proper orientation, usually a longitudinal view. It is possible to view a radial microsection of the holes (Fig. 23-3), but most analyses are of a longitudinal view (Fig. 23-4) since inner layer connections, inner layer registration, and land-to-hole throwing power can best be viewed; also, MIL-P-55110C specifies longitudinal sectioning. There is great information on drilling quality and other aspects of hole integrity to be observed with a radial microsection.

After the cast section has been removed from the encapsulating cup, it is first coarsely ground using No. 60 grit grinding disk on the grinder. This step should be carried out until the holes have been ground about half-way through. This should bare the hole walls longitudinally at the widest diameter of the

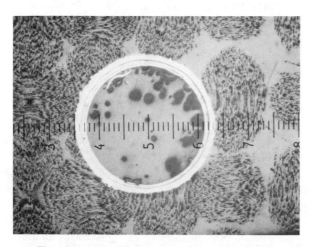

Fig. 23-3. Radial micro-section of plated through hole.

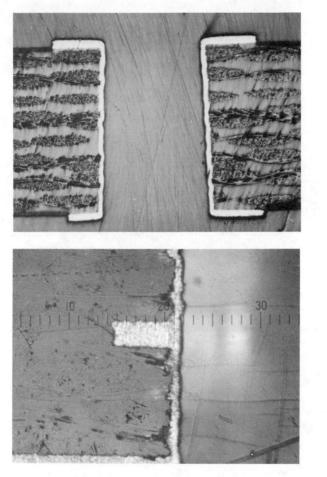

Fig. 23-4. (a) Longitudinal view of plated hole. (b) View of inner layer connection.

hole (Fig. 23-5). When using the grinder, it is good to view the holes and apply pressure so that the sides of the hole walls are parallel; nonparallel hole walls indicated that the grinding pressure is being applied unevenly.

The section is next sanded on strips of grit paper ranging sequentially from No. 240 to No. 600. When this has been completed, the viewing surface should be smooth, free of scratches, and dull. Another wheel is placed on the grinder; this wheel should contain polishing felt. This is used with an alumina slurry to polish the viewing surface to a high luster, free of all dullness, scratches, and voids. A cotton swab may now be dipped in etching solution and wiped over

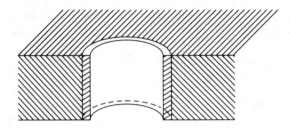

Fig. 23-5. Encapsulated section ground to center of holes.

the viewing surface to etch the metal. The section is now ready for viewing under the microscope.

MIL-P-55110C specifies that the sections shall be viewed for defects at 50 to 200 power, and that thickness measurements shall be taken at 200 to 400 power. This military specification also lists all pertinent criteria for determining acceptability of the section being viewed, and contains instructions for taking measurements for thickness of plated metal. Generally, at least four thickness readings are taken and averaged together, for each of the plated metals. Readings must be taken from both sides the the viewed hole wall (Fig. 23-6). A form should be drawn up for use in evaluating micro-sections (see chapter on quality assurance).

There are several manufacturers of microsectioning equipment. Two of the largest companies in this area are:

LECO CORPORATION
3000 Lakeview Avenue
St. Joseph, MI 49085
Tel: (616) 983-5531

BUEHLER LTD.
2120 Greenwood Street
Evanston, IL 60204
Tel: (312) 475-2500

SETTING UP YOUR OWN IN-HOUSE LABORATORY

When all is said and done there are three facts which tend to justify the in-house laboratory: (1) there are continuing savings over the weekly analyses performed by the outside laboratory service; (2) the laboratory is a strong sales tool; and (3) chemical and metallurgical analyses and information are available immediately, when they are needed. A medium sized printed circuit shop may spend $600–$800 per month for chemical analyses, and perhaps an equal

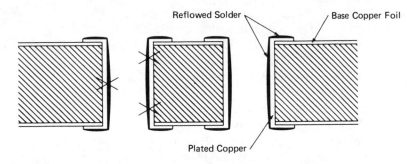

Reflowed Solder   Base Copper Foil

Plated Copper

Fig. 23-6. Minimum of three readings are averaged for thickness determination. Isolated thick and thin locations are ignored when taking readings. Use 200× to 400× magnification.

amount for Hull cell tests and micro-sections. This is a lot of money, enough to justify setting up a laboratory. A chemical analytical laboratory could be paid for in about 4–10 months, depending on the current outside lab bills.

The laboratory should be an integral part of any chemical processing business—and that is just what a printed circuit shop is. Since there are so many chemical processes involved, it is reasonable to expect that the customers and prospective customers of the printed circuit manufacturer would like to see an in-house laboratory. When the shop has a laboratory, it provides another opportunity for the shop to demonstrate its expertise in controlling its chemical processes.

Any complicated chemical processing business has certain processes which require careful evaluation by chemical means. When there is a need to examine quality of plated holes, and inner layer connections to decide whether or not to proceed with processing thousands of dollars worth of product, it is risky at best to do so without all the facts. The only reliable and timely means of gathering those facts is to have a chemical/metallurgical laboratory in house. Who can afford to wait several days for the test results which are needed now! It is not surprising that the more sophisticated buyers of printed circuits demand that their board vendors have this capability in house.

Requirements for a Laboratory

Setting up a laboratory requires that the following ingredients be assembled in one area:

1. A room which can be closed off from the rest of the plant and be equipped with:

a. Running water.
b. A sink that is acid resistant.

   c. A fume hood.

   d. Chemically resistant counters, shelves, and cabinets.

   e. Safety equipment such as acid shower, eye wash, and fire extinguisher.

2. Chemical analysis equipment and supplies:

   a. Balance, pH meter, glassware, and other labware.

   b. Standard reagents and indicators.

   c. An atomic absorption (A.A.) spectrophotometer is strictly optional.

   d. An oven; a small kiln would also be useful for characterizing prepregs.

3. Micro-sectioning equipment and supplies:

   a. Polisher/grinders and wheels.

   b. Hand sanding equipment.

   c. Diamond cutting saw.

   d. Plastic encapsulating supplies.

   e. A solder pot.

   f. A microscope, preferably with an inverted stage.

It also goes without saying that a technically competent person must be present to run the lab, or to train someone else to do so. If a particular printed circuit shop does not have such a person already on the payroll, one should be obtained; most college chemistry majors would be able to do the job with a little guidance from someone with greater experience. There is plenty of help available from suppliers of any of this equipment, and from technical sales representatives of vendors to the printed circuit industry. Most of the companies which supply the printed circuit industry have their own in-house research/sales support laboratories. These companies should be contacted; chances are they will be glad to offer assistance and training to the personnel of the printed circuit shop. Once the laboratory has been set up, even a part-time college student can be hired and trained to perform simple routine analyses. Platers are often very eager to learn the simple titrations needed to analyze most of their plating baths. Multilayer and quality assurance people are capable of learning micro-sectioning and the simple testing required for prepreg characterization. If a chemist or engineer is already on the staff, this type of person can administer the laboratory operation.

Once a technically qualified person is available the laboratory must be laid out and the laboratory furniture listed under No. 1 above purchased. Once a determination has been made for what chemical baths are to be analyzed a set of chemical analytical instructions must be obtained. It is generally best to use the analysis procedure found in the data sheet for the chemical bath. Almost all data sheets of a process contain the analytical information; those which do not, usually reference another document (or another data sheet) which does contain this information.

The M&T Chemical Company publishes a booklet entitled "Simple Methods for Analyzing Plating Solutions." This booklet contains well written instructions for many common analyses.

The most extensive book on analysis of plating baths is entitled *Analysis of Electroplating Baths* by Langford and Parker. This book contains excellent information on how EDTA (ethylenediaminetetraacetic acid) works, but many of the procedures given are more complicated than those which usually accompany data sheets. The book does contain excellent information on how to carry out analyses properly, and on chemical safety and other aspects of working in and running a laboratory. The book also contains chemical data on most compounds the plating chemist will have to work with. Unfortunately, it is published in Great Britain and difficult to obtain in the United States.

Modern Electroplating, edited by Lowenheim, has articles on electroplating, as well as the analytical procedures for the plating baths.

Handbook of Metal Finishing, published by *Metal Finishing* magazine also contains articles on plating together with chemical analyses. Like *Modern Electroplating,* most of the information is for the hardware plater, and there is a quantum jump between the needs of the printed circuit plater and the hardware plater.

Printed Circuit Handbook, edited by Donald Coombs, has analytical procedures together with a collection of articles on printed circuit plating, and other aspects of printed circuit processing.

When a set of analytical instructions has been decided upon, a list of chemicals, reagents and indicators must be drawn up; this list is derived from the analytical instructions. The required equipment, such as an analytical balance, pH meter, etc., is also derived from the instructions. Below is such a list, which is derived from the section on analytical procedures. Included in this list are some other items, such as a pipet stand, which are not part of a procedure, but are a necessary part of operating any laboratory. As a guide and reference, the suppliers for these supplies, their catalogue numbers and prices as of 1982 are also included.

Some chemicals, especially the indicators, are so expensive and used so sparingly that it would be wisest to contact any local laboratory (another printed circuit shop, or a college) and purchase a small quantity from their supply. There is no point in purchasing a 1 pound jar of phenolphthalein indicator for $73, when 0.1 gram in 100 ml of alcohol will last for years. Prices for chemicals and reagents vary greatly for laboratory quantities. It is wise to shop around. When purchasing normal or standard solutions, it is also wise to test the incoming lot for conformance to the standard on the label; manufacturers supplying chemicals at cut rate prices may not be supplying accurately standardized reagents.

ITEM	QUANTITY	PRICES		
		VWR	FISHER	AM. SCI. PRODS
1. Analytical balance (2432-AR)	1	$2250	$1890	$1800
2. pH Meter (Accumet Md 600)	1		525	
(Orion 301)	1	425		310
3. Hot plate/stirrer (Corning 10 × 10).	1	240		190
(Theramic 210)	1		185	
4. Auto zeroing buret 50 ml	1	72	63	73
5. Double buret stand	1	41	44	36
6. Dessication (aluminum 8″)	1	47	17	103
7. Drying oven (N8628-8)	1			260
(13-245-215G)	1		468	
(52348-310)	1	315		
8. Dessican (Aquasorb)	1 lb	47		33
(Drierite 8 Mesh)	1 lb		33	
9. Stir bar (1 × ⁵⁄₁₆)	1	1.10	2.1	1.35
10. Wash bottle (500 ml)	4 pK	7.66	5.55	
11. Aspirator	1	13	15	
12. Stir bar retriever	1	6	6	4
13. Gooch crucibles (50 ml Medium)	9/cs	98	104	76
14. Crucible tongs	1	8	9	3
15. Tweezers, 6″ stainless steel	1	7		4
16. Vacuum flask (1000 ml)	1	41	111	30
17. Gooch crucible holder	6 pk	18	18	18
18. Beakers (1000 ml)	6 pk	41	46	27
19. Beakers (250 ml)				
20. Beakers (400 ml)				
21. Erlenmeyer flask (250 ml)	12 pk	21	21	12
22. (1000 ml)	6 pk	22	13	13
23. Graduate cylinders (10 ml)				
24. (25 ml)				
25. (100 ml)				
26. Pipet (1 ml)	1	5	3	1
27. (2 ml)	1	5	3	1
28. (5 ml)	6 pk	29	46	7
29. (10 ml)	6 pk	33	40	9
30. (20 ml)	12 pk	48	61	26
31. (25 ml)	12 pk	51	66	28
32. (100 ml)	1	16	12	3
33. Disposal pipet	100 pk	17		3
34. Pipet support	1			61

Chemicals

ITEM	QUANTITY	VWR	FISHER	AM. SCI. PRODS
1. Murexide indicator	1 g	9	28	8
2. Sodium chromate	1 lb	38		18
	500 g		59	
3. Ferroin Indicator	4 oz.	14		6
	25 ml		5	
4. Iodine .1N	1 qt	8		5

ITEM	QUANTITY	VWR	FISHER	AM. SCI. PRODS
		PRICES		
5. Nitric Acid 69%	7 lb	19		11
	2.5 L		25	
6. Hydrochloric acid 0.1N	1 qt	12		10
	1 L		10	
7. Mercuric oxide	¼ lb	45		21
	100 g		37	
8. Methyl thymal blue	1 g	23	17	18
9. Ammonium chloride	1 lb	15		5
	500 g		11	
10. Bromophenyl blue	5 g	14	19	12
11. Sulfuric acid, concentrated	9 lb	28		13
	2.5 L		24	
12. Methyl orange indicator	1 oz	16		7
	25 g		8	
13. Ammonium hydroxide 28%	4 lb.	17		9
	2.5 L		19	
14. Sodium sulfite	1 lb	10		5
	500 g		9	
15. P.A.N. indicator	1 g	26		23
	5 g	78		
16. 35% hydrogen peroxide	—			
17. Disodium EDTA 1M	1 qt	16		8
18. Silver nitrate 0.1M	1 qt	100		87
19. Methyl red indicator	1 L	44		
20. Phenolphthalene	1 lb	73		

Laboratory Furniture

ITEM	VWR	FISHER	AM. SCI. PRODS
1. Fume hood, 28″ Labconco, with blower		1270	
2. Lab sink		98	
3. Goose neck faucet			134

Micro-Sectioning Equipment

ITEM	AM. SCI. PRODS
1. Sample-Klip supports (100)	$ 16.50
2. Sample-Kup molds	12.00
3. Cold mounting kit	31.00
4. Carbimet abrasive disks (8″)	84.00
5. Handimet strip grinder	415.00
6. Carbimet grit strips 240 grit	35.00
320 grit	35.00
400 grit	35.00
600 grit	35.00
7. Unitron inverted stage microscope with adjustable stage, eyepiece micrometer, and polaroid camera	5300.00
8. Economet III grinder/polisher	1395.00
9. Aluminum wheel	90.00
10. Mounting table	300.00
11. Slow cutting diamond saw	1800.00

All of the above will result in a fairly complete laboratory. There are added costs due to installation and the purchase of some items like cabinets and counters. Generally, a great deal of money can be saved by purchasing office or shop cabinets and counters, rather than laboratory cabinets and counters. It should be kept in mind that the prices given are as of mid-1982, and that prices tend to increase rapidly (25–100% every two years) on laboratory equipment and chemicals. No matter how fast they increase, however, the in-house laboratory will always remain a bargain.

DOCUMENTATION

It is not enough merely to perform an analysis and know that the bath is within tolerance, or in need of a chemical addition; the history of that plating bath should be developed. A log sheet on each plating bath must be kept. Ideally, there is no sudden change in the bath chemistry from one week to the next. However, the only way to know whether or not a chemical analysis seems reasonable is to have the previous analyses of that plating bath handy. The log sheet must also list any maintenance, such as carbolation, performed on the bath, and all the chemical additions made to it. When all this information is at hand, it is easy to tell whether or not an analysis is reasonable, or whether it should be performed over again. Just because an analysis does not fit in with previous analyses does not mean that it is unreliable. It is always possible that the wrong chemical addition had been made, or no addition at all, or the tank might have lost some plating bath.

Each plating tank should have its own section in the analytical logbook. The beginning of each section should contain a data sheet on each plating bath and all pertinent data relative to the bath make-up and operation (see the examples at the end of this chapter). When all the information is laid out and convenient, it becomes an easy matter to find critical information in a hurry. The data sheets should always include:

1. Bath concentrations, optimum and range.
2. Operating temperature.
3. Current density, anode and cathode.
4. Type of agitation recommended.
5. Tank capacity.
6. Bath make-up, exact amounts of each chemical.
7. Type of filters, and size.
8. Concentration of chemicals in each of the pure make-up chemicals.
9. Name of the plating bath (brand and generic).

The information developed in these logs will be well worth the effort when a plating problem does occur. There is nothing like a reliable chemical history to turn to when a problem arises. The analytical logbook is also an excellent location to store the data sheets on each process, supplied by the bath manufacturer.

M&T Copper Lume PC #1

1. Tank Capacity   475 Gallons
2. Bath Composition and Operating Conditions

	Optimum	Range
Copper	2.5 oz/gal	2 to 3
Copper Sulfate Pentahydrate	10.0	8 to 12
Sulfuric Acid	10% by vol.	9 to 12%
Chloride	50 ppm	40 to 100
Temperature	75°F	68 to 85°F
Current Density: Cathode	30asf	5 to 60
: Anode	15asf	5 to 30
Agitation	Low Pressure Air	
Tank Voltage	.5 to 6 volts	

3. Bath Make-Up

P2X Copper Sulfate	118.75 gallons
Sulfuric Acid	47.5 gallons
Chloride	202 ml HCl
PTH Brightner	2.85 gallons
D.I. Water	305.9 gallons

4. Pure M & T Concentrates

P2X Copper Sulfate	40.0 oz	10.0 Cu metal
Chloride (HCl)	1.5 fl. oz per 100 gal = 50 ppm Cl^- 4 ml = 1 ppm	

Hi-Thro Tin-Lead #1

1. Tank Capacity   500 Gallons
2. Composition and Operating Conditions

	Optimum	Range
Stannous Tin	2.02 oz/gal	1.61 to 2.69
Lead	1.34	1.07 to 1.88
Fluoboric Acid	53.7	47 to 67
Boric Acid	33.5	10 to Saturation
Temperature	75°F	60 to 100°F
Current Density	20asf	15 to 20asf
Agitation (Cathode Rod)	Mild	
Filtration	Constant	None to occasional
Anodes	60% Tin/40% Lead bagged with Dynel or Polypropylene	

3. Bath Make-Up

Tin Fluoborate	22.5 gallons
Lead Fluoborate	10.5
Fluoboric Acid	300.0
Boric Acid	93.0 lbs.
Peptone	8.5 gallons
Water	160.0

4. Pure Concentrates

Tin Fluoborate	109 oz/gallons	33.3 oz/gal Sn
Lead Fluoborate	119	65.0 Pb
Fluoboric Acid	89.9	
Peptone	40	

Orosene PC #1

1. Tank Capacity   63 Gallons
2. Composition and Operation

	Optimum	Range
Gold	1.0 tr. oz/gal	
pH	4.4 to 4.8	
Temperature	100°F	90 to 100°
Current Density	10asf	1 to 15
480 amp-min per troy ounce		
Baumē	8° minimum	

3. Bath Make-Up

Orosene P.C. Make-Up #1	63 lbs.
Orosene P.C. Make-Up #2	15.75 liters
Orosene P.C. Gold	63 tr. oz

ACR 404 Gold

1. Tank Capacity   20 Gallons
2. Composition and Operation

	Optimum	Range
Gold	1.0 tr. oz/gal	.75 to 1.5
pH	7.0	5.5 to 8.0
Temperature	160°F	120 to 180
Specific Gravity	1.074	1.074 to 1.14
Current Density	3asf	
259 amp-min per troy ounce		
1 tr. oz = 31.1 g		

3. Bath Make-Up
Supplied as liquid ready to go.

M & T Nickle Sulfamate #1

1. Tank Capacity   82 Gallons
2. Composition and Operation

	Optimum	Range
Nickle	10.25 oz/gal	7 to 12.5
Nickle Chloride	.8	.4 to 4.0
Boric Acid	5.0	4 to 6
pH	3.8	3.5 to 4.2
SN-1		½ to 1½ oz/gal
Anti-Pit #12		.1 to .3% by vol
Temperature	120°F	

3. Bath Make-Up

Nickle Sulfamate	42 gallons
Nickle Chloride	.68
Boric Acid	25.6 lbs.
SN-1	2.6 lbs.
Anti-Pit #12	.82 gal

4. M & T Pure Concentrates

Nickle Sulfamate	20 oz/gal Nickle metal
Nickle Chloride	96 oz/gal $NiCl_2 \cdot 6H_2O$

M & T Nickle Sulfamate #3

1. Tank Capacity   510 Gallons
2. Composition and Operation

	Optimum	Range
Nickle	10.25 oz/gal	7 to 12.5
Nickle Chloride	.8	.8 to 4.0
Boric Acid	5 oz/gal	4 to 6
pH	3.8	3.5 to 4.2
Temperature	120°F	
SN-1 Brightner		
Anti-Pit #12		.1 to .3% by vol

3. Bath Make-Up

Nickle Sulfamate	261.4 gal
Nickle Chloride	4.25 gal
SN-1	15.9 lbs.
Anti-Pit #12	.1 to .3% by vol.
Boric Acid	159 lbs.

4. M & T Pure Concentrates

Nickle Sulfamate	20 oz/gal Nickle metal
Nickle Chloride	96 oz/gal $NiCl_2 \cdot 6H_2O$

Adapted from *Printed Circuit Fabrication* magazine.

SECTION EIGHT
THE MARKETING PROGRAM

24
Sales Tools

What are sales tools? Sales tools are devices which can be used to market the company, its products, services, and expertise. Sales tools are used to inform, explain, demonstrate, and reaffirm that your company is a superior company to do business with. Ideally, everything about a company can be used as a sales tool. It is a well known fact that people do not buy printed circuit boards; they buy what is needed to fulfill their needs. Generally, except in the cases of rather outstanding marketing programs, they do not buy from a company, they buy from individuals. It is difficult to sell printed circuit boards; it is easier to sell the fulfillment of needs.

Some broad classifications of sales tools are:

1. Company brochures and advertisements.
2. The product line.
3. Certifications and recognitions.
4. The company personnel.
5. The plant facility.
6. Other services, such as assembly.
7. Quality, quick turnaround and on-time delivery.

THE COMPANY BROCHURE AND ADVERTISEMENTS

Obviously, personnel at electronics companies must be aware of the printed circuit shop and its ability to provide necessary products and services before they can do business. The salesman/saleswoman is the prime contact between the vendor (printed circuit shop) and the customer (electronics company). He/ she uses sales tools to present the ability of the vendor to supply products and services, and to fulfill needs. Advertising and a company brochure are very basic sales tools to aid the sales personnel.

The Brochure

Since the buyer or engineer is not going to drop everything to visit the printed circuit shop, the sales person must use the brochure to demonstrate the shop's abilities and capabilities. The brochure should not present a history of the company and the philosophy of the owners, it must present the types of printed circuits manufactured, the certifications to which it can build, and Underwriters Laboratories information. The presentation of this information is the number one requirement. After this has been accomplished, information on the facility can be presented, along with photographs of the personnel, product, and plant.

There are other items, or sales tools, which can be packed inside of the brochure, to make a complete brochure package; for example:

1. Facilities list; a listing of all equipment in the plant.
2. Letters of certifications from business, government agencies, and customers.
3. Favorable vendor performance ratings.
4. Copies of articles by plant personnel, or about the plant and its personnel.

THE PRODUCT LINE

The product line is the most important aspect of the company to sell, since it is the reason for its existence. Each type of product should be listed. It helps to say a few words about the function of each of the board types as well. Buyers and engineers can better identify with a board type when a description is provided; a name may have different meanings to different people. If any of these board types are built on more than one material, say so, If any of these boards can be built with Defense Electronic Supply Center (DESC) approval to MIL-P-55110C, say so.

Examples of board types include:

1. Multilayer:

- 3–12 layers.
- FR-4, FR-4/Teflon composit, and polyimide.
- Controlled impedance multilayers through 12 layers.
- Fine line to .005 inch traces and spaces.
- Buried vias.

2. Burn-in. Double sided and multilayer burn-in boards built on FR-5, Polyimide, and 1102 materials.
3. Flex circuits and rigid/flex harnesses.
4. Teflon circuits for microwave applications on woven and nonwoven fiberglass.
5. Soldermask over bare copper and alternatives:

- Soldermask over bare copper.
- Soldermask over tin, nickel, or tin-nickel; with solder plated in the holes.

6. Double sided computer and telecommunications boards.

7. Bondable gold on FR-4, 1102, and polyimide.

8. Backpanels with inserted pins.

9. Printed circuits built on ceramic, metal/epoxy, Kevlar/polyimide, or other exotic substrate.

CERTIFICATIONS

The two most important classifications in this area are UL (Underwriters Laboratories), and military. UL will test double sided and multilayer circuits for safety and flammability according to circuit parameters (trace width, etc.) and types of materials. Obtaining UL approval for numerous types of materials, fine line circuitry, and 94V-0 flammability are strong selling points for a salesman to have when dealing with customers.

Military certification is granted, pending successful test results, according to material types and according to whether the board is double sided or military. Military certification, granted by DESC, is perhaps an even stronger sales tool than a wide UL listing. Many commercial customers will cast a jaundiced eye on printed circuit shops which have not obtained DESC approval to build to MIL-P-55110C. Also, there is a lot of military business available to printed circuit manufacturers, especially in the multilayer area. DESC approval is granted only to shops which have fulfilled all the provisions of MIL-P-55110C; without this approval, the printed circuit shop cannot build printed circuits for military contracts.

Once DESC approval has been obtained, few military contractors will place business with a shop unless it meets documentation requirements as spelled out in MIL-I-45208 (see Chapter 5, The Quality Assurance Program). Shops which meet the requirements of this military specification have a competitive edge over other shops, and they probably have a quality and delivery edge also. In fact, when a commercial electronics company surveys a printed circuit shop, that survey is generally performed to MIL-I-45208. Printed circuit shops which qualify for MIL-P-55110C by DESC, and which qualify for MIL-I-45208 by military contractors, often will be asked to build commercial boards for the military contractor as well. It is often assumed by buyers of commercial boards that companies which are military qualified produce superior quality printed circuits than do shops which are not military qualified. This assumption may not be valid, but it does show the potency of DESC certification as a sales tool.

Customer Qualifications

This is actually another type of certification. For a printed circuit shop to do business with major corporations which are known for the excellence of their quality and technology, they must go through a stringent qualification process. In many instances this qualification process is tougher to pass than MIL-P-55110C testing, and maintaining that qualification is based strictly on performance.

PERSONNEL

Plant Personnel

Since people manufacture printed circuits, it is logical that some of the personnel at the shop should be used as sales tools. Since sales will come when the customer has been convinced that his/her needs can be met by dealing with the given printed circuit shop, various personnel can be of great value in demonstrating the printed circuit manufacturer's ability to meet those needs. From time to time, buyers and engineers will have need of information on printed circuits: materials, metals, tolerances, specifications, artwork, etc. When the need for this information comes to the attention of the sales person, it should be seen as an opportunity to demonstrate expertise by the shop. Perhaps the shop has people who present papers at trade shows, or publish articles or books. People of this type can be of great value in making sales calls with the sales person, especially when the sales call has been set up to have a variety of customer purchasing, quality, or manufacturing people present. Any opportunity for contact between customer and vendor personnel should be encouraged. The sales person should not hesitate, when there is an urgent need for information, to call plant personnel, even while in a meeting with a buyer. Plant personnel should be used to help staff exhibits at trade shows, as well.

Plant tours during customer surveys offer other opportunities for the sales person to let the customer meet the exceptionally talented people at the plant. As the survey progresses from one department or work area to the next, let the supervisors meet the customer and field questions which may be put to them by the customer.

Sales People

One of the most important persons which a company has for contacting the customer is the sales person. This person is generally the prime contact, and the customer knows the vendor through the sales person. As printed circuits become more complicated, the sales person will need to have more training. A technically qualified sales person can be a real asset to the printed circuit manufacturer. Also, technical knowledge will help the sales person to feel more confident when dealing with the customer. Some areas of knowledge about printed circuits which are helpful for sales people to have a grasp of are:

1. Reading the blueprints.

2. Understanding the factors which make a printed circuit easy or difficult to build.

3. Understanding common materials used to build printed circuits.

4. Certifications, such as military requirements and UL.

5. Grid, CAD, and auto insertion requirements.

6. Bare board testing.

7. Types of circuit:

    a. Multilayer.

    b. Burn-in.

    c. Microwave, Teflon.

    d. Controlled impedance.

    e. Soldermask over bare copper (SMOBC), and alternatives.

8. Tooling requirements:

    a. Drill tapes.

    b. Artwork (positives/negatives, soldermask and padmasters).

    c. Test fixtures.

This type of basic information, which is needed to deal effectively with purchasing and engineering personnel of the customer, will be discussed in Chapter 25, What Sales People Should Know About Printed Circuits. The sales people should also receive training in selling. The basic information on making sales calls, dealing with sales situations, handling complaints, closing, etc., should not be neglected.

Lastly, the customer should be handed a list of personnel to contact in the company to handle certain tasks, or to obtain information. The customer should always feel comfortable in knowing that he/she knows who to contact for information on various topics. Typical people who should be on any list are:

1. Sales person.

2. Production control and planning engineer.

3. Quality assurance manager.

4. General manager.

THE FACILITY

Obviously, the facility itself can be a strong sales tool. The facility is the tool used for manufacturing by the people who work for the company. The printed circuit sales person should understand something of printed circuit manufacturing, so that he/she can help the customer's buyer or engineer to understand why the facility is so well suited to manufacturing the types and quantities of boards which the customer needs; as well as why they can be built to such demanding quality levels as the customer requires.

Some of the important features about a facility which can be used to sell to a given customer are given below.

Extent of Facilities

Size, square footage, number of buildings or manufacturing locations, as well as the extensiveness of manufacturing equipment in quantity, quality, and function. The numbers provided for these can be used to emphasize committment to large production runs, prototype or quick turn-around, as well as to a wide variety of products, testing, and supporting services.

Also, estimated shipping volume, in dollars or panel square footage, is an important subject the sales person must be able to discuss with the customer. This information is critical in making decisions on the size of orders which can be placed at a facility. If a customer purchases $3 million annually of one part number, and a given vendor has monthly shipments of $350,000; this information is useful in helping the customer gauge how much that vendor is able to handle of his procurement requirements. Knowledgeable sales people must be able to discuss this information with buyers. The sales person must also be aware of what percentage of capacity the shop is at.

Some of the aspects about the facility which can be used to sell the customer are given below.

1. Drilling:

   a. Number of drilling machines and programming digitizers.
   b. Special considerations, such as:

      ● Reduced number of hits per drill bit.
      ● Special (aluminum) backup and entry material for drilling.
      ● Color coding of used drill bits, for resharpening purposes.
      ● Use of mylar copies of drilling first articles, and other drilling inspection steps.

2. Imaging:

   a. Availability of dry film photoresist as well as screen imaging.
   b. The number of screening stations.
   c. The availabillty of automatic screen printing equipment.
   d. Photo area:

- Size and layout of photo area.
- Fully equipped darkroom.
- Capability of using diazo, silver, and room light safe silver film.

e. Oven capacity for baking resist and soldermask.
f. Extensiveness of soldermasking capability:

- 1 part or 2 part soldermask.
- Ultra violet (UV) curable soldermask (or plating resist).
- Dry film soldermask.

g. Air conditioning and humidity control; for excellent artwork stability.

3. Plating.

a. Number and gallonage of each type of plating bath:

- Copper.
- Tin-lead.
- Nickel.
- Tin.
- Tin-nickel.
- Gold.

b. Automated plating lines.
c. Etching facilities.
d. Tin-lead fusing: hot oil and infrared; explain why both (one?) are used.
e. Advanced pollution control systems.
f. Presence of logbooks for each process.
g. Availability of metal thickness measuring equipment.

4. Fabrication (routing):

a. N/C router/drillers, and why they can be used to save manufacturing steps and to improve quality and throughput.
b. Hardware installation available.

5. Quality control:

a. Location of inspection operations in the manufacturing process.
b. Documentation/checklists used at these points.
c. Incoming and first article inspections.
d. Artwork inspection.

e. Quality feedback to manufacturing.
f. Thoroughness of inspection, and sample sizes.
g. Understanding of military, IPC, and customer acceptability standards by the inspectors.
h. Use and availability of in-house micro-sectioning facility.
i. Availability of bare board testing.

6. Production control and pre-production planning:

a. Excellence of control demonstrated in being able to track job movement, and quantities of each part number.
b. Extensiveness and care used in planning each job, and understanding of the requirements for each part number.
c. Excellence of document control.

7. Separate facility maintained to handle only prototypes; to assure the customer that small jobs, and jobs requiring rapid turnaround, will not get lost in the woodwork.

8. Laboratory. The presence of a laboratory is a strong sales tool. Printed circuit manufacturing is a chemical processing industry, and a laboratory should be used for process control:

a. Daily and weekly chemical analyses of all plating baths.
b. Hull cell testing.
c. Presence of all troubleshooting apparatus.

All of these items are selling points to be used in explaining the facility to the customer. Any one of them by itself can also be considered a selling point, or sales tool. The printed circuit sales person should know how to use them as such. The idea is not to explain how printed circuits are manufactured, but to explain how *you* manufacture printed circuits.

Process and Documentation Control

All potential customers are interested in the ability of their printed circuit vendor to maintain excellent control over documents and processes. If your shop has it, tout it. The sales person should be able to explain why good control is maintained. Common reasons for good controls are:

1. Written procedures for all manufacturing operations. This serves as an established work standard and a reference for all personnel, on all shifts to use.
2. Formal training sessions for operators.

3. Cross-training of operators, for versatility, backed up by formal training sessions.

4. In-house laboratory, equipped for:

a. Chemical analysis.
b. Hull cell evaluations of the plating baths.
c. Micro-sectioning of printed circuits, for product examination.

5. Quality circles approach to manufacturing.

6. High yields and minimal customer returns, hence few internal remakes to clog up the manufacturing schedule.

7. Knowledgeable supervisors, managers, and engineers.

8. Excellent supervision, leading to knowledgeable and dependable workers.

9. Logbooks for product and materials accountability.

10. Vendor performance records.

11. Planning and production control have written guidelines for handling customer documentation:

a. Incoming inspection of drawings and artwork, before manufacturing begins.
b. Separate control files maintained on each part number. Each file should contain all information about that part number, and a control number should be assigned to each file. Each revision of a part number should be assigned a separate control number and file.

The Lobby

The lobby of the printed circuit shop affords an excellent opportunity to display the manufacturing expertise of the printed circuit shop, rather like an ongoing trade show exhibit. Instead of pictures of landscapes, the walls can be decorated with:

1. Printed circuit boards.
2. Artwork.
3. Polyester overlays showing how printed circuits are designed or laminated.
4. Samples of materials, together with boards manufactured on those materials.
5. Photos of trade show exhibits.
6. Letters of certification from DESC or from customers.

Conducting Surveys

Most business of significant dollar volume is preceded by a survey by the customer of the printed circuit vendor's plant. The survey may be conducted by the buyer alone, or by a team from purchasing, quality assurance, and manufacturing. It is at this time that the benefits of the vendor, which have been touted by the sales person, can be seen at first hand by the customer. The survey must not be treated lightly by the printed circuit manufacturer or the sales/marketing team. Sadly, all too often a survey is treated as little more than "in the front door, a quick whisk around the plant, and out the front door again." There are a number of considerations which should not be overlooked in conducting surveys. Some of them are given below:

1. Managers and supervisors should be notified as soon as a survey date has been set, and again the morning of the survey.

   a. Every area of the plant should be clean and neat.
   b. Unsightly wall posters should be taken down.
   c. Any piles of scrap boards should be properly disposed of.
   d. All logbooks should be present and properly filled out and dated.
   e. All job work order travelers should be properly filled out and dated.
   f. Attention should be paid to any work flow bottlenecks. It can be embarassing to be conducting a survey and to enter a work area which is devoid of work. Even if work has to be brought in from another area, do it. Do not let the shelves in a work area sit barren when customers are being shown around the plant.
   g. People working in the inspection and touch-up areas must be wearing white cotton/polyester gloves.
   h. Just prior to the survey, check for food in the work areas. It is embarassing to be conducting a survey, only to find someone eating potato chips and drinking soda while handling the panels.
   i. Copies of the written work procedures should be prominently displayed.

2. Managers should be on hand, if possible, to greet the customer. It is best if several managers sit down in a conference room to talk with the customer. There should be an agenda of topics to discuss. This does not have to be written down and handed to the customer, but it should be in the minds of the sales/ marketing personnel and other managers who sit in on the meeting with the customer.

Obviously, some customers are going to be more organized and sophisticated than others. Some are going to have their own agenda and checklists of information which they want to obtain. A good management survey team will be aware of the information needed by the customer, and will channel the conversation in this direction. Being aware of these needs will set the customer at ease and make for a more productive visit. Good topics to start off the discussion, prior to plant tour, are:

a. What types of printed circuits is the customer interested in buying: multilayer, double sided, microwave, etc.?
b. What quantities is the customer interested in buying: prototype quantities or major production runs?
c. What is the need for bare board testing?
d. Will the boards require military MIL-P-55110C certification?

3. The plant tour. One of the best ways to demonstrate all the features which have so far been listed as sales tools is to show them off during the plant tour by the customer. The plant tour is your opportunity to show how *you* build printed circuits, not how printed circuits are built. Most members of a customer survey team will have been in printed circuit shops before. They may not, however, understand the manufacturing process, or how to tell a quality operation from a schlocky one. It will be helpful to have on hand such documentation as:

a. Letters of approval from various agencies and customers.
b. Favorable vendor rating letters.
c. Recent UL recognitions.
d. Articles published by company personnel.
e. Internal yield reports, to document excellent yield, turnaround, and quality.

The best place to start the tour is with Production Control and Planning. By starting here, the entire process can be explained. The rest of the tour can be according to the schedule as listed earlier. It is important to stress the control and quality aspects of the manufacturing operations. These are what you want to stick in the mind of the survey team members.

It is helpful to have an informal monologue developed, written, and memorized by the sales person or manager who conducts the survey. This is to ensure that all topics on manufacturing and quality will be presented to the customer, and in the best possible light. A video tape camera and recorder can be very useful in developing the survey format. There may even be instances where the video presentation can be made at the customer's plant, for example, as when the sales person is out of state on a sales call. All sales personnel, and all management personnel, should be capable of conducting the plant tour, or of leading the initial discussions with the customer when they visit the plant. Obviously, this can only be done with excellent customer survey and plant tour planning and training.

The content of the survey should be changed to reflect changes in the plant and to reflect what is learned about what customers want and need to know. Ideally, the survey leader will never get caught off guard without effective answers to questions posed by the customer during a survey. A good survey will supply most of the information needed by the customer to fill out his own checklists.

SUGGESTED LIST OF SALES TOOLS

1. Product line: make a listing of your company's product line.
   2. Certifications and recognitions:

   a. DESC approval for MIL-P-55110C
   b. Customer approval to MIL-I-45208
   c. Underwriters recognition for:

- Wide variety of materials.
- Multilayer.
- Fine line circuitry.
- High temperature ratings.

3. Manufacturing capabilities:

   a. Multilayer press: Punches for artwork and laminate registration.
   b. Extensive artwork processing abilities.
   c. Dry film photoresist imaging.
   d. Pin insertion for back panels.
   e. Hardware insertion.
   f. Bondable gold.
   g. Build on exotic materials.
   h. Soldermask over bare copper, and alternatives.
   i. Other unusual or exotic processes or products.

4. Supporting services:

   a. Assembly and other turnkey operations.
   b. Universal grid bare board testing for rapid and inexpensive testing and turnaround on small quantities.
   c. Group A or Group B testing to MIL-P-55110C requirements.
   d. Board repair: trace welding and brush plating.

5. In-house laboratory:

a. Chemical analysis.
b. Prepreg characterization for multilayer quality.
c. Hull cell evaluation of plating baths.
d. Micro-sectioning.
e. Excellent process control documentation.

6. Distinction within the printed circuit industry: reasons.
7. Engineering/Manufacturing expertise.
8. Distinguished customer list.
9. Favorable vendor performance letters.
10. High quality, documented.
11. Low customer return rate, documented.
12. Rapid turnaround on jobs, documented.
13. Excellent documentation control.
14. Written procedures for all manufacturing processes.
15. Video and other formal training for employees.
16. Excellence in preproduction planning.
17. Excellence in in-house documentation:

a. Printed company forms: travelers, reports, etc.
b. Yield reports.
c. Quality assurance manual.

18. Clean, modern, well equipped facility.
19. Demonstrated commitment to a given market:

a. Military.
b. Controlled impedance multilayer.
c. Microwave.
d. Flex, etc.

20. Knowledgeable sales staff.
21. Responsive sales staff:

a. Request for quotations and blueprints picked up.
b. Willingness to obtain specific information.
c. Promptly return phone calls.

22. Responsive management:

a. Willing to meet tight delivery goals.
b. Willing to work with customer to develop new products, processes, or
   materials.

   c. Willing to work with vendors to the printed circuit manufacturer to develop new products, processes, and materials.
   d. Rapid resolution of quality and delivery problems.
   e. Price and delivery adjustments as an aid in conducting customer relations.

23. Excellent customer/vendor communication:

   a. Sales response.
   b. Resolution and prevention of quality problems.
   c. Responsive production control personnel.
   d. Responsive planning engineer.
   e. Access to upper management.
   f. Keeping the customer posted on situations which will negatively impact delivery dates.

24. Customer surveys.
25. Help for the customer in avoiding common pitfalls when they switch to CAD and auto-insertion.
26. Sales and management staff which understands the importance of giving realistic due dates—and meeting them.
27. Computerized quoting system.
28. Advertisements:

   a. Company brochure.
   b. Facilities list.
   c. Magazine ads.
   d. Articles on the company/supplier spotlights.
   e. Promotional items.
   f. News releases.
   g. Mailing list.

29. Company literature:

   a. Reprints of articles on the company.
   b. Reprints by company personnel.
   c. Yield reports and newsletters.

30. Trade show exhibits.
31. Presentation of papers by company personnel, in conjunction with trade shows.
32. Published articles by company personnel.
33. Sample printed circuit boards which demonstrate manufacturing expertise, for sales people to make into a portfolio.

34. Sample printed circuits for sales portfolio which demonstrate printed circuits built on various materials.

35. Utilize company personnel to make sales calls, and even to make initial customer contacts, in special circumstances.

36. Form letters for company mailings which address specific aspects of the customer's needs.

37. Cut newsworthy items about the customer from newspapers, and send to buyer in a greeting card.

38. Training courses provided for sales people:

a. Printed circuit manufacturing.
b. Sales training.

39. Develop telephone ability for setting sales appointments with buyers.

40. Face-to-face contact with the customer:

a. Do not be afraid to hop on an airplane to visit a customer; whether to make a sales call, or to resolve a problem.
b. Plant visits to and by the customer, by sales, purchasing, and engineering personnel.
c. Luncheon meetings during sales calls or after plant tours provide an excellent opportunity to get to know the customer.

41. Quality assurance and planning personnel should be encouraged to visit the customer, and to invite customer personnel to the plant.

42. Learn to cover a sales territory effectively.

43. The lobby should be used as an in-house exhibit of manufacturing expertise.

44. A formal, and ongoing, marketing program.

25
What Sales People Should Know About Printed Circuits

The design and manufacturing requirements for printed circuits are becoming more precise and more demanding. The buyers of printed circuits must have a greater understanding of what they are buying, and printed circuit sales people must have a greater understanding of what they are selling. In the past printed circuits have served one function—to make electrical connections from point A to point B. No matter how fine the traces, and how densely packed, and no matter how many layers—a mere electrical connection from point A to point B was the only function. The concept of the role of the printed circuit has changed, and is still in a state of flux. The printed circuit is becoming more of a functioning component in the electronic system. Such considerations as those listed below have created the need for printed circuit manufacturers, sales personnel, and buyers to be fluent in more areas of printed circuit knowledge. Modern printed circuit sales people should be able to read blueprints and ask intelligent questions regarding tolerances, materials, and testing requirements—since all of these have drastic effects on price and delivery, and in some cases on the ability of any given printed circuit shop to manufacture the board.

1. Conformance of conductor dimensions to the design or artwork.
2. Fringing effect of conductor edges.
3. Uniformity and control of dielectric constant.
4. Thickness of plated metals.
6. Press fit connectors.
7. Grid, CAD, and auto-insertion.
8. Millimeter wavelength stripline circuitry.
9. High temperature burn-in boards.
10. The use of buried vias in mulilayer circuitry.
11. The use of buried resistor layers.
12. The use of surface mounted chip components.
13. The need for multilayer circuitry to pass Group A and Group B testing.
14. The use of flex and rigid-flex circuitry.
15. The use of greater varieties of materials.

PROCESSING AND MATERIALS

The first chapter of this book, How Printed Circuits Are Manufactured: Processes and Materials, deals with the basic steps and operations of manufacturing. There is a section on materials. The processing information should be read and understood. There are other ways for the sales person to gain an understanding of manufacturing considerations. The sales person should have an engineer or manager walk them through the process. This will afford the sales person ample opportunity to have each operation explained, while watching it being performed. This is a good time to take notes and ask questions. No sales person should be afraid to walk through the manufacturing area, meet the people, and ask questions. It may require several such trips through the plant before fully understanding each operation. There is a side benefit from learning the processes in this way: the sales person will be able to do a more credible job of conducting plant tours when customers visit for surveys.

The first chapter also deals with laminate materials upon which printed circuits are made. It is helpful for sales people if they can put together a portfolio of printed circuits made on as many different materials as they can obtain. When reading blueprints and talking with buyers/engineers, the sales person will at least be able to visualize the materials on which the printed circuits under discussion will be made.

BLUEPRINT READING AND DIFFICULTY FACTORS

The sales personnel should be able to look at blueprints and understand enough to do a quote, even if that quote will be reviewed before going to the customer. The information given below can be used to set up a checklist. The sales person can then study a series of blueprints and fill out the checklists. It is very helpful to have the planning engineer or other qualified person review the checklists and drawings for accuracy. Some of the important items which the sales person should be able to learn from reading the blueprint are given below:

Dimensions

Many times the board dimensions are referenced from two intersecting lines, which are labeled as .000. Each reference line is called a datum. All board dimensions in the X- and Y-axes start from the datum (Fig. 25-1). The datum may be a corner of the board, the center of a tooling hole, or some other location. When looking for the dimensions of the board, it is important to notice if the dimensions are given from a datum, or from the edges of the board. If the dimensions are given from a datum, the dimensions to the boards edges on both sides of the datum must be added together. It is a common error for novice blueprint readers to assume that a dimension is from edge-to-edge, when it is actually from a datum point to the edges.

Generally the only dimensions which are important for quoting the board price are the overall length and width.

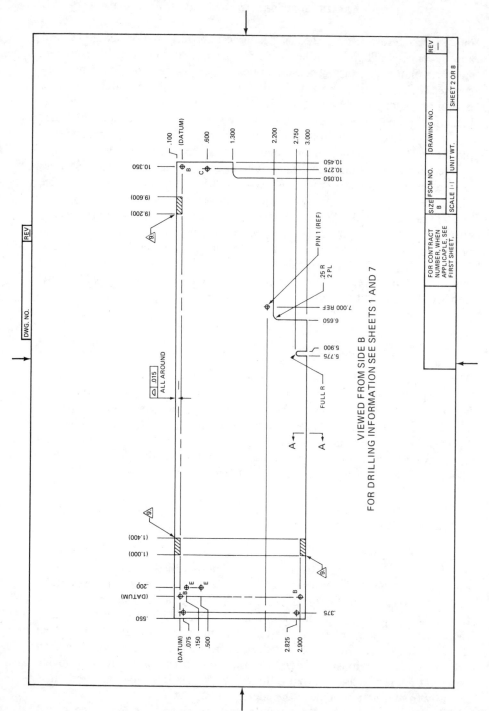

Fig. 25-1. The datum, or reference location, may be the corner of the board, the center of a hole, or some other location.

Dimensional Tolerances

These are often located in two places on the drawing: next to the title block, and next to a dimension.

1. Next to Title Block. There is a section labeled "tolerances." The tolerances will be displayed as,

$$.xx \pm .015''$$
$$.xxx \pm .005$$

The .xxx always reflects a tighter (more difficult to meet) tolerance than does .xx. Any tolerance listed as being tighter than ± 0.005 inches for a board dimension is unrealistic and should be questioned, if only to the planning engineer. This is generally considered to be the limit of the material for routing.

2. Next to the actual dimension on the drawing. When this is the case it is because the need for close tolerance is most critical at that dimension, and special attention should be paid to it during fabrication.

3. Dimensioned holes. Often, tooling holes are dimensioned in the blueprint. This is typically the case for tooling holes which are used to align the board for auto insertion of components. Auto-insertion tooling holes are toleranced within ± 0.002 inch, sometimes tighter. The presence of dimensioned holes, which are usually nonplated through, can affect the difficulty of manufacturing the board. Most nonplated holes can be drilled at the time the boards are N/C routed from the panel. Tightly toleranced dimensioned tooling holes, which are nonplated, cannot be second drilled during N/C routing and fabrication. These holes must be drilled first, then plugged with rubber stoppers during plating. This is time consuming and costly. If there are more than two of this type of hole per board, the sales person can ask which holes will be used for alignment; hence, only these two holes will have to be plugged, and the rest can be second drilled.

4. Board thickness. This may be stated during the material call out, such as ".062 1/1 FR-4." If this is the case, then the board thickness is obvious. Sometimes the board thickness is called out as shown in Fig. 25-2. The sales person should be on the lookout for unusual thicknesses, such as 0.080 inch, or 0.045 inch. These are not standard sizes and will require special purchase of laminate, at very high cost and with long lead times for non-multilayer boards.

Hole Sizes

There are plated through holes, and non-plated through holes. Holes can increase the difficulty of boards when:

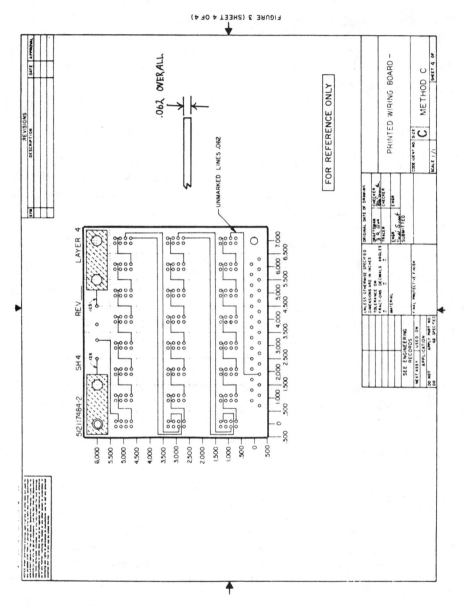

Fig. 25-2. One technique for denoting board thickness.

1. Hole diameter tolerance is less than ± 0.003 inches for plated through holes, and less than ± 0.005 for plated through holes larger than 0.100 inch. Because of the geometry of current density during electroplating, it is difficult to obtain a uniform distribution of current, hence it is difficult to obtain uniformity of plating thickness which will allow plating to less than ± 0.003 inch in hole diameter. To obtain ± 0.002 inch, often requested for press fit connectors, requires plating the panel at a very slow rate, for longer periods of time. Tightly toleranced hole diameters should be questioned. If the tight tolerance is absolutely required, the customer should be charged accordingly.

2. Copper plating thicker than 0.0015 inch. As the copper plating increases to much thicker than this, it becomes difficult to hold ± 0.003 inch diameter tolerance of plated through holes.

3. Aspect ratio more than 3:1. The aspect ratio is the thickness of the board divided by the diameter of the smallest plated through hole. Holes with high aspect ratios tend to have thin copper near the centers of the holes.

4. Diameters which are small, and must be accurately located. Small diameter holes, those around 0.025 inch or less, require more care when drilling. Outside of the fact that they break easily, the drill bits required tend to vibrate, since they are the thickness of a hair, or less. To place them accurately in an X-Y coordinate grid (the drilled panel) requires drilling with a slow table speed, so that excess vibration can be damped between hits of the drill bit. Drilling these types of holes can triple the time required to drill a panel of comparable hole quantity.

5. Excessive number of hole sizes. Generally, most printed circuits require no more than 3–5 hole sizes. There are times when a printed circuit shop must quote on boards with a dozen or more hole sizes. This adds greatly to drilling cost, since there may be excessive programming, excessive drill changes, and an excessive amount of time spent inspecting the hole sizes. Excessive numbers of different hole sizes should be avoided, since they also increase the chance of miscoding.

6. If nonplated holes can be second drilled during fabrication, as opposed to being first drilled, then plugged, their cost can be kept to a minimum. Companies which require large numbers of nonplated holes to be plugged will have to pay for them.

7. Drill tapes can be supplied by many customers; those who use CAD generation of their artwork. These tapes may or may not be of use to the manufacturer. If there are thousands of holes, perhaps it will be good to save the programming time. However, if there are a lot of hole sizes (eight or more), and only a few hundred holes the tape will not be of much value, since it will have to be spliced for automatic drill changing instructions.

Fabrication

There are a number of items which can add to fabrication (routing) costs:

1. Square interior cut-outs which require 90 degree corners, since router bits have a diameter.

2. Slots, typically placed between contact fingers for alignment and orientation purposes, add to the cost of manufacturing the board. Their cost is further increased if the bottom of the slot requires 90 degree corners.

3. Tightly toleranced dimensions. Dimensions which are tighter than 0.010 inch require great attention to maintain. If there is no functional requirement for ± 0.005 inch rather than $\pm.010$ inch, then the tolerance should be loosened. If the customer does not actually need to have an overtoleranced dimension held, the sales person should try and obtain a written waiver from the buyer or quality manager.

4. Recessed contact fingers (Fig. 25-3). All contact finger edges are beveled, to facilitate insertion into board connectors. The beveling is accomplished with a chamfering machine. However, when a customer has designed the board with recessed contact fingers (the contact fingers being recessed from the leading edge of the boad), the chamfering must be performed manually, with a file. This is time consuming and less reliable, and the customer should be charged accordingly.

6. Contact fingers on two or more edges or on adjacent edges (Fig. 25-4) are extremely labor intensive and time consuming. Several taping and plating operations are required.

7. Unusually shaped cut-outs or punch-outs require a specially made tooling die, which can cost several hundred to several thousands of dollars, plus the cost of punching.

8. Break-away configuration. Sometimes board users do not want the boards completely routed from the panel form. Boards are often routed so as to be still connected at a few locations. This way, an entire panel can be automatically loaded, and wave soldered. Only after soldering will the individual boards be broken apart from the panel. This form of fabrication should be little more expensive than regular N/C routing. However, it can only be performed with

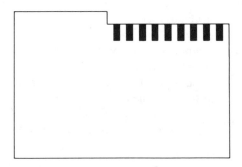

Fig. 25-3. Recessed contact fingers are difficult to chamfer (bevel).

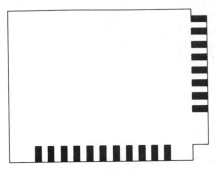

Fig. 25-4. Adjacent contact fingers require additional taping and de-taping operations.

N/C routing. If a given printed circuit shop has no N/C router, it must pay for the service—and this cost has to be tacked onto the cost of the board. Also, boards which have been scrapped out on a panel must be sent to the customer along with the good panels. This can be costly, since the gold on contact fingers cannot be reclaimed. Some customers will not acept panels with even one defective circuit. When this is the case, the entire panel is actually one board.

9. Edge plated boards are more expensive. These are boards which are partially routed from the panel, like break-away boards—only the routing operation is performed before electroless copper deposition. This is an entirely separate routing operation. The completed printed circuit will have plating all around the edges.

Contact Fingers

Generally these are gold plated over nickel. Even if the drawing or manufacturing specification does not call out for nickel, nickel should always be plated as a barrier coat between the gold and the copper. Gold is used for its excellent conductivity and ability to remain electrically conductive; if no nickel were plated between the gold and copper, the copper would migrate into the gold, tarnish it, and cause it to lose its conductivity. 30 micro-inches is adequate for most applications; 50 micro-inches is common. A square inch of gold, at a minumum requirement of 50 micro-inches, will cost around $.50 to plate at around $600 per troy ounce. Depending on the degree of control over the plating process, it might be slightly more costly.

Material Type

This is going to have a major impact on the cost of the board. Laminate can cost from $1.50 per square foot, to over $90 (Teflon). Hard materials, such as FR-5 and polyimide, require several times the number of drill bits/board than softer materials like FR-4 require. Also, a material such as Teflon will require special etching after drilling. This special etching operation is to roughen the drilled hole walls, to make them more receptive to an electroless copper deposit. This etching solution is hazardous (contains metallic Sodium) and expensive, at about $120 per gallon.

Hardware

If the boards require hardware, such as standoffs, turret lugs, and connectors, these will have to be factored into the cost on a per piece basis. This will also complicate the packaging operation.

Plating Thickness of the Conductor

Sometimes the plating thickness is called out as the thickness of metal inside the plated through hole, such as 1 mil, or as the thickness of the conductor, such as 1 ounce or 2 ounce. Conductor thicknesses of 1 ounce call for copper of 0.00144 inch thick. Copper foils for conductors called out as 2 ounce, actually will be made by using 1 ounce foil on the laminate, then plating another 0.0015 inch of copper to this by pattern plating.

Artwork

Generally, it is best if the sales person can obtain both the negatives and the positives of the circuit patterns. If the printed circuit manufacturer is going to use screen printing, negatives will be required. Dry film photoresist will require positives for outer layers and negatives for inner layers.

Definitions: *Negative.* The circuit pattern is shown by the clear part of the film.
Positive. the circuit pattern is shown by the dark part of the film.

1. Soldermasking. Sometimes only one piece of artwork is required, since it can be used for both sides. There are times when the pads on opposite sides of the board are shaped differently. When this is the case, two sets of artwork will be required. Artwork for soldermask should always be obtained from the customer, since it saves the Photo department much time in having to generate it from a drilled template. A padmaster can be used by expanding the pads photographically.
If the drawing is not labeled as to whether the side shown is component or solder side, inquiry should be made.
2. Epoxy Nomenclature. This is commonly called Legend by the screen printing department, and called Silk Screen on the drawing and by the buyer.

Soldermask Over Bare Copper (SMOBC)

This is becoming a popular method of building boards. The soldermask is over either bare copper or some non-solder metal such as tin-nickel, nickel, or tin. The solder is either plated in the holes and on the pads, or is applied by immersion in molten solder via hot air, or hot oil, leveling. Since there are extra steps involved, these types of circuits generally cost more than do regular solder circuits. Some printed circuit manufacturers refuse to build SMOBC, due to increased difficulty, cost, and manufacturability problems associated with solder stripping, solder mask bleed and hot air leveling of solder, company policy, obviously, must be known.

Fine Line Circuitry

Generally, any circuitry less than 0.008 inch is more difficult to manufacture than circuitry of 0.010–0.020 inch in trace width. The finer the line, the more difficult to reproduce, and the more difficult to touch up the artwork or imaged circuitry. Using screen printing, circuits with 0.010 inch can be produced satisfactorily in high quantity. For 0.008–0.005 inch traces, dry film photoresist will have to be used. Circuitry with traces finer than 0.010 inches will always cost more to manufacture, and be manufactured in lower yields, than will circuitry with wider traces.

Multilayers

Circuits of this type can be anywhere from fairly easy to extremely difficult to build. Most multilayers will fall between 4 and 12 layers. It is not the number of layers which makes the job difficult, as much as it is the degree of inner layer registration, the use of buried vias (electrical connections which are formed only between inner layers, with no hole drilled entirely through the board), and the need for controlled impedance.

MIL-SPEC PRINTED CIRCUITS

Many printed circuit shops advertise that they can build to MIL-P-55110C, without some of the documentation requirements required by that specification. This actually means nothing, as the sales person going after military contract type work will find out. Military customers will demand a copy of the letter of certification from DESC, and will survey the facility to see if it measures up to their understanding of MIL-I-45208.

Group A Testing

Group A testing is a battery of visual and microsectional tests which the circuits and the test coupons must pass. Some military customers will send a source inspector to the printed circuit shop for source visual inspection. Then they will allow the shop to ship the circuits with the test coupons. Final acceptance will result pending microsection results for Group A requirements. Other military customers will demand that the printed circuit shop perform all of the required visual and microsectional testing, and deliver the circuits with a letter of certification from the quality assurance manager. If the Group A testing is to be done by the printed circuit shop, the customer will have to be charged for that testing.

Group B Testing

These are further tests performed after the coupons have been subjected to severe environmental thermal shock. Only two tests per month are required by the printed circuit shop. However, some customers will want the shop to deliver only boards which have passed the Group B tests. For this service, the customer will have to be charged, and three to four weeks added on to the delivery date.

Ideally, military boards are no more difficult to build than commercial boards. Outside of the testing and documentation requirements, one of the most important reasons customers should be charged handsomely for military boards is the restriction on rework and buy-off through Material Review Boards. A broken trace can simply be welded before being shipped to the customer when it is on a commercial board. A broken trace on a military board is a cause of scrap. Where there is no requirement for passing thermal stress (10 second float on 550 degree Fahrenheit solder, then micro-section for cracked copper) testing on commercial boards, a failure of thermal stress will cause scrap for military boards. Only printed circuit manufacturers willing to commit to quality should be in the military market. But those who do make that commitment are entitled to receive significantly higher dollars for their products.

BARE BOARD TESTING

Printed circuits can be quite expensive. Even greater than the cost of the printed circuit is the cost of the components which will be loaded into it. If a short or open circuit develops in a loaded board, it is extremely difficult to isolate and remove the problem. For this reason, some customers will buy only printed circuits which have been electrically tested for opens and shorts. This is a service which the printed circuit sales person should be actively selling with the printed circuits. It is a good sales tool, since the customer gets only good boards, making quality and manufacturing engineers happy. Also, shops which are not equipped to sell this service will lose much business, and credibility, to shops which are so equipped.

Ideally, the test fixture and equipment is located at the manufacturer's plant. Bare board testing equipment will cost $30,000–$70,000, and the fixtures

$2,000–$7,000, depending upon the number of holes to be tested. Unfortunately, a minimum of 2–4 weeks is required to have a test fixture built. However, there are times when it is best to use an outside facility. Universal grid testing systems make it possible to build a test fixture and test the boards all within 48 hours. This can often be done for less than the cost of a discrete fixture alone. Universal grid systems cost between $100,000 and $400,000. The real value of this system, when using an outside service which owns it, is the capability of offering rapid and fairly inexpensive testing on small quantities, even prototypes.

Printed circuit manufacturers who pursue bare board testing as a service will find that they will have a slightly higher reject rate. Sometimes the defects can be reworked, other times not. When the manufacturer sells untested boards, the customer ends up paying for a portion of the boards which have electrical defects. However, when selling only tested boards, the cost of defective boards is borne entirely by the manufacturer. This added cost must be taken into account when quoting boards. However, the shop which aggressively goes after the testing market must also be willing to make a commitment to quality. Failure to do so will result in unhappy customers, loss of credibility, and loss of business. The printed circuit sales person should know where his/her company stands on these types of commitments.

APPENDICES

Appendix A
Yield Tracking: A Tool For Productivity

Although there are undeniable benefits to tracking yield in a manufacturing environment, many companies never get around to the task of setting up the simple mechanisms for doing so.

WHAT IS YIELD TRACKING, AND WHAT ARE SOME OF THE BENEFITS?

In its most basic form yield tracking is merely determining the ratio of product quantity accepted by Quality Control to the total product quantity processed. If a company sets out to manufacture 100 widgets; and during the process rejects 5 and completes 95 to the design requirements, it has achieved a yield of 95% on that job. That number may be interesting to know; but a little more effort spent during the process of making its determination can provide quantitative information on:

1. The ability of the company to manufacture according to specified requirements.
2. The success of each department in meeting productivity goals.   .
3. Identifying general and specific problems related to:

   - Incoming materials.
   - Documentation requirements.
   - Supervision, and employee skill level.
   - Company processes.
   - Electrical, mechanical, and chemical maintenance needs.

4. Customer returns as a percentage of product shipped.
5. Job turnaround time.
6. Percentage of jobs shipped late, and shipped short of requirements.

The availability of this information to those who must do planning and managing is enough to justify a yield tracking program. Other benefits become apparent once the program is underway.

WHAT ARE THE BASIC ELEMENTS OF THE PROGRAM?

The following is a brief description of one method to incorporate yield tracking into your manufacturing operation.

1. *Job Traveler.* This basic document (Fig. A-1) accompanies each job, from planning through shipping. It must list: (a) all materials and documentation requirements; (b) all manufacturing steps to be performed and the requirements at that step; (c) shipping quantity required, as well as the quantity in and out at each step in manufacturing.
2. *Inspection Reports.* These are checklists (Fig. A-2) against which the product is inspected and judged to be acceptable or not acceptable. Rejected product should be listed as to quantity and cause of rejection. When the job leaves the inspection point, the reject report becomes a part of the traveler.

WORK ORDER

PROCESS	QTY	DATE	EMP	REJ.
Rel. Date				
Film Insp.				
Program				
Process Film				
Blank				
Bake				
IL Image				
IL Etch				
Black Oxide				
Press				
Bake				
Drill N/C				
De Burr				
Cuposit				
Image S/S D/F				
Touch-Up				
Plate (Cu)				
Plate (Ni)				
Plate (SnPb)				
Plate (Special)				
Etch				
Tip Plate				
Reflow				
Inspection				
Clean				
Solder Mask				
Legend				
Fabrication N/C				
Fabrication Manual				
Champher				
Final Clean				
Final Inspection				
Electrical Test				
Hardware				
Shipping	Qty	Date	Invoice #	

Dwg.# _____ Rev. _____
A/W # _____ Rev. _____
A/W # _____ Rev. _____
Mst Dwg.# _____ Rev. _____

MATERIAL

Type: FR-4 _____ Color: Nat. _____
Thk: .031 .062 .093 .125 _____
Oz Sides .5/.5 1/1 1/0 2/2 2/0 _____
Size _____ x _____ 1 2 3 4 Deep
Run _____ Panels Bds Per Panel= _____

BOARD MARKINGS

UL Date Code Logo Lot #
Etched Screened Stamped

SCREENING

Solder Resist Over: Cu SnPb Ni _____ Sides
Type: Green Clear _____
Legend : Blk White _____

Pattern **PLATING** Contacts

Cu Ni Sn Pb Au/Cu Au/Ni Au/Au
Cu Thickness _____ Ni Thickness _____
Ni Thickness _____ Au Thickness _____
SnPb Thickness _____ Au Thickness _____

HOLE SIZE

Min	Max	Drill Size	Size After Plate
—	—	—	
—	—	—	
—	—	—	
—	—	—	
—	—	—	

TESTING

Micro Section
Porosity
Solder Sample
Source Insp.
Test Reports
Certs

NOTES: _____
_____
_____
_____
_____
_____
_____
_____

CUSTOMER:
QTY.:
P.O. NO.:
DUE DATE:
PART NO.:
ITEM:
REV.:
CONTROL #:
JOB NO.:

Fig. A-1.

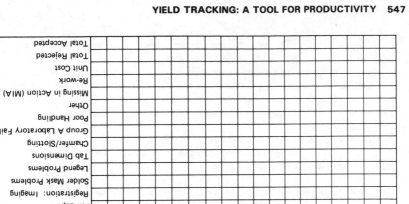

Fig. A-2.

3. *Inspection Stations.* The formal inspection station should be used as a repository for all product rejected at any point in the manufacturing operation.

4. All supervisors must train the people in their department to correctly fill out travelers: their initials, date processed, quantity processed, and quantity rejected. The reasons for the rejection or loss should be noted on the traveler; and the rejected or scrapped out product taken to the nearest inspection station. The deposited material must be identified according to customer, part number, and job number.

5. When the job leaves the final inspection area a copy of the traveler must be made and submitted to the quality supervisor along with the inspection reports which have accumulated during the manufacturing process.

When all of these requirements have been met, the company will have access to: (1) acceptable product ready to ship; (2) rejected product isolated and identified; (3) complete documentation on the history of each item processed on the job. All of this is the raw material for generating a meaningful yield report.

How is this information processed into a yield report? Before any report can be written, the quality supervisor must review the rejected product, as well as the inspection reports and travelers. It is good practice to look at everything that has been rejected during the manufacturing process. The quality supervisor must understand the reasons for the defects in order to understand the magnitude of problems which are detailed on the travelers and inspection reports; there is a difference between reading about a situation and seeing it with one's own eyes. Reviewing rejects also gives the quality supervisor an opportunity to set aside examples to be used for discussing quality issues with manufacturing people. Nothing speaks louder or is more undeniable than the real thing when you are trying to make a point. Also, if there has been an especially costly reject problem, the quality supervisor can dramatize it by having the evidence to show. Sometimes, product is rejected for invalid reasons or without sufficient cause. If this has been done, or if an item can be successfully reworked and shipped, the quality supervisor has the opportunity to correct the situation during a daily review of rejected product.

After reviewing the rejected material, the quality supervisor should review the traveler copies and the inspection reports attached to them. The information on them can be tabulated in a notebook form, such as Figs. A-3 and A-4. Quantities for processed, required and accepted product is listed on one page, together with turn-around time, yield and whether or not the shipping due date is met, for each job processed. The other page tabulates defects and quantities according to which department was responsible.

When organized like this, the information can be used to generate a report which provides an overview of most of the problems involved in manufacturing and how successfully the company is meeting their challenge.

(Figs. A-5, A-6, and A-7) This information can be used productively by managers, supervisors and hourly employees.

1. When the yield report is posted in each work area, employees tend to read it with interest. Operators want to see how their reject rate compares with other departments in the company, and most employees genuinely want to understand the problems of their department and the company as a whole.

2. At production meetings with supervisors the yield report serves as a weekly summary of quality related problems. It is a good tool managers can use to discuss areas needing attention with those closest to the problems and the solutions. A thorough discussion, on a weekly basis, of the yield report virtually assures that problems will receive the attention they deserve.

3. The yield report affords the Quality department a vehicle for discussing quality on a company wide manner once a week. Continual feedback on problems concerning incoming materials, documentation changes, product and process difficulties, handling and storage, workmanship,

CUSTOMER	PART NO./JOB	PROCESS	REQ'D	ACCEPT		COMMENTS	TURN-AROUND	LATE?	% YIELD

Fig. A-3.

DRILLING	IMAGING	PLATING	FAB	MATERIAL	HANDLING	UNACCOUNTED	COMMENTS

Fig. A-4.

YIELD REPORT FOR THE PERIOD ENDING AUGUST 30, 1981

BOARDS: PROCESSED   REQUIRED   ACCEPTED   %YIELD
        16211   14634   15663   96.6%
        CUSTOMER RETURNS
        1.6% FOR AUGUST

JOBS: LATE   CLOSED SHORT
     6.5%   3.5%

Causes for defects:
Drilling: 133 over/undersized holes, 88 missing, extra or misplaced holes, 25 burrs.
Imaging: 266 broken, nicked traces, pads and tips, 253 scrapped in touchup, 38 burned during
    Nome bake, 25 other nome and solder mask problems, 10 black contamination beneath
    the solder mask, 6 registration.
Plating: 165 holes too large due to thin plating, 139 over/under etching and exposed copper on
    surface, 122 pitting due to poor rinsing before copper plating, 60 burned and noduled
    plating, 54 resist breakdown, 12 dewetting, 10 copper peeling.
Fabrication: 100 N/C operator and machine problems, chamfer, slotting, and second drilling.
Material: 114 delamination, included particles, trace lifting, and stains. Also, 10 multilayers for
    delamination during reflow because of insufficient prepreg cure.
Handling: 54 scratched and nicked.
Missing: 126 unaccounted for.

Fig. A-5.

customer returns, excessive scrap, late deliveries, and equipment maintenance serve to heighten
employee awareness of these areas and how they affect that employee's job. Those areas which
do need attention stand out for all to see; once a company has done this a major roadblock to
improving productivity and quality has been overcome.

Special Notes

It is important that the Quality department make a survey of all rejected material before writing
the report; this is the only way to maintain credibility in the report. Many of the people reading

HANDLING—We lose many boards due to careless handling. Most of the defects are the result
    of scratches on the contact fingers of completed boards. Remember—one scratch on a
    $200 multilayer board makes $200 of scrap in one fell swoop.
NICKED AND CUT TRACES, TIPS AND PADS—Most of the time these come from imag-
    ing. During touchup, is excessive or repeating short or opens are noticed, both the screen
    and the artwork should be checked as a source of defects requiring attention.
PITTING DUE TO POOR RINSE BEFORE COPPER PLATING—When the boards are
    removed from the soak cleaner they require a great deal of rinsing, as the cleaner is very
    viscous and tenacious. If we cut short the spray rinse cycle: fail to rinse full length of
    panel, rinse two panels at once, not rinse long enough to remove suds—then pitting will
    occur on the edges of conductors and contact fingers. If the rinsing is poor enough, the
    pitting will also show up on the surface of the contact fingers and traces.
RESIST BREAKDOWN—This results in plating beneath the resist. The chief causes of this at
    XYZ Co. are poor job of neutralizing Cuposit with acid, and screening over finger prints
    and stains.

Fig. A-6.

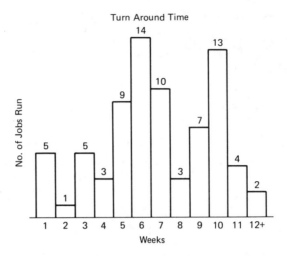

Fig. A-7.

the report are going to want to be certain that they did not have rejects inaccurately attributed to their department.

The degree of jobs shipped late and closed short, together with turnaround time, technically is not a matter for quality control. However, the yield rate affects a company's ability to deliver goods in a timely manner. For this reason, it is proper that employees be aware of how successfully the company is meeting its production commitments.

SUMMARY

A yield tracking system is not difficult to initiate. It can be kept basic and simple: or, with a little more effort it can be made to generate a wealth of information. By posting the report in all work areas the company will foster pride in workmanship and quality awareness. All this will help place the burden of quality and productivity where it belongs: on the shoulders of those who perform the work.

It may be best to include customer returns as a percentage of shipping on a monthly basis only, unless you have excessive returns.

This article appeared in *Quality* magazine.

Appendix B
CAD: Swift, Precise, Infinitely Repeatable, but Never Creative

By Jerrold Asher, Bishop Graphics, Inc., Westlake Village, CA 91359

The drafting and design professions are awash with the latest offspring of the computer industry: computer aided design or CAD. Computers now offer drafters and designers relief from the drudgery and boredom of repetitive drafting tasks. This is particularly applicable in printed circuit design where the symbols and many of the design rules are fixed and iterative. But part of any good PC design requires creativity and ingenuity to achieve highest component density and lower manufacturing costs. Only the human brain, so far, has proven best at innovating the PC design to its most cost-effective level. There are also some deep pitfalls in completing a successful CAD installation for PCB artworks. Not only must the hardware and the software capabilities be examined but the vagueries of the organization and the personalities involved must be considered in developing a successful CAD installation. This paper discusses some of the pros and cons of manual versus CAD generated designs, and the personal psyches of both management and drafting, which bear heavily on the sometimes illusive road to a successful installation.

A CAD System for Printed Circuit Drafting

In all areas where the omnipresent computer is first introduced, it is haled as a great cost-cutting, labor-saving, time-shrinking, error-eliminating benefactor to mankind. This certainly has not been minimized with the introduction of computer-driven plotting or computer aided drafting (CAD) for printed circuitry.

So much have the drafting/design professions been enamored with CAD that some of our industry countrymen would define the "D" in CAD as design and then have us believe that the CAD system can perform design tasks. Herein lies the rub. (Quote from RCA)

Computers aren't Creative

Certainly machinery just offering all the above benefits would be a boon. But there are those few misguided souls in our industry who feel that designing is a skill which is intuitively creative. And even the most erudite of the computer breed will generally agree, after not too many martinis, that creative tasks cannot be done by a blind, dumb beast which must be led from place to place by previously, carefully-prescribed instructions—in other words, programming.

The computer and its associate, the plotter, are indeed wondrous in their blinding operational speeds and meticulous accuracies. But when someone feels that they have the power to think, to design, which printed circuitry artworks require, that someone needs to switch channels from "Galactica 1980" to "Real People." Saying it another way, PC designing has to use imagination, something no computer/plotter yet possesses.

Designing in printed circuitry takes the form of careful component placement and orientation (positioning, if you will) to make for best electrical operations and minimum production assembly cost. For example, it should be kept in mind which components are to be closest to the circuit board's end connector (to reduce circuit path length and signal time delays), which components

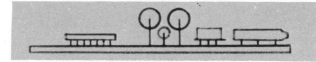

Fig. B-1.

need to be isolated from the others to allow heat dissipation or for reduction of RFI (radio frequency interference). Component placement must allow for potential removal of any component in case of its malfunction. The designer must make certain not to make smaller component removal impossible without always having to remove larger ones as shown in figure 1. Use of production automation, such as automatic component insertion equipment, requires the designer align all axial components on the same axis. And so it goes! There are many other considerations based on the type of function the PC board will perform, but all of the considerations require human thought process, imagination, and creativity, based on earlier assimilated knowledge.

I suggest that, though we document all the caveats and rules about PC circuitry layout and design, it is the cleverness and deft subtlety of the designer's *manipulation* of the rules which makes the better PC designs. Note emphasis on *using* the rules. Getting the rules all stuffed and stored in some easily retrievable fashion from computer memory will take some sophisticated programming effort, but it can be done. Simply knowing them all and being able to call them up in the tick of a nanosecond is baby step easy for any computer worth its integrated circuitry. Using the rules with skill is something computers will never learn. Or in simple words, good PC designers are *always* going to be in demand.

CAD Systems Need Human Helpers

Since the computer and it's buddy, the plotter, cannot think, i.e., excercise imaginative creativity, they must be assisted in their efforts to perform PC artwork generation. Two classes of CAD systems exist for this work. First is manual digitizing; and second is a combination of interactive designing with automatic routing. Both approaches use an automatic plotter, computer-driven, to produce the final phototooling artwork. Let me take up each individually for a little better understanding.

Manual Digitizing

With this approach, the layout is made manually just as it is when the entire artwork is made manually. Then, using the digitizer input to the CAD System, the X-Y coordinates of every critical point in the PC layout are identified to the computer, i.e., digitized.

Once digitized, the PC layout becomes a permanent part of the data base and is stored on some form of magnetic media-tape or disc. It can be reproduced as often as required in pen & ink plot or with a photoplotter onto light-sensitive photographic film. This method offers the first opportunity for errors. Should the digitizer operator not be careful to define what symbol each X-Y coordinate is locating, i.e., whether it is a pad, dual in-line package, or a change in conductor direction, the data base will suffer accordingly. To avoid this potential error source, many people like to structure the layout into the same layers or levels used in the CAD data base: a layer for the pad master, a second layer for solder side circuitry, a third layer for component side circuitry, etc. In this way, the digitizer operator will only have to define once per layer what type of symbol coordinates are being input. He or she can define for the digitizer the data following as all being pads of a specific OD and ID, dual in-lines of specific pin numbers and shape, solder-side circuitry, or component-side circuitry. Another advantage of using this layered approach is that less-skilled data processing personnel can be used as a digitizer operators in place of high-cost designers. The possible disadvantage to this method is that the layers must be carefully and precisely registered one to another in order to maintain the correct X-Y coordinates for elements of the circuitry in the same relative positions on the PCB but on different layers. It is of no value to have the world's most resolute digitizing device and accurate plotter, if the layers in the data base are not registered with the same degree of precision.

Interactive Design

This approach offers some of the real power of a computer aided system. As many of you are probably familiar and will observe on figure 2, the design of the PCB actually takes place on the face of a CRT. The designer calls from the computer data base the actual components which he or she will be using to mechanize the circuitry of the board. These components have been stored earlier in the CAD data base and appear on the face of the CRT in whatever position the designer desires. There are some frustrating limitations in this data base, but we'll address those later. As a further aid, the CAD system will allow the designer to move the component outlines around the screen or reorient their positions relative to the outline of the PCB.

Inputting the net list or some other form of data which tells the computer which pin numbers connect with which other pin numbers, is the only added data required for an interactive system to go to work.

Problems occur when the input typist makes errors in the net list. This turns out to happen more than occasionally. One major user reported a consistency of as many as 12 errors per 30 pages of data and is experimenting with very expensive, but more accurate voice-input terminals.

It should be remembered that these errors will not show up until the time and expense are taken to run a plot, an expense to be avoided if possible. And, don't forget the time and expense for someone to check the checkplot for errors.

With automatic routing, the computer now takes over and makes all of the interconnections between the components' pins . . . well almost, anyway. The estimates I have received are that

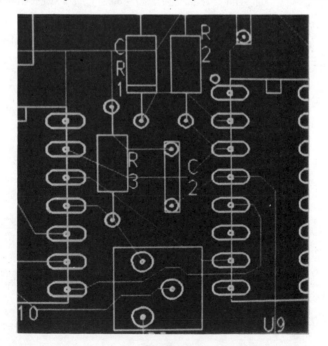

Fig. B-2.

the better router programs can do somewhere between 85 and 95% of all the connections. After the router's best effort at connections, it electronically "throws in the towel" and returns the last few connections it cannot make to the designer. Most automatic routers have a means of displaying either with color, dashed lines, or blinking lines those interconnections which it cannot make. Now the designer must take over and make these last few connections.

This is where potential problems creep in. The last few connections that the designer must make may require that he or she disconnect some of the previously-made router connections. Sometimes in fact, the designer must disconnect so many connections to make the last few unconnectables, that he or she may spend as much time rerouting as would have been needed to do the entire job manually. I admit this is the extreme case, but we hear of many instances where it happens.

To the best of my knowledge, automatic routers have been really effective only on digital circuitry. The more analog circuitry that is involved on the board, the less capable are the routers that are currently available. Anybody know of a good analog router? What percent completion?

Emitter-coupled logic and memory circuitry are still very poor candidates for automatic routers. It is my understanding that power supply boards are virtually impossible to do on any currently marketed router due to their irregularly-shaped components and oddly-shaped PC boards.

Routers have some other interesting logic which can never be as creative as that of a human designer. Here are a few examples of the kind of situation where the human designer can surpass the CAD system. In figure 3, section 1, we see first how the CAD system would handle the problem of avoiding the donut pad in its path, yet not run the conductor too close to the pads in the dual in-line rows between which it must pass. It chooses to neck down to a narrower conductor size to solve the problem and be able to keep going in an absolutely straight line. In section 2, we see how the human designer would solve the problem simply by taking a short curve around the obstruction and passing right down the center of the dual in-line for what is obviously a better production design.

Figure 4, shows how the CAD system has been programmed for minimal spacing to accommodate 4 conductors between the two rows of the dual in-line. When it is given the task of inserting only 2 interpad connectors, as in the middle view, it continues to maintain the same very narrow spacing. On the other hand, the human designer would spread out, as shown in the third section to take advantage of all the room that is available.

Fig. B-3.

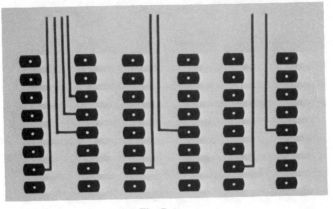

Fig. B-4.

Figure 5 shows a very difficult set of connections between 2 dual in-lines that are extremely close together. Notice how the human designer readily makes the connections by putting in simple curves of the conductors. The CAD system, however, goes through all kinds of intricacies to produce a pattern of connections which will be a virtual nightmare to produce in fabricating the circuitry.

Finally, in Figure 6, we show a condition commonly called taping "off the grid." Since the 1/10th inch grid system is commonly used, most CAD systems are programmed to operate on some fraction of 1/10th of an inch. Whatever this sub-multiple or fraction of 1/10th of an inch is on the CAD system, your circuitry must be fit onto that CAD system's gridding. You have no way of producing an off-grid pattern such as might occur in the figure we show here. In this case, even though the arrows depict an off-grid pattern which is not a sub-multiple of any known CAD system, the designer can easily manually tape this pattern by going off grid and placing tape in the pattern indicated by the arrows.

Figures 7, 8, 9 show some very difficult taping configurations where the manual designer actually improvised in order to make the tape fit. In figure 7, we see a combination of off-grid

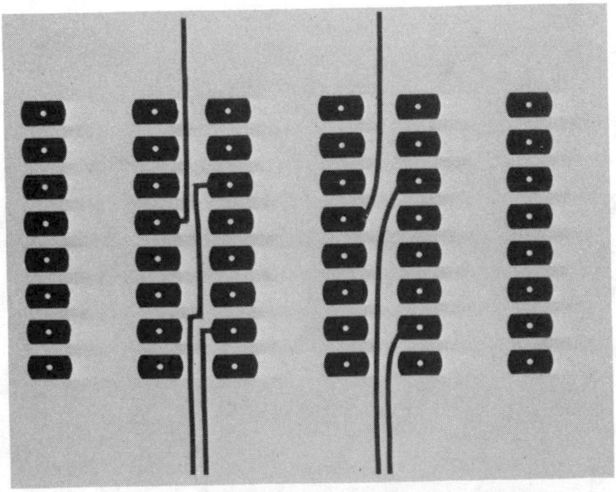

Fig. B-5.

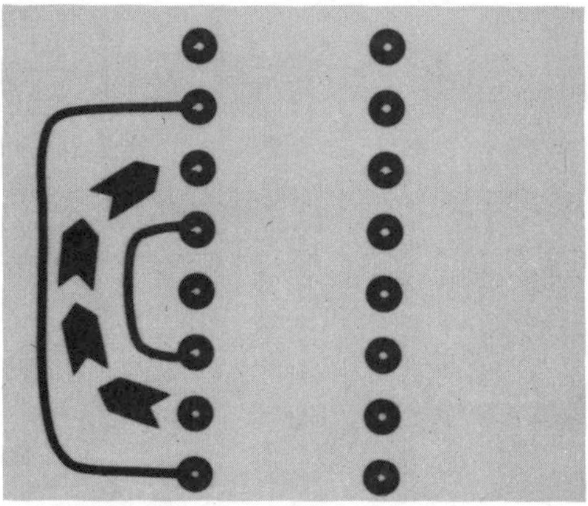

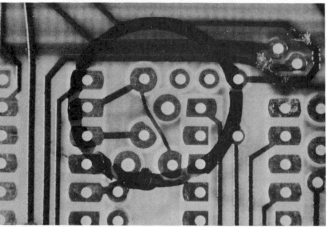

Fig. B-6.

Fig. B-7.

Fig. B-8.

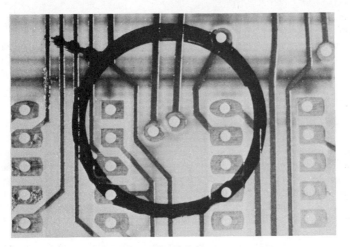

Fig. B-9.

taping, as well as the trimming of a pad in order to allow the conductor to slip through between the two pads in the center of the slide. In figure 8, we see a combination of pad trimming on both left and right side of the center pad, as well as some rather ingenious tape use which I suggest no CAD system could accomplish. Finally, we see on figure 9 the use of pad trimming on both the dual in-line on the left hand row of pads as well as some trimming of the two pads in the center of the slide in order to allow them to be fitted slightly closer together than their normal OD would really allow. The point being emphasized here is that with a manual designer, the designer rather than the system controls the design and can make the arbitrary decisions necessary when push comes to shove.

Seeing Is Not Always Believing

There are some optical disadvantages in using the interactive CRT design approach. The CRT is limited to a smaller size display than the typical C or D size drafting sheet on which we are accustomed to doing manual artworks. To compromise this disadvantage, the CRT uses pan and zoom techniques in order to allow both an overview of the entire board, as well as a closeup of any particular area where designing is taking place. However, these views are hard to work with for most of us. The problem is that we never get both total global view of the board simultaneously with an ability to zero in on a detailed section, as we can do with our eyeballs on a large piece of manual artwork. Using the CRT, we either get the global view with insufficient detail of a given area, or we have the zoom-in view with a very tight view of a given area but no feeling for how this incorporates with the overall design of the board. It's a compromise at best. As a matter of fact, I have found some CAD system users would rather go back and redigitize an artwork when they have a series of changes as opposed to trying to maintain some sense of proporation while redoing their design on the CRT.

Fractured Feelings, Injured Egos, and Other Pre-Installation Woes

Much of the problem of making a CAD system effective and productive is with personnel. The keystone of any attempt to put a CAD system to work is the absolute necessity of top management backing. This requires more than a simple commitment to corporate funding and setting aside a space for installation. The complete positive mental set that CAD is something beneficial for your firm must be shared throughout top management The slightest doubts or inklings of concern that this was not a wise decision on the part of management will invariably undermine the best intended CAD application.

Management must have the patience of at least one year before expecting productive results out of the system. Any results that are obtained short of the year, of course, are miraculous and all in one's favor. However, you must convince and be certain that management believes they are in for a minimum of a one-year start-up period.

The working designers and drafters must also be positive in their outlook toward the installation of the new system. How many of you or your people have fears that their jobs are at stake because the computer is taking over? How many of your people are fearful that they will have to learn a new skill? Here is something some of us don't particularly like to take on at a later age in life. If you have too many of either this type of mind set in your drafting group, you are in for a difficult to an impossible task in obtaining the cooperation and the enthusiasm that are required to make a CAD system work effectively. If there are too many of the older-type thinkers in the group, and there is no chance to change or replace them with others who are more positive, you are better off not even trying to put a CAD system into this environment. CAD, like any major departure from tradition, has installation birth pains. Everyone's enthusiasm is needed to minimize those painful moments.

One severe disadvantage to some designers—particularly we older guys—approaching CAD systems, is the need for a degree of regimentation. A fixed data base of symbols, sizes, widths, and other design elements is required when using a CAD system. You cannot simply reach into the PC drafting supply catalog and bring up whatever OD and ID donut size that you happen to like or whatever size trace width you'd like to use. You must use and stay with those design elements that are stored in the system data base. This puts some strong limitations on our more creative designers and may even frustrate them to the point where they take up strong drink or heavy smoking. I am not suggesting that this is necessarily a defeating element to a CAD system, but it certainly can be an area in which many designers will resent if faced with having to use a fixed set of design elements.

Hidden Costs

Some of the costs are not likely to be mentioned by your friendliest CAD sales representative and should be addressed in the very beginning. The most obvious one is maintenance. Besides the cost element, you must determine the source point from which the maintenance engineer will emanate: Is he around the corner? In the same city? In the same state? How many airline miles must he fly to reach your installation? All of this will impact your mean time to response when a service call is required.

You must make some judgement as to the quality of the service personnel who will be doing your system's maintenance. Our experience has been that a brand new service engineer is assigned to the newest installations. When he gets into trouble, which may be one, two or three *days* after he has tried to diagnose your problem, he finally calls a senior person who then will require additional time to extricate himself from some other duties and come over to see what is ailing your system. All of this time is costing you money, since you cannot be productive when your system is down. We talked with several of our own dealers who have used CAD systems and have found some who have literally thrown them out because of the lack of responsive service. These people are located in major metropolitan centers such as Chicago, Phoenix, and St. Louis.

On the other hand, I have visited several successful CAD installations which tend to rely on in-house personnel who are capable of system hardware maintenance and system software maintenance. Although I would not consider my information to be a representative sample of the total universe, based on what I have seen and heard, I have to conclude that in-house talent is the most effective way of keeping the system up and running.

I have noted an industry custom practiced in CAD systems which I observed in my earlier experience in data processing computer systems. That is to move any developed senior technician in field service into management or other forms of factory responsibility to take advantage of the senior level of experience. This leaves you, the user, with the newer, less-experienced technician, who has to do part of his on-the-job training on your system. You wind up with less than the most efficient service form and a longer mean time to repair.

Double Time Is Not Twice as Productive

An extensive cost area which most new CAD system installations miss is the double staffing, which will occur during the period the CAD system is being installed and brought on stream. This period of time we mentioned earlier can last anywhere from 9 months to a full year and represents the time during which the installation training takes place, the data base—of which I will speak in more detail later—is being constructed and input into the system, and useful production is slowly being obtained. During this period, the work must still get out. Therefore, you will have one group of your designers working at understanding and learning how to produce designs on the CAD system, while a second group of your designers will be doing the work— which has to get done manually. Since those of your staff learning and working on the CAD system obviously can not be turning out designs, you have some duplicate staff involved, so that the work keeps flowing. This means your overhead is going up extensively, and your management, whom you have hopefully gotten fully behind you and the CAD system, must be prepared for the extra costs the additional personnel will entail.

Don't make the mistake of trying to teach your entry level personnel how to operate the CAD system. The CAD operators should be your most senior and most productive PC board designers, especially if you use interactive design techniques. Using less than the most senior people on the CAD system is fraught with disadvantages. First, you will tie up the system for inordinately lengthy periods of time which may keep other departments or users from getting their work through your system. In a time-sharing mode, that is a system which has multiple terminals, an inexperienced person can take up so much time in inefficient design techniques as to cause the computer to slow down in serving the other terminals connected to it. Finally, a person who is not fully experienced or senior in designing PC boards will never be able to fathom the shortcuts and clever techniques that are available within the programming of the CAD system to help his or her design be done in a shorter time. One cardinal rule that everyone should follow is never use other than your very best designers on a CAD system. The temptation is always there, because it is generally believed that the CAD system can supply the experience level which a junior person has not had time to accumulate. On the contrary, the system does not supply a senior level of knowledge and actually suffers in the hands of a junior person attempting to operate it and generate PC designs.

Cockpit Errors

A potential error source is in the simple operating procedures for the system. There are operational steps which must be taken in setting up a photoplotter to do the final photographic negative. For example, as some of you probably know, you must prescribe certain scales, which aperture plates are to be placed in the carousel, what line widths are going to be used, etc. If these are not properly prescribed, or if the operator does not follow the instructions correctly and get the right aperture in the correct carrier on the carousel, or the programming doesn't call out the right carousel position, your artwork is going to turn out wrong. Unfortunately, this will not be known until the complete artwork has been run on the plotter and developed through a photographic development process When it does not come out correctly, the designer then must return to the beginning of the plotter queue and wait his or her turn for resubmission of the plotting job all over again. By the way, in manual taping, these kinds of errors are easy to correct: one simply dives into the middle of the artwork and changes the incorrect symbols to the correct ones without having to wait for his or her turn to use the expensive corporate asset which is in contention for use by several different individuals.

Another question that comes to mind is whether or not the pen plotter or photoplotter is out of scale or specification. We find most of the CAD systems that we meet in our travels have not been calibrated for as long as a year or even two. How can people operating a system of this magnitude, precision, and complexity not know whether their very elaborate electro-mechnical device is within specification? How many of you currently use CAD? How many know when your system was last calibrated? We don't understand this, but we find it to be common practice within this growing industry. If the plotter is not within calibration, you then find yourself running off a series of several hundred printed circuit boards which probably are not checked that closely at incoming inspection. They wind up on a production line in Hong Kong using automatic insertion equipment where the parts do not go into the holes for which they have been intended. At this point, it is a little late to remember or to worry that the CAD system has not been calibrated for a long number of months. Although I hate to put this into the "I told you so" department, a manual artwork is obviously readily calibrated at the time it is put down on drafting film. The symbols that are used are accurate to $\pm$ two 1000ths of an inch as guaranteed by the manufacturer, and the designer is using an underlying grid which is accurate to a similar set of specifications. There is no question about being in or out of specification when one does a manual artwork.

Savings: Now You See Them, Now You Don't

One of the most heavily-touted savings which CAD systems can provide according to the vendors, is in the area of CAM or computer aided manufacturing. We are told that the by-product punched tapes from the CAD design effort can be used to drive automatic insertion equipment and automatic inspection equipment. The concept here is that because the CAD system has computed or been fed the X-Y coordinates of every critical point on the PC board, it therefore can use these same data to produce the subsidiary tapes needed to run the manufacturing equipment. In principle, this is an excellent idea. In practice, however, it doesn't always work. This speaker recently attended a seminar on CAD systems and associated with members of drafting and design organizations from some of the larger electronic firms in the LA area. I was appalled to find that many of these people did not consider preliminary consultations with their manufacturing counterparts before or during their embarkation on a CAD system planning effort. Some of them did not even know who to contact or in what part of their organization cognizant manufacturing people existed. The seminar also took up a case history of a major computer peripheral manufacturer in our area who had installed a CAD system, was diligently turning out automatic insertion tapes, and was sending them properly over to the head of the manufacturing operation. Unfortunately, no one had ever contacted this latter, and he was not even using the tapes, because they did not fit the format acceptable to his insertion equipment. In other words, there appears to be a lack of communication between design and drafting departments and the manufacturing mangers. I, therefore, suggest that simply because a CAD system is capable of turning out manufacturing or CAM products does not necessarily mean that those products will even be used and that any cost savings available will derive to the company.

There's another case in which CAD developed manufacturing tapes do not necessarily become useful. This occurs where one uses an outside vendor to produce the PC boards from the artwork or negatives supplied by a design organization. Note if you will on figure 10 we have the same $4\frac{3}{4}'' \times 6''$ PC board arranged in three different patterns on the original raw material panel. As you can readily understand, the drill pattern for each of these three master panels would be different even though the same circuitry is involved. The pattern would depend on which arrange-

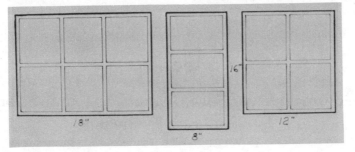

Fig. B-10.

ment the fabricator made of the PC board. Since the CAD system has no way of knowing in advance which arrangement the fabricator will use, the drill tape produced by the CAD system is of little value to the PC board fabricator. Even if the CAD system could be made known of the fabricator's arrangement in advance, there is no guarantee that the purchasing department is going to use the same fabricator everytime.

It is common knowledge that many PC board fabricators derive part of their income from "bomb-sighting" the drill tape each time. Therefore, they are just as likely to say that your CAD supplied drill tape will not work on their drilling equipment regardless of whether or not you are using the same panel layout as they are. In summary then, there are a number of unexplored areas of which anyone venturing into CAD systems should beware. The personnel involved and their attitudes toward the system, top management's attitude and support, or lack of, for the system, whether or not the by-products of the system will be as all pervasive and useful as they would appear to be on the surface, and finally some good procedures that must be followed to keep the system up and operating. If you can find satisfactory answers to all of these questions, you probably are going to have an excellent CAD system and will do well. If any of these areas hold significant doubts for you and your associates, you may do better not to plunge into a very expensive CAD installation but simply stay with the good old manual techniques which will get the job done and keep the work flowing out the door.

Appendix C

Soldermask Over Bare Copper: Alternatives and Manufacturing Techniques

There has always been interest in this type of circuit, but it has never been in greater demand than today. Current trends toward the following areas are responsible for this interest:

1. Higher circuit density.
2. Finer traces.
3. Tighter airgaps.
4. Smaller pads.
5. Wide bus traces and ground planes.
6. Higher reliability.
7. The desire for a more esthetically pleasing printed circuit.
8. The need for tighter control of impedance for stripline microwave and for ECL computer technology.

There are a number of alternatives to bare copper which the buyer and manufacturer of printed circuits should be aware of. These alternatives may result in greater yield and lower cost than the use of bare copper. The soldermask over bare copper type of circuit has always been more expensive and more difficult to manufacture—and it is likely to remain so. However, there are approaches which will reduce the burden significantly.

Before these alternatives are discussed, both users and manufacturers of printed circuits should understand the role of solder in printed circuit manufacturing. Solder is used by the manufacturer as an etch resist; it protects the copper circuitry during the etching operation. Solder also does a good job of protecting the solderability of the printed circuit for use by the buyer. It is convenient, easy to work with, and easy to maintain. To manufacture a printed circuit with no solder on the traces, but with solder on the pads and in the holes, is expensive and time-consuming from the manufacturer's view point. It involves chemically removing the solder and selectively reapplying it. It is the removal and reapplication steps which make soldermask over bare copper expensive to manufacturer.

ADVANTAGES OF SOLDERMASK OVER BARE COPPER

1. Solder slivers. The leading cause of electrical test failure is short circuits due to solder slivers. Solder slivers are caused by overhanging solder breaking off during a scrubbing operation and shorting between two conductors (Fig. C-1). These slivers can be virtually impossible to see, even with a magnification glass. If there is no solder, there can be no solder slivers. No matter how well the inspection process is carried out, printed circuits which have no solder on them must be inherently more reliable.

2. Solder is a reflowable metal; it melts at a relatively low temperature and has a surface tension. During the wave or dip soldering operation, the soldermask is actually floating on molten

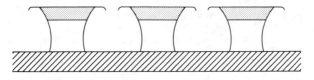

Fig. C-1. Solder slivers are the leading cause of short circuits.

solder. Molten solder wicks up beneath the solder mask during soldering. This causes a number of undesirable effects:

- Solder bridges often occur where two conductors are exposed at an opening in the solder-mask (Fig. C-2). This is especially common where there is little airgap between a hole pad (terminal area) and a conductor.
- The swollen solder on wide traces and land area is esthetically unappealing. The traces take on a crinkled, wrinkled surface.
- Large land areas can swell so much that the soldermask actually breaks. This is not a defect in the soldermask, or a defect in quality by the printed circuit manufacturer—it is a defect in the design, if anywhere at all. Many printed circuit designers avoid this problem by simply using a checkered pattern for the ground plane, instead of using a solid ground plane.

3. Visual inspection. Shiny solder surfaces are difficult to inspect, whether it is the manufac-turer or buyer doing the inspection. Reflection and glare reduce an inspector's ability to see defects, and contribute to eye fatigue. No matter how conscientious the inspector, the shiny sur-face must decrease the reliability of visual inspection.

4. Soldermask over bare copper, on the other hand, has no solder slivers; the soldermask is anchored to the copper and cannot float; solder cannot wick up beneath the soldermask; and the copper surface is less reflective than that of solder.

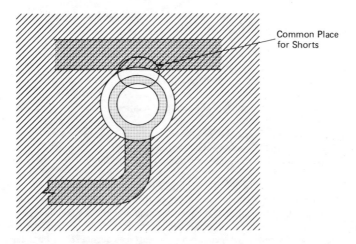

Common Place
for Shorts

Fig. C-2. Solder bridge.

DRAWBACKS TO SOLDERMASK OVER BARE COPPER

1. Added manufacturing operations mean added cost, and the possibility of lower yield.
2. Stripping solder from the printed circuit contributes greatly to waste disposal and pollution control problems of the manufacturer.
3. If there is a remnant of the solder, especially in the holes, the reapplied solder will not wet. The result is a board for scrap or rework.
4. Solder reapplication by immersion in molten solder is expensive. Either a very costly piece of capital equipment must be purchased, or the service paid for by having it subcontracted out.

- If the equipment is purchased, besides a large capital outlay, there is an increased maintenance burden.
- Subcontracting the immersion solder application to an outside service means abdicating scheduling and quality control.

5. Solder reapplication, via immersion in molten solder or by chemical deposition, may result in a printed circuit which is dewetted and which will perform poorly on being wave solderability tests:

- Copper builds up in all processes used to apply solder to copper surfaces. Electroplating is the only process for solder application where the copper content can be kept below 60 ppm, above which dewetting begins to occur.
- The holes on hot air leveled printed circuits may not solder nearly as well as the holes on printed circuits where the solder was plated.
- Slight bleeding of solder mask, very difficult to avoid from a manufacturing standpoint, typically results in scrap for SMOBC; while it will occur less and be acceptable on a traditional soldered circuit.

MANUFACTURING TECHNIQUES

Hot Air or Hot Oil Leveling

This is one of the most commonly used methods for applying solder to soldermask over bare copper type circuits. The basic process is as follows:

1. Image through hole plated printed circuit panels with a negative circuit pattern.
2. Electroplate copper and tin-lead. The tin-lead serves as the etch resist.
3. Strip the plating resist: screening ink or photoresist.
4. Etch copper foil.
5. Chemically strip the tin-lead from the copper circuitry.
6. Plate contact fingers.
7. Solder mask both sides of panels.
8. Tape off contact fingers with special hot air leveling tape.
9. Immerse panel in molten solder/hot air level; hot oil leveling may also be used.

One disadvantage of this method is the need for chemically stripping the solder. This requires extra manufacturing steps, uses great volumes of costly and hazardous metal stripper, and creates added waste hauling problems and expenses. Should any of the solder remain, especially in the holes, dewetting or non-wetting will occur. Nickel, tin-nickel, or tin cannot be used as a non-solder etch resist, as soldermask will peel from these surfaces during solder immersion.

The contact fingers must be plated prior to solder immersion. If the tape leaks, solder will adhere to the gold plated contacts. The taping, untaping, tape residue removal, and solder on the contact fingers result in labor intensive rework and touch-up operations.

Black Oxide as an Etch Resist

Black oxide provides a potentially viable method of using a non-solder etch resist. Black oxide is easy to apply, and strips cleanly with hydrochloric acid. The basic process is as follows:

1. Image through hole plated panels with negative circuit pattern. Vinyl screening resist must be used.
2. Electroplate copper only.
3. Immerse panels in black oxide, 3–5 minutes at 80+% potency.
4. Strip vinyl plating resist.
5. Etch copper foil.
6. Strip black oxide to 50 : 50 mixture of water and concentrated hydrochloric acid.
7. Plate contact fingers.
8. Soldermask both sides of panels.
9. Immerse panels in molten solder; hot oil or hot air level.

There are some considerations when using this process. The solder must be applied by immersion methods. However, there is no solder to strip. Vinyl resist is not as convenient to strip, as is the more common alkaline soluble screening resist. Care must be taken not to scratch the oxide through to the copper; it is fairly resistant to mild abrasion. By keeping black oxide bath potency up, immersion time is greatly reduced.

Plate Solder in the Holes with Nickel as an Etch Resist

This technique is one of the oldest methods used for selectively applying solder to the holes and pads. It uses a non-solder electroplated etch resist. This method uses a double screening method; once for the circuit pattern, and once for the hole pattern. The basic process is as follows:

1. Image through hole plated panels with a negative circuit pattern.
2. Electroplate copper and nickel (or tin-nickel or tin).
3. Rescreen the electroplated panels using the padmaster or soldermask artwork; the first layer of resist image is not stripped. This will result in panels with only the holes exposed for solder plating.
4. Electroplate tin-lead in the holes and on the pads.
5. Strip both layers of plating resist.
6. Etch copper foil.
7. Plate contact fingers.
8. Reflow electroplated tin-lead.
9. Soldermask both sides of the panels.

This techniqiue is straight forward. There is no need to strip solder from the circuitry, and no need to immerse the panel in molten solder; and no need for an added taping operation to protect the contact fingers. The only disadvantage, outside of the double screening, double plating operations, is that possiblity of slivers has not been totally eliminated. There may be nickel, instead of solder slivers. However, the soldermask will not float, peel or crinkle during wave soldering; and the esthetic appearance of the board is excellent.

Plate Solder in the Holes with Immersion Tin as an Etch Resist

This technique, like the nickel resist method just described, uses double screening and double plating. However, the etch resist is not electroplated and does not cause slivers, and soldermask adhesion is excellent. The basic process is as follows:

1. Image through hole plated panels with negative circuit pattern.
2. Electroplate copper.
3. Rescreen image panels using padmaster or solder mask artwork. This will result in panels with only the pads and holes exposed for tin-lead plating.
4. Electroplate tin-lead in the holes and on the pads.
5. Strip the plating resist.
6. Etch copper foil.
7. Plate contact fingers.
8. Reflow tin-lead plated in the holes, and on the pads.
9. Soldermask both sides of panels.

This technique also is straightforward. Alkaline soluble screening resist should be used for imaging. The chief requirement is that the immersion tin be maintained at about 80% bath strength or better. This can easily be accomplished by tracking panel square footage, and making frequent additions, for example, 12.5 pounds of tin salts in a 100 gallon tank per 100 panels. The only problems which detract from this method result by allowing tin content to slip below 60–70%. The end product is one which has all the features of soldermask over bare copper, and none of the drawbacks of using an electroplated etch resist, such as nickel. Immersion tin is not reflowable, solder mask adhesion is excellent, and so is board appearance.

It is also possible to use immersion tin as an etch resist, then strip it and apply molten solder with hot oil or hot air leveling. The immersion tin is much more easily and economically stripped than is tin-lead.

Immersion Tin-Lead for Solder Application

It is possible to apply solder to the bare copper in the holes and on the pads using immersion chemistry. There are several specialty chemical manufacturers producing baths which will deposit up to 150 micro-inches (0.00015 inches) of tin-lead. The deposit can then be reflowed, the crest of the deposit having the potential of reaching 300 micro-inches. The basic process is as follows:

1. Image through hole plated panels with negative circuit pattern.
2. Electroplate copper and tin-lead.
3. Strip plating resist.
4. Etch copper foil.
5. Strip tin-lead from copper circuitry.
6. Plate contact fingers.
7. Soldermask both sides of panels.
8. Immersion plate tin-lead in holes and on pads.
9. Reflow deposited tin-lead.

The immersion bath must be maintained at a high concentration. This may be difficult, as the bath may require up to four addition agents, yet analysis is provided for only one component. It is good to leach the deposit in hot water for several minutes after tin-lead immersion. This will help to assure removal agents from the immersion bath which may co-deposit with the tin-lead. There is always the possibility of poor quality deposit, so deposit quality must be monitored by inspecting the reflowed tin-lead. Still, the operation does work, and is used.

Panel Plate, Print and Etch

This would appear to be an ideal method. After all, inner layers and single sided circuits are produced this way. The basic process is as follows:

1. Electroplate copper to the bare through-hole-plated panel.
2. Flood the holes with plating resist, and apply a positive circuit pattern: a print-and-etch pattern. Alternatively, use dry film photoresist to apply a positive circuit pattern. The photoresist will be used to "tent" over the plated through holes to keep out the etchant.
3. Etch copper foil. The etchant and etching resist must be compatible.
4. Strip etching resist.
5. Plate contact fingers.
6. Soldermask both sides of panels.
7. Apply solder by hot oil or hot air leveling.

This, too, is a straightforward method. The problems result from the requirement that resists either completely plug 100% of the plated through holes, or, in the case of photoresist, reliably "tent" 100% of the holes. Should there be only one hole per circuit which voids because of insufficient resist coverage, that circuit will be scrap. Photoresist does not adhere well to electroplated copper, and must have 0.010 inch or larger annular ring to provide enough grip for tenting. This means that circuit and artwork would almost have to be laid out with this process in mind.

Plate Holes, Print and Etch

This sounds like another straightforward technique. but has reliability problems. The basic process is as follows:

1. Use padmaster to image through-hole-plated panels. This will result in panels where only the holes and pads are exposed for plating.
2. Electroplate copper and tin-lead in holes and on the pads.
3. Strip plating resist. What results is bare copper panel with copper/tin-lead plated holes.
4. Screen image the panels with a positive circuit pattern.
5. Etch copper foil.
6. Plate contact fingers.
7. Reflow tin-lead in the holes and on the pads.
8. Soldermask both sides of panels.

The advantages are obvious: no metals need be stripped, and there is no need to immerse panels in molten solder. However, the reliability of flooding the fillet formed at the hole/copper surface junction (Fig. C-3) with resist is low. Generally, a coarse screen must be used, such as 150 mesh

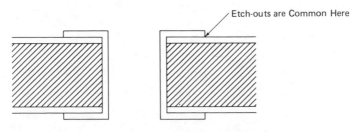

Fig. C-3. Fillets must be flooded with resist.

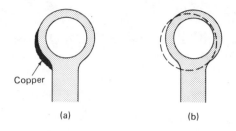

Fig. C-4. Misregistration effects. (a) Unetched copper due to resist coverage. (b) After misregistration during second screening, resist prevents clean etching at pads.

polyester. When this happens, the ability to produce fine line circuitry is diminished. If there is misregistration, which is very likely to occur, excess and unwanted copper will remain around the pads after etching [Fig. C-4(a)]. Plated circuitry tends to increase in width, as the plating creeps over the resist. This actually reduces registration difficulties [Fig. C-4(b)] as the plated pad will be larger in diameter than the pad on the artwork.

SUMMARY

Printed circuit manufacturers increasingly receive requests for quotation on soldermask over bare copper today. There are numerous alternatives, some of which are quite viable and will meet the goals of the customer. These alternative techniques should be explored by more manufacturers, as an aid in reducing the manufacturing burden placed on them by soldermask over bare copper. At the same time, the manufacturer must understand the manufacturing difficulties and should decline SMOBC if it will add unwanted manufacturing difficulties.

Appendix D
Control and Operation of Printed Circuit Plating Baths with the Hull Cell

All too often plating baths in the printed circuit shop operate at less than optimum performance. Even the most common, avoidable, and easy to manage problems lead to scrap, down time, and missed due dates, when action is taken without understanding the problem. The most valuable tool the plater, or process engineer, has to evaluate his plating baths is the Hull cell.

What is the Hull cell, and what can you use it for? "The Hull cell is a miniature plating unit designed to produce a cathode deposit that records the character of electroplate obtained at all current densities within the plating range."[1] Simply put, it is a small plating tank (267 ml) you can use to quickly test the performance of your electroplating baths.

A brass panel is situated at an angle to the anode. This gives you plating from high current density, nearest the anode, to low current density, furthest from the anode. Plating tests are run at a constant amperage (typically 1 or 2 amps for printed circuit applications) for 5–10 minutes.

Once the brass panel has been plated, it is compared to the Hull cell ruler. The Hull cell ruler lets you see what the current density is at any location on the panel. What can this information be used for? Most plating bath constituents can be determined by chemical analysis or some other measurement: metals, acids, pH, specific gravity, temperature, surface tension, etc. But certain additives are difficult to determine chemically or by other measurement: brighteners, leveling agents, stress reducers, grain refiners; and other desirable and undesirable contaminants, organic and ionic. If a constituent has an effect on plating (almost all do) its presence or absence can usually be detected by a Hull cell.

Plating used to be referred to as Black Magic. Problems would arise and neither their cause nor their cure would be understood by the platers. In more recent years the view of Black Magic has subsided as plating quality became critical to the electronics industry. The best way to keep practitioners of Black Magic out of your plating shop is to read your bath manufacturers' literature and to use the Hull cell on a regular basis in your operation.

Do any of these problems sound familiar?

Copper

1. Dull, burned, or step plated copper; with or without pitting or nodules, or both.
2. Localized dullness or haziness or fish eying at plated through holes.
3. Dull plating along edges of traces and ground planes.
4. Hazy black or discolored copper accompanied by high rectifier voltage and low amperage.
5. Large pits along traces or contact fingers. This may show up as "half pits" along the edges of contact fingers or traces—and resemble overetching.
6. Copper-to-copper peelers.

This article appeared in *Printed Circuit Fabrication* magazine.
[1] *Directions for Hull Plating Tests*, R. O. Hull & Co., Inc.

Tin-Lead

1. Dark or streaky tin-lead.
2. Gassing at the cathode in tin-lead, with or without treeing (tin-lead burning).
3. Thin plating or failure to cover in whole or part, no matter how often you check rack and panel connections.
4. Large weekly addition of tin and lead fluoborate required to maintain bath concentration.
5. Dewetting, nonwetting, or gopher burrowing (bumps) in the reflowed solder.
6. Dark, murky, cloudy tin-lead plating bath with visibility only a few inches.

Nickel

1. Pitting.
2. Nodules.
3. Step plating.
4. Dull, hazy plating.
5. Large additions of nickel sulfamate or sulfate.
6. Uneven plating from rack to rack.
7. Nickel-copper peelers or "bubbles" beneath the plating.
8. Gold peelers off the nickel.

These are all common problems that plague printed circuit shops from time to time. Have you ever seen a plater try adding "a shot of brightener," or a quart of peptone to a bath, and still the problem remained? Have you ever dumped a plating bath in frustration or thinking it was actually the only thing to do? Your options for identifying and resolving and preventing problems such as these are greatly increased by using the Hull cell.

In most printed circuit shops the big work horses are acid copper, tin-lead, and nickel. This discussion will be limited to these baths. However, if you run pyro copper, bright acid tin, tin-nickel, gold, silver, rhodium, or any other electroplating bath, the Hull cell is equally valuable.

It is difficult to get optimum performance, month after month, out of any plating system. Good plating procedures and well trained platers help, but problems do arise. When the Hull cell is used as part of your routine, a developing problem can be detected before it gets out of hand. When a problem gets to the point that it impacts production, the Hull cell allows you to zero in on the cause and determine what course of action will eliminate it.

The Hull cell gives you a greater perspective on your plating bath performance—as long as you understand what the test panel is telling you. Often, if you are having a plating problem, several test panels must be run in order to draw valid conclusions. The following information can serve as guidelines on how and why to start monitoring you plating baths with the Hull cell. A good point to keep in mind: do the Hull cell testing in house. It is neither difficult nor expensive to set up and perform, and Hull cell testing is far too important to farm out to an analytical service. Almost all bath vendors recommend using the Hull cell, and most publish charts and pictures of what the test panel will look like for a variety of conditions. Remember, the more your use it, the more you will learn from it, and the more you will understand your own plating baths.

Generally, the Hull cell should be evaluated against the actual plating coming from the bath. The person performing the Hull cell tests should get in the habit of pulling panels from the plating tank and examining them. It is important to understand the correlation between actual plating and what the same condition looks like on a Hull cell panel.

ACID COPPER

1. Acid copper baths need brightener added throughout the day. Additions are usually made on the basis of milliliter per amp-hour. However, there are many factors affecting the rate at which

the brightener is actually consumed. The net result of these factors is that brightener is usually on the increase or on the decrease in your bath. If concentration gets too high or too low, you will have detrimental plating.

    a. Extended periods of bath shutdown causes anode film to loosen. If the air agitation is turned on prior to dummy plating, extra brightener will be consumed in rebuilding that anode film.

    b. Temperature affects the ability of brightener to do its job. A colder bath will require more brightener for a degree of leveling than does the same bath at a higher temperature.

    c. Anode to cathode ratio affects the throwing power and the ability of an anode to properly corrode and hold the anode film.

    d. Large additions of anode area (replenishment) will also contribute to increased brightener consumption.

2. Plating personnel also play a major roll in determining the rate of brightener consumption. It often occurs that a plater will forget to add on schedule, or will make an unscheduled addition.

3. High and low metal and acid concentrations have some of the good and bad aspects of varying brightener concentration. In many cases, you can operate with less brightener by running copper concentration far above what the manufacturer recommends. This condition results from an excess of anode area; the anode actually dissolves into the plating bath and increases concentration of copper.

4. Contamination may occur through a variety of channels/ including the normal operation of the plating bath; it is inevitable. It is not, however, necessary to produce scrap because of it. Whether it happens gradually, or overnight, contamination can be detected with the Hull cell. Furthermore, you can characterize the nature of the contamination and how to eliminate it by performing tests on a few hundred milliliters of bath. It is practical to do a purification on 500 mls of bath and run a Hull cell to determine effectiveness—before you touch the whole bath. This is one of the major benefits of Hull cells.

5. It is common for an electrical defect to cause problems that look like a chemical condition related to the bath: poor connections, and reversed polarity, and no electricity flowing at all, or excess current.

6. Lack of agitation may also cause problems that appear to be chemical in nature.

Any of these conditions can cause problems. Yet, when the problem occurs it is difficult to say with any degree of certainty "*This* is the cause." When a problem does arise, there is a definite sequence of action you may take to determine the cause and the cure. For example: Dull, step plated copper. *Don't* add brightener to your plating bath until you:

    a. Run a Hull cell panel and see how it looks. If the sample you ran looks optimum per your experience and the manufacturer's literature (see Fig. D-1), look for a nonchemical reason for the plating problem and stop running Hull cells until you've checked bath temperature, looked at the plating tank, and talked with the platers.

    b. If the test panel doesn't look optimum, add a small amount of brightener (equivalent to 1 quart in the plating tank) to the test sample and run another test panel in the Hull cell. If the plating quality has improved, chances are good that your only problem was low brightener. You can perform more tests with more brightener until you achieve optimum results, before you add a drop of brightener to your plating tank.

    c. If the brightener added had only marginal results on the test panel, look for other causes of your problem:

1. Check the most recent analysis, or analyze it immediately. Is everything near optimum? Have all the adds been made?

Hull Cell Control of CopperLume P.C.
Operating Conditions:
Temp. 75°F
Current 2 Amps
Time 10 Min.
Air Agitation

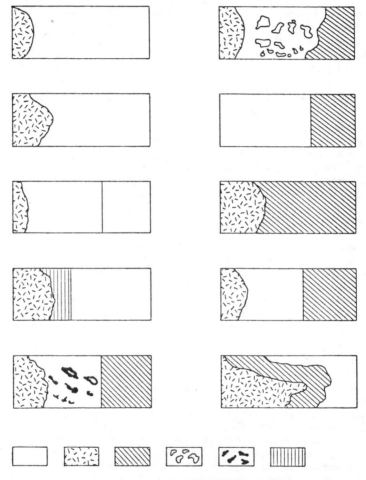

Fig. D-1. Courtesy of M & T Chemical Co.

2. Is the bath temperature within range? If not, heat a sample and run another Hull cell test.
3. Chloride can change drastically in a few days. Take 1 ml of concentrated hydrochloric acid and dissolve it into 40 ml of water. Take 1 ml of this and add to the Hull cell; this will raise chloride by about 35 ppm. Do another Hull cell test on this.
4. Always read manufacturer's literature, and never take drastic action until you have evaluated the possible results with a Hull cell.

Acid Copper Hull Cell Conditions

The author generally uses 2 amps with mechanical agitation for 5 minutes to monitor brightener; use at least 10 minutes for any other purpose; such as investigating anode polarization.

TIN-LEAD

Tin-lead plating also requires Hull cell use for optimum performance. Most of the common problems of tin-lead can be caused by the plating itself, the copper it is plated upon, or other processes the tin-lead will see down the line—resist stripping, etching, scrubbing, fusing, and the preparation steps for fusing. Just about any problem associated with tin-lead can be caused before, during, or after the plating. Because of this, you need a reliable method for distinguishing and eliminating possible causes. The Hull cell is most valuable here.

1. Too much peptone, or other organic contaminant, will cause outgassing. This shows up as bumps or gophers burrows in the fused metal. But outgassng may be due at least in part to other factors.
2. Dewetting, too, may be due to copper in the bath, dirty or oxidized copper substate, or ill-prepared tin-lead prior to fusing.
3. Thin tin-lead may be due to poor connections, lack of anode area, low amperage, low peptone and throwing power, low agitation and metal content, or too great a pH in an ammonia etch.
4. Treeing and gassing at the cathode may be due to peptone level, metal content, amperage, poor connections, or lack of agitation.

Copper gets into the bath every time you place a panel in the tank for plating, and every time the bus bars are rinsed off. Peptone should be added soley on the basis of Hull cell test.

Treeing will show up on the Hull cell panel as "burning" along the left edge of the panel. Treeing is all right, as long as it does not extend into the plating range, as measured with the Hull cell ruler. When it does, the need to add a small amount of peptone is indicated—the least volume required to eliminate treeing in the plating range. Excess peptone will show up on the Hull cell panel as an S-shaped curve in the right side of the panel. The deposit will be lighter on the left of the curve, than the right side. When you see this on your Hull cell panel, you will see it in the plated and fused printed circuit board—unless you carbolate or carbon treat immediately.

Higher copper will show up as a dark streak along the right hand (low current density) edge of the Hull cell panel. If the tank is dummy plated for a few hours, at 2–3 amps per square foot, the problem should go away on the next Hull cell panel.

Tin-Lead Hull Cell Conditions

The author generally uses 1 amp with no agitation for 5 minutes to monitor peptone level; use 10 minutes for any other reason.

Agitation is not recommended because it is easier to identify contamination with reduced agitation. Most tin-lead baths do not have a great deal of agitation anyway.

Nickel

Nickel baths are easier to run with the routine use of a Hull cell. Lack of brightener and the presence of metallic contamination show up readily on the test panel. Low brightener will show as dullness extending from the left edge of the panel into the plating range. Metallic contamination shows as haziness in the plating range. Patchy dullness may indicate the presence of organic contamination. It is easy to confirm any possibility merely by dummy plating, or carbon treating small samples, and running another Hull cell test.

GETTING SET UP FOR RUNNING HULL CELLS

You will need to buy the following:

1. Acrylic plastic Hull cell. These are available in several sizes, the smallest being the 267 ml, and the one I recommend. They are also available with and without heater and air sparger (for which you need an aquarium air pump). I recommend the basic Hull cell without heater and without air sparger.

2. D.C. current source; a small rectifier works nicely.

3. Agitation system. You can use a stirring rod or buy a convenient mechanical agitator. The mechanical agitator is very nice to have, especially for running large numbers of Hull cells; and you probably get slightly more consistent results.

4. Anodes. You need one anode for each bath you wish to test with the Hull cell.

5. Brass Hull cell panels.

RUNNING THE HULL CELL TEST

1. Place the desired anode in the Hull cell.

2. Fill the Hull cell with the desired plating bath, corresponding to the anode.

3. Clean the brass Hull cell panel:

- 5 minute soak in alkaline cleaner.
- Water rinse.
- Dip into 50% sulfuric acid.
- Deionized water rinse—should have good water break.

4. Place panel in cell, connect NEGATIVE lead to panel, and POSITIVE lead to anode.

5. Begin agitation.

6. Turn on and set rectifier and timer.

7. At completion of time:

- Turn off rectifier, timer and agitation.
- Rinse panel with deionized water and alcohol to dry free from spots.

8. Compare panel to Hull cell ruler.

WHERE TO BUY HULL CELL EQUIPMENT

R. O. Hull & Company
3202 W. 71st Street
Cleveland, Ohio 44102
Tel.: (216)651-7300

Kocour Company
4800 S. St. Louis
Chicago, Illinois 60632
Tel.: (312) 847-1111

You can buy the equipment from the above sources or from most plating supply houses.

SPECIAL NOTES

The back of the Hull cell can give you a measure of the throwing power of your plating bath: the ability of your bath to cover your cathode. Sometimes it is difficult to get the backside of the brass panel clean enough to give reliable results. For this reason, I generally recommend using brass panels made by the R. O. Hull Co.

There are other configurations and models of the Hull cell; you might want to investigate these for your own application.

Electroplating addition agents are expensive. Moreover, if their concentration in the plating bath gets out of hand, the result is apt to be down time, scrap boards, and missed due dates. There are so many factors operating which affect the quality of plating, that the well run printed circuit plating shop must possess the means of monitoring plating quality. It is just as vital to be able to trouble shoot plating problems without performing drastic measures to the entire bath. The Hull cell is your most valuable tool in meeting these objectives.

Appendix E
Troubleshooting the IR Fusing Process

Mike Palazzola, Chemelex Technical Service Chemelex, Division of RBP
Chemical Corporation

In printed circuit board manufacturing, many problems remain undetected until the final fabrication phase of IR fusing. These problems may occur at one specific point or be a culmination of little things throughout the entire process. Regardless of the problem source, the board at this stage is near final completion and has accumulated considerable material and labor expense. For this reason, it is necessary to properly identify these problems as quickly as possible, and go back in process and remedy the situation.

In the following paragraphs, I would like to discuss various problems which come to light after IR fusing, and offer advice on solving them. While some of the suggestions may appear to be basic, it is important that you do not overlook the obvious when troubleshooting the IR fusing process.

DEWETTING

One of the most common faults noticed after IR fusing is dewetting, yet it is not necessarily a fault of fusing (Fig. E-1). Dewetting appears as a ball of solder, exposing a thin silver-gray surface (total non-wetting exposes a copper surface). In most cases, this problem is caused by the improper cleaning of copper surfaces prior to tin-lead plating. Normally, this residue is a copper oxide which can be easily removed with a simple acidic cleaner. If a check of pre-plating methods proves cleaning to be adequate, an evaluation of the tin-lead plating bath itself should be made.

A check should be made on the level of inorganic contaminants, such as copper (dragged in from an overused fluoboric acid tank) or iron (caused by an accidental dropping of a nut, bolt, etc. into the tank), which can alter bath make-up. If an evaluation does prove the bath to be at fault, two courses of action are available:

1. The entire bath can be replaced (which can prove to be expensive).
2. Dummy plate: entire panels plated at a lower current density (3–5 amps/square foot) in order to plate out trace metal contaminants.

Dewetting should not be confused with another surface condition in which plating, after fusing, has a gritty or granular appearance (similar to gritty sand). This is not a dewetting condition but is attributable in most cases to excessive grain refiner in the tin-lead bath which leads to excessive plate-out of the grain refiner with the tin-lead. The grain refiner also breaks down and contains a certain level of impurities. As this level builds up, it will co-plate and the resultant will be an improperly fused board. (Fig. E-2). If this condition exists, the solder bath must undergo carbon treatment (which removes all traces of organic matter) and be properly replenished. To a lesser degree, this can also be caused by the excessive presence of particulate matter,

This article appeared in *Printed Circuit Manufacturing* magazine.

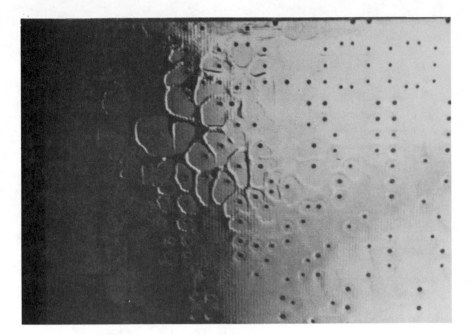

Fig. 1. Dewetting.

Fig. 2. Organic impurities; sand finish.

such as dust, carbon particles (improper bath filtration after carbon treatment), filter fibers, and sludge. In most cases, continuous filtration through a 10–20 micron filter will remove particulate matter.

POOR COSMETICS

Another problem which plagues may PC fabricators is unacceptable board appearance. Ideally, after passing the fusing and rinse cycles, the board should display a shiny, mirror-like surface (Fig. E-3). Often overlooked is the fact that the alloy composition of the solder bath can have a remarkable affect on end-product cosmetics. Grain structure can range from a spangled, galvanized appearance for a high tin make-up (Fig. E-4) to shiny, reflective cosmetics at eutectic (63% tin, 37% lead) to a gray, dull surface as characterized by high lead.

One of the more common causes for off-eutectic plating is improper bath composition (assuming the bath is one with good throwing power). This condition can be monitored through available analytical techniques. In addition, this condition may also be the result of a couple of other variables:

1. Excessively high or low current densities.
2. Thickness of solder plate (if tin-lead plating is less than 0.00030 inches thick, solder results will be very similar in appearance to high lead plating—dull and gray.

WHITE RESIDUES

White residues on the board after fusing can be, and often are, considerable problem areas appearing on the metal, the laminate, or both (Fig. E-5). White residue on the metal is generally attributable to excessive tin-lead oxides prior to fusing, improper rinsing of the flux after fusing, or improper use of the flux itself.

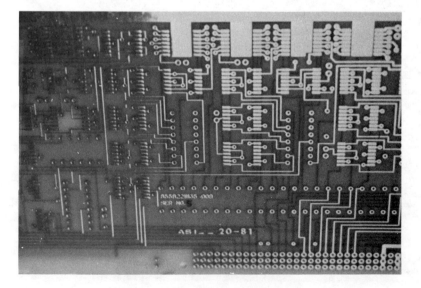

Fig. 3. Properly reflowed tin-lead.

Fig. 4. High tin content; galvanized.

Fusing fluids can break down a degree of oxidation, but when excessive, the oxide film will remain unfused and intact. This film will float on the melted solder, and recondense when the board cools, thus leaving residual matter in the fused areas.

Even if the board is properly pre-cleaned, an improper rinsing procedure following fusing will create troubles. Because most proprietary products contain some level of chloride or halogenated

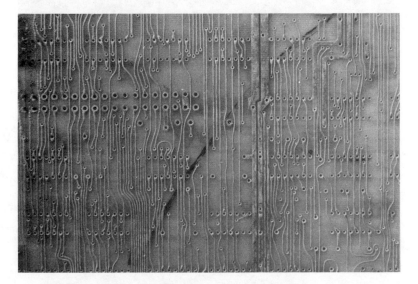

Fig. 5. Residue on laminate and metal surface.

acid, incomplete removal will result in tin-lead attack, causing a marred appearance or, worse, poor solderability. Therefore, proper rinsing techniques should utilize warm, running water with a light, nylon scrub brush (water temperature and exposure time required will vary, depending upon which proprietary flux is used).

It should be kept in mind that not all water-soluble fluxes are, in fact, water soluble. With particular proprietary products, a breakdown occurs in which, after fusing, insoluble particles remain on the board and require additional cleaning aids or procedures for thorough cleaning.

Improper use of fluxes can also leave residue problems if flux reuse is abused. While additional mileage can be achieved by replenishing flux with isopropyl alcohol, flux reuse will pick up foreign matter (inorganic contaminants such as iron, copper, tin-lead, and organic breakdown products), which will lead to rinsing problems.

In regard to flux usage, it is also recommended that specific gravity be maintained. If ignored, the flux will become highly concentrated, again leading to rinsing difficulties. Several other factors which require attention when removing white residue from metal are the water quality and amount of time elapsing after IR fusing before rinsing. The water used for rinsing should be consistent and as mineral-free as possible. Boards, after fusing, should be rinsed thoroughly when panel temperatures after fusing are 200 degrees F. or less, because a warmer board aids solubility of residue removal. Also, if boards after fusing are allowed to stand for any length of time, or cool to room temperature, residuals become difficult to remove.

White residues on the laminate (as opposed to on the metal) are a less common occurrence and are a result of an interaction between the fusing fluid and improperly cured, or poor quality laminate (Fig. E-6 and E-7). The flux itself will make a difference as to the level of interaction. While most flux manufacturers make use of the same chemical type, specific types and the additives that are utilized will make a difference. This type of residue can be removed with solvents, such as acetone, methylene chloride, trichlorethane, etc. However, in this situation it is best to consult your laminate supplier.

Fig. 6. Dye absorption of laminate.

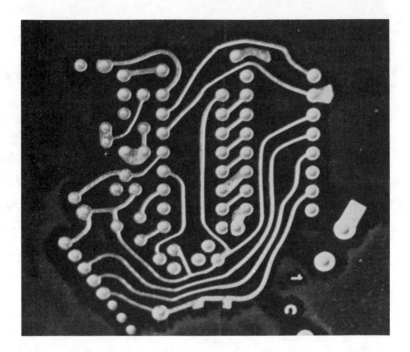

Fig. 7. White residue on laminate.

POOR EDGE COVERAGE

Proper edge coverage is critical to PC board performance and is a primary reason for fusing. Poor edge coverage can generally be eliminated by reviewing two areas: neutralizing and rinsing of the ammoniacal etchant, and effectiveness of the acid used in your fusing fluid. The acid level in the fusing fluid will eliminate most copper oxides that form after etching; however, the fusing fluid is not designed to act as a copper cleaner. Ammoniacal etchant residues need to be neutralized after etching and properly rinsed off. If this does not happen, you can expect further line degradation and poor coverage after fusing.

Because acid levels can vary considerably from flux to flux, the fabricator has a choice concerning activity level. A lower acid formulation will be less effective in oxide removal and will require a more stringent post-etch/pre-fusing clean cycle. However, final rinsing will be easier. With a higher activity flux, you may be able to get by without a pre-clean procedure, yet post-fusing rinsing will be more critical. In addition, the higher the acid content, the greater the attack on equipment. The trade-offs exist regarding high versus low activity flux usage, and the decision is up to the fabricator.

GENERAL PROBLEMS

Do your boards look good on the surface but have white residue in the holes (Fig. E-8)? Also, do your boards look significantly better when hot oil fused versus infrared fused? When using hot oil, you employ a mechanical rinsing effect of immersing the board into a hot oil, causing a

Fig. 8. White residue in hole.

flushing effect upon removal from the solution. Also, there is a more even heat distribution across panel surface and through hole.

When using infrared, the board is exposed to infrared rays on the surface, and the flux is used as the carrier to heat the board evenly, through-hole and across the surface. Since no mechanical flushing is achieved, the solder solidifies, thus entrapping certain residuals which will be pronounced through hole. Sometimes these residuals are nothing more than tin-lead oxides, which good precleaning techniques will take care of. At other times these residues may be your co-plated grain refiner. This indicates an excess grain refiner or possibly the wrong grain refiner is being used.

When infrared fusing, do you sometimes experience a need for more heat? If this occurs, you should check the alloy composition. Remember, nothing is constant, and don't overlook the obvious.

Eutectic solder fuses at the minimum temperature; any variations in the eutectic point will require more heat to fuse. Also, continued use of the same flux may require an increase of heat. Again, most fluxes may be reused, but these solutions are not infinite.

Excessive smoke is a nuisance for operators, and generally can be eliminated. Assuming adequate exhaust exists there are several other problem sources. Boards, if run at too high a temperature, or at too slow operating speed, will generate smoke. Panels should be run at optimum heat and speed at all times, not only to eliminate smoke but also to prevent other mishaps, such as flash fires, delamination, and poor reflow.

A final problem area is in regard to dryness of the boards after fusing. If this occurs, several reasons may exist: the flux may be thinned excessively or applied too sparingly; excessive temperatures may accelerate flux evaporation or excessive air flow from exhaust may evaporate the fusing fluid.

Various aspects of infrared fusing problems, their respective causes, and solutions have been looked at. Many problems that are noticed after fusing are generally a result of previous deficiencies in process. When problem solving, never make assumptions; sometimes the obvious is not so obvious. Also, take one variable at a time. Process of elimination may take more time, but its results are always beneficial.

Appendix F
Preserving Solderability with Solder Coatings

James P. Langan, Federated Metals

Tin and tin lead coatings are employed to provide and preserve solderability of printed circuit boards and component leads. Solderability indicates the ease with which molten solder will wet the surface of the metals being joined. Wetting is a surface phenomenon which depends, among other factors, on surface cleanliness. When molten solder leaves a continuous permanent film on the metal surface, it is said to wet the surface. A bond is formed by atomic attraction between the solder base metals. The reaction may also involve some diffusion of the solder metal or alloy into the base metal or vice versa. Tin and tin lead coatings provide corrosion protection of a base metal which has been rendered solderable by a cleaning process prior to plating or fluxing process prior to hot coating.

These coatings can be applied by electroplating or hot coating. The advantage of electroplating is that it can perform a three fold purpose, namely to act as a metal etch resist, preserve solderability, and provide corrosion protection. To do this effectively, a quality plating is required.

Quality plating is impossible without good process control which is essential to produce a deposit that is low in porosity, low in co-deposited solderable surface. In the case of solder plate, it is essential to deposit the correct alloy composition. When it is required to reflow solder plate, control closer than the 50–70% tin permitted by Mil Std 275D is usually needed.

Today, most solder plated printed circuit boards are reflowed to insure elimination of the solder sliver created by undercutting during etching. (See figures 1, 2, 3). This is done to avoid the possibility of this sliver breaking off from vibration and causing a short. Reflowing also guarantees that the cleaning process prior to plating is adequate. It also produces a slightly denser deposit with less porosity and improves the appearance of the coating. To reflow properly, the alloy composition should be between 57%–67% tin.

Hot coated solder is an excellent process for preserving solderability. However, it requires an organic etch resist which is capable of protecting the plated through holes. Dry film resists can be used or an etch resist. Air leveling equipment is required to insure sufficient thickness and guarantee hole clearance. This equipment is expensive and requires good maintenance.

An advantage of this process is that solder mask can be applied over copper which eliminates the problem of solder resist wrinkling.

TIN-LEAD ALLOY PLATING

The deposition of tin-lead is one of the simplest alloy plating system operations since the standard electrode potentials of the two metals differ by only 10 mV and are closer than any other two metals except silver and mercury. In addition, the potentials of lead and tin are only slightly less noble than hydrogen and, since the metals have a high hydrogen overvoltage, the alloys that

Reprinted from California Circuits Association Proceedings, San Mateo, California, October 15–16, 1980

Fig. F-1. Conductor pad after etching where tin-lead is the etch resist material. Note that under-cutting causes solder sliver (667× magnification).

deposit out of the acid solutions have a cathode current efficiency of 100%. Unlike brass where deposit color is an indication of alloy composition and an approximation of the alloy composition can be obtained by matching it to color standards, tin-lead alloys do not differ enough in appearance from one composition to another and must, therefore, be controlled by analytical methods.

Tin-lead alloys can be plated from several types of solutions, but the fluoborate is the one

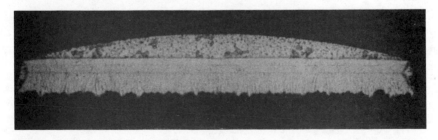

Fig. F-2. Tin-lead deposit after reflow to remove solder sliver (200× magnification).

Fig. F-3. Side of conductor pad after reflow. Fusing flux cleaned side of pad, causing pad solder coating (750× magnification).

universally used. The alloy can be deposited from sulfamate, chloride, fluosilicate and pyrophosphate solutions, but up to now, none has gained wide acceptance.

Tin-lead fluoborate solutions most often are prepared from purified liquid concentrates. They can be made from reacting hydrofluoric acid with boric acid and then reacting this mixture with lead carbonate to form lead fluoborate. Tin fluoborate can be prepared by reacting stannous oxide with fluoboric acid. Unless one has the proper facilities and know-how, it is advisable to buy the purified liquid concentrate.

CONTROL OF PLATING SOLUTION

In order to assure highly reliable and economical tin-lead plating, the alloy must be applied over a clean basis metal.

Controlling the alloy composition depends on the following factors:

1. The stannous tin to total metal content, Stannous/Stannous + Lead.
2. Current density.
3. The addition agents used and their concentrations.
4. The alloy composition of the anodes.
5. Solution contamination.

CO-DEPOSITED IMPURITIES

The major metallic impurity that affects the quality of tin and solder is copper. This is why it is essential to only use chemicals and anodes that are manufactured to rigid quality specifications.

Copper is lower than both tin and lead in the electro motive series and thus will readily co-deposit with tin and lead. As little as 20 ppm of copper will cause a darkening of the deposit in low current density areas. A tin lead fluoborate solution where 139 ppm of copper was deliberately added resulted in 0.25% copper in the deposit. Copper can cause oxidation of tin in solution and has a passivating effect on the anode. Most of the early problems when reflowing of tin lead was specified in MIL-STD-275 D were caused by copper impurities in the solution. The major sources of copper contamination were plating salts and impure anodes.

Fortunately, once the source of the copper is removed, the copper level in the solution can quickly be reduced by low current density electrolysis. In the late sixties and early seventies, when reflowed tin lead started to become a requirement for most Printed Circuit Boards, careful attention was paid to assure the purchase of high quality raw materials. In 1972, the cost of tin and lead started to escalate and in many cases more emphasis was placed on price rather than quality.

One of the contributing factors is that there is no Federal Specification governing the purity of raw materials employed in this plating process. Quite often, Federal Specification QQS571E is used as a guideline. This particular specification is for wire solder, bar solder, and paste solder. Since it has no reference to electroplating, blind adherence to the "letter of the law" could result in serious plating problems. This specification permits 0.08% copper which, if allowed in anodes, reflow and soldering problems would result. Antimony is 0.2% minimum and 0.5% maximum, and at these levels a heavy black anode sludge would develop which would cause a lowering of anode efficiency.

The following is a suggested specification for tin and solder anodes:

	Sn/Pb	Sn
Ag	0.004	0.004
Al	0.001	0.001
As	0.01	0.02
Au	0.001	0.001
Bi	0.01	0.01
Cd	0.001	0.001
Cu	0.005	0.005
Fe	0.005	0.005
Pb	Balance	0.05
S	0.001	0.001
Sb	0.01	0.01
Sn	$60 \pm .05\%$	99.9 Min
Zn	0.002	0.002

In solder plating, it is extremely important for alloy control that the tin content of the anode be at least 0.5% of the nominal. Low impurity levels are essential in order to reduce sludging, to maintain anode efficiency close to 100%, and to aid in close alloy control of the deposit. Low tin in anodes can result in excessive additions of tin fluoborate which is costly and complicates alloy control.

Plating chemicals should also meet rigid specifications. The following is a suggested specification for fluoborate chemicals:

	Tin Fluoborate	Lead Fluoborate	Fluoboric Acid
Stannic Tin	1.0% Max	NA	NA
Chloride	0.05 Max	0.05	.01
Sulfate	0.03 Max	0.03	.02
Copper	0.005 Max	.002	.005
Iron	0.01 Max	.01	0.005
Lead	0.005 Max	Major	0.003
Nickel	0.001 Max	.001	0.005
Zinc	0.001 Max	.001	0.001

Fluoborate chemicals are liquid concentrates that are approximately 50% solutions. The approximate concentration can easily be checked by specific gravity.

TYPICAL CONCENTRATES

	$Sn(BF_4)_2$	$Pb(BF_4)_2$	HBF_4
Wt. %	51%	51%	49%
Metal equivalent	20.7	27.7	NA
Specific gravity	1.6	1.72	1.37
lbs./gal.	13.3	14.6	11.4

OCCLUDED ORGANIC IMPURITIES

Solder plating and bright tin deposits require the use of organic additives in order to produce a smooth fine grain deposit and for increasing throwing power. Organics in a plating solution usually undergo some decomposition during electrolysis so that it is difficult to determine what exact compounds are present in the plating solution. Organics are absorbed into the deposit during electrolysis and the amount occluded in the deposit is a function of concentration in the solution

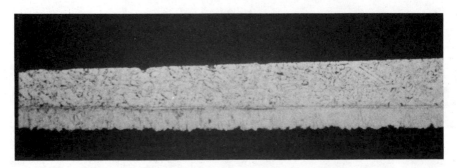

Fig. F-4. Normal reflowed tin-lead deposit (80× magnification).

Fig. F-5. Reflowed tin-lead with excess occluded organics (80× magnification).

and the time of plating. Organics if present in excessive amounts can cause reflow and soldering problems because of out gasing during heating (See figures 4, 5 and 6). It is important that organics in solution be kept to the minimum required to produce a sound deposit. Excess metallic contamination usually requires more addition agents to produce a smooth deposit and maintain throwing power.

It is suggested that tin and solder plating solutions be carbon treated at least 4 times per year to insure removal of breakdown products and to avoid excessive buildup from indiscriminate additions. The solution is treated with 4–5 pounds of activated carbon per 100 gallons and filtered until a clear solution results. After treatment, fresh addition agents are added to bring the concentration up to operating level. Consult suppliers for recommended additions after treatment. In the case of bright tin solutions, plating at lower operating temperatures usually reduces the amount of addition agents required to produce a sound deposit and improves the deposit's ability to preserve solderability. It becomes a question of practicality of how low a solution can be cooled since below 60°F can be expensive especially during the summer months.

SELECTION OF TYPE OF DEPOSIT

Bright tin and solder plate can be employed as a metal etch resist and both can be effective coatings for preserving solderability. Solder plate has the advantage in cost, since tin is $8.68/

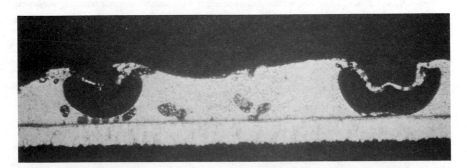

Fig. F-6. Reflowed tin-lead with excess occluded organics causing outgassing (80× magnification).

lb. and lead is $0.42/lb.* Alloying with 40% lead sharply reduces metal cost. However, its main advantage is that it can be reflowed after plating which provides a test for solderability, removes solder slivers produced during the etching process and produces a denser deposit. Solder plate is not susceptible to whisker growth and tin transformation at low temperature does not take place when tin is alloyed with lead.

The main reason for using bright tin deposits is that this deposit is slightly better as a contact surface to replace gold. Tin plated from the sulfate solution can be deposited with a bright surface that is less susceptible to tarnishing from handling and is a better contact surface than solder plate.

INORGANIC CONTAMINATION

Solder and tin plate both melt and dissolve during the soldering process and the solder bond is to the metal underneath. The same impurities that are detrimental in a solder bath in wave soldering can also cause soldering problems when co-deposited in a tin or tin lead electrodeposit. These impurities are copper, iron, zinc, cadmium and gold. Copper is the contaminant most often found since reclaimed anodes are usually manufactured from copper bearing metal and drag in from previous plating operations can contain copper. Periodic low current density electrolysis during non-working hours with corrugated cathodes will quickly reduce contaminants such as copper from a plating solution.

A solder or tin solution that consistently produces quality work will have a copper level of less than 10 parts per million. On printed circuit boards, tin and solder deposits also must function as an etch resist metal. Co-deposited metallics and occluded organics can effect these deposits' ability to withstand the etching solutions. Tarnishes formed during etching also lower solderability.

HOT SOLDER COATED PRINTED CIRCUIT BOARDS

Hot solder coatings have been considered as the ultimate for preserving solderability on printed circuit boards. Up to recently, techniques for clearing the holes have been unsuccessful in clearing the plated-through holes and keeping sufficient thickness on the board. Earlier attempts to do this by hot oil or wax under pressure left insufficient solder to do an adequate job of preserving solderability. These coatings were under 50 microinches and in time most of the coating alloyed with copper, forming an unsolderable intermetallic compound. Coatings of tin or solder under 50 microinches have limited shelf life.

For years the hope of improved techniques to produce a thicker hot solder coating with cleared plated-through holes, prompted extensive development efforts. Several years ago, air leveling techniques proved successful in clearing the holes and leaving sufficient thickness for adequate shelf life.

Air leveling requires a high purity oxide-free solder and a flux that is designed for this process.

FLUX CONSIDERATIONS

Fluxes that are used in wave soldering are not always adequate for air leveling. To date, organic acid water soluble fluxes have proven to be the most successful. These fluxes should be composed of a non-volatile vehicle to reduce generation of smoke. It is also important that it be easily removed by water cleaning to reduce processing costs and, its residues are biodegradable.

The flux should have heat stability and be capable of producing a thick adherent flux coating on the board to act as a thermal barrier protecting the laminate from heat damage while in the

---

Metals Week 9/9/80

solder pot. The flux should be of a consistency that it remains on the boards after they are withdrawn from the solder pot. This flux film aids during the air leveling step by reducing the solder surface tension, thereby preventing solder smear or adherence of solder balls on exposed laminate or solder resisted areas.

It is helpful if the air leveling flux does not have a flash point for greater safety and fewer OSHA related storage problems.

SOLDER ALLOYS

Most air leveling units will use either 63/87 or 60/40 tin-lead. The eutectic alloy is the most widely used. It has been found that high purity oxide-free is essential for best results.

Solder that has been treated to remove oxides, sulfides and other non-metallics is helpful. The following is the recommended purity level of solder for this process:

Ag	0.015	Cd	0.001
Al	0.005	Zn	0.005
As	0.03	Sn	63 ± 0.5
Bi	0.05	Pb	Balance
Cu	0.01	Sb	0.2 - 0.5
Fe	0.005		

The main contaminant in used solder is copper. The copper level increases rapidly in the beginning but starts to slow down at 0.15. At 0.2 the dissolution of copper diminishes but will increase slowly. The alert level for copper contamination is 0.25% and the pot should be changed at 0.32. A solder bath will dissolve a small percentage of the metal surface being soldered. In time, this will cause contamination of the solder bath. A regular analysis will detect this quickly and often corrective action can be taken before rejects.

To analyze solder properly, a well equipped laboratory with sophisticated equipment is essential as well as a highly trained chemist or technician. If this is not available, it is better to contract with an independent laboratory that performs this type of analysis on a regular basis. This will assure accurate and reliable results.

A regular analytical program provides a complete record of all solder baths and enables one to know when solder is approaching dangerous contamination levels so that solder bath can be changed before problems arise.

The table that accompanies this article lists suggested guidelines for monitoring contamination levels. The "alert" level is where the pot should be changed if the product appearance is unsatisfactory or if there are known soldering problems. The "danger" level is the level at which it is recommended that the pot be changed even if there are no apparent signs of a problem.

Percentage

Material	Alert	Danger
Copper	0.25	0.3
Gold	0.1	0.2
Zinc	0.005	0.008
Cadmium	0.01	0.02
Iron	0.015	0.02
Aluminum	0.005	0.008
Copper & Gold	0.30	0.4

CLEANING

Following soldering and air leveling the boards should be quickly cleaned in water. Warm or hot water will clean faster and more efficiently than cold water. In some cases, neutralizers are employed to insure complete removal of flux residues.

SUMMARY—HOT SOLDER COATINGS

The major advantage of hot solder air leveled coatings is that dewetting will occur if the base metal is not properly prepared for solder and this will be readily seen. This process is essential on additive circuitry when long shelf life is required. The other advantage is that it can be used in conjunction with solder mask to limit the coating only to the areas to be soldered. This permits solder mask over copper which eliminates wrinkling problems of solder mask over tin-lead. Today, some processes strip tin-lead plating after etching and then solder mask over copper and hot air level solder coat.

SUMMARY—PLATED COATING

The ability of tin and solder deposits to preserve solderability is in large measure dependent on the purity of the deposit, and in the case of solder plate, alloy composition is extremely important. The use of high purity raw materials used in plating, close process control and frequent purification treatments are essential for producing a consistent quality deposit. Reflow of solder plate has provided an operation that can be used as a 100% quality control procedure. It also eliminates solder slivers and forces close control of the plating process.

Bright tin can provide a deposit that can also be used as a contact surface which eliminates the use of gold on contact fingers.

Both processes require high purity deposits in order to produce consistent quality that will assure solderable surfaces. The need to minimize both organic and inorganic solution contamination cannot be overemphasized.

Appendix G
Screen Printing is the Answer

Robert M. Nersesian, Tetko, Inc., Elmsford, NY

The secret to this precise application of screen printing is understanding and controlling the variables.

Screen printing is a process with many variables—variables which must be controlled if a cost-effective process is to be incorporated into a manufacturing operation. In the production of circuit boards, the greatest demands are placed upon screen printing, and it is necessary to understand each of its components, their interrelationships, and the way to which they can be controlled.

Within the past couple of years, screen printing has gone through dramatic changes. The technology of screen printing has grown in direct proportion to the demands of the industry. It is possible today to produce the high-quality product needed in large quantities, in the most economical way. The resultant quality of the product is directly related to the quality of the process.

The key to screen printing's success has been attributed to the efforts made in the areas of screen printing process control. Screen printing techniques and procedures are no longer left for chance "scientific cookbook procedures." Along with the necessary sophisticated major equipment and proccess-control tools, they are now a reality.

Screen printing, being a process made up of many variables, demands that each variable be controlled before a quality cost-effective process is realized. To achieve the gratification of a cost-effective process, knowledgeable manpower combined with the necessary technology and equipment are required. New equipment alone cannot guarantee the results of a process. In the same respect, knowledgeable manpower alone cannot work miracles with outdated and inferior equipment.

Today, equipment is manufactured to meet the demands of the industry. Since the demands of the industry are constantly being changed, so must the equipment. Whereas two years ago a piece of sophisticated equipment was a luxury, today it is a necessity.

However, the state-of-the-art equipment available today is not a necessity for all companies. An individual evaluation of your process and what is needed in the final product from that process must be conducted. Then, the correct selection of equipment needed to best suit that process can be made.

Of primary importance to the screen printer is the screen itself. Most effort should be made in upgrading and fine-tuning the screen making operations. This includes correct fabric or wire cloth selection, understanding their characteristics, and the way they should be used.

In screen printing, the screen is the printing plate. Any prints taken from the screen can be no better than the screen. If the artwork is the original for determining quality of the stencil image, the stencil image is the original for controlling the quality of the print. For this reason, it is imperative to make all efforts to produce the finest screens possible.

The two major decisive areas of screen making are tensioning and imaging. Each is very critical in determining the final print. It is not difficult to make a good screen look bad on a press,

---

This paper was given at CIRCUIT EXPO, 1982, Worcester, MA.

so correct press setup procedures are a must. However, it is extremely difficult to make a bad screen look good.

With the advent of precise hand-screening tables, it is possible to get the same quality print from a hand table as with a semi or fully-automatic press. The only advantage an automatic press will present is the consistency from the first print to the last print.

When buying equipment, do as much research into the venture as possible. Consult all the manufacturers and compare and contrast the claims of each. Do not be afraid to spend a little more if you think you'll get a little more.

When looking for new equipment or using new equipment, use your suppliers as much as possible. It is their job to keep abreast of the latest technology; their input is invaluable. Follow manufacturers' recommendations, so if a problem arises, a standard checklist can be used to troubleshoot the problem.

Of major concern to manufacturers of printed circuit boards using the screen printing process is registration. Screen printing, an off-contact printing process, makes it virtually impossible to get a one-to-one relationship between the stencil and the print. The key to holding registration is to keep the off-contact distance to a minimum during printing. The lower the off-contact, the less elongation and image distortion. Allowing registration to be our major topic, a discussion of the variables and their effect on the final print will follow.

Screen Frame

The primary function of the screen frame is to act as a support for the screen fabric or wire cloth. The material and profile used has a great bearing on the dimensional stability of the screen media

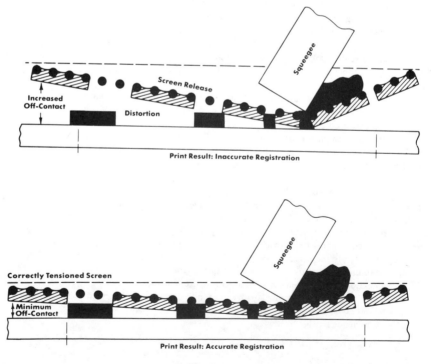

Fig. G-1.

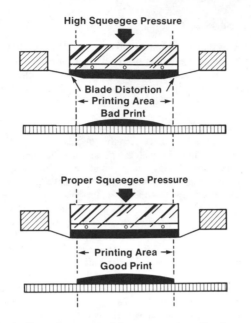

This illustration shows what happens when the frame is too small for the squeegee. There is insufficient clearance between the squeegee ends and the frame sides, so that high squeegee pressure must be applied to overcome screen tension. This causes the squeegee blade ends to distort and leads to uneven ink deposits, hence poor print quality.

To avoid this condition, allow more space between the squeegee ends and the frame. One way of judging the proper frame size is to provide a screen area at least double the image area. Another method is to allow 5″ to 6″ clearance between each end of the squeegee and the frame edge. Sufficient squeegee pressure can then be applied to overcome screen tension without distorting the blade ends.

Fig. G-2.

and stencil. Due to the continuous influence of high fabric tension and exposure to solvents and cleaning agents, the frame components must be made of materials that resist mechanical and chemical influences.

The selection of the frame size should be based on the image size. For minimum off-contact, optimum print detail, longer stencil and screen service life, and ease of printing, the inner area of the frame on all sides must be at least five to six inches larger than the corresponding image size. This will guarantee distortion-free printing and minimum elongation for consistent registration.

The functions of various screening media are numerous. As a support for the stencil, the fabric has a direct influence on the durability and service life of the screens. Edge definition and print detail are greatly affected by the selection of mesh counts and thread diameter.

Stainless Steel Wire Cloth

Stainless steel wire cloth is the most common screening medium used in the circuit industry for primary imaging. Its low initial elongation makes it ideal for close-tolerance registration. The high tension achieved with stainless steel makes possible minimum off-contact, minimum squeegee pressure, and minimum screen and stencil wear—hence, minimum elongation for tight, consistent registration. Stainless steel's dimensional stability is unequalled by any other screening medium.

One disadvantage of stainless steel is that it is subject to continual elongation during printing, ultimately resulting in screen fatigue. Theoretically, the first print made from a stainless steel screen is the best; each succeeding print will deteriorate until it is unusable. This is due to two factors: One, stainless steel wire contains little or no elasticity; and two, under the influence of squeegee drag and off-contact, the mesh is subject to a continual elongation. This results in material slackening and increased off-contact—therefore, image growth, poor registration and lower screen and stencil life. Another factor contributing to a loss in register with stainless steel wire cloth is the flattening out or the crossing points or "knuckles" of the wire during printing.

The number of prints that can be made with stainless steel until the screen is unusable is dependent on the procedures used to prepare the screen for printing. If handled correctly, a stainless steel screen should perform with accuracy for an entire length of run. A stainless steel screen can print a job of 10,000 with no problem. It is being done presently in a production situation. If warranted, it is estimated that such a screen could print an additional 10,000 easily. This success can be achieved only through correct screen-making procedures. That includes tensioning and imaging; and between the two, tensioning is the key.

Tensioning of the Screening Medium

Correct tensioning of the screen medium is probably the single most important aspect of the screen-making process and is an absolute prerequisite for high-quality printing. The tension of the screening medium has a direct influence on the registration, image quality, ink deposit, and service life of the fabric (or wire cloth) and stencil. As much time as possible must be allowed for the tensioning sequence. The highest possible dimensional stability cannot be retained if the fabric or wire cloth is not tensioned in stages.

There are numerous ways of tensioning screening media today. Most stretching devices are divided into two main groups: devices with mechanical stretching mechanisms and devices with pneumatic stretching mechanisms. The type of quality printing you wish to achieve will determine the method of tensioning used for the screening media.

More important than the type of stretching mechanism is the type of fabric-holding device; i.e. rigid versus laterally moving fabric-holding devices. The most common device used in the circuit industry is the self-stretching chase. These chases work by locking the screening medium into a rigid fabric-holding device on a frame, and with the use of tensioning bolts, the fabric is tensioned right on the frame.

Stainless steel wire cloth is most commonly used in the self-stretching chases. The advantage of this type frame is after a number of images are printed and the screening medium has lost some tension (due to elongation or fabric slippage), it is easily retensioned to the original tension values.

A major disadvantage is the self-stretching chases, in addition to being very heavy, is that the screening medium is held in place by a rigid fabric-holding device (which will not move laterally during stretching), causing overstretching at the corners and possible premature tearing of the screening medium. It is good practice to consciously place the fabric a bit looser into the holding device near the corners to minimize its detrimental effects.

Due to the rigid fabric-holding device, warp and weft threads will not run parallel and perpendicular to one another. Also, tension is not uniform due to the tensioning bolts creating high-tension centers across the area of the screening medium. These conditions will result in faulty registration, poor print detail, skipping, and lower fabric and stencil life.

A stretching device must be able to tension the screening medium in each direction independently. Also, movable fabric-holding devices are more desirable than rigid holding devices due to their resulting even distribution of tension. For the most uniform tension from any stretching device, the fabric-holding devices should be parallel to the frame sides and warp and weft directions should be straight and parallel to the frame. Take as much time as necessary to place the screening medium into the fabric-holding device uniformly—any irregularities in doing so will result in irregular tension.

Overstretching or understretching will adversely affect registration and the service life of the screen and stencil. It is extremely important to handle precision woven screening media according to the manufacturer's recommendations. These tension recommendations are made in order to obtain the longest life and best possible printing characteristics from the screening medium.

In order to establish the correct tension on the screen, tensiometers are used to monitor the tension placed on the screening medium. There are electronic and mechanical devices available, but the best ones will make possible readings in both the warp and weft directions and anywhere on the screening medium. With self-stretching chases, this type of device is a necessity for obtaining the most uniform screen possible.

What are the correct tension values for stainless steel wire cloth? Each mesh count will have a different recommended value which correlates to the thread diameter used for that particular mesh count. The breaking points, yield points, and plastic deformation points of the threads are taken into consideration when arriving at the recommended tension value. It is recommended to use a tensiometer that is calibrated to read in the same values in which the tension values are derived—N/cm^2.

Special tensioning instructions for stainless steel wire cloth include tensioning in warp direction first and taking as much time as possible to tension the screen. During weaving, the warp threads exhibit a greater crimp at the "knuckles" or crossing points. Due to this exaggerated crimp in the warp direction, a greater percent elongation is required to tension the warp threads correctly. Gradual tensioning also is required to flatten out the knuckles or crossing points. A minimum of two to three days is required to flatten out the knuckles and produce a stable screen for optimum performance.

For example, if you require a tension of $32 \ N/cm^2$ for your 325.0011 inch stainless steel, the first day a tension of $22 \ N/cm^2$ should be placed upon the screen. The second day the screen should be brought up to $28 \ N/cm^2$ and the third day the screen can be brought up to $32 \ N/cm^2$. A noticeable difference in print quality, registration, and screen and stencil life should be pleasantly realized.

If possible, it also is recommended to print in the weft direction of the wire cloth. During weaving, the warp wires are under relatively little strain as compared to the weft wires. During printing, the cross-squeegee direction is under the greatest strain. Therefore, it is recommended to have the most stable wire (warp) in the cross-squeegee direction. This will aid in obtaining mesh stability for close-tolerance printing.

Because of the many variables, it is impossible to recommend off-contact distances (at least I do not like to). Whatever screen is being used, the minimum possible off-contact distance for that screen should be used. If the tension on the screen or the ink type or substrate only allows a minimum of .050 inch off-contact for good screen "snap-off" behind the squeegee, take the minimum.

When procedures have been set for screen making, it is good practice to set up a checking system. When a correctly made screen is set up on the press with its minimum off-contact distance and the job is completed with no major problems, measure the off-contact distance. Make a shim or a wedge-block that can be used by the pressman to set up all screens at the prearranged off-contact distance. This will speed up and insure consistent set up in addition to maintaining the optimum performance from the screen.

Stainless steel, due to its precision during weaving, is excellent for close-tolerance printing and printing requiring controlled thickness of deposit. Stainless steel is not cheap. Use stainless steel wire cloth correctly and let the "proof be in the printing."

Synthetic Screening Media

Many of the boards printed do not require extremely tight, close-tolerance registration (print-n-etch, flexible circuitry, touch panels). Because of this, stainless steel wire cloth is not required. However, many people still use stainless steel wire cloth for their "non-critical" printing requirements. Synthetic screening media (monofilament polyester and metalized polyester) can be used for producing the majority of the non-critical boards.

Synthetic screening media have a unique quality which is not found with stainless steel wire cloth. If tensioned correctly to manufacturer's recommendation, synthetic screening media will retain their registration integrity for the entire life of the screen.

Synthetic screening media possesses a "memory." Theoretically, this means that if tensioned correctly, the first print will be identical to the last print. Why hasn't this revolutionized the circuit printing industry?

Although synthetic screening media are ideal for prints requiring registration accuracy, the initial print may be out of register. Synthetic screening media exhibit greater elasticity than stainless steel wire cloth. Also, the recommended tension values for synthetic screening media are approximately half that of stainless steel wire cloth. Simply, synthetic screening media are not as tight as stainless steel—hence, higher initial off-contact, greater elongation, and greater image distortion.

Even though it is difficult to predict the amount of "out of register" with synthetic screening media, it is safe to assume that if prepared correctly, they will give a consistent print. This does not help if the "out of register" is beyond the tolerances. If the tolerances are not exceeded with the use of synthetic screening media, one can enjoy the benefits of greater exposure control (dyed fabrics), greater speed in screen making, lower cost, and greater cost-effective screen printing.

Many shops producing print-n-etch boards, flexible circuitry, and/or touch panel boards would benefit from an evaluation of monofilament polyester. One area monofilament polyester is best suited for is solder masking. Stainless steel wire cloth cannot be used for solder masking. The reason for this is that stainless steel does not have the necessary elasticity to conform to the irregular surface structure of the circuit lines and pads. The most common mesh counts used to solder mask are monofilament polyester 110 T, 156 T, 195 T and 240 T. With the growing popularity of UV curable solder mask, mesh counts closer to the 240 T range should be evaluated.

The major problem with solder masking is skipping. This can be caused by too hard a squeegee (60 to 70 durometer is recommended), too fast a press speed (speed should be relatively slower than primary imaging), and incorrect frame size versus squeegee size versus image size (allow five to six in. from end of squeegee and print stroke to frame).The squeegee should not overhang the image by more than one inch on either side. Other causes are incorrect screen tension, fabric not "peeling" behind squeegee, warped panels, and/or overplating.

Tension control and uniformity are important with any screening medium and prove to be invaluable when screening with synthetic meshes, as with monofilament polyester and metalized polyester. As with stainless steel, take as much time as possible to tension the synthetic meshes. It is recommended to allow three to four hours per screen, but again, if you can only spare 45 minutes, please take all of the 45 minutes.

Metallized Polyester

Metallized polyester is a monofilament polyester fabric which is nickel-plated, which results in a screening medium with half the elasticity of monofilament polyester. As compared to regular monofilament polyester, the metallized polyester will make possible a tighter screen, lower off-contact, and better registration.

In addition to having half the elasticity of regular monofilament polyester, metallized polyester exhibits excellent ink passage. This is due to the nickel-plating which encapsulates the knuckles or crossing points, producing a flatter fabric which minimizes squeegee drag and eliminates the crossing points or knuckles from interfering with the ink flow. With the limited elasticity of stainless steel and the memory of polyester, metallized polyester is worth the investigation.

As with any screening medium, tensioning of monofilament polyester and metallized polyester is very critical. A stretching device with laterally-moving clamps is most desirable. When tensioning the synthetic meshes and bonding them to a rigid aluminum frame, it is advisable to use a profile of 1½ × 2½ inches with a minimum of ⅛ inch wall thickness, no matter what the screen size. A 2½ inch bonding surface will add greatly to the dimensional stability of the fabric and stencil.

Synthetic screening media show a definite cost savings when compared to stainless steel wire cloth. However, an evaluation of your tolerances will determine if an investigation is warranted of synthetic screening media.

Recommended Screen Tension Values

SCREEN MEDIA	MESH COUNT	RECOMMENDED TENSION VALUE IN N/CM	
Silk	2XX–30XX	7–9 N/cm	
Polyester Multifilament	2XX–8XX	15–18 N/cm	
	10XX–14XX	13–15 N/cm	
	16XX–25XX	12–13 N/cm	
		THREAD TYPE	
	MESH COUNT	T	HD
Nylon Monofilament	Up to 195	12–14	13–15 N/cm
	195–240	10–12	11–13 N/cm
	260–305	10–11	11–12 N/cm
	330–355	9–10	10–11 N/cm
Polyester Monofilament	90–195	15–16	16–18 N/cm
	195–240	14–15	15–16 N/cm
	260–305	13–14	13–15 N/cm
	330–355	12–13	12–14 N/cm
	390–470	12–13	—N/cm
Metalized Polyester Monofilament	123–186	18–20	—N/cm
	195–240	16–18	—N/cm
	260–355	65–17	—N/cm
	390–470	14–16	—N/cm

Recommended Screen Tension Values

SCREEN MEDIA	MESH COUNT	RECOMMENDED TENSION VALUE IN N/CM
Stainless Steel Fine Wire Cloth	180 Mesh, .0018″ wire dia. and coarser	34–40 N/cm
	200 Mesh, .0016″ wire dia. to 270 Mesh, .0016″ wire dia	32–34 N/cm
	270 Mesh, .0014″ wire dia. to 300 Mesh, .0012″ wire dia	30–32 N/cm
	325 Mesh, .0011″ wire dia. to 400 Mesh, .0010″ wire dia.	28–30 N/cm

A correctly made screen is not the answer to everything. Make certain that correct press procedures are incorporated into your process. Again, always remember that you can make a good screen perform poorly on the press due to incorrect press set-up procedures (too high off-contact, too much squeegee pressure, flood bar driving ink into mesh, etc), but you cannot make a poor screen perform well on a press.

Appendix H
Forecasting Waste Treatment Requirements[1]

Marshall I. Gurian, Ph.D.
Advanced Systems, Inc.

The printed circuit board manufacturing process presents certain challenges in the strategy for the optimum treatment of process rinse waters. The industry consists of shops of widely differing size and financial strength, ranging from large captive shops of the major electronic and computer companies to some very small prototype service shops. Regardless of facility size, the main pollutants are heavy metals, particularly copper and lead. Analysis and selection of optimum treatment techniques are difficult due to the number of chemical formulators who provide proprietary process chemicals containing complexers (including chelates) and whose formulations are not available to the end user or to his treatment advisers. Thus, although the general process flow is well known, there are many variations in the actual combinations of chemistries.

The printed circuit industry, like many others, is faced with more rigidly enforced waste treatment regulations, as well as increasing expenses for process water supply, sewer charges, and chemical waste disposal charges. The conventional precipitation, flocculation, and settling methods can prove to be operationally and financially burdensome to the chemically unsophisticated shop management. The sizing of such plants as well as evaluation of potential alternative treatment methods requires detailed flow and contaminant analyses which are costly and often beyond the in-house capability of the circuit producer. Thus, it was desirable to develop an analysis strategy which would be as generally applicable as possible, minimizing the expense of detailed process surveys, and having the flexibility to accommodate a variety of chemical systems.

In order to properly survey the operation of any chemical process area, the conventional method requires a very detailed sampling survey involving timed composite samples as well as many individual analyses and flow measurements. Even when these measurements are completely and competently performed, they do not allow easy extrapolation for the cases of increased production, varying process loading, and changes in type or supplier of process chemistry.

This method, which was originally developed as a simplified analytic tool for ion exchange application, is adaptable to general waste stream loading analysis. In addition, it is established in a format which allows computer estimates of dragover contributions in various configurations of production so that a detailed model can be developed. The method involves straightforward information input, and is easily adapted to process variation and production loading changes.

PROCESS CHARACTERIZATION

The printed circuit board processes can be divided into several convenient categories for both production and waste treatment purposes. The following is representative of one class of produc-

[1]This article is reproduced with the permission of *Printed Circuit Fabrication* Magazine where is appeared in Vol. 5 No. 5 (May 1982), p. 32. This article is also reproduced with the permission of California Circuits Association, where it appeared in the 1981 Fall Symposium Proceedings (p. 10).

tion, namely the two sided, pattern plated, conventional subtractive technology. This is certainly the most prevalent technology, and it is suitable for representative analysis. However, the variations within this technology and the alternative technologies available in some shops make comprehensive generalization difficult.

Without attempting to describe the details of all manufacturing, it is important to examine each of these processes to identify the treatment problems presented. Since the process categories form localized production centers in the typical plant layout, this inspection can also point up specialized treatment opportunities where ease of separation of chemical systems occur. In order of their manufacturing sequence, these process categories are described briefly with their pollution contributions summarized.

1. *Mechanical Fabrication.* Deburring and scrubbing abrasive machinery produces fine copper particulates which can dissolve when mixed with process chemicals from other areas.

2. *Electroless Chemical Deposition.* Chemical cleaners, followed by tin-palladium surface activators and electroless copper reduction bath. Typically six or seven process and rinse stages comprise this area with effluent only moderate in copper metal content. There are problem complexers in these baths.

3. *Image Formation.* Sometimes includes particulate formers similar to section 1. Typically single chemical rinse from dry film photoresist process containing mild caustics and some organics, negligible metal problem.

4. *Pattern Electroplating.* Usually simple acid cleaners followed by acid copper electroplating and fluoborate tin-lead alloy electroplating. Rinses moderately high in metal content and relatively free of chelates.

5. *Stripping and Etching.* Stripping of pattern resist material usually high in caustic and organics, low in metals. Etching usually high in copper and ammonia. Sometimes solder activator present adding some small amounts of lead, tin, and organics.

6. *Contact Finger Plating.* This is generally a small volume series of tanks to remove solder, and electroplate nickel and gold only on the contact finger area of the boards. Gold dragout reclamation usually economically demanded. Some tin, lead, and nickel contamination, but usually a very minor effluent contributor.

7. *Solder Reflow.* Typically contains a chemical clean, and a water cleaning stage after flux application and infrared heating. Some tin and lead content, but generally detergent and organic effluent.

8. *Final Fabrication.* Generally no wet process content.

The treatment problems presented by these types of process chemistry are further magnified by the nature of the process cycles. For example, electroless copper plating requires an immersion time of 20–40 minutes. Thus, one rack of boards is delivered to the rinse in a period of 10–40 minutes depending on the process approach. The concentration of contaminant in the rinse water has therefore a cyclical nature of the type illustrated in Fig. H-1. In most production lines, this curve is somewhat modified by the cascade rinsing procedures utilized to conserve water. The result of the combination of many processes of noncoincidental cycles is that the actual analytic picture of even a well scheduled operation is not easily defined. A typical water analysis study of a "regular conventional" printed circuit shop would therefore have to consider the time variation of many chemical species over a relatively long sampling period in order to properly specify the parameters for treatment planning. Estimates of from $5,000 to $10,000 for analysis only are not unusual. Sometimes these reports include the treatment recommendations, and sometimes they only include the process survey information. A further limitation to the rigorous analytic survey approach is that modeling changes of chemical formulations and production flows is extremely difficult, and the data reflects only a composite of what occurred during the sampling period, with limited expansibility.

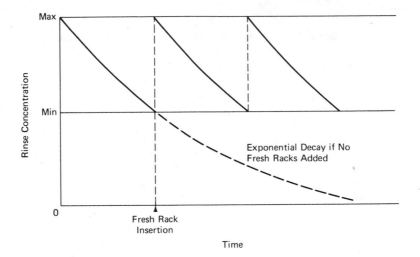

Fig. H-1. Concentration profile of flowing rinse with periodic product insertion.

ANALYTIC METHOD

The conventional survey approach represents a very prohibitive cost for the small to medium shop. It was desirable to formulate a method which could be used to estimate the water contamination and treatment options with techniques which could be implemented with a minimum of equipment and technical expertise. A further desire was to establish a simple model which could be used to estimate the result of process and production changes as well as forecasting new facility waste stream loading.

A brief description of the estimation technique is as follows:

1. Categorize all processes by type of rack used.
2. Measure dragover per square foot for each type.
3. Characterize the dilution characteristic of each chemical bath.
4. Estimate the production rate.
5. Measure the flow rate of all rinses (special techniques for conductivity controls).
6. Develop the numerical algorithm for estimation.
7. Calculate individual contributions
8. Average (flow weighted) for gross estimate.

The implementation of the technique for a specific case can be found in summary in Table H-1. For illustrative purposes, a descriptive expansion of the above technique follows.

1. There are basically two racking types used in most process lines. The electroless deposition line typically utilizes a "basket" type rack into which several boards 10–12 typically) are inserted vertically on approximately ½ in. centers. The electroplating racks are planar with clamping arrangements for current transfer to the (2–6) workpieces. Each company usually standardizes on its own design of the racks, and typical dragover characteristics may be attributed to each type of rack. The assumption is made that the racked panel array has a repeatable dragover regardless of which chemistry is being utilized. Obviously, there is a possibility for error, but it is not critically significant as long as there is not an unusually viscous process solution

Table 1
Process Rinse Analysis—Medium Production Shop
Conductivity

Rack Dragout (R):
    A. Flat Rack (5.25 sq.ft. contained)—12 ml/sq.ft.
    B. Basket Rack (78 sq.ft. contained)—34 ml/sq.ft.
Production Rate (X): 70 sq.ft./hour (5000 sq. ft./week)

PROCESS	FLOW (F) l/hr	RACK	DR. OUT (RX) ml/hr	STD COND (M) *	AVGE COND (RXM/F) *	WTD COND (RXM) **
Acid Cu Plt	340	FLT	840	1600	3950	1.34
Sn/Pb Plate	340	FLT	840	1700	4200	1.43
Acid Clean	230	FLT	840	250	913	.21
Rev.Ct. Cln	450	FLT	840	1300	2425	1.09
Pre Etch	450	BSK	2450	1200	6350	2.94
Cat/Accel	680	BSK	2450	100	360	.24
Eltrlss Cu	570	BSK	2450	250	1074	.61
Final Etch	1360	BSK	2450	1100	1981	2.70

Total Flow: 4420 l/hr (19.5 gal/min)
Flow Averaged Conductivity: 2390 u mho/cm

*units of micromho/cm
**units of mho/cm
Note: Rinses included in this table represent the candidates for reuse, and not all rinses in the shop. This facility represents a low water usage shop. By implementation of dragover and spray rinses, a concept was developed to reduce the flowing stream conductivity by nearly a factor of five. The pre etch and final etch were immersion sulfuric-peroxide.

being used. Most of the common processes are consistent enough for this approximation to be relatively valid.

2. The easiest method for dragover estimation is to analyze the concentration of process liquid and of a nonflowing rinse immediately following. By removing the rack from the process solution per normal procedures and immersing into a measured volume of clean rinse water, the increase in concentration of measurable contaiminants in the rinse can quickly be converted into volumetric dragover. For similicity this can be converted to a ml per square foot factor which aggregates both the workpiece geometry and rack contributions. Measured values are usually in the range of 10–15 ml/sq ft for the flat rack and 30–50 ml/sq ft for the basket rack. The dragout in either case can be significantly reduced by as little as 5 seconds drain off or "hang" time over the process tank before submersion in the rinse. Of course, spray rinsing and static dragover rinsing are extremely advantageous in limiting contamination of the flowing rinses. If these methods are used, their effect should also be estimated by sampling dragover into a following clean rinse.

3. The dilution factor for each bath can be measured in a very straight forward manner. For the simple concentration of chemical elements, a dilution of 1 ml of sample to 1 liter with distilled water is easily tested for chemical components as well as for conductivity. This gives a single analytical reference for a concentration representative of the probable range of rinse waters. Of course, the chemical results are a quantitative dilution of the bath concentrations. However, the conducitivity reading is significant at this ratio. For most of the chemistries involved, Fig. H-2 shows that the conductivity is also a simple dilution function in a two order of magnitude range.

This is specifically interesting in the analysis of contamination levels for purposes of ion exchange treatment, where the total ionic content is the treatable criterion, not merely the metals or other specific components. Furthermore, as a criterion for treatability with ion exchange, the linearity of the two decade dilution plot of conductivity implies a simple ionic character. Where there are dissociating species and colloidal character, the plots are specifically not linear or the slope is significantly off of the "45 degree" simple dilution slope. For example, the "electroless tin" line in Fig. H-3.

The simple rule resulting from this discussion is that the single point (1000/1) dilution factor may be used in most of the treatable cases as a representative number for estimation purposes. "Non-ideal" cases can be treated in a slightly more cumbersome yet direct manner. The dilution curve is estimated from several points and this curve referred to specifically in the ongoing procedure.

4. The production rate of the shop is merely the processed area of laminate per hour of operation. The gross square feet of laminate as purchased is used rather than using the actual wetted area for direct association with known production.

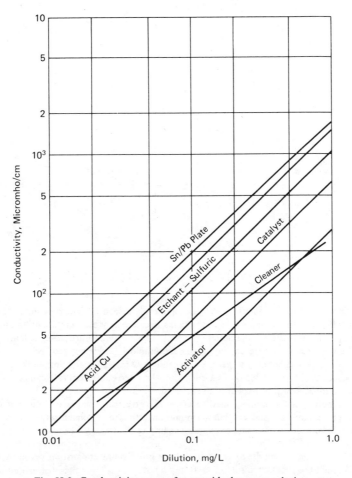

Fig. H-2. Conductivity curves for near-ideal process solutions.

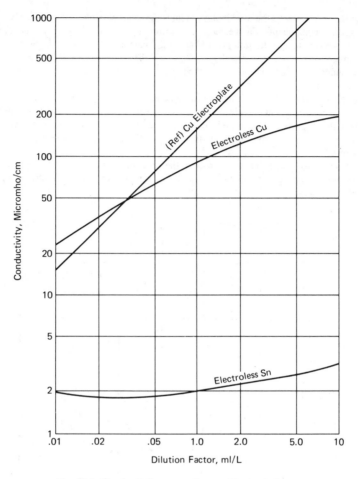

Fig. H-3. Conductivity curves for non-ideal solutions.

5. The flow rate of each of the rinses can then be simply measured. The problem in many shops is the difficulty of access to the outlet of the tanks because they are permanently plumbed into the drain system. In this case, the outlet or overflow can be temporarily plugged, and the level increase rate measured. Where cascade rinses are employed, it is not required to analyze the dynamics of each tank, but only to get the aggregate flow rate from the final tank.

Where conductivity controlled rinses are used, a calculated equivalent rate can be used instead of measurement. This will be explained in section 6.

6. The numerical algorithm is a method of easily correlating the basic factors of dragover, rinse flow, production rate and concentration into an estimate of concentration or conductivity in the treatable stream. Consider the following example:

- Let R equal the racked dragover constant for a given configuration (ml per gross sq. ft.).
- Let X equal the production rate in sq. ft. per hour.
- Let F equal the rinse flow in liters per hour.

- Let M equal the concentration at a dilution of 1 ml per liter of water.
- The average process dilution is then RX/F ml/l.
- The average rinse concentration is MRX/F expressed in units consistent with the units of M (mg per liter of particular metals, conductivity, TDS, etc.)

Where the flow rate is controlled on a conductivity basis, there is an equivalent calculation method. If Q is the conductivity controller setting, and M_c is the conductivity at the 1 ml per liter dilution, then the flow rate is calculated by M_cRX/Q. Thus, by knowing the dilution factor of the conductivity and the copper content, for example, the flow rate and the output copper content can be estimated. Where the rinse is fed by tap water rather than demineralized water, the baseline conductivity of the source water must be subtracted from the conductivity control set point to obtain the "working" control parameter Q.

7. The above algorithm may then be systematically employed in the analysis of each of the treatable rinses in the process line. The tabular approach in Tables H-2 and H-3 illustrates these calculations.

8. By summarizing the results in a weighted averaging procedure, multiplying the concentration by the flow rates, the relative contributions of each of the rinses may be examined. If the weighted result shows an unacceptably large treatment consideration, particularly important in ion exchange economics, process modifications may be possible.

At first appearance, the above procedure may seem obvious and not particularly beneficial. However, in actual plant terms, the simplicity and "Aggregate" averaging can help a great deal

Table 2
Prcess Rinse Analysis—Medium Production Shop
Copper Concentration

Rack Dragout (R):
    A. Flat Rack (5.25 sq.ft. contained)—12 ml/sq.ft.
    B. Basket Rack (78 sq.ft. contained)—35 ml/sq.ft.
Production Rate (X): 70 sq.ft./hour (5000 sq. ft./week)

PROCESS	FLOW (F) l/hr	RACK	DR. OUT (RX) ml/hr	STD CONC (M) *	AVGE CONC (QRXM/F) *	WTD CONC (RXM) **
Acid Cu Plt	340	FLT	840	18.8	46.4	15.8
Sn/Pb Plate	340	FLT	840	—	—	—
Acid Clean	230	FLT	840	3.0	11.0	2.5
Rev.Ct. Cln	450	FLT	840	3.8	7.1	3.2
Pre Etch	450	BSK	2450	30.0	163.	73.4
Cat/Accel	680	BSK	2450	—	—	—
Eltrlss Cu	570	BSK	2450	3.0	12.9	7.4
Final Etch	1360	BSK	2450	45.0	81.	110.2

Total Flow: 4420 l/hr (19.5 gal/min)

Flow Averaged Copper Concentration: 48.1 mg/l

*units of mg/l
**units of g/l

Note: Rinses included in this table represent the candidates for reuse, and not all rinses in the shop. This facility represents a low water usage shop. Both the pre etch and the final etch were batch process sulfuric peroxide. This represents the same shop as Table 1.

Table 3

Process Rinse Estimate—Small Production Shop

Rack Dragout (R):
    A. Flat Rack—120 ml/sq.m.
    B. Basket Rack—200 ml/sq.m.
    C. Unracked, Conveyorized—60 ml/sq.m.

Production Rate (X): 1.43 sq.m./hour (2000 sq.m.year)

PROCESS	FLOW (F) l/hr	RACK	DR. OUT (RX) ml/hr	STD COND (M) *	AVGE COND (RXM/F) *	WTD COND (RXM) **
Alk Clean	230	BSK	286	18	23	5
Sul/Prx Etch	460	BSK	286	1200	746	343
Catalyst	460	BSK	286	660	410	189
Accelerator	230	BSK	286	100	124	29
Eltrlss Cu	230	BSK	286	250	310	72
Acid Cu Pit	345	FLT	172	1600	800	275
Sn/Pb Plt	345	FLT	172	1700	850	292
Acid Clean	230	FLT	172	25	19	4
Develop	230	CON	86	30	11	3
Sul/Prx Etch	460	CON	86	1200	224	103
Brighten	230	CON	86	100	37	9

Total Flow: 3400 l/hr (15 gal/min)
Flow Averaged Conductivity: 389 micromho/cm

*units of micromho/cm
**units of millimho/cm
Note: These estimates are aggregates; the actual operation showed typical 450 micromho content with a baseline of 40 micromho in the reuse waters. Thus, the gross estimate produced above was within 10 percent of actual experience.

in the practical estimation of treatment requirements. The tedium of repetitive sampling, either manually or with mechanical sampling devices, is eliminated along with the analytic expense of testing a great many samples. In addition, the analysis can be performed by relatively inexpensive techniques mastered in most cases by in-plant personnel even where lack of chemical sophistication exists.

Another benefit is that the dynamics of the processes are smoothed. There might be a basis for argument that the possible maximum treatment cases could be overlooked, and that it would be therefore possible to overload a system designed only for the average loading. The question of system consistency must be addressed on a shop by shop basis. The better managed production flows attempt to balance product flow throughout the plant. On the contrary, all shops at some times (and some shops routinely) run odd lots, emergency remakes, and other cases which make process loading variations important. This must be a judgment in the system specification decision, with the clearly economic "trade-off" of sizing for a peaking load versus the risk of overload with some frequency. It is important to realize that the parameters of the peak loading may be introduced into the same algorithm and the degree of overloading thus predicted.

SUMMARY

The usefulness of the method is only limited by the creativity and ability of the user. Once some very basic and easily obtained test information is deduced about the chemicals and process parameters, evaluations of plant changes in chemical formulations, rinse flows, production loading, and other factors can be routinely made. If the data were stored in tabular or computerized form, the quick evaluation of changes and their potential impact upon the treatment process can be routinely calculated.

Of course, this type of analysis cannot completely replace the comprehensive water survey, but in many cases it can give significant results in the design and specification of the treatment plant. It also affords the shop operator a means within his own control to check on the reasonableness of a proposed system and treatment advice given by design consultants and regulatory personnel.

ACKNOWLEDGEMENT

This work is an expansion of part of a paper presented by the author at the Water Reuse Symposium II, August 25, 1981, Washington, D.C. sponsored by the AWWA Research Foundation, American Water Works Association.

Appendix I
Wave Soldering of Discrete Chip Components[1]

Ralph Woodgate

The term chip, or chip component, is used to describe several types of devices and therefore let us start by defining exactly what we are discussing. In this paper, the word "chip" denotes the tiny discrete components, resistors, capacitors and the surface mounted SOT semiconductors. These all have two common features. They are extremely small and they are designed to be mounted on the surface of the printed wiring board. They have some other less obvious characteristics. Except for the SOT's they have no leads; therefore, there are two fewer connections for each device, which increases the overall reliability. They are designed to pass through the solder wave and are resistant to this temperature increase. The parts can be made to a closer tolerance. Compare the size of a chip resistor with a ⅛ watt leaded device.

The size advantage is obvious over a standard component, as is the tighter packaging possible with chip components. When the relative costs are compared, the chip component is lower in cost when the total placement and soldering processes are considered. There are no wire leads to tangle, and the chip components are ideally suited to automatic assembly processes.

A review of a sample board using these components will quickly show the packaging density that can be expected and also indicate the problems that exist in soldering these dense boards. Wave soldering, however, is an essential part of the overall process if the advantages of using chip components is to be realized at a cost which is acceptable for the majority of the products that are manufactured. Correctly designed wave soldering is capable of producing defect free, high quality, low cost joints.

Let us then go back and look at the overall assembly process. Parts are usually packaged in reels, which can be stacked into the placement machines. There is some discussion of packaging in other forms, tubes, even loose using vibratory feed. However, the reel package appears to be the most common method in use today. As we have mentioned, the parts are small, light and are easy to move and control during assembly. One machine vendor uses an air transport, blowing the devices through small tubes at high speeds, over quite long distances.

The parts are held onto the board by dots of an adhesive, usually a quick setting epoxy. Once cured, the board is soldered by passing over the wave. This not only solders the chip components in position, but also joins the leaded components which have previously been mounted and the leads clenched.

The adhesive is placed in one of three ways. Single dots can be placed by a second head on the machine immediately prior to placing the chip, or all of the dots of adhesive can be placed at one operation, by using an array of pins to transfer adhesive from a container to the board. Where the chips can be placed on the board before the leaded parts are assembled, or where the board has only surface mounted devices, the adhesive can be applied through a silk screen or a metal stencil.

---

[1]This paper was presented at Circuit Expt '82, held in Worcester, Mass. Ralph Woodgate is Executive Vice-President of Electrovert Consulting Service, and the author of *Handbook of Machine Soldering*, published by John Wiley & Sons.

Today, however, most chips are used in conjunction with leaded components and if these latter are machine inserted, then the chips must be placed later, or they will foul the cut and clinch heads and either prevent them functioning correctly, or the chips will be damaged or knocked off the board. In this case the protrusion of the leads, of course, prevent the use of stencil or silk screen.

As a view of the soldered board suggests, soldering is not easy; the connections are very close together. However, it is also quite obvious that packaging of this kind is almost impossible to inspect for solder joint defects and extremely difficult to rework if joint problems are found.

It is therefore most important that the process is configured to produce zero defects. Many of the basic principles of normal wave soldering apply, but there are also some features particular to soldering chip devices. These are leaching, solder shorts because of the very close spacing and solder skips, or joints which do not have any solder on them at all. Solder shorts are no different from the problem seen whenever joints are close together and will be discussed later. Leaching is the loss of material from the contact ends of the chips caused by the tin in the solder dissolving the metallization. This occurs to some extent whenever components are soldered, but because of the very small volume of metal available to form the connection, in soldering these chips, the small amount of metal removed becomes of considerable importance.

The component manufacturer specifies that after six seconds exposure to 60/40 solder at 240°C, the connecting metallization on the chip will not be reduced to less than 70% of the original metal. This means that for safety, the parts should not be exposed to the molten solder for more than three or four seconds. This in turn means that the board must be preheated adequately to permit a fast conveyor speed and the shortest possible soldering time. Preheat becomes a very important factor in the correct soldering of discrete chip components.

To solder at high conveyor speeds, the preheater will become inconveniently long unless a high intensity heat source is provided. Electrovert has developed such a preheater which utilizes quartz tube elements. The tubes have integral reflectors, are fast in heating, and can provide adequate energy for all soldering purposes in a very short length. In addition, the tubes can be individually switched to provide preheating only over the width of the board to be processed. This effectively conserves energy consumption in the system.

Because of the fast response time these heaters are not suitable for thermocouple control and a totally new electronic open loop control system has been developed to operate with this preheater.

Solder skips were originally believed to be due to gasses evolved from the flux which prevented the solder from reaching the joints. Indeed at one time it was proposed that a hole be drilled in the board at each component site to permit these gasses to escape. It may well be that there is a certain amount of truth in this theory, but at Electrovert we studied the effect of the solder wave when soldering these components.

The standard Lambda wave proved to be effective in soldering about 98% of the joints on the board, but in every case there were a few joints which did not solder at all. We found that there were consistent patterns to the failed joints which tied in to their position relative to the surrounding joints. When this pattern of defects was examined, we found it to be consistent in that the same joints failed to be soldered in every board of the same type. We concluded then that the super smooth laminar flow of the Lambda wave was being diverted around some joints by adjacent components. We experimented by deliberately maladjusting the Lambda and causing a turbulent flow.

The incidence of skips was reduced, but the incidence of bridges increased. Now one of the reasons for producing the very smooth Lambda wave is to be able to provide bridge-free joints with extremely small spaces between circuit pads. This is done by the slow flowing rear section where the board lifts at a very small angle from a smooth solder surface which is traveling at the same speed as the board. The natural forces of surface tension and gravity can then act to remove all surplus solder. When the wave was made turbulent, these conditions no longer applied.

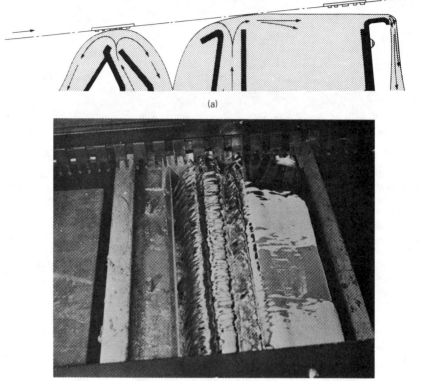

(a)

(b)

Fig. I-1. (a) Schematic of the Electrovert chip wave soldering nozzle. (b) Dual wave configuration of the Electrovert CHIPPak∠ nozzle. The initial wave (left) is a narrow, somewhat turbulent flow that accelerates in the direction counter to board travel, insuring proper wetting. The second wave is smooth, laminar and based on the Electrovert Lambda concept. It provides a finishing operation.

We therefore decided to split the solder wave into two sections. A narrow turbulent wave which would effectively wet all the joints, followed immediately by a smooth Lambda type wave to remove all the excess solder from the board. This proved to be an extremely effective wave design for soldering chip devices and performed equally well on the discrete wire ended components usually mixed in with the chips.

Tests run on boards loaded with rectangular chips, SOTs and the cylindrical MELFS, have proved that this preheat and solder wave system can produce defect-free soldered joints, without problems of leaching, solder skips, or bridges.

In summary then, for successful chip soldering, there are certain guidelines.

Although the new waves have reduced the concern for trapped gasses generated during soldering, it is wise to optimize the adhesive to prevent any generation of gas bubbles during exposure to the solder wave. Similarly, it is wise to use a flux with the minimum solids content that will produce satisfactory joints. This should not be so reduced that the board looks "dry" after soldering. In fact there should be an obvious but minimum layer of flux remaining after emerging from the wave.

Pre-heat is important and should never be less than 100°C as measured on the top of the board as it enters the first wave. Soldering speed should be adjusted to produce 3–4 seconds exposure to the solder. A tempered glass plate is useful in measuring the contact length. Also to reduce leaching, the solder should be held at the lower end of the normal range of temperatures.

Finally, the solder wave must be one designed specifically for chip soldering. It must be carefully and correctly adjusted according to the vendor's instructions. The other precautions taken during normal soldering are also of vital importance—solderability of parts, clean flux at the right density, clean solder, and good housekeeping and handling. With all of these things put together it is possible to wave solder these new packages quickly, efficiently and without defective joints.

Index*

Aluminum, clad laminate use of, 14
Amperage determination
 contact finger plating, 384
 pattern plating, 289
Amp-hour, definition, 41
Analytical laboratory procedures,
 489
Annular ring,
 definition. *see* Pad 41
Anode
 considerations, 344
 definition, 41
Artwork
 definitions, 42, 143
 emulsion orientations, 149
 generation, 250
 inspection, 101, 250
 multilayer, 151
 phototool problems, 178
 processing, 139
 registration systems, 159
 soldermask, 150
 spread and choke, 155
Autotype calculator, 222

Base, 42
Black oxide
 process, 452
 purpose, 8
Blueprint reading, 531
Booking, for lamination, definition, 8
Breakdown, of resist, 42
Brightener, 42
Brush plating, 42
Bus bar, 42

CAD artwork
 advantages, 16
 example of, 24
 generation, 249

Calibration
 program, 95
 requirements, 63
Carbon treatment, 361
Chemical analysis, 489
Chemical names and formulas, 363
Chemical storage and waste, 356
Collimation, 43
Computer Aided Design (CAD)
 basic process, 19
 definition, 15
 electronic table, *20*
 net list, example of, 22
 software, 18
Contact finger plating
 amperage determination, 384
 appearance of plated metals
  gold, 381
  nickel, 383
 automatic line, 367, *371*
 gold plating, 372
 manual line, *368*, 372
 nickel plating, 378
 procedures, 385
 sequence in manufacturing operation, 5
Copper. *see* Pattern plating
 anodes, 344
 plating, amperage determination, 289
 tank design, 342
 thickness determination, 283
 through hole plating. *see* electroless copper
 throwing metal, 360
Current density, definition, 43
Customer relations, 69
 customer returns, 132
 customer surveys, 524

Definitions, 41
 artwork, 104
 digital printed circuit design, 32

---

*Italics = photograph.

617